D0161510

Approaching Quantum Computing

Dan C. Marinescu and Gabriela M. Marinescu

Upper Saddle River, New Jersey 07458

Library of Congress Cataloging-in-Publication Data

Marinescu, Dan C.
Approaching quantum computing / Dan C. Marinescu, Gabriela M. Marinescu.
p. cm.
Includes bibliographical references and index.
ISBN 0-13-145224-X
1. Quantum computers. I. Marinescu, Gabriela M. II. Title.

QA76.889.M36 2004
004.1—dc22 2004053113

Vice President and Editorial Director, ECS: *Marcia J. Horton*
Vice President and Director of Production and Manufacturing, ESM: *David W. Riccardi*
Executive Managing Editor: *Vince O'Brien*
Managing Editor: *David A. George*
Developmental/Production Editor: *Rose Kernan*
Director of Creative Services: *Paul Belfanti*
Art Director: *Jayne Conte*
Managing Editor, AV Management and Production: *Patricia Burns*
Art Editor: *Gregory Dulles*
Manufacturing Manager: *Trudy Pisciotti*
Manufacturing Buyer: *Lisa McDowell*
Senior Marketing Manager: *Holly Stark*

Pearson Prentice Hall
Pearson Education, Inc.
Upper Saddle River, New Jersey 07458

Printed in the United States of America

10 9 8 7 6 5 4 3 2 1

ISBN 0-13-145224-X

Pearson Education Ltd., *London*
Pearson Education Australia Pty, Ltd., *Sydney*
Pearson Education Singapore, Pte. Ltd.
Pearson Education North Asia Ltd., *Hong Kong*
Pearson Education Canada, Inc., *Toronto*
Pearson Educación de Mexico, S.A. de C.V.
Pearson Education—Japan, *Tokyo*
Pearson Education Malaysia, Pte. Ltd.
Pearson Education, Inc., Upper Saddle River, *New Jersey*

Contents

APPENDICES

Preface

RATIONALE

Tremendous progress has been made in the areas of quantum computing and quantum information theory during the past decade. Thousands of research papers, a few solid reference books, and many popular science books have been published in recent years on this subject. The growing interest in quantum computing and quantum information theory is motivated by the incredible impact these disciplines are expected to have on how we store, process, and transmit data and knowledge in this information age.

Computer and communication systems using quantum effects have remarkable properties. Quantum computers enable efficient simulation of the most complex physical systems we can envision. Quantum algorithms allow efficient factoring of large integers with applications to cryptography. Quantum search algorithms considerably speed up the process of identifying patterns in apparently random data. We can guarantee the security of our quantum communication systems because eavesdropping on a quantum communication channel can be detected with high probability.

It is true that we are years, possibly decades, away from building a quantum computer requiring little, if any power at all, filling up the space of a grain of sand, and computing at speeds that are unattainable today even by covering tens of acres of floor space with clusters made from tens of thousands of the fastest processors built with current state-of-the-art solid-state technology. All we have at the time of this writing is a 7-qubit quantum computer capable of computing the prime factors of a small integer, 15 [139]. To break a code with a key size of 1024 bits requires more than 3000 qubits and 108 quantum gates [82]. Building a quantum computer presents tremendous technological and theoretical challenges. At the same time, we are witnessing a faster rate of progress in quantum information theory than in quantum computing. Applications of quantum cryptography seem ready for commercialization. Recently, a successful quantum key distribution experiment over a distance of some 100 km has been announced.

It is difficult to predict how much time will elapse from the moment of a great discovery until it materializes into a device that profoundly changes our lives or drastically affects our understanding of natural phenomena. The first atomic bomb was detonated in 1945, less than 10 years after the discovery of nuclear fission by Lise Meitner and Otto Hahn [91]. The first microprocessor was built in late 1970s, some 30 years after the creation of the transistor on December 23, 1947 by William Shockley, John Bardeen, and Walter Brattain. Francis Harry Compton Crick and James Dewey Watson discovered the double helix structure of the genetic material in 1957 and the full impact of their discovery will continue to reverberate for years to come.

The authors believe that the time to spread the knowledge about quantum computing and quantum information outside the circle of quantum computing researchers and students majoring in physics is ripe. Students and professionals interested in information sciences need to adopt a different way of thinking than the one used to construct today's algorithms. This certainly presents tremendous challenges, since, for many years, computer science students have been led to believe that they can get by with some knowledge of discrete mathematics and little understanding of physics at all.[1] We are going back to the age when a strong relationship between physics and computer science existed.

TOPICS, PREREQUISITES, AND CHAPTER DESCRIPTIONS

This text is devoted to quantum computing. We treat the quantum computer as a mathematical abstraction. Yet, we discuss in some depth the fundamental properties of a quantum system necessary to understand the subtleties of counterintuitive quantum phenomena such as entanglement. Chapter 1 introduces the reader to the quantum world by way of several experiments. Chapter 2 provides the most basic concepts of quantum mechanics and of the supporting mathematical apparatus. Chapter 3 introduces the qubit and hints at simple physical realizations of a qubit. Chapter 4 is devoted to quantum gates and quantum circuits. Chapter 5 presents quantum algorithms. The last chapter, Chapter 6, introduces the reader to quantum teleportation, quantum key distribution, and dense coding, and then presents reversible computations. The text is intended to be self-contained; concepts, definitions, and theorems from linear algebra necessary to develop the mathematical apparatus of quantum mechanics are introduced in Chapter 2. Appendix A introduces basic algebraic structures. Appendix B presents modular arithmetic necessary for understanding the factoring algorithms. Appendix C is devoted to the Walsh–Hadamard transform, and Appendix D summarizes basic concepts related to the Fourier transform. *Approaching Quantum Computing* is intended as a textbook for a one-semester first course in quantum computing. The time table we suggest for covering the material is: five weeks for Chapters 1 and 2, five weeks for Chapters 3 and 4, four weeks for Chapter 5 and Appendices A, B, C and D, and two weeks for Chapter 6. Any graduate or undergraduate student with a solid background in linear algebra, calculus, and physics should be able to do well in the class.

Features

This volume combines a qualitative presentation with a more rigorous, quantitative analysis. Whenever possible, we attempt to avoid the sometimes difficult mathematical apparatus, the trademark of quantum mechanics. In his marvelous book *A Brief History of Time* [71], Stephen Hawking, the astrophysicist, who

---

[1] This seems to be a perennial problem. When James II, the king of Great Britain, insisted that a Benedictine monk be given a degree without taking any examinations or swearing the required oaths, Isaac Newton, who was the Lucasian professor at Trinity College at Cambridge, wrote to the vice-chancellor "Be courageous and steady to the Laws and you cannot fail." The vice-chancellor took Newton's advice and was dismissed from his post.

is now the Lucasian professor, shares with his readers the warning he got from his editor: "Expect the sales to be cut in half for every equation in your book." There are k° —102 equations in this series of lectures and $2100 \approx 100010$ is a very large number. Detailed presentation and step-by-step analysis to illustrate the behavior of quantum circuits are given, along with numerous examples that will guide the reader in solving the problems at the end of each chapter. A solutions manual for instructors who adopt the book is available through the publisher.

ACKNOWLEDGMENTS

We are indebted to Tom Robbins, our thoughtful and supportive publisher, who communicated with us his vision and helped shape the book. Our appreciation and thanks go to production editor Rose Kernan for her timely and thorough supervision of the project on a day-to-day basis, as well as to managing editor David George, and senior marketing manager Holly Stark of Prentice Hall, whose professionalism turned our manuscript into this book. Many thanks are extended to George Dima, an accomplished artist and a fine violinist, who created the drawings for the cover of this book and for the opening page of each chapter. Peter Shor has made several constructive suggestions and signaled some of the problems in an early version of the manuscript. P.K. Aravind, Dan Burghelea, Octavian Carbunar, Robert E. Lynch, John Hayes, Boris Zel'dovich, and Helmut Waldschmidt have gone over more evolved versions of the manuscript with a fine-toothed comb. Andrei Marinescu's comments helped improve the presentation of several chapters. Erol Gelenbe encouraged us to undertake this project. We are greatly indebted to all of them.

DCM extends his thanks to former students and post-doctoral fellows who have stimulated his thinking with their inquisitiveness: Ladislau Böloni, Marius-Cornea Hasegan, Jin Dong, Kyungkoo Jun, Ruibing Hao, Yongchang Ji, Akihiro Kawabata, Christina Lock-Black, Ioana Martin, Mihai Sirbu, K.C. van Zandt, Zhonghyun Zang, Bernard Waltsburger, and Kwei Yu Wang. He is indebted to his friends and collaborators throughout the years: Timothy S. Baker, Dan Burghelea, Franz Bush, Hagen Hultzsch, Chuang Lin, Robert Lynch, Veron Rego, John Rice, Michael Rossmann, H. J. Siegel, and Wojciech Szpankowski. He acknowledges the support of the National Science Foundation, which has funded his research for many years. The students enrolled in the quantum computing class taught by the authors at UCF during the Spring and Fall 2003 semesters have signaled a fair number of typos and other errors. Of course, the authors are responsible for all of the remaining errors and are anxious to be advised of their presence and to correct as many as possible in future printings. We also welcome suggestions for improving the presentation and for inclusion in future printings. A second volume, under preparation, *Approaching Quantum Information Theory,* is devoted to quantum information theory.

DAN C. MARINESCU
GABRIELA M. MARINESCU

Notations

c	The speed of light in vacuum: $c = 3 \times 10^{10} cm\ s^{-1}$.
h	Planck's constant: $h = 6.6262 \times 10^{-34}\ J\ s$.
$\hbar$	Reduced Planck's constant: $\hbar = \frac{h}{2\pi} = 1.054 \times 10^{-34}\ J\ s$.
k_B	Boltzmann's constant: $k_B = 1.3807 \times 10^{-23}\ J\ K^{-1}$.
G	The universal gravitational constant: $G = 6.672 \times 10^{-8}\ cm^3\ g^{-1}\ s^{-2}$.
$a \in \mathcal{A}$	Element a *in* (belongs to) set $\mathcal{A}$.
$\forall a \in \mathcal{A}$	For all elements a *in* set $\mathcal{A}$.
$\exists a \in \mathcal{A}$	There exists an element a in set $\mathcal{A}$.
$\mathbb{R}$	The field of real numbers.
$\mathbb{C}$	The field of complex numbers.
$\mathbb{Z}_n$	The finite field of integers modulo n with n a prime number.
$i = \sqrt{-1}$	Imaginary number. Square root of -1.
$\alpha_0, \alpha_1, \ldots$	Complex numbers: $\alpha_j = \Re(\alpha_j) + i\Im(\alpha_j)$.
$\alpha_0^*, \alpha_1^*, \ldots$	Complex conjugates: $\alpha_j^* = \Re(\alpha_j) - i\Im(\alpha_j)$.
$\mid \alpha_j \mid$	The modulus of the complex number α_j: $\mid \alpha_j \mid = \sqrt{[\Re(\alpha_j)]^2 + [\Im(\alpha_j)]^2}$.
$e = 2.71828\ldots$	Euler's number.
$e^{i\alpha}$	Euler's formula: $e^{i\alpha} = \cos\alpha + i \sin\alpha$.
$\mathbb{C}^n$	n-dimensional vector space over the field of complex numbers.
$\mathcal{H}_n$	n-dimensional Hilbert space.
$\mid \psi\rangle, \mid \varphi\rangle$	`ket` vector in Dirac's notation (e.g., $\mid \psi\rangle, \mid \varphi\rangle$ column vectors in $\mathbb{C}^4$) $$\mid \psi\rangle = \begin{pmatrix} \alpha_0 \\ \alpha_1 \\ \alpha_2 \\ \alpha_3 \end{pmatrix}, \quad \mid \varphi\rangle = \begin{pmatrix} \beta_0 \\ \beta_1 \\ \beta_2 \\ \beta_3 \end{pmatrix}.$$
$\langle\psi \mid$	`bra` vector in Dirac's notation—the dual of $\mid \psi\rangle$ (i.e., $\mid \psi\rangle^\dagger$). The row vector $\langle\psi \mid$ is the transpose of the complex conjugate of $\mid \psi\rangle$: If $\mid \psi\rangle = \begin{pmatrix} \alpha_0 \\ \alpha_1 \\ \alpha_2 \\ \alpha_3 \end{pmatrix}$ then $\langle\psi \mid = \mid \psi\rangle^\dagger = (\mid \psi\rangle^*)^T = \begin{pmatrix} \alpha_0^* & \alpha_1^* & \alpha_2^* & \alpha_3^* \end{pmatrix}$.

$\langle \psi \mid \varphi \rangle$ — The scalar (inner) product of $\mid \psi \rangle$ and $\mid \varphi \rangle$; it is a complex number:

$$\langle \psi \mid \varphi \rangle = \mid \psi^* \rangle^T \mid \varphi \rangle = \begin{pmatrix} \alpha_0^* & \alpha_1^* & \alpha_2^* & \alpha_3^* \end{pmatrix} \begin{pmatrix} \beta_0 \\ \beta_1 \\ \beta_2 \\ \beta_3 \end{pmatrix},$$

$$\langle \psi \mid \varphi \rangle = \alpha_0^* \beta_0 + \alpha_1^* \beta_1 + \alpha_2^* \beta_2 + \alpha_3^* \beta_3.$$

$\mid \psi \rangle \otimes \mid \varphi \rangle$ — The tensor product of $\mid \psi \rangle$ and $\mid \varphi \rangle$; it is a vector:

$$\mid \psi \rangle \otimes \mid \varphi \rangle = \mid \psi \rangle \mid \varphi \rangle = \begin{pmatrix} \alpha_0 \\ \alpha_1 \\ \alpha_2 \\ \alpha_3 \end{pmatrix} \otimes \begin{pmatrix} \beta_0 \\ \beta_1 \\ \beta_2 \\ \beta_3 \end{pmatrix} = \begin{pmatrix} \alpha_0\beta_0 \\ \alpha_0\beta_1 \\ \alpha_0\beta_2 \\ \alpha_0\beta_3 \\ \alpha_1\beta_0 \\ \alpha_1\beta_1 \\ \alpha_1\beta_2 \\ \alpha_1\beta_3 \\ \vdots \\ \alpha_3\beta_0 \\ \alpha_3\beta_1 \\ \alpha_3\beta_2 \\ \alpha_3\beta_3 \end{pmatrix}.$$

$\mid \psi \rangle \langle \varphi \mid$ — The outer product of $\mid \psi \rangle$ and $\mid \varphi \rangle$; it is a linear operator or a matrix:

$$\mid \psi \rangle \langle \varphi \mid = \mid \psi \rangle (\mid \varphi \rangle^*)^T = \begin{pmatrix} \alpha_0 \\ \alpha_1 \\ \alpha_2 \\ \alpha_3 \end{pmatrix} \begin{pmatrix} \beta_0^* & \beta_1^* & \beta_2^* & \beta_3^* \end{pmatrix}$$

$$= \begin{pmatrix} \alpha_0\beta_0^* & \alpha_0\beta_1^* & \alpha_0\beta_2^* & \alpha_0\beta_3^* \\ \alpha_1\beta_0^* & \alpha_1\beta_1^* & \alpha_1\beta_2^* & \alpha_1\beta_3^* \\ \alpha_2\beta_0^* & \alpha_2\beta_1^* & \alpha_2\beta_2^* & \alpha_2\beta_3^* \\ \alpha_3\beta_0^* & \alpha_3\beta_1^* & \alpha_3\beta_2^* & \alpha_3\beta_3^* \end{pmatrix}.$$

$\| \mid \psi \rangle \|$ — The norm of vector $\mid \psi \rangle$:
$\| \mid \psi \rangle \| = \sqrt{\langle \psi \mid \psi \rangle} = \sqrt{\mid \alpha_0 \mid^2 + \mid \alpha_1 \mid^2 + \mid \alpha_2 \mid^2 + \mid \alpha_3 \mid^2}$.

$\mathbf{A}$ — Linear operator.

$[\mathbf{A}, \mathbf{B}]$ — The commutator of operators **A** and **B**:
$[\mathbf{A}, \mathbf{B}] = \mathbf{AB} - \mathbf{BA}$.

$\{\mathbf{A}, \mathbf{B}\}$ — The anti-commutator of operators **A** and **B**:
$\{\mathbf{A}, \mathbf{B}\} = \mathbf{AB} + \mathbf{BA}$.

$\partial \mathbf{A} / \partial a_i$ — Partial derivative of operator **A**.

A	The matrix associated with the linear operator $\mathbf{A}$.
I_n	The $n \times n$ identity matrix. All its diagonal elements are equal to 1 and all non-diagonal elements are equal to 0. $\mathbf{I}$ is the identity operator.
$\mathrm{Tr}(A)$	The trace of matrix A; the sum of its diagonal elements.
$\det(A) = \mid A \mid$	The determinant of matrix A.
M^A_{ij}	Minor obtained by eliminating row i and column j from A.
$GL(n, \mathbb{R})$	n-linear group of matrices over $\mathbb{R}$. The set of $n \times n$ matrices with elements from $\mathbb{R}$ and with non-zero determinant form a group.
$GL(n, \mathbb{C})$	n-linear group of matrices over $\mathbb{C}$. The set of $n \times n$ matrices with elements from $\mathbb{C}$ and with non-zero determinant form a group.
$\mathcal{G}$	A matrix group, any linear subgroup of $GL(n, \mathbb{R})$ or $GL(n, \mathbb{C})$.
ρ	An n-linear representation of a group G, which preserves the identity: $\rho : G \mapsto \mathcal{G}$ such that $\rho(e) = I_n$.
$\chi_\rho(g)$	The character of a matrix representation of a group G: $\chi_\rho(g) = Tr[\rho(g)], \forall g \in G$.
A^T	Transpose of matrix A_{mn}; row $i, 1 \leqslant i \leqslant m$ of A_{mn} becomes column i of A^T_{mn}.
A^*	Complex conjugate of matrix A_{mn}: $a_{ij} \rightarrow a^*_{ij}, 1 \leqslant i \leqslant m, 1 \leqslant j \leqslant n$.
$A^\dagger$	Hermitian conjugate, or dual of A: $A^\dagger = (A^*)^T$.
δ_{ij}	Kronecker's delta function: $\delta_{ij} = \begin{cases} 0 & \text{if } (i \neq j) \\ 1 & \text{if } (i = j) \end{cases}$.
$\Delta X \, \Delta P_X \geq \frac{\hbar}{2}$	A formulation of Heisenberg's Uncertainty Principle; ΔX and ΔP_X are the indeterminacies in particle position and momentum along direction X, respectively.
$i\hbar \frac{d}{dt}\Psi = \mathbf{H}\Psi$	Schrödinger's Equation.
$\mathbf{H}$	Hamiltonian operator.
$p = h/\lambda$	de Broglie's Equation. p is the momentum of a particle and λ is the wavelength of the wave associated with it.
$\mid \psi_i \rangle$	The state of a quantum system.
$\langle A \rangle$	The expected value of observable A of N independent copies of a quantum system in state $\mid \psi_i \rangle$: $\langle A \rangle = (1/N) \sum_{i=1}^{N} \langle \psi_i \mid A \mid \psi_i \rangle$.
ρ	The density matrix of N independent copies of a quantum system in state $\mid \psi_i \rangle$: $\rho = (1/N) \sum_{i=1}^{N} \mid \psi_i \rangle \langle \psi_i \mid$.

$I, \sigma_x, \sigma_y, \sigma_z$ — The identity (I) and the Pauli matrices:

$$I = \sigma_0 = \begin{pmatrix} 1 & 0 \\ 0 & 1 \end{pmatrix}, \quad \sigma_x = \sigma_1 = X = \begin{pmatrix} 0 & 1 \\ 1 & 0 \end{pmatrix},$$

$$\sigma_y = \sigma_2 = Y = \begin{pmatrix} 0 & -i \\ i & 0 \end{pmatrix}, \quad \sigma_z = \sigma_3 = Z = \begin{pmatrix} 1 & 0 \\ 0 & -1 \end{pmatrix}.$$

H, S, T — The `Hadamard` (H), `phase` (S) and $\pi/8$ (T) matrices for one qubit gates:

$$H = \frac{1}{\sqrt{2}} \begin{pmatrix} 1 & 1 \\ 1 & -1 \end{pmatrix} \quad S = \begin{pmatrix} 1 & 0 \\ 0 & i \end{pmatrix} \quad T = \begin{pmatrix} 1 & 0 \\ 0 & e^{i\pi/4} \end{pmatrix}.$$

G_{CNOT} — The matrix of the `controlled`-NOT, (`CNOT`), a two-qubit gate:

$$G_{\text{CNOT}} = \begin{pmatrix} 1 & 0 & 0 & 0 \\ 0 & 1 & 0 & 0 \\ 0 & 0 & 0 & 1 \\ 0 & 0 & 1 & 0 \end{pmatrix}.$$

G_{Toffoli} — The matrix of the `Toffoli` gate, a three-qubit gate:

$$G_{\text{Toffoli}} = \begin{pmatrix} 1 & 0 & 0 & 0 & 0 & 0 & 0 & 0 \\ 0 & 1 & 0 & 0 & 0 & 0 & 0 & 0 \\ 0 & 0 & 1 & 0 & 0 & 0 & 0 & 0 \\ 0 & 0 & 0 & 1 & 0 & 0 & 0 & 0 \\ 0 & 0 & 0 & 0 & 1 & 0 & 0 & 0 \\ 0 & 0 & 0 & 0 & 0 & 1 & 0 & 0 \\ 0 & 0 & 0 & 0 & 0 & 0 & 0 & 1 \\ 0 & 0 & 0 & 0 & 0 & 0 & 1 & 0 \end{pmatrix}.$$

W_j — The Walsh-Hadamard Transform of j qubits:
W_j is defined recursively as:
$W_1 = H, \quad W_{j+1} = H \otimes W_j, j > 1.$

$u \oplus v$ — Sum modulo 2 (`XOR`). Given binary n-tuples
$u = (u_0, u_1, \ldots, u_{n-1})$ and $v = (v_0, v_1, \ldots, v_{n-1})$ then
$u \oplus v = (u_0 \oplus v_0, u_1 \oplus v_1, \ldots, u_{n-1} \oplus v_{n-1})$
$u_i \oplus u_i = 1$ if ($u_i = 0$ and $v_i = 1$) or ($u_i = 1$ and $v_i = 0$),
$u_i \oplus u_i = 0$ if ($u_i = 0$ and $v_i = 0$) or ($u_i = 1$ and $v_i = 1$).

$\lfloor a \rfloor$ — Largest integer smaller than or equal to the real number a, floor (a).

$\gcd(a, b)$ — The greatest common divisor of integers a and b. The largest integer dividing both a and b.

$\mid a \mid$ — The absolute value of integer a.

$\text{lcd}(a, b)$ — The least common multiple of integers a and b.

$\mid \mathcal{A} \mid$ — The cardinality of the set $\mathcal{A}$. The number of elements of $\mathcal{A}$.

F^* — The set of non-zero elements of a field F, $\{F - \{0\}\}$.

$\text{ord}(\alpha)$	The order of the element $\alpha \in F$, with F a finite field.
$a^{-1} \bmod n$	The multiplicative inverse of integer a modulo n: $a \cdot a^{-1} \equiv 1 \bmod n$.
$\mathbb{Z}_p[x]$	Polynomials in x with coefficients from the finite field $\mathbb{Z}_p$.
$[g(x)]$	The equivalence class containing the polynomial $g(x)$.
(a_0, a_1, a_2, a_3)	Vector representing the polynomial $g(x) = a_0 + a_1x + a_2x^2 + a_3x^3$.
$V_k(F)$	Vector space of k tuples over the field F.
$S = k_B \ln(\Omega)$	The thermodynamic entropy S. k_B is the Boltzmann's constant. Ω is the number of microstates.
X	Discrete or continuous random variable.
$p_X(x)$	The probability density function of the random variable X.
$p_{X,Y}(x, y)$	The joint probability density function of random variables X and Y.
$p_{X\mid Y}(x \mid y)$	The conditional probability density function of X conditioned by Y.
$F_X(x)$	The cumulative distribution function of the random variable X. $F_X(x) = \text{Prob}(X < x) = \int_{-\infty}^{x} p_x(t)dt \rightarrow$ *continuous case.*
$\mathbf{E}\,[X]$	The expected value of the random variable X: $\mathbf{E}[X] = \begin{cases} \sum_{i=1}^{n} x_i p_X(x_i) & \rightarrow \textit{discrete case} \\ \int x p_X(x)dx & \rightarrow \textit{continuous case.} \end{cases}$
$\text{Var}[X]$	The variance of the random variable X: $\text{Var}[X] = \begin{cases} \sum_{i=1}^{n} (x_i - \mathbf{E}[X])^2 p_X(x_i) & \rightarrow \textit{discrete case} \\ \int_{-\infty}^{+\infty} (x - \mathbf{E}[X])^2 p_X(x)dx & \rightarrow \textit{continuous case.} \end{cases}$
$H(X)$	The Shannon entropy of random variable X: $H(X) = \begin{cases} -\sum_{i=1}^{n} p_X(x_i) \log_2 p_X(x_i) & \rightarrow \textit{discrete case} \\ -\int p_X(x) \log_2 p_X(x)dx & \rightarrow \textit{continuous case.} \end{cases}$
$H(X, Y)$	The joint entropy of X and Y: $H(X, Y) = \begin{cases} -\sum_{x_i} \sum_{y_j} p_{X,Y}(x_i, y_j) \times \log_2 p_{X,Y}(x_i, y_j) \\ -\int_X \int_Y p_{X,Y}(x, y) \times \log_2 p_{X,Y}(x, y). \end{cases}$
$[n, M]$-code	Block code with M codewords each of length n.
(n, k)-code	Linear code with k information symbols and blocks of length n.
(n, k, d)-code	Linear code with k information symbols, blocks of length n, and distance d.
$C^{\perp}$	Orthogonal element of code C.
$\{n, k, d_1, d_2\}$	Quantum code where n qubits are used to store or transmit k bits of information and allow correction of up to $\lfloor (d_1 - 1)/2 \rfloor$ amplitude errors and simultaneously up to $\lfloor (d_2 - 1)/2 \rfloor$ phase errors.

CHAPTER 1
Introduction

Two of the greatest scientific discoveries of the twentieth century—quantum mechanics and the general theory of relativity—target physical phenomena which we rarely, if ever, experience in our daily lives. After all, no human being has ever traveled at a speed approaching the speed of light, nor are we often in the position to observe quantum effects on Earth.

Quantum is a Latin word meaning "some quantity." In physics it is used with the same meaning as the word *discrete* in mathematics; it refers to a quantity or variable that can take only sharply defined values, as opposed to a continuously varying quantity. The concepts of *continuum* and *continuous* are known from geometry and calculus.

It is extremely difficult to accept the non-determinism governing the quantum world. Some of the greatest minds of our time, including Albert Einstein, had doubts about such unsettling ideas. More troubling for most of us is the concept of *non-locality*: the fact that two quantum objects can influence one another, even when separated by a distance measured in light years, and that the change of state of one determines an instantaneous change of state of the other.

Most laws governing the physical phenomena studied by quantum physics are counterintuitive and can only be presented with the aid of a sophisticated mathematical apparatus applied to a mathematical model of the physical system. We have to free our thinking from sensory information and step into a very different world. For example, one would expect the same result when first measuring the position of an object and then its velocity, or first measuring the object's velocity and then its position. This is true for a car, an airplane, a rocket, or a bullet in flight, but it is false for quantum particles such as electrons or photons. Measuring the position of a quantum particle first makes it impossible to determine its velocity with an arbitrary high level of accuracy. The reverse is also true; measuring the velocity first makes it impossible to determine the position of the particle with an arbitrary high level of accuracy. This strange phenomenon can be understood if we construct a mathematical model of the process of measuring a physical property of a quantum system. The explanation of this effect has its roots in a trivial property of the matrix multiplication: the product of two matrices is non-commutative.[1] Measuring a property (e.g., the velocity or the position) of a quantum particle corresponds to a mathematical operation of applying the operator associated with that property (observable) to the vector describing the state of the system. A linear operator in quantum mechanics has a matrix associated with it; therefore, the non-commutativity of the product of two matrices implies that the order of the measurements is important.

The laws of physics, and, in particular, the laws of quantum mechanics limit our ability to process information increasingly faster and cheaper using present day solid state technologies. The question addressed by the new discipline of quantum information processing is whether the strange world of quantum phenomena can be harnessed and eventually put to "good use." It turns out that communication and computer systems using quantum effects have remarkable properties. A quantum communication channel transmits information using quantum effects (e.g., when the information is encoded into the spin of an electron, or in the polarization of a photon). Eavesdropping on a quantum communication channel can be detected with a very high probability. Quantum computing enables efficient simulation of the most complex physical systems that we can envision. Quantum algorithms have been discovered that allow efficient factoring of large integers. This is of immense practical interest because efficient factoring algorithms would allow us to decrypt with ease communication based upon today's state of the art encryption techniques. Parallel search quantum algorithms are equally promising for data mining and other important applications.

This chapter starts with a discussion of the laws of physics that limit our ability to build faster computers using current technologies. Following this introduction we present several experiments that provide some insight regarding quantum effects along with a practical application of quantum information theory. To conclude the chapter we outline the milestones leading to the new discipline of quantum computing.

1.1 COMPUTING AND THE LAWS OF PHYSICS

Computers are systems subject to the laws of physics. One of these laws—the finite speed of light—limits the potential reliability of future computing systems. The components of a computer exchange information among themselves (e.g.,

[1] If A and B are two arbitrary matrices, then their product is non-commutative, $AB \neq BA$.

the processor reads and writes information from/to memory, data is transferred from the internal registers to the Arithmetic and Logic Unit (ALU) and so on). Transmission of information is associated with a transport of energy from the source to the destination. The speed of this transmission is limited; it takes one nanosecond (10^{-9} second) for an electromagnetic wave to travel a distance of 30 cm in vacuum and about 20 cm in a metallic conductor. Therefore, the speed of a computer is limited by the size of its components. It is inevitable that in our quest to increase the speed, the components of a computer will approach atomic dimensions.[2] When this happens, switching (the change of the state of a component) will be governed by Heisenberg's Uncertainty Principle.[3] It follows that we may not be able to determine the state of that component with absolute certainty and the results of a computation, while carried out by a very fast computer, will be unreliable.

The technology enabling us to build smaller and faster computing engines encounters other physical limitations as well. We are limited in our ability to increase the density and the speed of a computing engine. Indeed, the heat produced by a super dense computing engine is proportional to the number of elementary computing circuits, thus, to the volume of the engine. To prevent the destruction of the engine we have to remove the heat through a surface surrounding the device (see Figure 1.1(b)). Let us assume that we pack the logic gates as densely as possible in a sphere of radius $r \geqslant 1$. The heat dissipated is then proportional to the number of gates and thus, to the volume of the sphere, $V = (4/3)\pi r^3$. At the same time, the heat can only be removed through the surface of the sphere $A = 4\pi r^2$. Therefore, *the amount of heat increases with the cube of the radius of the sphere, while our ability to remove heat increases as the square of the radius of the sphere.*

If a small, but finite amount of energy, ϵ, is dissipated to perform an elementary operation, then the increase in speed of the computing device requires at least[4] a linear increase of the amount of energy dissipated by the device (see Figure 1.1(a)). The computer technology vintage year 2000 requires some $\epsilon = 3 \times 10^{-18}$ Joules per elementary operation. Even if this limit is reduced, say 100-fold, we shall see a 10-fold increase in the amount of power needed by devices operating at a speed 10^3 times larger than the speed of today's devices. Moreover, the heat will be dissipated within a much smaller volume and our ability to cool the system will be diminished.

---

[2]The atoms are $(1 - 2) \times 10^{-10}$ m in radius.

[3]According to Heisenberg's Uncertainty Principle we cannot determine both the position and the momentum of a quantum particle with arbitrary precision. In his Nobel Prize lecture on December 11, 1954 Max Born said of this fundamental principle of Quantum Mechanics: "... It shows that not only the determinism of classical physics must be abandoned, but also the naive concept of reality which looked upon atomic particles as if they were very small grains of sand. At every instant a grain of sand has a definite position and velocity. This is not the case with an electron. If the position is determined with increasing accuracy, the possibility of ascertaining its velocity becomes less and vice versa."

[4]In fact the power is proportional to the cube of the frequency.

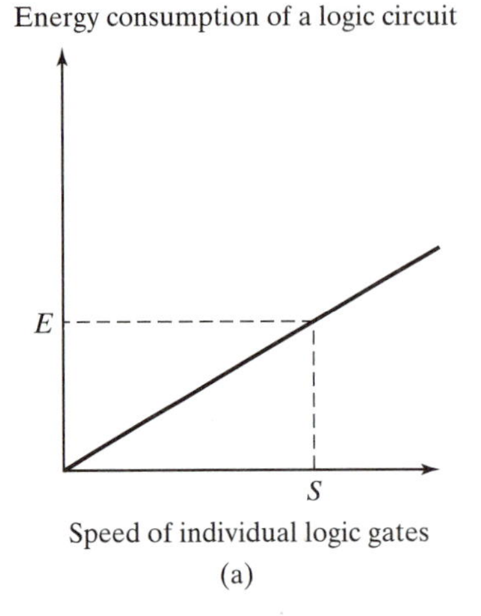

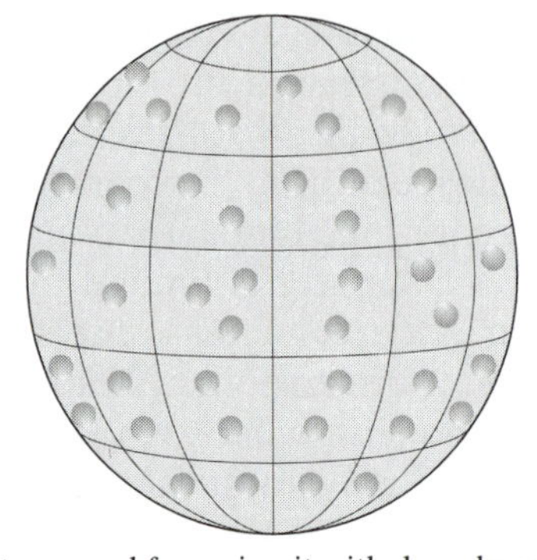

FIGURE 1.1: Energy consumption and heat dissipation of classical logic gates. (a) If there is a minimum amount of energy ϵ necessary to perform an elementary logic operation, the best we can hope for is a linear increase of the amount of energy used by the device as the speed of the logic gates increases. Thus, the energy required at speed S is $E = \epsilon S$. (b) To increase the speed of the device we need to pack the logic gates as densely as possible, say in a sphere of radius $r \geqslant 1$. The heat dissipated will be proportional with the number of gates and therefore with the volume of the sphere, $V = (4/3)\pi r^3$. However, the heat can only be removed through the surface of the sphere $A = 4\pi r^2$.

In 1992, Ralph Merkle from Xerox PARC calculated that a 1 GHz computer operating at room temperature, with 10^{18} gates packed in a volume of about 1 cm^3 (the size of a sugar cube) would dissipate 3 MW of power [29]. A small city with 1000 homes each using 3 kW would require the same amount of power; a 500 MW nuclear reactor could only power some 166 such circuits.

The laws of classical physics limit our ability to compute faster and cheaper, requiring us to look elsewhere for an answer and to consider a revolutionary rather than an evolutionary approach to computing. One of the directions to be explored is that of quantum physics.

Does quantum theory play only a supporting role by defining the limitations of present day physical systems used for computing and communication? The answer is a resounding no. Quantum properties such as uncertainty, interference, and entanglement form the foundation of a new brand of theory, the quantum information theory, where computational and communication processes rest upon fundamental physics.

Quantum computing is not fiction. We stand a reasonable chance of building quantum computers during the next several decades and we expect amazing results.

1.2 QUANTUM INFORMATION

Until recently, information and computation models have had a feeble connection to physics. Complexity theory addressed the time and space complexity of algorithms. Time and space are physical attributes, thus a connection with physical reality is still maintained. Information theory was concerned with entropy as a measure of the uncertainty associated with a random event and its relationship with information transmission over communication channels. It is fair to say that our information and computation models lacked physical awareness and required little understanding of the basic laws of physics. The laws of quantum mechanics have been viewed more as an annoyance than a necessity; they may be useful at a distant point in the future when the physical systems used to store and transfer information will consist of a few atoms only.

This view is now challenged by significant results streaming from the new discipline of *quantum computing*. Quantum computing and quantum information theory are concerned with the transmission and processing of quantum states and the interactions of such quantum information with the "classical" one.

A quantum bit, or *qubit* for short, is a quantum system used to store information. As opposed to a classical bit, which can be in one of the two states 0 and 1, a qubit can exist in a continuum of states. Moreover, *we can measure the value of a bit with certainty without affecting its state, while the result of measuring a qubit is non-deterministic and the measurement alters its state.* A bit in state 0 measured in an existing computer yields an answer of 0, and remains in this state following the measurement.

A classical communication channel allows electromagnetic waves to propagate and is characterized by its capacity, the maximum quantity of information that can be transmitted through the channel per unit of time. A *quantum communication channel* is a physical system capable of delivering quantum systems more or less intact from one place to another.

The path from classical information to quantum information is a process of extension, refinement, and completion of our knowledge. It follows the evolution of our thinking in other areas of science. Consider the example of number theory. One started with a concept inspired by the physical reality, positive integers; then one realized that the additive inverse of a positive integer must be defined, and negative integers were born. Soon one discovered that the multiplicative inverse of an integer is needed and the rational numbers were introduced. After a while irrational numbers, and then real numbers, were added to the family of numbers.

1.3 QUANTUM COMPUTERS

Let us take a closer look at computing machines to catch a glimpse of the potential advantages of their quantum incarnation. A classical computing engine is a deterministic system evolving from an input/initial state to a final/output state. The input state and all states traversed by the system during its dynamic evolution have some canonical labeling. Moreover, the label of each state can be measured by an external observer. The label of the final state is a deterministic function f of the label of the input state; we say that the engine computes a "function" f. Two classical computing engines are *equivalent* if they compute the same function f given the same labeling of their input and output states [42].

Quantum computers are stochastic engines because the state of a quantum system is uncertain and, therefore, a certain probability is associated with any possible state the system can be in. The output states of a stochastic engine are random: the label of the output state cannot be discovered. All we can do is to label a set of pairs consisting of an output state of an observable and a measured value of that observable. In layman's terms, *observable* means a characteristic, or attribute; in quantum mechanics we say that each pair consists of an eigenstate of a Hermitian operator and its eigenvalue, as we shall see in Chapter 2.

There is an asymmetry between the input and the output of a quantum engine reflecting the asymmetry between preparation and measurement in quantum mechanics. We put more information into the preparation of a quantum system than we can get back out in a single measurement of that system.

Quantum mechanics is a mathematical model of the physical world. This model allows us to specify states, observables, measurements, and the dynamics of quantum systems. As we shall see later, a Hilbert space, a space of n-dimensional complex vectors is the center stage for quantum mechanics.

A Hilbert space is indeed a very large space and we need every bit of it. Let us revisit the statement made earlier that numerical simulation of physical processes was, and continues to be, the motivation for the development of increasingly more powerful computing engines. Whether scientists look up at the sky and try to answer fundamental questions related to the evolution of the universe or try to decipher the structure of matter, they need to simulate increasingly more complex systems. One measure of the complexity of a system is the number of states the system can be in. For example, the theory of black hole thermodynamics predicts [42] that a system enclosed by a surface with an area A has a number $N(A)$ of observable states given by:

$$N(A) = e^{Ac^3/4\hbar G}$$

with $c = 3 \times 10^{10}$ cm/second, the speed of light;
$\hbar = 1.054 \times 10^{-34}$ Joules $\times$ seconds is Planck's constant; G the gravitational constant, $G = 6.672 \times 10^{-8}$ cm^3/(gram second^2). Even for a relatively small spherical object with a radius of 1 km, $N(A) \approx e^{80}$, which is a very large number.

Let us now turn our attention to the concept of numerical simulation of a physical system. In order to simulate the behavior of a physical system we first construct a model of the system. We then have to design a program that reflects the essential properties of the physical system captured by the model.

A model is an abstraction of a physical system. Therefore, the result of the simulation can only be as good as the model. For example, an atomic model of a virus structure describes the geometric position and the type of each atom. The atomic model of a virus structure is useful for virological and immunological studies and for the design of antiviral drugs. Antiviral drugs block the binding site of a virus, and prevent it from infecting healthy cells. Typically a macromolecular structure such as a virus has several million atoms. If instead of describing the position and the type of each atom we describe only groups of atoms (e.g., groups of 10^2 atoms), then we reduce the complexity of the model (instead of 10^6 objects we describe only 10^4), while making it less accurate. If the model of the virus is less accurate, then the antiviral drugs designed based upon this model are less likely to be effective.

Suppose the model of the system has n states. The simulation program must contain a description of each state, making the space complexity of the simulation program at least $\mathcal{O}(n)$. The model should also describe the dynamics of the system. This implies that the simulation program should describe the set of transitions among the n states and the action(s) associated with every transition. Assuming that each state is reachable from every other state, the dynamics of the system requires the description of the action(s) taken for each of the $\mathcal{O}(n^2)$ transitions. Multiple actions may be associated with a transition from a state s_i to a state s_j, depending upon the past history of the system, reflected by the path followed by the system to reach state s_i.

We conclude that the time (and the space) complexity of a simulation algorithm is at least $\mathcal{O}(n)$, where n represents the number of states of the model. To carry out a numerical simulation we need a physical computer with sufficient resources (mainly CPU rate and primary memory) to complete the simulation in a time frame of hours or days rather than years.

We have to balance the accuracy of the model against the feasibility of the numerical simulation. Today's computers cannot simulate a system with a very large state space making an accurate simulation of a physical system rarely possible. Quantum mechanics allows us to accommodate an extremely large state space, and may lead to an exact simulation of a physical system.

Mathematically, the state, $| \psi \rangle$, of a *quantum bit* (*qubit*) is represented as a vector in a two-dimensional complex vector space.[5] A vector in this space has two components and the projections of the state vector on the basis vectors are complex numbers.

While a classical bit can be in one of two states, 0 or 1, the qubit can be in states $| 0 \rangle$, and $| 1 \rangle$ called *computational basis states* and also in any state that is a linear combination of these states. This phenomenon is called *superposition*.

Consider now a system consisting of n such particles whose individual states are described by vectors in the two-dimensional vector space. In classical physics the individual states of particles combine through the Cartesian product.

The possible states of the quantum system of n particles form a vector space of 2^n dimensions; given n bits, we can construct 2^n n-tuples and describe a system with 2^n states. Individual state spaces of n particles combine quantum mechanically through the tensor product. If X and Y are vectors, then their tensor product $X \otimes Y$ is also a vector, and its dimension is $\dim(X) \times \dim(Y)$. For example, if $\dim(X) = \dim(Y) = 10$, then the tensor product of the two vectors has dimension 100.

The state space of a quantum system having n qubits has 2^n dimensions. There are 2^n *basis states* forming a computational basis and there are *superposition states* resulting from the superposition of basis states. The catch is that even though

[5]A detailed explanation of the terms used now can be found in Chapter 2.

one quantum bit—a system with 2^1 basis states—can be in one of infinitely many superposition states, when the qubit is measured the measurement changes the state of the quantum system to one of the two basis states. *From one qubit we can only extract a single classical bit of information.*

In quantum systems *the amount of parallelism increases exponentially with the size of the system*, thus it increases exponentially with the number of qubits. This means that the price paid for an exponential increase in the power of a quantum computer is a *linear increase in the amount of matter and space* needed to build the larger quantum computing engine. Adding a single qubit doubles the power of a quantum computer. For example, a quantum computer with 11 qubits has twice the computational power of a quantum computer with 10 qubits.

Access to the results of a quantum computation is restrictive because any interaction of an outside observer with a quantum system disturbs the quantum state. The process of disturbing the quantum state through the interaction with the environment is called *decoherence*. This is a major problem we have to overcome in designing quantum algorithms.

Two photons or two electrons can be in a superposition state of close coupling with each other, in an intimately fused state known as an *entangled state*, a state with no classical analogy. *Entanglement* is the exact translation of the German term *Verschränkung* used by Schrödinger who was the first to recognize this quantum effect. It means that the state of a two-particle quantum system cannot be written as a tensor product of the states of the individual particles. The state of an entangled system cannot be decomposed into contributions of individual particles.

Two quantum particles may be in an entangled state even if they are not in close proximity to each other. Even when the entangled particles are separated from one another, a change of state of one of the entangled particles instantaneously affects the other particle and determines its state. For example, the *singlet state*[6], the antisymmetric state of a pair of electrons with antiparallel spins is an entangled state. The two electrons find themselves in the same orbital (three quantum numbers are identical) but, according to the *Pauli exclusion principle*, they have their spins oriented in opposite directions: $\mid 1 \uparrow\rangle \mid 2 \downarrow\rangle$ or $\mid 1 \downarrow\rangle \mid 2 \uparrow\rangle$. The notation $\mid 1 \uparrow\rangle \mid 2 \downarrow\rangle$ means that the first electron has its spin "up" and the second one has its spin "down." If one of the electrons is made to change the orientation of its spin as a result of an experiment, then a simultaneous measurement of the other finds it in a state with an opposite spin.

It is very difficult to provide an intuitive and simple explanation of this subtle phenomenon which has no classical analogy. With the risk of oversimplification consider an example of entanglement taken from our daily life. When Alice and Bob, two characters from many cryptography texts, get married to each other, their lives become entangled. After a few months they find out that

[6]Since their spins are antiparallel, the two electrons in a singlet state $\mid \uparrow \downarrow\rangle - \mid \downarrow \uparrow\rangle$ have different spin quantum numbers $+1/2$ and $-1/2$. The total spin of the state is zero.

Alice is pregnant. Shortly afterwards, Bob takes off for an intergalactic voyage to Andromeda. The very moment Alice gives birth to their child, Bob's state changes instantly, despite the fact that he may be half a light-year away. Bob becomes a father. An external observer could see the baby and decide that Bob's state has changed.[7]

Entanglement is a very puzzling phenomenon. Richard Feynman writes [53]: "A description of the world in which an object can apparently be in more than one place at the same time, in which a particle can penetrate a barrier without breaking it, in which widely separated particles can cooperate in an almost psychic fashion, is bound to be both thrilling and bemusing."

In a recent paper, Charles Bennet and Peter Shor [16], two of the pioneers of quantum computing and quantum information theory, discuss the similarities and dissimilarities between classical and quantum information. They point out that "classical information can be copied freely, but can only be transmitted forward in time to a receiver in the sender's forward light cone. Entanglement, by contrast cannot be copied, but can connect any two points in space-time. Conventional data-processing operations destroy entanglement, but quantum operations can create it, preserve it, and use it for various purposes, notably speeding up certain computations and assisting in the transmission of classical data (quantum superdense coding) or intact quantum states (teleportation) from a sender to a receiver."

These facts are intellectually pleasing, but bring to mind two questions. Can such a quantum computer be built? Are there algorithms capable of exploiting the unique possibilities opened by quantum computing?

The answer to the first question is that *only a seven-qubit quantum computer has been built so far* [139]; several proposals to build quantum computers using nuclear magnetic resonance, optical and solid state techniques, and ion traps are being studied. The answer to the second question is that problems in integer arithmetic, cryptography, or search problems have surprisingly efficient solutions in quantum computing.

We have a reasonable hope that quantum computers can be built. When the time comes, quantum computers will be able to solve computational problems that are unsolvable today. These problems will remain unsolvable using classical computers, even if we assume that Moore's law[8] governing the rate of increase of the speed of classical devices will continue to hold for the next few decades.

There are skeptics who have serious doubts about the future of quantum computers. One of them, Dyakonov, contrasts the excitement generated by recent results in quantum information theory with the enormously difficult

[7]There is no physical change in Bob's state but only a (somewhat abstract) change in his family status as a result of his wife's (very physical) birthing process. Bob learns of his new state only when a message sent through a classical communication channel reaches him, but this is besides the point.

[8]Moore's law states that the speed of microprocessors doubles every 18 months.

problems posed by the actual physical realization of a quantum computing device[9] [48].

1.4 THE WAVE AND THE CORPUSCULAR NATURE OF LIGHT

Sophisticated quantum phenomena such as interference and entanglement are critical to understanding quantum computing and quantum information theory. While we do not need quantum mechanics to explain most of the phenomena we observe in our daily life, there are many phenomena involving light that cannot be explained unless we accept the fact that light comes as corpuscules, or finite grains of energy. Look through a window; besides the objects in your garden you may see faint reflections of yourself. This happens because some of the photons, the discrete particles carrying light, bounce off the glass.

The nature of light was a constant source of interest for philosophers and physicists. Aristotle believed that white light was a basic single entity. An early treatise on the subject, Kepler's "Optics", did not challenge this idea. Isaac Newton's first work as the Lucasian Professor at Cambridge was in optics; in January 1670 he delivered his first lecture on the subject. The chromatic aberration[10] in a telescope lens convinced Newton that Aristotle's hypothesis was false. Newton argued that white light is a mixture of many different types of rays that are refracted at slightly different angles, and that each type of ray produces a different spectral color. Newton's "Opticks" appeared in 1704; it dealt with the theory of light and color covering the diffraction of light and "Newton's rings." To explain some of his observations he had to use a wave theory of light in conjunction with his corpuscular theory.

At the end of the nineteenth century, James Clerk Maxwell constructed a consistent theory of electromagnetism and showed that light is a form of electromagnetic radiation. Yet, as Maxwell theory was gaining universal acceptance, in

---

[9] Talking about the universal quantum gate that can transform an arbitrary state of the quantum computing engine into another state via a unitary transformation, Dyakonov tells an old joke. "During the first months of World War II, an inventor kept telling anybody willing to listen that he had an idea of extreme military value. Eventually, the rumor reached Stalin who decided to listen to the inventor.

INVENTOR: I propose that you should have three buttons on your desk colored green, blue, and white. When you push the green button, all enemy ground forces shall disappear; when you push the blue button, all enemy U-boats shall sink to the bottom of the oceans; and, when you push the white button, all airplanes shall blow up in smoke.

STALIN: How will it work?

INVENTOR: Well, it is up to your scientists and engineers to figure that out. I just gave you the idea."

[10] A lens does not focus different colors in exactly the same place because the focal distance depends on refraction. The refractive index for blue light which has a short wavelengths is larger than that of red light, which has a long wavelengths. The chromatic aberration of a lens is seen as fringes of color around an image.

1905 Einstein used Planck's new quantum theory to explain the photoelectric effect.

Einstein's explanation, involving photons, seemed to signal a return to the Newtonian model of light. Photons are like any other particles, though massless. They carry both momentum and energy and act like billiard balls. When colliding with the electrons at the surface of a metal the photons can knock them free and leave behind a positive charge. This is the simplified explanation for the photoelectric effect. The photoelectric effect is often used to measure the intensity of light. A device called a *photomultiplier* allows us to detect very low intensity light, down to individual photons.

In 1923, Louis de Broglie proposed that a wave should be associated with every kind of particle. He provided an equation that linked the momentum p of a particle to the wavelength λ of the wave associated with it and to Planck's constant h:

$$p = \frac{h}{\lambda}.$$

For example, electrons have a finite mass and exhibit a corpuscular behavior and, at the same time, are diffracted like waves (see the double slit experiments discussed in Section 2.13). Louis de Broglie's assumption opened a new chapter in quantum physics and in our understanding of the atomic and subatomic worlds.

A number of experiments involving light yield very intriguing results that cannot be explained without invoking the principles of quantum mechanics. Here we discuss several experiments to put us in the frame of mind needed for understanding quantum information. First, we present a simple experiment revealing the granular nature of light. Next, we discuss a simple model to address the non-deterministic effects. We challenge this model to explain the results of successive measurements performed upon a beam of quantum particles using different bases. Finally, we augment the model with the superposition probability rule for events that may occur on alternative paths and discuss several experiments whose results are consistent with this rule.

1.5 DETERMINISTIC VERSUS PROBABILISTIC PHOTON BEHAVIOR

Contemporary theory regards light as a flux of photons, while, at the same time, supporting the fact that light exhibits wave properties. Consider a device called a *beam splitter*, a half-silvered mirror (see Figure 1.2(a)). A beam of light falling on a beam splitter is split into two components, one transmitted and one reflected. A beam splitter may transmit a larger or a smaller fraction of the incident light depending upon the characteristics of the silver deposition. For the experiments discussed in this section the two components are of equal intensity. The color (i.e., the wavelength) of the light is not altered by a beam splitter, a behavior consistent with a wave.

If we decrease the intensity of the incident light we are able to observe the granular nature of light. Imagine sending a single photon. Either detector `D1` or detector `D2` in Figure 1.2(a) senses the photon. Repeating the experiment

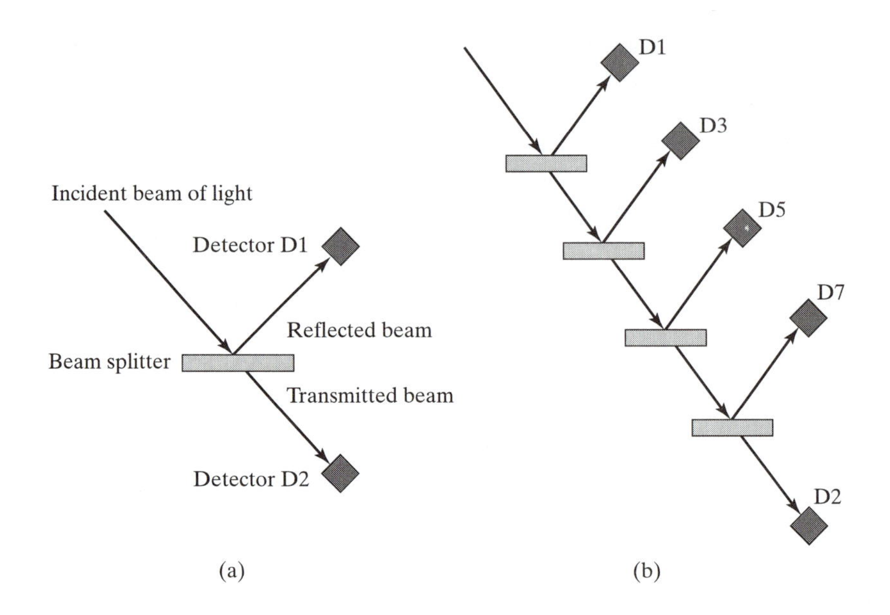

FIGURE 1.2: (a) One beam splitter. We send a single photon and observe that either detector D1 or detector D2 senses the photon. Repeating the experiment many times, we observe that each one of the two detectors records about the same number of events. (b) Cascaded beam splitters. All detectors have a chance of detecting the single photon sent in one experiment. The cascaded beam splitter experiment convinces us to dismiss the idea that a photon carries a "gene" and exhibits a deterministic behavior.

involving a single photon over and over again, we observe that each one of the two detectors records about the same number of events (one event is the detection of a photon).

This is puzzling: Could there be hidden information that controls the behavior of a photon? Does each photon carry a "transmit" or a "reflect" gene so that a photon with a "transmit" gene continues and reaches detector D2, while another photon with a "reflect" gene ends up at D1 [94]? If this is true, the two genes should have an equal probability of occurrence throughout the entire population of photons.

It is not difficult to dismiss this *genetic view of photon behavior*: Consider the setup in Figure 1.2(b) with a cascade of beam splitters. As before, we send a single photon and repeat the experiment many times, counting the number of events registered by each detector. According to our theory we expect the first beam splitter to decide the fate of an incoming photon; the photon is either reflected by the first beam splitter or transmitted by all of them. Thus, only the first and last detectors in the chain are expected to register an equal number of events.

The experiment shows that *all the detectors have a chance to register an event.* This result discredits our theory of deterministic behavior caused by a gene and leads us to seek another possible explanation of the strange effects revealed by this experiment, one based upon a probabilistic behavior of quantum particles.

In this context the term "probabilistic behavior" means that multiple outcomes of an experiment are possible and there is a certain probability of occurrence for each outcome. For example, we may assume that a photon tosses a coin when reaching a beam splitter, with the result of the toss determining the fate of the photon emerging from that beam splitter.

If a photon tosses an unbiased coin the probability of flipping four heads in a row to reach detector D2 is $p_{D2} = (1/2)^4$. Thus, if we repeat this experiment 1,000,000 times, then detector D2 will register about $1{,}000{,}000/16 = 62{,}500$ events. For this discussion we considered independent events and assumed that a photon is transmitted by a beam splitter if it flips a head and is reflected if it flips a tail.

1.6 STATE DESCRIPTION, SUPERPOSITION, AND UNCERTAINTY

Let us now try to construct a simple mathematical model for the state of a two-dimensional quantum system that can explain what we have described so far. When measured by an external observer the state should reveal one of two values that are opposite to each other, 0 or 1. Other names for the two opposite values[11] are possible (for example, "transmitted" and "reflected"; "black" and "white"; "hard" and "soft").

Let us consider that the state is represented by a vector $\mathbf{v}$ connecting the origin of the coordinates system to a point on the periphery of a unit circle. The two axes are labeled 0 and 1 and the square of the projection of the vector on each of them gives the probability of observing the outcome 0 or the outcome 1, respectively, see Figure 1.3.

The projections of the vector are called *probability amplitudes*. The probability of an outcome is the square of the number giving its probability amplitude. The two possible outcomes of the experiment occur with probabilities $p_0 = q^2$ and $p_1 = 1 - q^2$. The normalization condition requires that $p_0 + p_1 = 1$; this restricts us to vectors of length one.

In Figure 1.3(a) we illustrate the case when the two projections are equal: $q = 1/\sqrt{2}$ and thus $p_0 = p_1 = 1/2$. In Figure 1.3(b) we observe that one of the projections is larger than the second, $p_1 > p_0$, thus we are more likely to observe the outcome 1 than 0.

This very simple model has profound implications, as we shall see in the next chapter. The model is able to explain experimental results that classical physics fails to explain. A salient feature of the model is that the state represented by the vector $\mathbf{v}$ is a *superposition* of two possible basis states $| 0\rangle$ and $| 1\rangle$. Here $\mathbf{v}$ is the actual state of the system before we make an observation. To measure (observe) the system means to project the state vector on the two basis states.

The outcomes corresponding to projections on these basis states are all we can learn about the system as a result of a measurement, leaving us uncertain about the actual state of the system before we made our observations. However,

[11]We often use the mathematical term *orthogonal states* when talking of opposite valued outcomes for reasons that shall become apparent later.

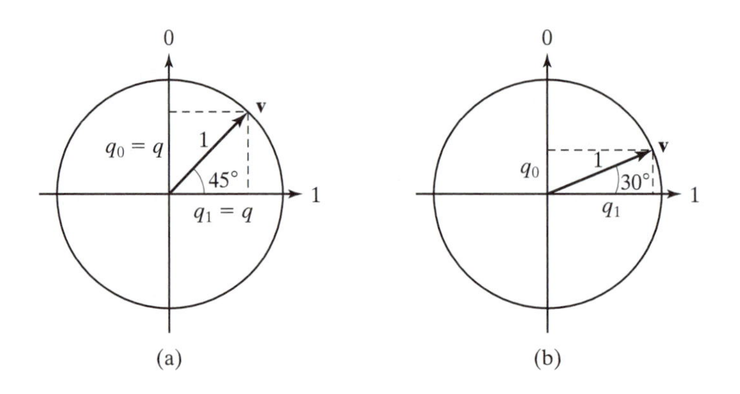

FIGURE 1.3: A model for the measurement of a quantum system in a superposition state; the binary quantum state is represented by a vector **v** of unit length. The square of the projection of the vector gives the probability of an outcome. (a) The two possible outcomes have equal probabilities, $p_0 = p_1 = q^2$ with $q = 1/\sqrt{2}$. (b) One of the outcomes, 1, has a larger probability than the other, 0, $p_1 = q_1^2 > p_0 = q_0^2$ because $q_1 > q_0$.

after we performed our observation, the system is no longer in an uncertain state; rather, it is precisely in one of the two possible states, $| 0\rangle$, or $| 1\rangle$.

This model explains the behavior we noticed in the cascaded beam splitter experiment presented in Figure 1.2(b). We use Dirac's notation[12] for the state vector of an incoming photon

$$| \psi\rangle = \alpha_t \, | t\rangle + \alpha_r \, | r\rangle$$

with α_t the probability amplitude of the photon to be transmitted and α_r the probability amplitude of the photon to be reflected. In our case, $\alpha_t = \alpha_r = 1/\sqrt{2}$, thus the probability of a photon being transmitted is equal to the probability of it being reflected, $p_t = p_r = (1\sqrt{2})^2 = 1/2$. All cascaded beam spitters are identical and the probability of the two observable events—transmission and reflection—are equal.

In this case, the quantum behavior is consistent with the classical probabilistic behavior discussed in Section 1.5. If we repeat the experiment 1,000,000 times, then `D1` will register about 500,000 events, `D3` will register about 250,000 events, `D5` will register about 125,000 events, and `D7` and `D2` will register about 62,000 events each.

[12] A common notation in linear algebra for, say, three-dimensional vectors having components a_1, a_2, a_3 is $\mathbf{v} = (a_1, a_2, a_3)$ which is equivalent to $a_1(1, 0, 0) + a_2(0, 1, 0) + a_3(0, 0, 1)$, where $(1, 0, 0)$, $(0, 1, 0)$, $(0, 0, 1)$ denote basis vectors. Sometimes the notation $\mathbf{v} = a_1\mathbf{i} + a_2\mathbf{j} + a_3\mathbf{k}$ is used, where **i, j, k** are unit vectors along x, y, z axis, respectively. Beginning with Dirac, physicists have been using a different notation for vectors in quantum mechanics. In the expression above $| \psi\rangle$ represents a vector, $| t\rangle$ and $| r\rangle$ denote unit vectors along a pair of orthogonal axes, and α_t and α_r are complex numbers, the projections of $| \psi\rangle$ along these two axes. Dirac's notation discussed in more detail in Section 2.5 simplifies the formalism for more intricate functions and makes the relationship among various functions and operators easier to understand.

There are other models that could explain the behavior we have seen so far, but it turns out that this model was not selected accidentally; it is consistent with other Gedanken (thought) experiments, as well as with real experiments in quantum physics.

We shall see in the next chapter that uncertainty and probabilistic behavior are quintessential properties of quantum systems. In fact, one of the pillars of the quantum model of the physical world is Heisenberg's Uncertainty (Indeterminacy) Principle which states that one cannot measure both the position X and the momentum P_X along direction X of a particle with arbitrary precision. The uncertainty in the measurement of the position, ΔX, and the uncertainty in the measurement of the momentum, ΔP_X, must satisfy the following inequality

$$\Delta X \times \Delta P_X \geqslant \tfrac{1}{2}\hbar$$

with $\hbar = h/2\pi = 1.054 \times 10^{-34}$ Joules $\times$ second and
$h = 6.626 \times 10^{-34}$ Joules $\times$ second.[13]

This inequality reflects the impossibility to know precisely the complete state of a quantum system. It also reflects the fact that any measurement disturbs the system being measured.

1.7 MEASUREMENTS IN MULTIPLE BASES

Let us now test the model we have developed in a more complex setting. We discuss the case when we measure properties of a quantum system using distinct vector bases.

Assume that we have a quantum particle with two properties, "color" and "hardness" [4]. The "color" of a particle may be either "white" or "black" and a particle may be "hard" or "soft." The two properties are attributed to a particle in a totally random fashion, about 50% of the particles are "white" and about 50% are "black"; similarly about 50% are "hard" and about 50% are "soft."

Imagine that we can construct a "color separation system", a box with a slit on the left-hand side, one on the right-hand side and one at the top, see Figure 1.4(a). A beam of particles enters the box through the slit on the left and the beam is split by a color-based beam splitter, the "white" particles continue along their path and exit the box through the right slit, while the "black" particles are deflected by an angle of 90° and exit through the upper slit. We construct also a "hardness separation system" (see Figure 1.4(b)) similar to the "color" separation system. This time the beam splitter lets the "soft" particles continue and deflects the "hard" ones.

When we use the "color separation system" exactly 50% of the particles, the "white" ones are allowed to continue; when we use the "hardness separation system" only 50% of incoming particles, the "soft" ones, are allowed to continue.

Now we design a more sophisticated experiment involving a sequence of three separation systems; first color, then hardness, and finally one more color separation system (see Figure 1.4(c)). They are aligned, allowing the beam of

[13]For practical purposes, Planck's constant can be expressed in units of Joules/Hz.

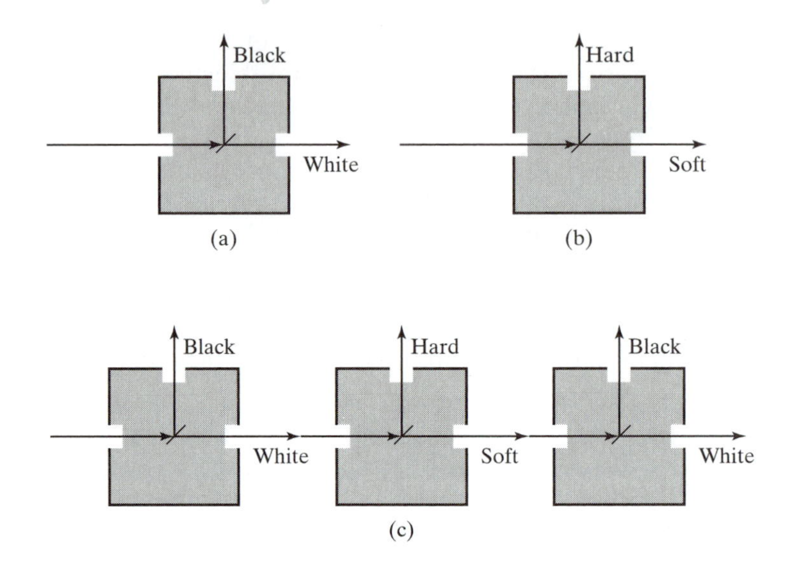

FIGURE 1.4: (a) A "color separation system." (b) A "hardness separation system." (c) Three measurement systems in cascade. We observe an equal number of "white" and "black" particles emerging from the third box though we expected that no "black" particles enter the second box and the second box cannot fabricate "black" particles.

"white" particles emerging from the first box to enter the second separation box and then the "soft" particles emerging from the second box to enter the third one. We observe an equal number of "white" and "black" particles emerging from the third box. It appears that the hardness separation system is able to "fabricate" "black" particles, even though we have taken all possible precautions in building the hardness box so that it only deflects the "hard" particles and lets the "soft" ones continue their path.

Can we explain this strange behavior using our model? The answer is yes. Recall from the description of our model in Section 1.6 that each measurement represents a projection of the state vector on the vectors forming the basis of a coordinate system. We perform three measurements in distinct bases rotated in respect to one another, with each base corresponding to one of the two separation systems, see Figure 1.5.

Let us move closer to the traditional notations used in quantum mechanics and denote the state vector $\mathbf{v}$ as $| \psi\rangle$. Assume that when we use the first separation system associated with property, $\mathcal{P}_1$, we perform the measurement on a basis formed by two orthogonal vectors $|\leftrightarrow\rangle$ and $| \updownarrow \rangle$. Call α_0 and α_1 the projections, or the "amplitudes" of $| \psi\rangle$ on these basis vectors.

Recall that the result of a measurement can only be one of the two basis vectors and assume that the result of the first measurement is the state vector $| \updownarrow \rangle$. Now we use the second separation system associated with property $\mathcal{P}_2$ and we measure on a new basis formed by the vectors $|\swarrow\!\!\nearrow\rangle$ and $|\nwarrow\!\!\searrow\rangle$. Call β_0 and β_1 the projections of $| \updownarrow \rangle$ on the new basis. Assume that the second measurement yields the result $|\swarrow\!\!\nearrow\rangle$. This result is observed with probability β_0^2 while $|\nwarrow\!\!\searrow\rangle$ is observed

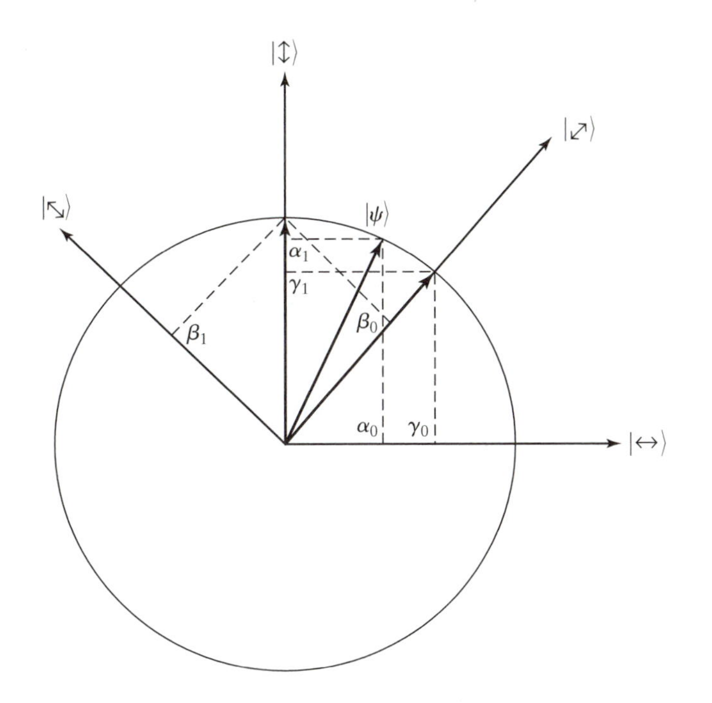

FIGURE 1.5: We perform three successive measurements and use two vector bases. The first has $|\leftrightarrow\rangle$ and $|\updownarrow\rangle$ as basis vectors, while the second has $|\swarrow\!\!\nearrow\rangle$ and $|\nwarrow\!\!\searrow\rangle$ as basis vectors. The first measurement represents the projection of the original state vector $|\psi\rangle$ on the bases formed by $|\leftrightarrow\rangle$ and $|\updownarrow\rangle$. The projections are $|\alpha_0\rangle$ and $|\alpha_1\rangle$. As a result of the first measurement we obtain either $|\leftrightarrow\rangle$, or $|\updownarrow\rangle$. Let us assume that we obtained $|\updownarrow\rangle$. The second measurement means to project the result of the first measurement, $|\updownarrow\rangle$, onto the second coordinate system formed by $|\swarrow\!\!\nearrow\rangle$ and $|\nwarrow\!\!\searrow\rangle$. Now we get two distinct projections β_0 and β_1. Assume that we observe the result $|\swarrow\!\!\nearrow\rangle$. For the third measurement we project $|\swarrow\!\!\nearrow\rangle$, the result of the second measurement, on the bases of the first coordinates system, formed by $|\leftrightarrow\rangle$ and $|\updownarrow\rangle$. The two projections are γ_0 and γ_1.

with probability β_1^2. Now we perform a third measurement using the separation system associated with property $\mathcal{P}_1$; this time the state vector $|\swarrow\!\!\nearrow\rangle$, is projected on the basis vectors $|\leftrightarrow\rangle$ and $|\updownarrow\rangle$ and its projections are γ_0 and γ_1. Thus we have non-zero probabilities to observe both $|\leftrightarrow\rangle$ and $|\updownarrow\rangle$ states. The model explains why the hardness separation system seems to fabricate "black" particles.

No refinement of the experimental setup discussed in this section can possibly alter this unexpected outcome. We have identified a practical manifestation of the fact that the quantum state is a superposition of state projections on a set of basis vectors. The state of the particles emerging from the second box is a superposition of projections on "white" and "black" basis vectors.

We have to free our thinking from the notations used and grasp the essence of the phenomena. We have a vector describing the state and we use the traditional notation of quantum mechanics introduced by Dirac [46]

$$|\psi\rangle = \alpha_0 \,|\, 0\rangle + \alpha_1 \,|\, 1\rangle$$

with α_0 and α_1 complex numbers and with $| 0\rangle$ and $| 1\rangle$ two vectors forming an orthonormal basis for the vector space.

We use the words "measurement" and "observation" rather loosely to describe an interaction of a quantum particle with its environment. Indeed, a beam splitter, or a color separation system, implies an interaction with a photon or another particle, rather than just an observation or measurement in the classical sense.

1.8 MEASUREMENTS OF SUPERPOSITION STATES

We now reinforce what we have just learned in the context of an experiment emphasizing the superposition states of light particles, photons. The concepts discussed here are analyzed in depth in most physics texts, our favorite being Feynman's [53].

Light is a form of electromagnetic radiation.[14] As an *electromagnetic radiation*, light consists of an electric and a magnetic field which are perpendicular to each other and, at the same time, perpendicular to the direction that the energy is transported in by the electromagnetic (light) wave. The electric field oscillates in a plane perpendicular to the direction of flight. The way the electric field vector travels in this plane defines the polarization of the light. We say that the light is *linearly polarized* when the electric field oscillates along a straight line. The light is *elliptically polarized* when the end of the electric field vector moves along an ellipse. The light is *circularly polarized* when the end of the electric field vector moves around a circle. If the light comes toward us and the end of the electric field vector moves around in a counterclockwise direction, we say that the light has *right-hand polarization*; if the end of the electric field vector moves in a clockwise direction, we say that the light has *left-hand polarization*. A *polarization filter* is a partially transparent material that transmits light of a particular polarization.

This experiment is discussed in [112]. We have a source S capable of generating randomly polarized light and a screen E on which we measure the intensity of the light. We also have three polarization filters, A, for vertical polarization, $\updownarrow$, B for horizontal polarization, $\leftrightarrow$, and C, for polarization at $45^\circ \nearrow$.

Using this experimental setup we make the following observations:

1. Without any filter the intensity of the light measured at E is I, see Figure 1.6(b).
2. If we interpose filter A between the source S and the screen E, then the intensity of the light measured at E is $I' \approx I/2$, see Figure 1.6(c).
3. If between filter A and the screen E in the previous setup we interpose filter B, then the intensity of the light measured at E is $I'' = 0$, see Figure 1.6(d).
4. If we interpose filter C between filters A and B in the previous setup, the intensity of the light measured at E is $I''' \approx I/8$, see Figure 1.6(e).

[14] The wavelength of the radiation in the visible light spectrum varies from 780 nm (red) to 390 nm (violet) (1 nm $= 10^{-9}$ m).

The photons emitted by the source S have random polarizations; this means that the vectors describing the polarization of the photons have random orientations, see Figure 1.6(a).

Let us denote by $| \psi \rangle$ a two dimensional vector representing the random polarization of a photon.[15] The vector $| \psi \rangle$ can be expressed as a linear combination of a pair of orthogonal basis vectors. We can use different orthonormal bases and we choose one denoted as $| \updownarrow \rangle$ and $| \leftrightarrow \rangle$ to represent the vertical and the horizontal polarization, respectively:

$$| \psi \rangle = \alpha_0 \, | \updownarrow \rangle + \alpha_1 \, | \leftrightarrow \rangle, \quad | \alpha_0 |^2 + | \alpha_1 |^2 = 1.$$

Measuring the polarization of a photon is equivalent to projecting the random vector $| \psi \rangle$ onto one of the two basis vectors. The measurement performed by a vertical polarization filter, $| \updownarrow \rangle$, provides an answer to the question: "Does the incoming photon have a vertical polarization?" Similar statements can be made about horizontal polarization filters, 45° polarization filters, or filters with any other polarization.

After the measurement in Figure 1.6(c), the superposition state $| \psi \rangle$ is resolved as one of the basis states, $| \updownarrow \rangle$, or $| \leftrightarrow \rangle$, a photon is forced to choose either the vertical, $| \updownarrow \rangle$, or the horizontal, $| \leftrightarrow \rangle$ polarization state. The probability that a photon in state $| \psi \rangle$ with random polarization is forced to choose the $| \updownarrow \rangle$ state is

$$p_{\updownarrow} = | \alpha_0 |^2.$$

The probability that the photon is forced to choose the $| \leftrightarrow \rangle$ state is

$$p_{\leftrightarrow} = | \alpha_1 |^2.$$

The sum of the two probabilities must be 1, each photon is forced to choose one of the two basis states

$$p_{\updownarrow} + p_{\leftrightarrow} = 1.$$

Once the choice is made, only those photons which have made a choice agreeing with the polarization of the filter are allowed to pass. This explains the results of the observations (2) and (3) above, see Figure 1.6(c) and 1.6(d). Indeed, due to the random polarization of the photons emitted by the source only about 50% of them emerge as vertically polarized from filter A. None of the photons can make it through filter B; all of them have a vertical polarization, $| \updownarrow \rangle$, thus the projection of their polarization on $| \leftrightarrow \rangle$ basis vector is zero. The probability of each photon to reach the screen E is zero.

Up to now everything seems clear and reasonable. The interesting fact is the introduction of filter C with a 45° polarization between filters A with vertical polarization and B with horizontal polarization. Filter C measures the quantum

---

[15] A randomly polarized photon is described by its density matrix as discussed in Chapter 2. The state vector representation is only valid when the phases of the coefficients α_0 and α_1 are random with respect to each other.

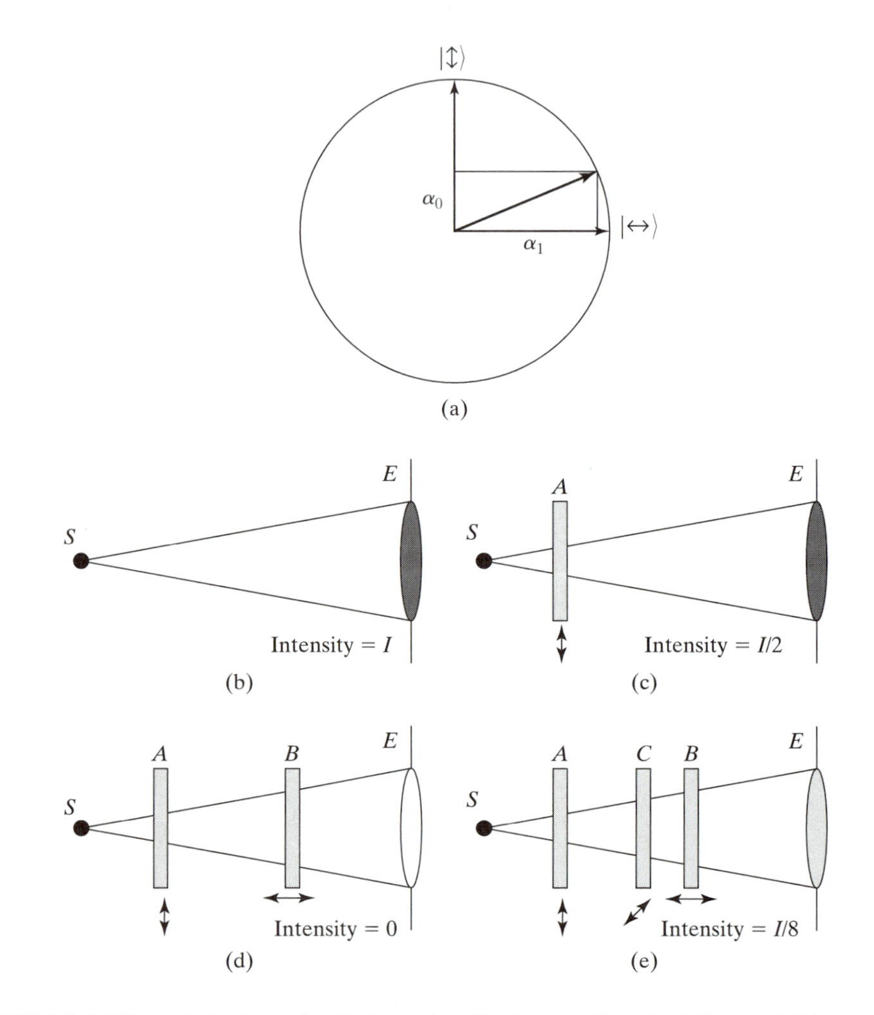

FIGURE 1.6: (a) The polarization of a photon is described by a unit vector $|\psi\rangle = \alpha_0 \mid \updownarrow\rangle + \alpha_1 \mid\leftrightarrow\rangle$ on a two-dimensional space with basis $\mid \updownarrow\rangle$ and $\mid\leftrightarrow\rangle$. Measuring the polarization is equivalent to projecting the random vector $\mid\psi\rangle$ onto one of the two basis vectors. The two projections are α_0 and α_1. Once the choice is made, only those photons which have made a choice agreeing with the polarization of the filter are allowed to pass. (b) Source S sends randomly polarized light to the screen E; the measured intensity is I. (c) The filter A with vertical polarization is inserted between the source S and the screen E. Now the intensity of the light measured at the screen E is about $I/2$. (d) Filter B with horizontal polarization is inserted between filter A and the screen E. The intensity of the light measured at E is now 0. (e) Filter C with a 45° polarization is inserted between filter A with vertical polarization and filter B with horizontal polarization. The intensity of the light measured at E is about $I/8$.

state with respect to a different basis than filters A and B; the new basis consists of vectors at 45° and 135°, given by:

$$\left\{\frac{1}{\sqrt{2}}(\mid \updownarrow\rangle + \mid\leftrightarrow\rangle),\quad \frac{1}{\sqrt{2}}(\mid \updownarrow\rangle - \mid\leftrightarrow\rangle)\right\}.$$

Filter C with a 45° polarization forces incoming photons to choose between the two basis states, 45° and 135°. About 50% of the photons emerge from filter C with a 45° polarization and continue along their path. The other 50% end up with a 135° polarization and are stopped by filter C.

Recall that only about 50% of the photons emitted by the source reach C because of the filtering done by A, therefore only about 25% of the photons emitted by the source reach filter B in Figure 1.6(e).

Now, once again the basis used to measure the polarization of the photons changes, filter B has a horizontal polarization and forces a measurement using the original basis vectors, $| \updownarrow \rangle$, or horizontal, $| \leftrightarrow \rangle$. Again the incoming photons with a 45° orientation are forced to make a choice, based upon the projection on the new base. As before, only about 50% make it through. Thus only about $1/8 = 12.5\%$ of the original number of photons make it to the screen, and all have horizontal polarization.

What did we learn from this simple experiment? First, we identified a candidate to store and transport information; we can use the polarization of a photon to store a bit of information. But this bit is unusual; it may not only take the values 0 and 1 but an infinite set of values. Second, there is something strange about the measurement process; once we measure a bit we affect its state. Even though this bit may take infinitely many values prior to the measurement, our interaction with it during the measurement process forces it to take one of the two possible values, 0 or 1.

1.9 AN AUGMENTED PROBABILISTIC MODEL—THE SUPERPOSITION PROBABILITY RULE

Now let us augment our model with an additional rule to describe even more complex experiments when an outcome may be reached through several different paths. As before, we consider a state vector

$$| \psi \rangle = \alpha_0 \, | 0 \rangle \; + \; \alpha_1 \, | 1 \rangle$$

describing a superposition of two basis state vectors, denoted as $| 0 \rangle$ and $| 1 \rangle$. Here α_0 and α_1 are the probability amplitudes corresponding to the two possible outcomes. The probability of an outcome is the modulus of the complex number defined as its probability amplitude.

The behavior of quantum systems in such cases is governed by the *superposition probability rule*: *if an event may occur in two or more indistinguishable ways then the probability amplitude of the event is the sum of the probability amplitudes of each case considered separately.*

To illustrate the new probability rules consider the experiment in Figure 1.7(a). A photon emitted by `S1` or by `S2` could be either reflected (R) or transmitted (T) by `BS1` and `BS2` and will eventually be detected by either `D1` or `D2`. In certain conditions, we observe experimentally that a photon emitted by `S1` is always detected by `D1` and never by `D2`, while the photon emitted by `S2` is always detected by `D2` and never by `D1`.

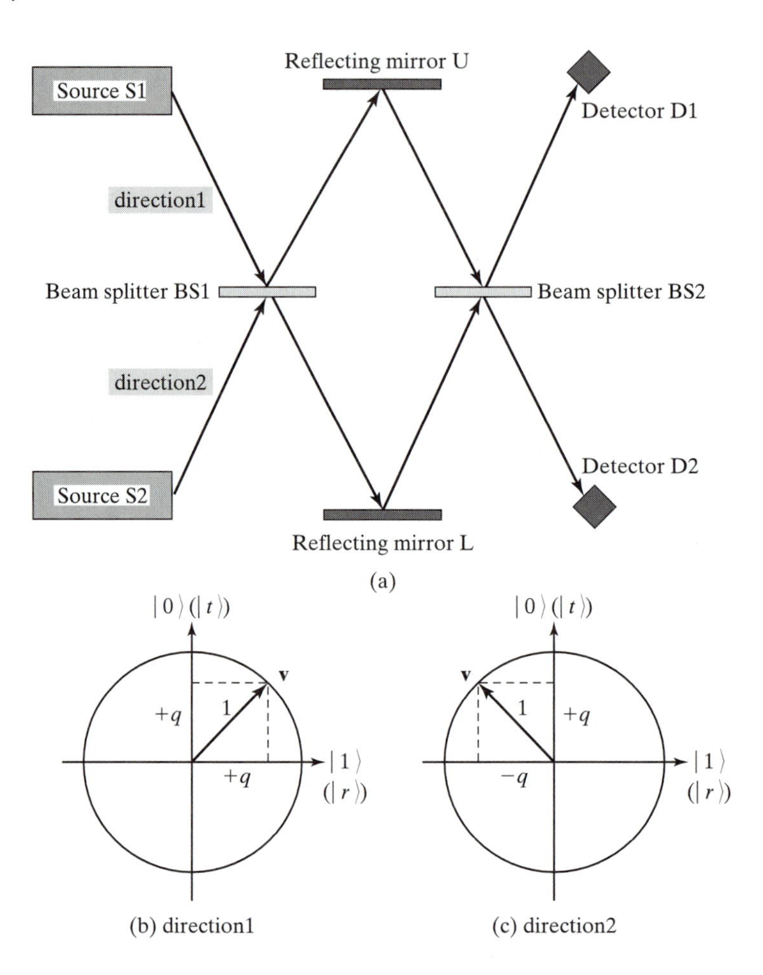

FIGURE 1.7: (a) Two sources, S1 and S2, generate photons, one at a time. There are two beam splitters, BS1 and BS2, and two detectors, D1 and D2. The beam splitters BS1 and BS2 are constructed so that the probability of a photon being reflected is equal to the probability of the photon being transmitted, $q = 1/\sqrt{2}$. We observe experimentally that all photons generated by S1 are detected by D1, none reach D2. Conversely, all photons generated by S2 are detected by D2. The experimental observations are consistent with the superposition probability rule. (b) The state vector of a photon coming to one of the beam splitters from `direction1` is described by the vector $|\psi_1\rangle = \alpha_0 \mid 0\rangle + \alpha_1 \mid 1\rangle$. The projection of this state vector on $|0\rangle$ is the probability amplitude of the event that the photon is transmitted, $\alpha_0 = +q$. The probability amplitude of a photon being reflected is $\alpha_1 = +q$. (c) The state vector of a photon coming to one of the beam splitters from `direction2` is described by the vector $|\psi_2\rangle = \alpha'_0 \mid 0\rangle + \alpha'_1 \mid 1\rangle$. The projection of this state vector on $|0\rangle$ is the probability amplitude of the photon being transmitted, $\alpha'_0 = +q$. The projection on $|1\rangle$ is the probability amplitude of a photon being reflected, $\alpha'_1 = -q$.

A photon incident on one of the beam splitters may come from either `direction1` or from `direction2`. The state vector of a photon coming to one of the beam splitters from `direction1` is described by the vector

$$|\psi_1\rangle = \alpha_0 \mid 0\rangle + \alpha_1 \mid 1\rangle,$$

while the state vector of the same photon coming to one of the beam splitters from direction2 is described by the vector

$$| \psi_2 \rangle = \alpha_0' \mid 0 \rangle + \alpha_1' \mid 1 \rangle .$$

The two state vectors must be different. Indeed, the phenomenon we describe must be reversible, implying that we should be able to trace back each photon to its source.

A photon emitted by one of the sources (S1 or S2) may take one of four different paths (see Figure 1.8) depending on whether it is transmitted or reflected by each of the two beam splitters. For example, the possible paths followed by a photon emitted by S1 are:

TT—transmitted by both BS1 and BS2,

RR—reflected by both BS1 and BS2,

TR—transmitted by BS1 and reflected by BS2, and

RT—reflected by BS1 and transmitted by BS2.

A photon emitted by S1 will reach D1 following either the *TT*, or the *RR* path. Similarly, a photon emitted by S2 will reach D2 following either the *TT* or the *RR* path. A photon emitted by S1 will reach D2 following either the *TR* or the *RT* path. Similarly, a photon emitted by S2 will reach D1 following either the *TR* or the *RT* path.

The probability amplitude of a photon coming from direction1 being transmitted by BS1 is $+q$, while the probability amplitude of a photon coming from direction2 being transmitted by BS2 is also $+q$ with $q = 1/\sqrt{2}$. Thus, the event *TT* has a probability amplitude q^2, as shown in Figure 1.8(a). Note that in this case the photon arrives at the two beam splitters from different directions.

Now let us examine the *RR* case. This time the photon arrives at the two beam splitters from the same direction. The probability amplitude of a photon coming from direction1 being reflected by BS1 is $+q$ and the probability amplitude of a photon coming also from direction1 being reflected by BS2 is also $+q$. Thus, the probability amplitude of the event *RR* is again $(+q)(+q) = q^2$, as shown in Figure 1.8(b).

The probability amplitude of a photon emitted by S1 to reach D1 is the sum of the two amplitudes of *TT* and *RR*, thus it is $q^2 + q^2 = 2q^2$. Therefore, the probability of a photon emitted by S1 reaching D1 is
$p_{S1 \rightarrow D1} = (2q^2)^2 = 4q^4 = 4(1/\sqrt{2})^4 = 1$.

Now the probability amplitude of a photon originating from direction1 being transmitted by BS1 is $+q$ and the probability amplitude of a photon originating from direction2 to be reflected by BS2 is $-q$. Consequently, the event *TR* has a probability amplitude $(+q)(-q) = -q^2$, as shown in Figure 1.8(c). The probability amplitude of a photon coming from direction1 to be reflected by BS1 is $+q$ and the probability amplitude of a photon originating from direction1 being transmitted by BS2 is $+q$. It follows that the event *RT*

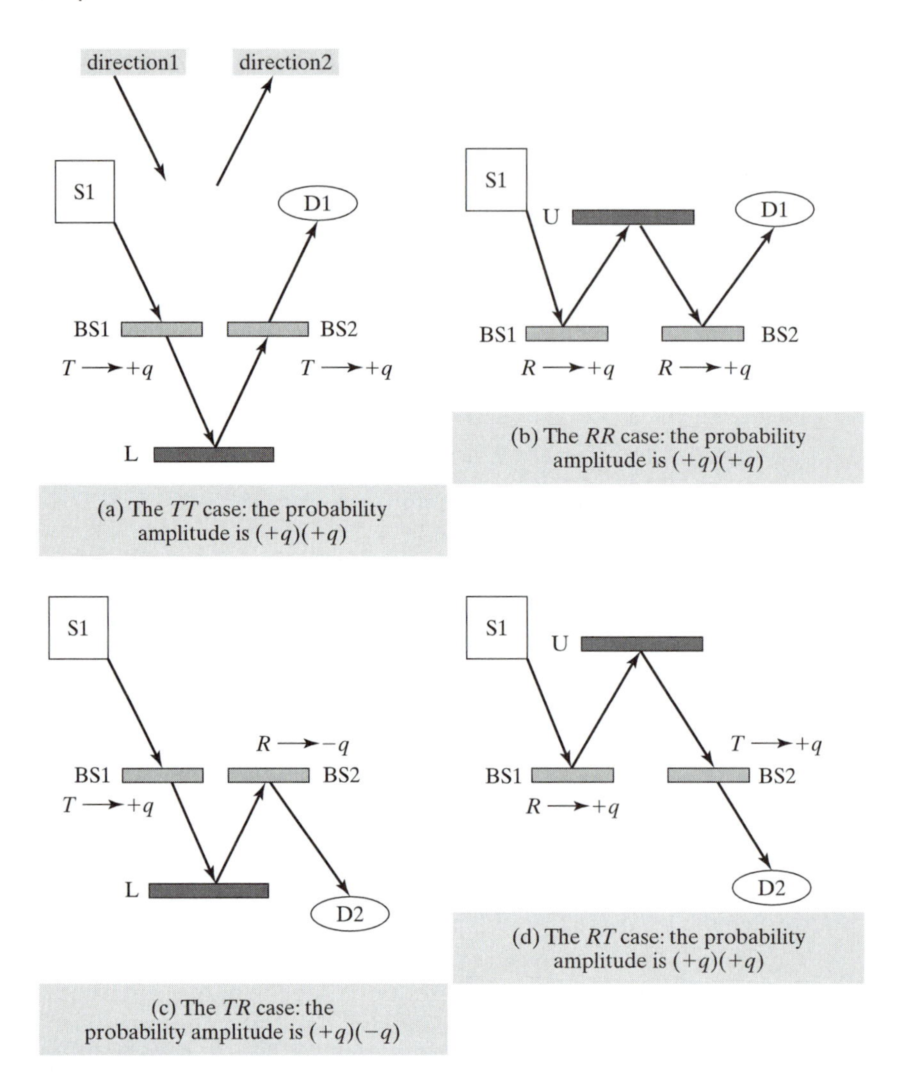

FIGURE 1.8: There are four possible events when a photon is emitted by S1. (a) The probability amplitude of a photon coming from direction1 being transmitted by BS1 is $+q$ and the probability amplitude of a photon coming from direction2 being transmitted by BS2 is also $+q$. Consequently, *TT*—the event that the photon is transmitted by both beam splitters has a probability amplitude $(+q)(+q) = q^2$. (b) The probability amplitude of a photon coming from direction1 being reflected (by BS1) is $+q$ and the probability amplitude of a photon coming also from direction1 being reflected by BS2 is also $+q$. It follows that the probability amplitude of *RR* is again $(+q)(+q) = q^2$. (c) *TR*—the photon is transmitted by the first beam splitter and reflected by the second with a probability amplitude $-q^2$. (d) *RT*—the photon is reflected by the first beam splitter and transmitted by the second with a probability amplitude q^2. As before, $q = 1/\sqrt{2}$. The probability amplitude of a photon emitted by S1 reaching D1 is the sum of the two amplitudes of *TT* and *RR*, it is equal to $2q^2$. The probability of a photon emitted by S1 reaching D1 is $p_{S1 \rightarrow D1} = (2q^2)^2 = 4q^4 = 4(1/\sqrt{2})^4 = 1$.

has a probability amplitude $(+q)(+q) = q^2$, as shown in Figure 1.8(d). The probability amplitude of a photon emitted by S1 reaching D2 is the sum of the two amplitudes of *TR* and *RT*, thus it is $q^2 - q^2 = 0$. Therefore, the probability of a photon emitted by S1 reaching D2 is $p_{\texttt{S1}\rightarrow\texttt{D2}} = 0$.

Similar arguments can be provided for a photon emitted by S2. It follows from this discussion that a beam splitter is characterized by four numbers, the probability amplitudes of a photon coming from either `direction1` or `direction2`, being either transmitted or reflected.

The superposition probability rule enables us to show that

$$p_{\texttt{S1}\rightarrow\texttt{D1}} = 1 \qquad p_{\texttt{S1}\rightarrow\texttt{D2}} = 0$$

and

$$p_{\texttt{S2}\rightarrow\texttt{D2}} = 1 \qquad p_{\texttt{S2}\rightarrow\texttt{D1}} = 0$$

and to explain the results of the experiment presented in this section. We can now comprehend the meaning of the term "indistinguishable" in the formulation of the superposition probability rule. Once a photon reaches detector D1 we know that it originated from S1, but we have no idea whether it followed path *TT* or *RR*; the two paths are indistinguishable. The system is *reversible* because a photon detected by D1 can always be traced back to its source, S1, regardless of the path followed.

Finally, we stress the fact that probability rules are different for classical systems when the Bayes rules for conditional probabilities apply. Assume that it is known that event $\mathcal{A}$ occurred, but it is not known which one of the set of mutually exclusive and collectively exhaustive events $\mathcal{B}_1, \mathcal{B}_2, \ldots, \mathcal{B}_n$ has occurred. In this case, the conditional probability that one of these events, $\mathcal{B}_j$ occurs, given that $\mathcal{A}$ occurs, is given by

$$P(\mathcal{B}_j \mid \mathcal{A}) = \frac{P(\mathcal{A} \mid \mathcal{B}_j)P(\mathcal{B}_j)}{\sum_i P(\mathcal{A} \mid \mathcal{B}_i)P(\mathcal{B}_i)}.$$

$P(\mathcal{B}_j \mid \mathcal{A})$ is called the *a posteriori probability*. For example, assume that there are two different routes to reach the summit of Everest, call them $\mathcal{B}_1$ and $\mathcal{B}_2$. Denote the probability of reaching the summit by $P(\mathcal{A})$ and the probability of taking route i, $i = \{1, 2\}$ by $P(\mathcal{B}_i)$. Then, according to Bayes rule:

$$P(\mathcal{A}) = P(\mathcal{A} \mid \mathcal{B}_1) \times P(\mathcal{B}_1) + P(\mathcal{A} \mid \mathcal{B}_2) \times P(\mathcal{B}_2).$$

with $P(\mathcal{A} \mid \mathcal{B}_i)$ the conditional probability of reaching the summit via route i. In the general case when there are n different alternatives:

$$P(\mathcal{A}) = \sum_{i=1}^{n} P(\mathcal{A} \mid \mathcal{B}_i) \times P(\mathcal{B}_i).$$

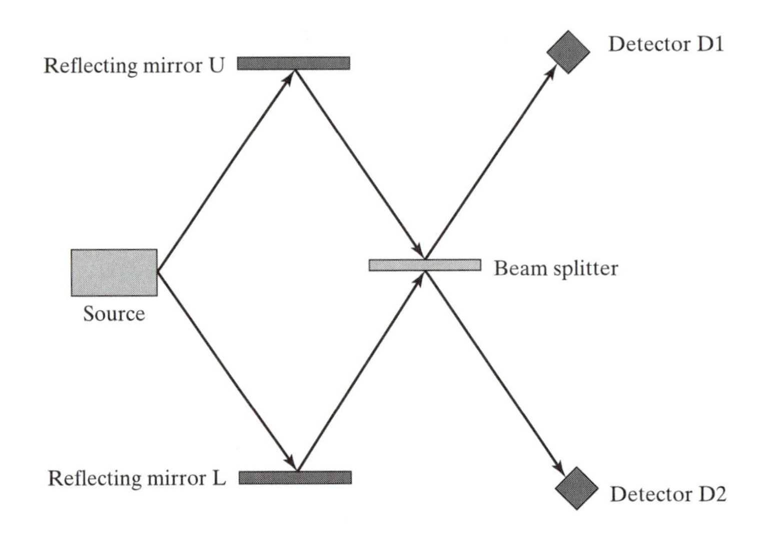

FIGURE 1.9: A source of light generates two photons simultaneously. Each photon is reflected by a mirror and directed to a beam splitter. Each photon has a 50-50 chance of being transmitted or reflected when reaching the beam splitter. The outcome of the experiment is recorded by two detectors. Both photons may reach the same detector as one of the photons is reflected and the other is transmitted. Alternatively, both photons may be reflected or both may be transmitted by both beam splitters. The latter case, when each photon reaches a different detector, is called a coincidence, and is never observed.

1.10 A PHOTON COINCIDENCE EXPERIMENT

Another experiment is illustrated in Figure 1.9. A group at the University of Rochester used the phenomenon of photon *parametric down-conversion*[16] to generate two photons simultaneously in two separate beams. This process is by no means trivial; however, a detailed explanation of this process is beyond the scope of this text. Each beam is directed by a reflecting mirror to a beam splitter [93]. Upon reaching the beam splitter each photon has a 50-50 chance of being transmitted or reflected.

In the realm of classical physics there are two possible outcomes:

1. Each of the two photons has a different fate; one is reflected, while the other is transmitted by the beam splitter. Both photons end up either at detector `D1` or at detector `D2`. When the photon coming from mirror U is reflected by the beam splitter and the one coming from mirror L is transmitted, both

[16]Parametric down conversion is an elementary quantum process of decay of a photon of frequency ω_p into two new photons of lower frequencies ω_1 (*signal* photon) and ω_2 (*idler* photon), so that $\omega_p = \omega_1 + \omega_2$; the emission of the two highly correlated (entangled) photons (the photons have correlated polarizations) is simultaneous. This decay process appears when a photon from an incident laser beam interacts with a nonlinear medium, for example a crystal of ammonium dihydrogen phosphate (ADP), and splits into two lower frequency *signal* and *idler* photons.

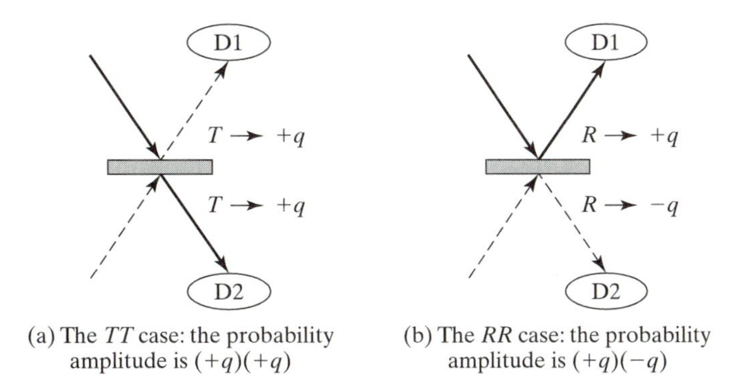

(a) The *TT* case: the probability amplitude is $(+q)(+q)$

(b) The *RR* case: the probability amplitude is $(+q)(-q)$

FIGURE 1.10: (a) The probability amplitude for the *TT* case is q^2. (b) The probability amplitude for the *RR* case is $-q^2$. There are two indistinguishable ways for a coincidence to occur, *TT* and *RR*, thus the probability amplitude of this event is zero. Here $q = 1/\sqrt{2}$.

photons end up at detector D1. When the scenario is reversed both photons end up at detector D2.

2. Both photons have identical fate; they are either reflected or transmitted by the beam splitter. In this case detectors D1 and D2 record one photon each and signal a coincidence. If both photons are transmitted by the beam splitter, the one coming from mirror U ends up at detector D2 and vice-versa. When both photons are reflected by the beam splitter the one coming from mirror U ends up at detector D1 and the one coming from mirror L ends up at detector D2.

The second outcome, which represents the coincidence, is very hard to observe! But now we can explain why: the beam splitter is characterized by the four numbers giving the probability amplitudes for the two photons to be transmitted and reflected $((p_T, p_R)_1, (p_T, p_R)_2) = (+q, +q, +q, -q)$. Call *TT* the event that both photons are transmitted by the beam splitter, *RR* the event that both are reflected. Figure 1.10(a) shows that the probability amplitude for the *TT* event is $1/2$ and Figure 1.10(b) shows that the probability for the *RR* event is $-1/2$. The two photons are indistinguishable after the beam splitter; thus, the probability amplitude of being both transmitted or both reflected is the sum of the two probability amplitudes, and it is equal to zero. In our Gedanken experiment, in which we assume the probability amplitudes given above, we could never observe a coincidence.[17] When only one photon is emitted there is an equal chance that it is detected by either D1 or D2, as expected.

Milburn [93] calls this a quantum two-up experiment. He describes a game of chance "Australia's very own way to part a fool and his money." Two fair and identical coins are tossed up in the air. Four outcomes are possible: two heads (*HH*) with probability $p_{HH} = 1/4$, two tails (*TT*) with $p_{TT} = 1/4$, or one

[17] In reality, in a carefully designed experiment, coincidences can be detected with a probability of about 10^{-4}.

head and one tail (HT or TH) with $p_{HT} = p_{TH} = 1/4$. The spinner aims to toss the HH combination three times before tossing either a TT, or five consecutive odds (HT or TH). If (s)he succeeds (s)he wins, if not (s)he loses.

1.11 A THREE BEAM SPLITTER EXPERIMENT

Let us now test our model augmented with the superposition probability rule in a slightly different setup involving three beam splitters, as shown in Figure 1.11.

We want to show that a photon emitted by `S1` has an equal probability of being sensed by `D1` or `D2`. Then, it is trivial to show that the same holds true for a photon emitted by `S2`.

As before, we assume that we have two particles with different state vectors. To make things easier to understand and demystify our notations, we use $| t\rangle$ and $| r\rangle$ instead of using the basis vectors $| 0\rangle$ and $| 1\rangle$, where t stands for "transmitted" and r for "reflected." The two particles moving in `direction1` and `direction2`, respectively, have the following state vectors

$$| \psi_1\rangle = q \,| t\rangle + q \,| r\rangle$$

and respectively

$$| \psi_2\rangle = q \,| t\rangle - q \,| r\rangle$$

with $q = 1/\sqrt{2}$ the probability amplitude for either transmission, or reflection. We assume 50-50 beam splitters, thus the two probability amplitudes are equal.

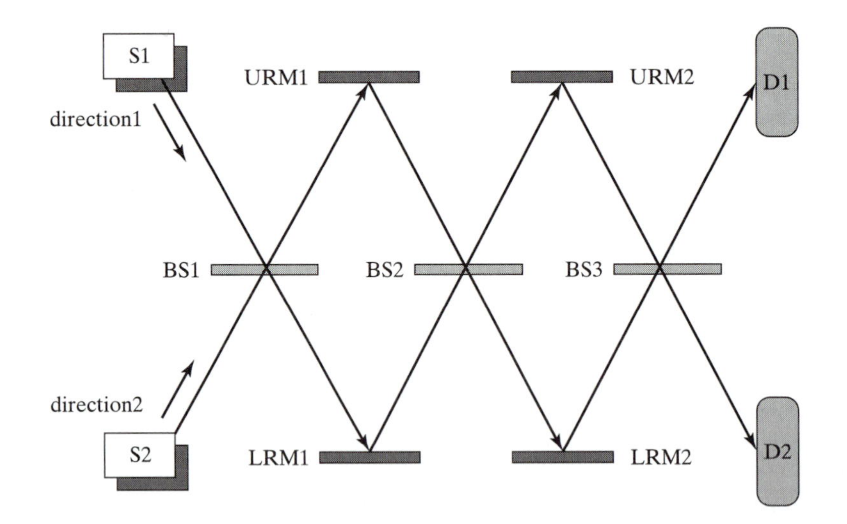

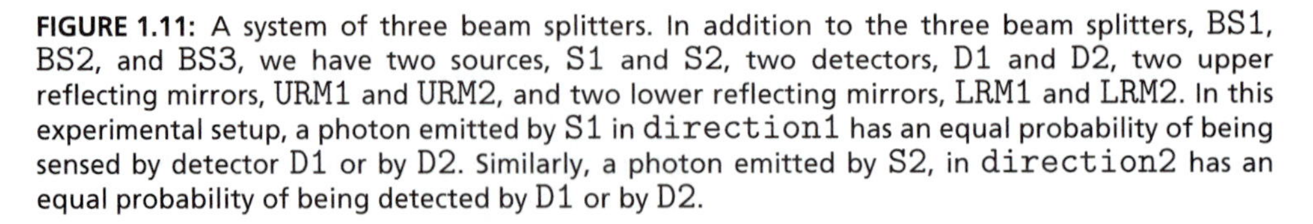
FIGURE 1.11: A system of three beam splitters. In addition to the three beam splitters, BS1, BS2, and BS3, we have two sources, S1 and S2, two detectors, D1 and D2, two upper reflecting mirrors, URM1 and URM2, and two lower reflecting mirrors, LRM1 and LRM2. In this experimental setup, a photon emitted by S1 in `direction1` has an equal probability of being sensed by detector D1 or by D2. Similarly, a photon emitted by S2, in `direction2` has an equal probability of being detected by D1 or by D2.

This time we use a three-tuple to denote the events at the three beam splitters. RTR means that a photon is reflected (R) by BS1, transmitted (T) by BS2 and reflected (R) by BS3. A probability amplitude of $(+q)(+q)(-q)$ associated with path RTR means that the probability amplitude of reflection at the first beam splitter is $(+q)$, the probability amplitude of transmission at the second beam splitter is $(+q)$, and the probability amplitude of reflection at the third beam splitter is $(-q)$. The following table summarizes the eight possible paths taking a photon from S1 to D1 or to D2, the probability amplitude (PA) of transmission or reflection at each beam splitter, the total probability amplitude for each path and the probability of the photon reaching D1 or D2.

$S \mapsto D$	Path	Individual PAs	Path PA	Total PA	Probability
$S1 \mapsto D1$	TTR	$(+q)(+q)(+q)$	$+q^3$		
	RRR	$(+q)(+q)(+q)$	$+q^3$		
	TRT	$(+q)(-q)(+q)$	$-q^3$		
	RTT	$(+q)(+q)(+q)$	$+q^3$	$2q^3 = 1/\sqrt{2}$	$p_{S1 \rightarrow D1} = 1/2$
$S1 \mapsto D2$	TTT	$(+q)(+q)(+q)$	$+q^3$		
	RRT	$(+q)(+q)(+q)$	$+q^3$		
	TRR	$(+q)(-q)(-q)$	$+q^3$		
	RTR	$(+q)(+q)(-q)$	$-q^3$	$2q^3 = 1/\sqrt{2}$	$p_{S1 \rightarrow D2} = 1/2$

It is left as an exercise for the reader to prove that an experimental setup with an odd number of beam splitters, $1, 3, 5, \ldots (2k + 1) \ldots$ yields the result presented in this section

$$p_{S1 \rightarrow D1} = p_{S1 \rightarrow D2} = 1/2 \qquad p_{S2 \rightarrow D1} = p_{S2 \rightarrow D2} = 1/2.$$

When we have an even number of beam splitters, $2, 4, 6, \ldots (2k) \ldots$ we get

$$p_{S1 \rightarrow D1} = 1 \quad p_{S1 \rightarrow D2} = 0 \qquad p_{S2 \rightarrow D1} = 0 \quad p_{S2 \rightarrow D2} = 1.$$

1.12 BB84—THE EMERGENCE OF QUANTUM CRYPTOGRAPHY

Before closing this introductory chapter we want to stress that quantum computing and quantum information theory have very important applications, some of which are feasible even today. We have selected computer security to illustrate the power of quantum information theory.

The explosive developments in the areas of communication and computer networks have affected many aspects of human activity including, but not limited to, economic, defense, scientific, educational, and social. The Internet supports financial transactions, enables distance learning and collaborative research, and allows large companies to share databases of material properties and various design tools. Specialized networks support defense applications while telemedicine allows physicians to share medical knowledge. In all these examples confidential information such as credit card numbers, personal records, proprietary code and data, medical records, and highly classified defense data, flow through different communication media.

To ensure confidentiality, data is often encrypted. The most reliable encryption techniques are based upon *one time pads*, whereby the encryption key is used for one session only and then discarded. These techniques create the need for reliable and effective methods for distributing encryption keys. The challenge faced by encryption key distribution lies in the physical difficulty associated with detecting the presence of an intruder when communicating through a classical communication channel. To date, secure and reliable methods for cryptographic key distribution have largely eluded the cryptographic community in spite of considerable effort and ingenuity.

Could quantum information help solve the problem of secure distribution of encryption keys once and for all? The answer is provided by an idea published by Bennett and Brassard in 1984, hence the name BB84 [13]. Let us consider communication using photons prepared in different states of polarization. Figure 1.12(a) shows photons with vertical/horizontal (VH) and diagonal (DG) polarization at 45° and 135°. A calcite crystal is used to separate photons of different polarization, as shown in Figure 1.12(b).

We already know from the previous experiments that the vertical—horizontal and diagonal (45°–135°) states of the photon polarization are pairs of orthogonal states. In other words, *if the measuring system is made to distinguish states in one basis, it cannot reliably distinguish the states corresponding to the other basis*. In our case the crystal is oriented so that it separates photons with vertical polarization from those with horizontal polarization. The crystal cannot distinguish between photons with 45° polarization and those with 135° polarization. If we position the crystal to separate photons with vertical polarization from those with horizontal polarization and photons with diagonal polarization arrive, their state will be randomly altered by the crystal; recall that any measurement alters the quantum state.

This preliminary discussion indicates that Alice and Bob can communicate using polarized photons and detect whether Eve, the perennial eavesdropper, has attempted to intercept the exchange of information-carrying photons. In our setup Alice and Bob are also connected via a classical channel, say a telephone line, in addition to the quantum communication channel (see Figure 1.12(c)). Eve is able to intercept all communications on both the quantum and the classical channel. The quantum key distribution protocol is fully described in Section 6.6; here we present only an outline of the protocol.

First, let us assume that Eve is very persistent and unsophisticated, intercepting all photons sent by Alice to Bob, as well as their exchanges over the voice line. Alice and Bob communicate using the following algorithm:

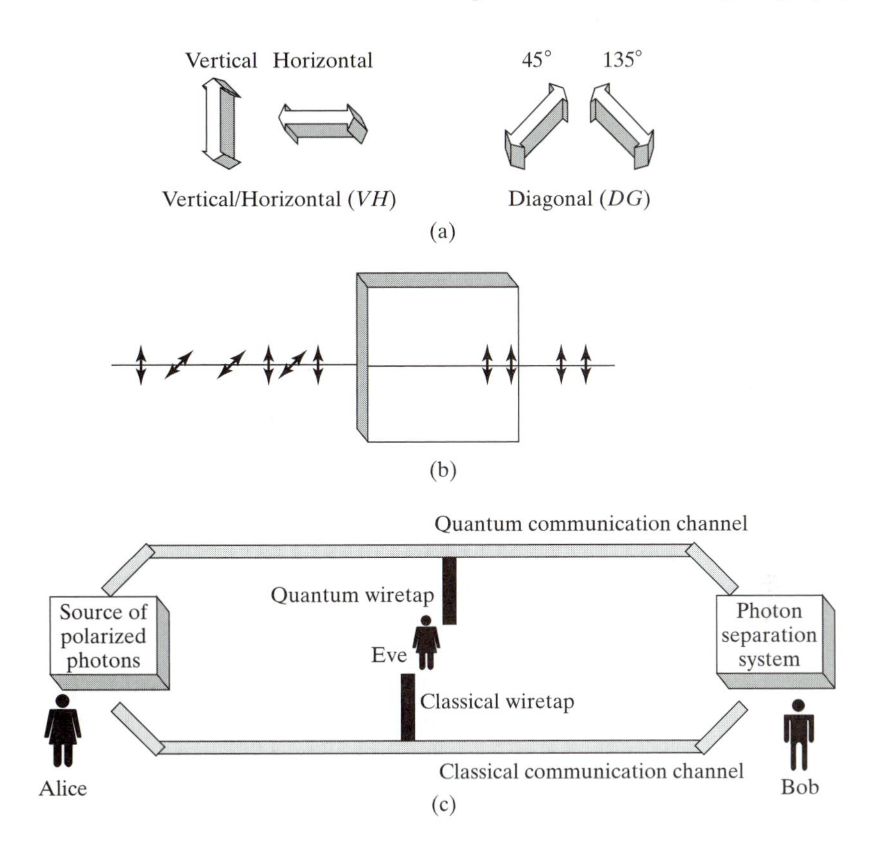

FIGURE 1.12: The quantum key distribution algorithm of Bennett and Brassard. (a) Photons prepared by Alice may have vertical/horizontal (VH) or diagonal polarization (DG). The photons with vertical/horizontal (VH) polarization may be used to transmit binary information as follows. A photon with vertical polarization may transmit a 1 while one with a horizontal polarization may transmit a 0. Similarly, photons with diagonal (DG) polarization may transmit binary information, 1 encoded as a photon with 45° polarization, and 0 encoded as a photon with a 135° polarization. (b) Bob uses a calcite crystal to separate photons with different polarizations. Shown here is the case in which the crystal is set up to separate vertically polarized photons from the horizontally polarized ones. To perform a measurement in the DG basis the crystal is oriented accordingly. (c) Alice and Bob are connected via two communication channels, one quantum and one classical. Eve eavesdrops on both communication channels.

1. Alice selects n as the approximate length of the encryption key. She generates two random strings a and b, each of length $(4 + \delta)n$. By choosing δ sufficiently large, Alice and Bob can ensure with a very high probability that the number of bits kept is close to $2n$.

   A subset of the bits in string a will be utilized as the encryption key and the bits in string b will be used by Alice to select the basis VH or DG for each photon sent to Bob.
2. Alice encodes the binary information in string a based upon the corresponding values of the bits in string b. For example, if the i-th bit of string b is 1 then Alice selects vertical-horizontal (VH) polarization; if the i-th bit

is 0, then she selects diagonal (DG) polarization. If VH is selected, then a 1 is sent in the i-th position of string a as a photon with vertical polarization, and a 0 is sent as a photon with horizontal polarization. If DG is selected, a 1 is sent as a photon with a 45° polarization and a 0 is sent as a photon with 135° polarization. Both Alice and Bob use the same encoding convention for each of the bases.

3. In turn, Bob randomly picks up $(4 + \delta)n$ bits to form a string b'. He uses one of the two bases for the measurement of each incoming photon in string a based on the corresponding value of the bit in string b'. For example, a 1 in the i-th position of b' implies that the i-th photon is measured in the DG basis, while a 0 implies that the photon is measured in the VH basis. As a result of this measurement Bob constructs the string a'.
4. Bob uses the classical communication channel to request the string b, to which Alice responds with b on the same channel. He then sends her string b' on the classical channel.
5. Alice and Bob keep only those bits in set $\{a, a'\}$ for which the corresponding bits in set $\{b, b'\}$ are equal. Let us assume that Alice and Bob keep only $2n$ bits.
6. Alice and Bob perform several tests to determine the level of noise and eavesdropping on the channel. The set of $2n$ bits is split into two sub-sets of n bits each. One sub-set will consist of the *check* bits used to estimate the level of noise and eavesdropping, while the second sub-set consists of the *data* bits used for the quantum key. Alice selects n check bits at random and sends the positions and values of these bits to Bob over the classical channel. Next, they compare the values of the check bits. If more than t bits disagree, where t is an integer determined by the level of noise on the channel, they decide that an eavesdropper has intercepted the key and abort the current attempt. Then they run the protocol again.

For example, if the sequences sent by Alice and the one used by Bob are

Photon#	**1**	**2**	**3**	**4**	**5**	**6**	**7**	**8**	...
Alice $\gg$	VH	VH	DG	VH	VH	DG	DG	VH	...
Bob $\ll$	VH	DG	DG	VH	DG	VH	DG	DG	...
Match(Y/N)	Y	N	Y	Y	N	N	Y	N	...

then Alice tells Bob that photons numbered 1, 3, 4, 7, ... were sent and received using the same basis. Alice consults her log to find out which polarization was selected for each of these photons. In turn, Bob examines the results of his measurements performed on these photons and determines the polarization of each photon. The measurements are performed in the same basis that photons are prepared in; therefore the results of Alice's and Bob's investigations should be the same. For example, if Alice uses vertical polarization for photon #1,

135° polarization for photon #3, horizontal polarization for photon #4, and 45° polarization for photon #7, then the binary string transmitted by Alice, and received correctly by Bob is 1001.

In absence of eavesdropping, roughly 50% of the photons are measured correctly. Initially, Alice selects each of the two basis randomly with a probability of $p_{VH}^{Alice} = p_{DG}^{Alice} = 1/2$. Bob independently selects the basis with probabilities $p_{VH}^{Bob} = p_{DG}^{Bob} = 1/2$. Four events are possible when selecting the basis, $(VH)(VH)$, $(VH)(DG)$, $(DG)(VH)$, and $(DG)(DG)$; the encoding of events reflects the selection of Alice's and Bob's selection; $(VH)(DG)$ means that Alice has selected VH and Bob DG. The probabilities of the four events are:

$$
\begin{aligned}
p_{(VH)(VH)} &= p_{VH}^{\text{Alice}} \times p_{VH}^{\text{Bob}} = 1/4 \\
p_{(DG)(DG)} &= p_{DG}^{\text{Alice}} \times p_{DG}^{\text{Bob}} = 1/4 \\
p_{(VH)(DG)} &= p_{VH}^{\text{Alice}} \times p_{DG}^{\text{Bob}} = 1/4 \\
p_{(DG)(VH)} &= p_{VH}^{\text{Alice}} \times p_{DG}^{\text{Bob}} = 1/4
\end{aligned}
$$

It follows that the probabilities of Alice and Bob selecting the same basis, or selecting different basis, respectively are

$$
\begin{aligned}
p_{\text{agreement}} &= p_{(VH)(VH)} + p_{(DG)(DG)} = 1/2 \\
p_{\text{disagreement}} &= p_{(VH)(DG)} + p_{(DG)(VH)} = 1/2.
\end{aligned}
$$

When Eve measures every single photon, she alters the state of all photons for which she had used the wrong basis. Arguments similar to those presented previously show that the polarization of approximately 50% of the photons sent by Alice is altered by Eve. Under these conditions Bob and Alice may agree on only half of the photons whose state was not altered by Eve, meaning on some 25% of the total number of photons. This is observed only after Alice and Bob perform the test (6) to determine the level of noise and intrusion.

Eve could be less intrusive and measure only a small fraction of the photons passing through the quantum communication channel. This method lowers her probability of her being detected by Alice and Bob, but it also lowers her chances of getting enough information about the encryption key. Alice and Bob could preempt this by including a number of parity check bits in their message. For example, Alice may add a parity check bit to each group of say 15 bits to ensure that each group of 16 bits contains an even number of 1s. In this case, regardless how subtle Eve's intrusion is, it cannot go unnoticed.

Many research teams work diligently on practical applications of quantum cryptography. In June 2003, Andrew Shields and colleagues revealed a record-breaking cryptographic link over a distance slightly larger than 100 km [146]. In view of the facts discussed in this section it is not surprising that quantum cryptography is the first example of a quantum information theory application ready for commercialization.

1.13 A QUBIT OF HISTORY

Quantum computing is the result of a marriage between two great discoveries of the twentieth century, quantum mechanics and the general-purpose computer.

It all started more than one hundred years ago when, in 1900, Max Planck proposed an amazing solution to a puzzling problem, the so called *ultraviolet catastrophe*. Contrary to experimental evidence and to common sense, classical physics predicted that the intensity of radiation emitted by a hot body increases without any limit as the frequency of radiation increases. A hot blackbody[18] in equilibrium would radiate an infinite amount of energy. Since this is a physical impossibility, it followed that thermal equilibrium was impossible and that conclusion was absurd. Planck calculated the so-called *blackbody radiation* spectrum assuming that the body emitted energy in discrete packets called *quanta*. His calculations agreed with the experiment, avoiding the classical theory contradiction. Shortly afterwards, in 1905, Albert Einstein used Planck's quantum hypothesis to explain what happens when light shines on a negatively charged metal plate, the phenomena known as the *photoelectric effect*. Then, in 1913, Niels Bohr proposed a quantum model of the atom. Twelve years later, in 1925, Werner Heisenberg developed an astounding new formulation of what was first called "quantum mechanics" by Max Born. Heisenberg's work was followed, in 1926 by Erwin Schrödinger's Wave Equation. In collaboration with Heisenberg and Jordan, Born published investigations on the principles of quantum mechanics (sometimes called matrix mechanics) during the years 1925 and 1926, and soon afterwards, his own studies on the statistical interpretation of quantum mechanics. Six of these scientists got the Nobel Prize in physics for their revolutionary discoveries: Planck in 1918, Einstein in 1921, Bohr in 1922, Heisenberg in 1932, Schrödinger in 1933, and Max Born in 1954. The Nobel Prize lectures of these great scientists make for fascinating reading, and can be found at the Nobel Prize Web site: `http://www.nobel.se/physics`.

In 1935, an eccentric young fellow (don) of King's College at Cambridge by the name of Alan Turing dreamed up an imaginary typewriter-like device called the Universal Turing Machine. Turing conceived the principle of the modern computer: He described an abstract digital computing machine consisting of a limitless memory and a scanner that moves back and forth through the memory, symbol by symbol, reading what it finds and writing additional symbols. The actions of the scanner are dictated by a program of instructions that is stored in the memory in the form of symbols. This is Turing's stored-program concept.

The *Universal Turing Machine* embodies the essential principle of a computer: a single machine that can be turned to any well-defined task by being supplied with the appropriate program. Turing envisioned a machine that could do anything with a few simple instructions, an idea we take for granted today. He believed that an algorithm could be developed for almost any problem and that the most difficult task was to break down a problem into a sequence of simple instructions that the machine could perform.

---

[18]A blackbody is an idealized object which absorbs and emits radiation of all possible frequencies.

Turing had reached the conclusion that "every function which can be regarded as computable can be computed by a universal computing machine." He reached this conclusion at about the same time that the work of the American logician Alonzo Church was published. Turing's paper, "On Computable Numbers with an Application to the Entscheidungsproblem",[19] referred to Church's work, and was published in August 1936 [138]. In his paper Turing makes a triple correspondence between logical instructions, the action of the mind, and a machine that could in principle be embodied in a practical physical form. He was the first to use a concept referred to in modern language as an *algorithm* as the "definite method."

Almost 10 years later, in the fall of 1945, the world's first general purpose computer, the brainchild of J. Presper Eckert and John Mauchly, became operational after several years of development at the Moore School of the University of Pennsylvania. The ENIAC (Electronic Numerical Integrator and Calculator) could perform 5000 addition cycles a second and do the work of 50,000 people computing by hand; it could calculate the trajectory of a projectile in 30 seconds, compared to the 20 hours required with a desk calculator. The ENIAC required 174 kW of power for the 17,468 vacuum tubes, 70,000 resistors, and 10,000 capacitors. The cost of electricity required to keep the filaments of the vacuum tubes heated and the fans running to dissipate the heat was about $650 per hour, even when the computer was not operating [90].

Arguably, computer simulation of physical systems and phenomena in the mid 1940s was the motivating force for the development of general-purpose computers. It is not a coincidence that general-purpose computers are based on the so called *von Neumann architecture*, named after the renown mathematician and physicist. In the early 1940s, scientists associated with the Manhattan project at Los Alamos, including John von Neumann and Richard Feynman, were feverishly developing a fission device. Sophisticated calculations were necessary, and the hundreds of "human calculators" employed to solve large numerical problems were simply not sufficient. The Manhattan Project first resorted to the use of tabulating machines and calculators before having access to the ENIAC. Nicolas Metropolis and Stanley Frankel, two other physicists from the Manhattan Project, had the honor of running the first test programs once the ENIAC became operational; the problem they solved remains classified even today. Later that year, Stanislaw Ulam, who doubted Edward Teller's design of a thermonuclear device, used the ENIAC to compute the results of a thermonuclear reaction at increments of one ten-millionth of a second.

Some question the paternity of the ideas presented in the 1946 report, co-authored by John von Neumann, [30] which proposed the development of a new computer called the EDVAC (Electronic Discrete Variable Automatic Computer). This report may well be the reason why we talk today about von Neumann architecture, rather than Eckert-Macauly architecture; however this

---

[19]German term for the "decision problem" posed by Hilbert: "Could there exist, at least in principle, a definite method or process, by which it could be decided whether any mathematical assertion was provable?"

is beside the point for our topic. Rather, we would like to stress that numerical simulation became a new investigative tool in the mid twentieth century. Numerical simulation complements the two traditional exploratory methods of science: experimental work and theoretical modeling.

In 1948, Claude Shannon published "A Mathematical Theory of Communication" in the Bell System Technical Journal. This paper founded a new discipline, information theory, and proposed a linear model of a communications system. Shannon considered a source of information that generates words composed of a finite number of symbols transmitted through a channel; if x_n is the n-th symbol produced by the source, then x_n is a stationary stochastic process [118].

The first commercial computer, UNIVAC I, capable of performing 1900 additions/second was introduced in 1951; the first supercomputer, the CDC 6600, designed by Seymour Cray, was announced in 1963; IBM launched System/360 in 1964; while a year later DEC unveiled the first commercial minicomputer, the PDP 8, capable of performing some 330,000 additions/second. A large percentage of the cycles of all these systems were devoted to numerical simulation. In 1977, the first personal computer—the Apple II—was marketed. The IBM PC rated at about 240,000 additions/second was introduced four years later in 1981.

Today's computers are very different from the ENIAC. In 2003, a high-end PC had two 2.5-GHz CPUs, 2 GB of memory, an 80-GB disk, and a 100-Mbps Ethernet interface. What about the future?[20] Changes in the very large scale integration (VLSI) technologies and computer architecture are projected to lead to a 10-fold increase in computational capabilities over the next five years and 100-fold increase over the next 10 years. By 2011, the CPU speed is projected to be 11.5 GHz, the main memory is expected to increase to 16 GB.

By 2003 the minimum feature size[21] for integrated circuits reached 0.09 μm and, according to the Semiconductor Industry Association Roadmap [147], is expected to decrease to 0.05 μm by 2011 (see Table 1.1). As a result, the density of memory bits is expected to increase 64-fold during this period, while the cost per memory bit is expected to decrease 5-fold. It is projected that the density of transistors will increase 7-fold during this period, while the density of bits in logic circuits will increase 15-fold, and the cost per transistor will decrease 20-fold.

While solid-state technology continued to improve at a very fast pace, making our computers faster and cheaper, theoretical physicists became restless, asking questions about the physical limitations of our computational models. They reasoned that if there is a minimum amount of power dissipated for the execution of a logical step, then the faster computers become, the more power is needed for their operation. The amount of power dissipated increases as computers become faster, making it harder to deal with the heat

[20]Predicting the future is a practice involving considerable risks and potential ridicule. Cases in point: in 1943, discussing the future of the computer industry, Thomas J. Watson, the chairman of IBM corporation said: "I believe that there is a market for maybe five computers."; in 1949, a widely circulated popular science magazine speculated: "Computers in the future may weigh no more than 1.5 tons."

[21]The minimum feature size is the width of the smallest line or gap constructed on a chip in the manufacturing process.

TABLE 1.1: Projected Evolution of VLSI Technology

Year	2003	2005	2008	2011
Minimum feature size (μm, 10^{-6} meter)	0.10	0.08	0.07	0.05
Memory—Bits per chip (billions, 10^9)	1	2	6	16
Logic—Transistors per cm^2 (millions, 10^6)	24	44	109	269
Microprocessor—Transistors per chip (millions, 10^6)	95.2	190	539	1523

generated during the execution of instructions. This conclusion motivated Rolf Landauer to investigate the heat generation during the computational process, starting from the basic laws of thermodynamics. His results were published in 1961 [87].

Following in Landauer's footsteps, Charles Bennett began studying the logical reversibility of computations in 1973. This concept is discussed in Section 6.17. In layman's terms *logical reversibility* means that once a computation is finished, one can retrace every step and reconstruct the data used as input for every step. Bennett argued that Turing Machines and any other general purpose computing automata are logically irreversible [11]. A device is said to be irreversible if its transfer function does not have a single-valued inverse. Bennett developed a theoretical framework that proved that reversible general-purpose computing automata can be built and that their construction makes the possibility of building thermodynamically reversible computers plausible. There is no positive lower bound on the energy dissipated per logical step by a thermodynamically reversible computer; in principle, such a device could compute dissipating little, if any, energy.

Widespread interest in quantum computing likely stemmed from the contributions of Richard Feynman. In 1981 he gave a talk titled "Simulating Physics with Computers" at a meeting held at MIT [29]. Feynman argued that in traditional numerical simulations such as weather forecasting or aerodynamic calculations, computers model physical reality only *approximately*. He advanced the idea that physics was computational and that a quantum computer could do an *exact simulation* of a physical system, even of a quantum system. Feynman identified quantum mechanics as the most important ingredient for constructing computational models of physics.

Feynman speculated that computation can be done more efficiently in many instances by using quantum effects [57]. His ideas were inspired by the previous work of Bennett [11, 12] and Benioff [9]. Starting from basic principles of thermodynamics and quantum mechanics, Feynman suggested that problems for which polynomial time algorithms do not exist could be solved; computations for which polynomial algorithms exist could be speeded up considerably and even made reversible.

Ed Fredkin, Thomasso Toffoli, and Norman Margolus, who were associated with the Laboratory for Computer Science at MIT, also contributed to the field of quantum computing. Fredkin and Toffoli gates are some of the most common building blocks of quantum circuits.

In 1985, David Deutsch reinterpreted the Church-Turing conjecture as "every finitely realizable physical system can be perfectly simulated by a universal model computing machine operating by finite means" and conceived a universal quantum computer [42]. In 1993, Charles Bennet, Giles Brassard, Christian Crepeau, Richard Jozsa, Asher Peres, and William K. Wooters discovered quantum teleportation.

In 1994, Peter Shor developed a clever algorithm for factoring large numbers [119] and generated a wave of excitement for the newly founded discipline of quantum computing. A year later Robert Calderbank, Peter Shor, and Andrew Steane addressed the problem of reliability of quantum computing and communication and developed quantum codes. A summary of the milestones in quantum physics, quantum computing, and quantum information theory is presented in Table 1.2.

1.14 SUMMARY AND FURTHER READINGS

In this chapter, we discussed the physical limitations of classical electronic circuits used in modern computer systems. Heat dissipation restricts our ability to compute faster and cheaper, while quantum effects ultimately will limit the reliability of future computing systems built with current technologies. We conclude that fundamentally new ideas are necessary for building systems that are able to store and process information with higher speed and at a lower cost. Systems using quantum particles, which are governed by the laws of quantum mechanics, seem ideally suited to storing and processing large volumes of information using little, if any, energy.

Next, we presented the milestones in the evolution of quantum ideas and the unprecedented developments in quantum computing and quantum information theory we witnessed during the past two decades. We introduced the reader to the world of quantum effects by way of several experiments involving beams of photons, or light in layman's terms. These experiments were meant to put us in the frame of mind necessary for understanding quantum information. First, we presented a simple experiment revealing the granular nature of light. Next, we discussed a simple model for non-deterministic effects. Then, we challenged this model to explain the results of successive measurements performed on a beam of quantum particles using different bases. We augmented the model with the superposition probability rule for events that may occur on alternative paths, and discussed several experiments with results consistent with this rule.

The sometimes strange and non-intuitive outcomes of such experiments are relatively easy to grasp by using elementary probability arguments. We concluded the section with an application of quantum information theory to computer security, namely the quantum key distribution.

TABLE 1.2: Milestones in Quantum Physics, Computing, and Quantum Computing

Year	Milestone
$\approx$ 1800	Thomas Young (1773–1829) conducts the "double-slit experiment."
1900	Max Planck presents the blackbody radiation theory; *quantum theory* is born.
1905	Albert Einstein develops the theory of the photoelectric effect.
1911	Ernest Rutherford develops the planetary model of the atom.
1913	Niels Bohr develops the quantum model of the hydrogen atom.
1923	Louis de Broglie relates the momentum p of a particle with the wavelength λ of the wave associated with it, $p = h/\lambda$.
1925	Werner Heisenberg formulates the *matrix quantum mechanics.*
1925	Max Born and Pascual Jordan use infinite matrices to represent basic physical quantities developing a complete formalism for quantum mechanics.
1926	Erwin Schrödinger proposes the equation for the dynamics of the wave function.
1926	Erwin Schrödinger and Paul Dirac show the equivalence of Heisenberg's matrix formulation and Dirac's algebraic formulation with Schrödinger's wave function.
1926	Paul Dirac and, independently, Max Born, Werner Heisenberg, and Pascual Jordan obtain a *complete formulation of quantum dynamics.*
1926	John von Neumann introduces Hilbert spaces to quantum mechanics.
1927	Werner Heisenberg formulates the *Uncertainty Principle.*
1927	Davisson and Germer observe the diffraction of electrons by a crystal and confirm the wave character of an electron and thus, also de Broglie's theory.
1928	Paul Dirac develops the *relativistic quantum mechanics.*
1932	John von Neumann publishes *Mathematical Foundations of Quantum Mechanics.*
1936	Alan Turing envisions the *Universal Turing Machine.*
1936	Alonzo Church publishes a paper asserting that "every function which can be regarded as computable can be computed by a universal computing machine".
1945	ENIAC, the world's first general purpose computer, and brainchild of J. Presper Eckert and John Macauly becomes operational.
1946	John von Neumann co-authors a report outlining the principles of the program-stored computer, the *von Neumann architecture.* Proposes EDVAC.
1946	John Wheeler makes the assumption that the two photons produced through electron-positron annihilation have opposite polarizations.
1948	Claude Shannon publishes *A Mathematical Theory of Communication.*

(continued overleaf)

TABLE 1.2: (*continued*)

1949	Madame Chien-Shiung Wu and Irvin Shaknow confirm Wheeler's hypothesis and produce a pair of *entangled photons* for the first time.
1951	UNIVAC I, the first commercial computer is delivered.
1961	Rolf Landauer decrees that computation is physical and studies heat generation.
1973	Charles Bennet studies the logical reversibility of computations.
1981	Richard Feynman suggests that physical systems including quantum systems can be simulated exactly with quantum computers.
1982	Peter Benioff develops *quantum mechanical models of Turing machines*.
1984	Charles Bennet and Giles Brassard introduce *quantum cryptography*.
1985	David Deutsch reinterprets the Church-Turing conjecture.
1993	Charles Bennet, Giles Brassard, Christian Crepeau, Richard Jozsa, Asher Peres, and William K. Wooters discover *quantum teleportation*.
1994	Peter Shor develops a clever algorithm for factoring large numbers.
1996	Robert Calderbank, Peter Shor, and Andrew Steane develop *quantum codes*.

We learned from the experiments discussed in this chapter that a photon can be used to store and transmit information. Measuring the polarization of the photon in order to retrieve the classical information (0 or 1) from the quantum information reflected by its quantum state produces a reliable outcome if and only if the same basis used to encode the information is used for the measurement. Moreover, the measurement process alters the state of a quantum system; the state vector is projected on one of the two basis state. This effect is exploited by the quantum key distribution algorithm of Bennett and Brassard to detect the presence of an intruder.

Roland Omnes points out in [100]: "When a theory is so strange that it must be interpreted, whether it be relativity theory or quantum mechanics, the aim of this interpretation is to reconcile the fundamental, outrageously abstract concepts with plain empirism." This is precisely the reason why in this book we discuss several experiments revealing quantum effects, and, whenever possible, we provide intuitive explanations and present analogies, even when they fail to convey the full complexity of the physical phenomena.

We avoided the sometimes difficult mathematical apparatus throughout this section, the trademark of quantum mechanics. The only mathematical concepts used were the vectors and the tensor product. The readers unfamiliar with these concepts are encouraged to refresh their knowledge of linear algebra and consult a text such as *Lectures on Linear Algebra* by J. M. Gelfand [62].

A fair number of popular science books attempt to provide intuitive explanations of some of the phenomena of interest to quantum computing and quantum information, with minimal references to mathematical equations. The list should start with Richard Feynman's *QED—The Strange Theory of Light and Matter* [56]. David Deutsch's *The Fabric of Reality* should also be high on the reading list of the curious reader [44]. A book by A. D. Aczel provides a very vivid encounter with some of the great physicists of the last century and an entertaining presentation of entanglement, one of the most puzzling and non-intuitive physical phenomena [2]. Closer to the subject of quantum computing are J. Brown's *The Quest for Quantum Computer* [29] and G. J. Milburn's *The Feynman Processor* [93] and *Schrödinger's Machines* [94].

Richard Feynman's *Lecture Notes on Computation* should be a required reading for anyone interested in the subject of quantum computing and quantum information theory [57]. We found the discussion of reversibility, the presentation of quantum gates, and information theory particularly impressive. Charles Bennett's paper entitled *Quantum Information and Computation* provides a very insightful discussion of some of the critical aspects of the subject [15]. A more advanced and rigorous, yet comprehensible introduction to quantum computing is E. Rieffel and W. Polack's paper [112].

However, the most authoritative text to date is M. A. Nielsen and I. L. Chuang's *Quantum Computing and Quantum Information Theory* [98]. Several lecture notes are available on the Internet. Among them, the ones of John Preskill stand out [111].

1.15 EXERCISES AND PROBLEMS

1.1. Read the Nobel lectures of Max Planck, Albert Einstein, Niels Bohr, Werner Heisenberg, Erwin Schrödinger, Max Born, and Richard Feynman and discuss how each of them viewed his contributions to science as reflected in his lecture. These lectures can be found at *http://www.nobel.se/physics/laureates*.

1.2. How did the political turmoil in the 1930s and 1940s in Europe influence the relations between the great physicists Max Planck, Albert Einstein, Niels Bohr, Werner Heisenberg, and John von Neumann, as well as their lives (according to [2])?

1.3. Present and critically analyze the main ideas related to "positivism", "reductionism", and "holism", as approaches to explaining physical phenomena, see [44] pages 4–28.

1.4. What is the justification of the statement in [44], page 197, "Complexity theory has not yet been sufficiently well integrated with physics to give many quantitative answers?"

1.5. Critically analyze Fredkin's idea that the universe is a gigantic digital computer processing information [58]. Note that the physicist Philip Morrison remarked "the only reason Fredkin thought that the universe was a computer was because he was a computer scientist, in the same way that if he had been a cheese maker, he would have claimed that it was made out of cheese" [29], page 59.

1.6. Discuss the benefits of computer simulation as an alternative method to science and engineering, complementing the traditional approaches, theoretical modeling and experiments. Give examples from your own area of interest and outline the benefits of computer simulation for your specific examples.

1.7. Relate the number of states of a system subject to computer simulation to the execution time and the space complexity of the algorithms used for simulation.

1.8. Is it feasible to simulate a quantum system using a classical computer? Justify your answer.

1.9. Consider the multiple beam splitter experiment discussed in Section 1.4. Assuming that a beam splitter tosses a fair ternary coin to decide if a photon should be transmitted, reflected, or absorbed, what are the probabilities of the five detectors in Figure 1.2(b) detecting a photon? If the experiment is carried out 100,000 times, how many counts will each detector register?

1.10. Draw the four diagrams showing the path of a photon emitted by S2 in Section 1.9, and calculate the probability of a photon from S2 to reach D1 and D2.

1.11. Consider the experiment discussed in Section 1.9 and illustrated in Figures 1.7 and 1.8. Assume that a beam of electrons crosses the paths of the photons coming from sources S1 and S2 before they reach the beam splitter BS1. How will the outcome of this experiment be affected?

1.12. Write a Java program to simulate the system discussed in Section 1.9 and presented in Figures 1.7 and 1.8.

1.13. Write a Java program to simulate the BB84 protocol. You have to simulate:

(a) a source of polarized photons that randomly picks up either a vertical/horizontal (VH) polarization and then a vertical polarization, or a horizontal polarization, or a diagonal polarization (DG) and then a 45° or 135° polarization;

(b) the photon separation system that randomly picks up a vertical/horizontal (VH) orientation, or a diagonal (DG) orientation for the crystal, and then measures an incoming photon without any error if its basis is identical with the crystal orientation;

(c) a quantum communication channel transporting a photon from the sender to the receiver with a possible eavesdropping component; and

(d) a classical noiseless communication channel allowing the sender and the receiver to exchange binary information.

1.14. Consider a setup similar to that in Sections 1.9 and 1.11. Show that for the case of an experimental setup with an odd number of beam splitters, $(1, 3, 5, \ldots, 2k+1, \ldots)$, the result is that presented in this section

$$p_{S1\rightarrow D1} = p_{S1\rightarrow D2} = 1/2 \qquad p_{S2\rightarrow D1} = p_{S2\rightarrow D2} = 1/2.$$

For the case of an even number of beam splitters, $(2, 4, 6, \ldots, 2k, \ldots)$, show that

$$p_{S1\rightarrow D1} = 1 \quad p_{S1\rightarrow D2} = 0 \qquad p_{S2\rightarrow D1} = 0 \quad p_{S2\rightarrow D2} = 1.$$

1.15. Consider an experimental setup similar to that in Section 1.9 and assume that the state vectors of photons originating from `direction1` and `direction2`, respectively, are

$$| \psi_1 \rangle = +q \,| 0 \rangle + q \,| 1 \rangle \qquad | \psi_2 \rangle = -q \,| 0 \rangle + q \,| 1 \rangle.$$

Show that

$$p_{S1\rightarrow D1} = 0 \quad p_{S1\rightarrow D2} = 1 \qquad p_{S2\rightarrow D1} = 1 \quad p_{S2\rightarrow D2} = 0.$$

CHAPTER 2

Quantum Mechanics—A Mathematical Model of the Physical World

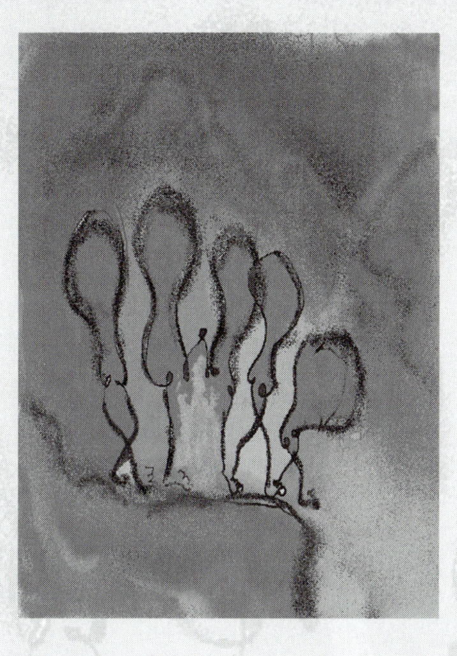

The material discussed in this chapter represents a minimal set of concepts necessary to understand the physical phenomena the new discipline of quantum computing is based upon. A number of experiments, some of them discussed in the previous chapter, have shown that the classical theory cannot explain some of the phenomena involving light, as well as atomic and subatomic particles. During the first decades of the twentieth century it became obvious that we needed a more refined model of the physical world. Quantum mechanics represents a mathematical model of the physical world able to explain phenomena, such as the energy spectra of atoms, which classical mechanics fails to explain. The predictions of quantum mechanics have been confirmed by a plethora of experimental evidence collected during the last hundred years.

We first introduce the basic definitions and concepts necessary to understand the mathematical apparatus of quantum mechanics. First, we introduce basic concepts from linear algebra. We present vector spaces, Hermitian operators, Hilbert spaces, as well as inner products, tensor products, and outer products of vectors in a Hilbert space.

Then, we present some of the milestones in the evolution of quantum ideas and briefly discuss two of the pillars of quantum mechanics, Schrödinger's Equation and Heisenberg's Uncertainty Principle. We present in some depth the double slit experiment imagined by Young to demonstrate the interference phenomena associated with the wave-like behavior of light and the Stern-Gerlach experiments revealing the spin.

The postulates of quantum mechanics are then introduced gently. We discuss the representation of quantum states and observables, outline the properties of quantum operators and the spectral decomposition of an operator, and conclude with a section devoted to the measurement of observables.

2.1 VECTOR SPACES

To define the concept of a vector space we first need to introduce two basic algebraic structures, the group and the field. A *group* is a set G with one binary operation "·", called *multiplication*, which satisfies three conditions:

1. Associative law: $\forall (a, b, c) \in G \quad a \cdot (b \cdot c) = (a \cdot b) \cdot c$.
2. Identity element: There is an identity element $e \in G$ such that $a \cdot e = e \cdot a = a, \forall a \in G$.
3. Inverse element: $\forall a \in G, \exists a^{-1}$ such that $a \cdot a^{-1} = a^{-1} \cdot a = e$.

A group G whose operation satisfies the commutative law $a \cdot b = b \cdot a$, $\forall a, b \in G$ is a *commutative*, or *Abelian*, group.

A *field* is a set F equipped with two binary operations, *addition* and *multiplication*, with the following properties:

1. Under addition, F is an Abelian group with the identity (or neutral) element 0 such that $0 + a = a, \forall a \in F$.
2. Under multiplication, the nonzero elements form an Abelian group with identity element 1 such that $1 \cdot a = a, \forall a \in F$, and $0 \cdot a = 0, \forall a \in F$. The additive and multiplicative identity elements are different, $0 \neq 1$.
3. The distributive law holds: $a \cdot (b + c) = a \cdot b + a \cdot c$.

A *vector space* $\mathcal{A}$ assumes three objects:

1. An Abelian group $(V, +)$ whose elements are called "vectors" and whose binary operation "+" is called *addition*,
2. A field F (usually $\mathbb{R}$, the real numbers, or $\mathbb{C}$, the complex numbers), whose elements are called "scalars", and
3. An operation called "multiplication with scalars" and denoted by "·", which associates to any scalar $c \in F$ and vector $\alpha \in V$ a new vector $c \cdot \alpha \in V$. The operation of multiplication of a vector with a scalar has the following properties:

$$c \cdot (\alpha + \beta) = c \cdot \alpha + c \cdot \beta$$
$$(c + c') \cdot \alpha = c \cdot \alpha + c' \cdot \alpha$$
$$(c \cdot c') \cdot \alpha = c \cdot (c' \cdot \alpha), \quad 1 \cdot \alpha = \alpha$$

where $\alpha, \beta \in V$ and $c, c' \in F$.

Observations:

(a) Often we omit the "·" symbol and write the product of two scalars as cc' instead of $c \cdot c'$ and the product of a scalar c with a vector α as $c\alpha$ instead of $c \cdot \alpha$.

(b) In this volume we are only concerned with either $\mathbb{R}$, the field of real numbers, or $\mathbb{C}$, the field of complex numbers.

Given n scalars $\{c_1, c_2, \ldots, c_n\} \in \mathbb{R}$, then the set of n vectors $\{\alpha_1, \alpha_2, \ldots, \alpha_n\} \in \mathbb{R}^n$ are *linearly independent* if

$$c_1\alpha_1 + c_2\alpha_2 + \ldots + c_n\alpha_n = 0 \implies c_1 = c_2 = \ldots = c_n = 0.$$

Vectors that are not linearly independent are called *linearly dependent.*

A *subspace* $\mathcal{S}$ of a vector space $\mathcal{A}$ is a subset of $\mathcal{A}$ which is closed with respect to the operations of addition and scalar multiplication. This means that the sum of two vectors in $\mathcal{S}$ is in $\mathcal{S}$. For any vector and any scalar, the product of a vector with a scalar is also a vector, $\forall\alpha \in \mathcal{S}$ and $\forall c \in \mathbb{R}$ then $c\alpha \in \mathcal{S}$.

Examples of subspaces:

1. The set of polynomials of degree at most m is a subspace of the vector space of all polynomials which is a subspace in the vector space of complex valued continuous functions $C_R(\mathbb{R})$.
2. The set of all continuous functions $f(x)$ defined for $0 \leqslant x \leqslant 2$ is a subspace of the linear space of all functions defined on the same domain.

The set of all linear combinations of any set of vectors of a vector space $\mathcal{A}$ is a subspace of $\mathcal{A}$. Given $c', c'_1, c'_2, \ldots, c'_n \in F$ the following two identities allow us to prove this statement

$$(c_1\alpha_1 + c_2\alpha_2 + \ldots + c_m\alpha_m) + (c'_1\alpha_1 + c'_2\alpha_2 + \ldots + c'_m\alpha_m)$$
$$= (c_1 + c'_1)\alpha_1 + (c_2 + c'_2)\alpha_2 + \ldots + (c_m + c'_m)\alpha_m$$
$$c'(c_1\alpha_1 + c_2\alpha_2 + \ldots c_m\alpha_m) = (c'c_1)\alpha_1 + (c'c_2)\alpha_2 + \ldots(c'c_m)\alpha_m$$

The subspace consisting of all linear combinations of a set of vectors of $\mathcal{A}$ is the smallest subset[1] containing all the given vectors. The set of vectors *spans* the subspace.

A linearly independent subset of vectors which spans the whole space is called a *basis of a vector space.* A vector space $\mathcal{A}$ is *finite dimensional* if and only if it has a finite basis.

If $\{b_1, b_2, \ldots, b_m\}$ and $\{c_1, c_2, \ldots, c_n\}$ are two bases for a vector space and m and n are finite then $m = n$.

The minimum number of vectors in any basis of a finite-dimensional vector space $\mathcal{A}$ is called the *dimension of a vector space*, $\dim(\mathcal{A})$. For example, the

[1] The smallest subset is the subset with the smallest cardinality.

ordinary space $\mathbb{R}^3$ can be spanned by three vectors, $(1, 0, 0)$, $(0, 1, 0)$, and $(0, 0, 1)$.

2.2 n-DIMENSIONAL REAL EUCLIDEAN VECTOR SPACE

Consider now an n-dimensional real vector space, $\mathbb{R}^n$. If for every pair of vectors $\alpha, \beta \in \mathbb{R}^n$ we have an associated real number (α, β) such that the following four conditions are satisfied

1. $(\alpha, \beta) = (\beta, \alpha)$
2. $(c\alpha, \beta) = c(\alpha, \beta)$, if $c \in \mathbb{R}$
3. $(\alpha + \gamma, \beta) = (\alpha, \beta) + (\gamma, \beta)$, $\forall \gamma \in \mathbb{R}^n$
4. $(\alpha, \alpha) \geqslant 0$ and $(\alpha, \alpha) = 0$ if and only if $\alpha = 0$

then we say that we have an n-dimensional *Euclidean space* and that (α, β) is the *inner product* of the vectors α and β. All the facts known from Euclidean geometry can be established in an Euclidean space. The *length of a vector* α in an Euclidean space is defined to be the real number

$$|\alpha| = \sqrt{(\alpha, \alpha)}.$$

The *angle between two vectors* α and β is the real number θ:

$$\theta = \arccos \frac{(\alpha, \beta)}{|\alpha|\,|\beta|} \quad \Longrightarrow \quad \cos\theta = \frac{(\alpha, \beta)}{|\alpha|\,|\beta|}.$$

If $(\alpha, \beta) = 0$, where $\alpha \neq 0$ and $\beta \neq 0$, then $\theta = \pi/2$, and $\cos\theta = 0$. Two vectors α and β in an Euclidean space are *orthogonal* if

$$(\alpha, \beta) = 0.$$

A set of n vectors $\mathcal{E} = \{e_1, e_2, \ldots, e_n\}$ is an *orthogonal basis* in an n-dimensional Euclidean space if the vectors in the set are pairwise orthogonal. If, in addition, each vector has unit length, then the vectors form an *orthonormal basis* and satisfy the condition

$$(e_i, e_j) = \delta_{i,j} = \begin{cases} 0 & \text{if } i \neq j \\ 1 & \text{if } i = j \end{cases} \qquad 1 \leqslant (i, j) \leqslant n.$$

with $\delta_{i,j}$ the *Kronecker delta function.*

It is relatively easy to show [62] that every n-dimensional Euclidean space contains orthogonal bases. Given an orthonormal basis $\{e_1, e_2, \ldots, e_n\}$ of an n-dimensional Euclidean space we can express any two vectors as

$$\alpha = a_1 e_1 + a_2 e_2 + \ldots + a_n e_n$$

and

$$\beta = b_1 e_1 + b_2 e_2 + \ldots + b_n e_n.$$

Then, we use the fact that $(e_i, e_j) = \delta_{i,j}$ to show that

$$(\alpha, e_i) = a_i$$

and that

$$(\alpha, \beta) = \sum_{i=1}^{n} a_i b_i.$$

The simplest functions defined on vector spaces are the linear transformations. The function $A(\alpha)$ is a *linear form (function)* if

$$A(\alpha; \beta) = A(\alpha) + A(\beta)$$

and

$$A(c\alpha) = cA(\alpha).$$

Consider two vector spaces $\mathcal{A}$ and $\mathcal{B}$ over the same field F. Let $\alpha \in \mathcal{A}$, $\beta \in \mathcal{A}$, $c \in F$, and $A(\alpha) \in \mathcal{B}$. We are given a function A which maps vectors in $\mathcal{A}$ to vectors in $\mathcal{B}$.

The function $A(\alpha; \beta)$ is said to be a *bilinear form (function)* of vectors α and β if

1. For any fixed β, $A(\alpha; \beta)$ is a linear function of α,
2. For any fixed α, $A(\alpha; \beta)$ is a linear function of β.

This implies that, if $\mathcal{A}$ and $\mathcal{B}$ are two vector spaces over the same field F and we consider variable vectors $\alpha, \alpha' \in \mathcal{A}$ and $\beta, \beta' \in \mathcal{B}$ and scalars $a, b, c, d \in F$, the function $A(\alpha, \beta)$ with values in F is a *bilinear function* if

$$A(a\alpha + b\alpha'; \beta) = aA(\alpha; \beta) + bA(\alpha'; \beta)$$

and

$$A(\alpha; c\beta + d\beta') = cA(\alpha; \beta) + dA(\alpha; \beta')$$

A bilinear function is *symmetric* if

$$A(\alpha; \beta) = A(\beta; \alpha).$$

The inner product of two vectors in an Euclidean vector space is an example of a symmetric bilinear function.

If $A(\alpha; \beta)$ is a symmetric bilinear form, then the function $A(\alpha; \alpha)$ is called a *quadratic form*. A quadratic form $A(\alpha; \alpha)$ is *positive definite* if for every vector $\alpha \neq 0$

$$A(\alpha; \alpha) > 0.$$

The bilinear form $A(\alpha; \beta)$ is called the *polar form associated* with the quadratic form $A(\alpha; \alpha)$. It can be shown that $A(\alpha; \beta)$ is uniquely determined by its quadratic form [62].

Now we can provide an alternative definition of an Euclidean vector space as *a vector space with a positive definite quadratic form $A(\alpha; \alpha)$. The inner product (α, β) of two vectors is the value of the bilinear form $A(\alpha; \beta)$ associated with $A(\alpha; \alpha)$.*

2.3 LINEAR OPERATORS AND MATRICES

A rectangular array of elements of a field F with m rows and n columns, A, is called a *matrix*

$$A = \begin{pmatrix} a_{11} & a_{12} & \dots & a_{1n} \\ a_{21} & a_{22} & \dots & a_{2n} \\ \vdots & \vdots & \ddots & \vdots \\ a_{m1} & a_{m2} & \dots & a_{mn} \end{pmatrix}.$$

Matrix A can be interpreted as a linear map from the vector space of dimension n, F^n, to the vector space of dimension m, F^m, equipped with the canonical base. If $n = m$, then A is a linear map from F^n to itself.

Let $\alpha_1 = (a_{11}, a_{12}, \dots, a_{1n}), \alpha_2 = (a_{21}, a_{22}, \dots, a_{2n}), \dots,$ $\alpha_m = (a_{m1}, a_{m2}, \dots, a_{mn})$ be a set of vectors spanning a subspace of dimension m of a vector space $V_n(F)$ called the *row space* of the $m \times n$ matrix A.

The elementary row operations on a matrix are:

1. Interchanging of any two rows,
2. Multiplication of all the elements of a row by a constant $c \in F$, and
3. Addition of any multiple of a row to any other row.

Two matrices are *row-equivalent* if one is obtained from the other by a finite sequence of row operations.

When $n = m$ the *identity matrix* $I = [a_{ij}]$ is the matrix with $a_{ii} = 1$ and $a_{ij} = 0$, if $i \neq j$. For example, the identity matrix in a vector space of dimension 8 is

$$I_8 = \begin{pmatrix} 1&0&0&0&0&0&0&0 \\ 0&1&0&0&0&0&0&0 \\ 0&0&1&0&0&0&0&0 \\ 0&0&0&1&0&0&0&0 \\ 0&0&0&0&1&0&0&0 \\ 0&0&0&0&0&1&0&0 \\ 0&0&0&0&0&0&1&0 \\ 0&0&0&0&0&0&0&1 \end{pmatrix}.$$

A *permutation matrix* P is an identity matrix with rows and columns permuted.

The *determinant* of an $n \times n$ matrix $A = [a_{ij}]$ is the polynomial

$$\det(A) = |A| = \sum_{\phi} \operatorname{sgn}(\phi) a(1, 1\phi) a(2, 2\phi) \dots a(n, n\phi).$$

ϕ denotes one of the $n!$ different permutations of integers $1, 2, \dots, n$. If ϕ is an even permutation then $\operatorname{sgn}(\phi) = +1$. For an odd permutation $\operatorname{sgn}(\phi) = -1$.

The determinant can be written as:

$$\det(A) = A_{i1}a_{i1} + A_{i2}a_{i2} + \ldots + A_{in}a_{in}.$$

Here the coefficient A_{ij} of a_{ij} is called the *cofactor* of a_{ij}. A cofactor is a polynomial in the remaining rows of A and can be described as the *partial derivative* $(\partial|A|/\partial a_{ij})$ of A. The cofactor polynomial contains only entries from an $(n - 1) \times (n - 1)$ matrix M_{ij} (also called a "minor") obtained from A by eliminating row i and column j.

If we permute two rows of A then the sign of the determinant $|A|$ changes. The determinant of the transpose of a matrix is equal to the determinant of the original matrix:

$$|A^T| = |A|.$$

If two rows of A are identical then:

$$|A| = 0.$$

A square $n \times n$ matrix is *triangular* if all entries below the diagonal are zero. The determinant of a triangular matrix is the product of its diagonal elements. To compute the determinant $|A|$ one should perform elementary row operations on matrix A and reduce it to a triangular form.

The *characteristic polynomial of matrix A* is:

$$c(\lambda) \equiv \det(A - \lambda I) \quad \text{or} \quad c(\lambda) \equiv |A - \lambda I|.$$

The *trace of matrix A* is the sum of its diagonal elements.

$$\mathrm{Tr}(A) = \sum_i a_{ii}.$$

The *trace of an operator* **A** is the trace of A, the matrix representation of **A**.[2] Given any two matrices A and B over F and a scalar $c \in F$ it is easy to show that the trace has the following properties:

1. It is cyclic

$$\mathrm{Tr}(AB) = \mathrm{Tr}(BA).$$

2. Linearity

$$\mathrm{Tr}(A + B) = \mathrm{Tr}(A) + \mathrm{Tr}(B) \quad \text{and} \quad \mathrm{Tr}(cA) = c\mathrm{Tr}(A).$$

3. Invariance under the *similarity transformation*, S

$$\mathrm{Tr}(SAS^\dagger) = \mathrm{Tr}(S^\dagger SA) = \mathrm{Tr}(A).$$

[2]We make a distinction between the linear operator **A** and A—the matrix representation of **A**—only when the context requires such a distinction.

The determinant, the trace, and the characteristic polynomial are quantities associated with any linear map from a finite dimensional vector space into itself.

Now we can express a bilinear form $A(\beta; \gamma)$ in terms of the projections of the two vectors, β and γ, namely $b_1, b_2, \ldots, b_n$ and $c_1, c_2, \ldots, c_n$ on the orthonormal basis $e_1, e_2, \ldots, e_n$.

$$A(\beta; \gamma) = (b_1 e_1 + b_2 e_2 + \ldots + b_n e_n; c_1 e_1 + c_2 e_2 + \ldots + c_n e_n).$$

But A is a bilinear function, thus

$$A(\beta; \gamma) = \sum_{i,j=1}^{n} A(e_i; e_j) b_i c_j.$$

Here $A(e_i; e_j)$ is a real number that we denote by a_{ij}. Then

$$A(\beta; \gamma) = \sum_{i,j=1}^{n} a_{ij} b_i c_j.$$

The matrix with elements a_{ij}, $A = [a_{ij}]$, $1 \leqslant (i, j) \leqslant n$ is called the *matrix of the bilinear form* $A(\beta; \gamma)$ relative to the basis $e_1, e_2, \ldots, e_n$.

2.4 HERMITIAN OPERATORS IN A COMPLEX n-DIMENSIONAL EUCLIDEAN VECTOR SPACE

Later in this chapter we shall use vectors to describe the state of a quantum system. Quantum systems evolve in time and, in order to capture the dynamics of a system in a mathematical model, we have to study transformations of states represented as mathematical operators applied to vectors. Note that the concepts "form", "transformation", and "operator" are used interchangeably throughout this chapter.

It is also necessary to consider vector spaces over fields other than $\mathbb{R}$, the field of real numbers. $\mathbb{C}^n$, the n-dimensional vector space over $\mathbb{C}$, the field of complex numbers is of particular interest to us.

The concepts of linear and bilinear transformations introduced for an n-dimensional Euclidean vector space over the field of real numbers, extend to finite-dimensional vector spaces over other fields. For example, *the inner product* in a vector space $\mathbb{C}^n$ over a field $\mathbb{C}$ of complex numbers is a bilinear map which for every $\alpha, \beta \in \mathbb{C}^n$, in addition to bilinearity property, satisfies three conditions

$$(\alpha, \beta) = (\beta, \alpha)^*$$

$$(\alpha, \alpha) \geq 0$$

and

$$(\alpha, \alpha) = 0 \implies \alpha = 0.$$

Recall that $c = (\alpha, \beta)$ is a complex number, $c = a + ib$, and $c^* = a - ib$ with $i = \sqrt{-1}$.

The inner product in a finite-dimensional vector space induces a norm, but a norm may exist even if an inner product is not defined. A finite dimensional vector space with a norm is a *Banach space.*

The inner product permits us to measure the "length" of a vector and the "angle" between two vectors. If $\alpha, \beta \in \mathbb{C}^n$ and $c \in \mathbb{C}$, then the norm is a non-negative function with the following properties

$$\|\alpha + \beta\| \leq \|\alpha\| + \|\beta\|$$

$$\|c \cdot \alpha\| = |c| \, \|\alpha\|$$

and

$$\|\alpha\| = 0 \quad \text{if and only if} \quad \alpha = 0.$$

We define orthogonality and orthonormal vector bases in an n-dimensional Euclidean vector space over the field of complex numbers similarly to the ones defined for an n-dimensional Euclidean vector space over the field of real numbers.

We now introduce Hermitian operators, a concept analogous to the symmetric bilinear form in a real n-dimensional Euclidean space. If $\alpha, \beta \in \mathbb{C}^n$ then the bilinear form $\mathbf{A}(\alpha; \beta)$ is called *Hermitian* if

$$\mathbf{A}(\alpha; \beta) = \mathbf{A}^*(\beta; \alpha).$$

A bilinear form has a matrix associated with it and the necessary and sufficient condition for an operator $\mathbf{A}$ to be Hermitian is that its matrix $A = [a_{ij}]$ relative to some basis $e_1, e_2 \ldots, e_n$ satisfies the condition

$$a_{ij} = a_{ji}^*.$$

The *adjoint operator* associated with a bilinear form $\mathbf{A}$ is denoted as $\mathbf{A}^\dagger(\alpha; \beta) = \mathbf{A}^*(\beta; \alpha)$. If $\alpha \in \mathbb{C}^n$ then by definition $\alpha^\dagger = (\alpha^*)^T$.

If A is the matrix representation of the linear operator $\mathbf{A}$ and we want to obtain the adjoint matrix, we first construct the complex conjugate matrix A^* and then take its transpose

$$A^\dagger = (A^*)^T.$$

For example, the adjoint of matrix

$$A = \begin{pmatrix} 1 - 5i & 1 + i \\ 1 + 3i & 7i \end{pmatrix}$$

is

$$A^\dagger = \begin{pmatrix} 1 - 5i & 1 + i \\ 1 + 3i & 7i \end{pmatrix}^\dagger = \left[\begin{pmatrix} 1 - 5i & 1 + i \\ 1 + 3i & 7i \end{pmatrix}^*\right]^T.$$

Thus,

$$A^\dagger = \begin{pmatrix} 1 + 5i & 1 - i \\ 1 - 3i & -7i \end{pmatrix}^T = \begin{pmatrix} 1 + 5i & 1 - 3i \\ 1 - i & -7i \end{pmatrix}.$$

The adjoint of the matrix

$$I_3 = \begin{pmatrix} 1 & 0 & 0 \\ 0 & 1 & 0 \\ 0 & 0 & 1 \end{pmatrix}$$

is

$$I_3^\dagger = \begin{pmatrix} 1 & 0 & 0 \\ 0 & 1 & 0 \\ 0 & 0 & 1 \end{pmatrix}$$

and the adjoint of matrix A with real elements, $(a, b, c, d, e, f, g, h, i, j, k, l, m, n, o, p) \in \mathbb{R}$

$$A = \begin{pmatrix} a & b & c & d \\ e & f & g & h \\ i & j & k & l \\ m & n & o & p \end{pmatrix}$$

is

$$A^\dagger = \begin{pmatrix} a & e & i & m \\ b & f & j & n \\ c & g & k & o \\ d & h & l & p \end{pmatrix}.$$

An operator $\mathbf{A}$ is *normal* if

$$\mathbf{A}\mathbf{A}^\dagger = \mathbf{A}^\dagger\mathbf{A}.$$

If $\mathbf{A}$ is a Hermitian (self-adjoint) operator, then it is also a normal operator.

An $n \times n$ matrix U is *unitary* if

$$U^\dagger U = I_n$$

where I_n is the identity matrix in an n-dimensional vector space

$$I_n = \begin{pmatrix} 1 & 0 & 0 & \dots & 0 \\ 0 & 1 & 0 & \dots & 0 \\ 0 & 0 & 1 & \dots & 0 \\ \vdots & \vdots & \vdots & \dots & \vdots \\ 0 & 0 & 0 & \dots & 1 \end{pmatrix}.$$

Unitary operators preserve the inner product of vectors. Let $(\alpha, \beta) \in \mathbb{C}^n$, then

$$(U\alpha, U\beta) = (\alpha, \beta).$$

Usually we write the inner product using the "·" symbol. With this convention the previous equation becomes

$$U\alpha \cdot U\beta = \alpha \cdot \beta.$$

Given two operators **A** and **B**, the *commutator* of the two operators is

$$[\mathbf{A}, \mathbf{B}] = \mathbf{AB} - \mathbf{BA}.$$

The operator **A** commutes with **B** if

$$[\mathbf{A}, \mathbf{B}] = 0.$$

The *anti-commutator* of **A** and **B** is

$$\{\mathbf{A}, \mathbf{B}\} = \mathbf{AB} + \mathbf{BA}.$$

The operator **A** anti-commutes with **B** if

$$\{\mathbf{A}, \mathbf{B}\} = 0.$$

2.5 *n*-DIMENSIONAL HILBERT SPACES—DIRAC'S NOTATION

Traditionally, a Hilbert space is defined as an infinite-dimensional vector space with an inner product and its associated norm [85]. Note that not any norm comes from an inner product but any inner product induces a norm.

A norm, and therefore an inner product, allows us to define "convergence." A pair consisting of a vector space and an inner product is called a *Hilbert space* if the vector space is complete with respect to the norm induced by the inner product. "Complete" means that any Cauchy sequence[3] is convergent. If a vector space is finite dimensional it is automatically complete, hence it is a Hilbert space.[4] If not, one can add elements and complete it to a Hilbert space.[5] The relationship between Hilbert and Banach spaces is summarized in Figure 2.1.

The infinite-dimensional vector spaces that are important in quantum mechanics are analogous to the finite-dimensional vector spaces and can be spanned by a countable basis.[6] They are called *separable Hilbert spaces* [92].

The quantum computing literature [98] has inherited the convention of calling an n-dimensional complex Euclidean vector space an n-dimensional Hilbert space, $\mathcal{H}_n$. We shall follow this convention throughout this book.

With this convention an *n-dimensional Hilbert space $\mathcal{H}_n$ is an n-dimensional vector space over the field of complex numbers with an inner product and associated norm*. The elements of $\mathcal{H}_n$ are n-dimensional vectors. These vectors can be added together, or multiplied by complex numbers and the results of

---

[3]A sequence $\{a_n\}$ is Cauchy if for any $\epsilon > 0$ there exists N so that $\|a_k - a_r\| < \epsilon$ for $k, r > N$.

[4]A vector space which is complete and has a norm is called a Banach space. A Hilbert space is a space with an inner product. This space is a Banach space with respect to the induced norm. A finite dimensional vector space equipped with a norm is always a Banach space. An infinite dimensional vector space can be completed to a Banach space.

[5]The procedure is similar to the completion of rational numbers to real numbers.

[6]The set of vectors forming an orthonormal basis is countable.

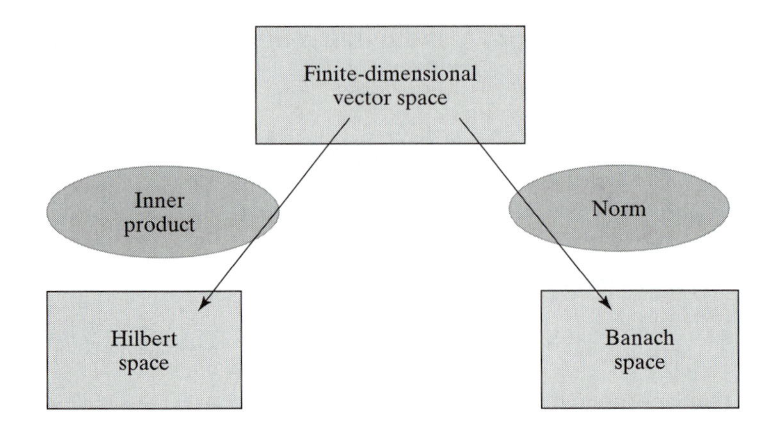

FIGURE 2.1: A finite-dimensional vector space with a inner product is a Hilbert space. A finite-dimensional vector space with a norm is a Banach space.

these operations are also elements of the Hilbert space. $\mathcal{H}_n$ is isomorphic with $\mathbb{C}^n$. An n-dimensional Hilbert space $\mathcal{H}_n$ is also called a *unitary space*.

All the concepts discussed in the previous section apply to $\mathcal{H}_n$. A collection of vectors $\{e_1, e_2, \dots, e_n\} \in \mathcal{H}_n$ is called an orthonormal basis if the inner product of any two of them is zero, $(e_i e_j) = 0, \forall (i, j) \in \{1, n\}$, and the inner product of any of them with itself is one, $(e_i, e_i) = 1, \forall i \in \{1, n\}$.

A set of n orthonormal vectors is an *orthonormal basis* in $\mathcal{H}_n$. For the following discussion we select one of many possible choices of orthonormal basis vectors. We use the traditional notation of quantum mechanics for vectors in $\mathcal{H}_n$ introduced by Dirac and denote the vectors of this particular orthonormal basis as `kets`

$$\{| 0\rangle, | 1\rangle, \dots, | i\rangle, \dots, | n - 1\rangle\},$$

or, as `bras`

$$\{\langle 0 |, \langle 1 |, \dots, \langle i |, \dots, \langle n - 1 |\}.$$

In matrix representation each `ket` vector $| i\rangle$ is expressed as a column vector with a 1 in the $i^{th} + 1$ row and 0 in all the others. For example,

$$| 0\rangle = \begin{pmatrix} 1 \\ 0 \\ \vdots \\ 0 \\ \vdots \\ 0 \end{pmatrix}, | 1\rangle = \begin{pmatrix} 0 \\ 1 \\ \vdots \\ 0 \\ \vdots \\ 0 \end{pmatrix}, \dots, | i\rangle = \begin{pmatrix} 0 \\ 0 \\ \vdots \\ 1 \\ \vdots \\ 0 \end{pmatrix}, \dots, | n - 1\rangle = \begin{pmatrix} 0 \\ 0 \\ \vdots \\ 0 \\ \vdots \\ 1 \end{pmatrix}.$$

In turn, each `bra` vector $\langle i \mid$ is expressed as a row vector with 1 in the $i^{th} + 1$ position and 0 in all the others:

$$\langle 0 \mid = (\ 1\ 0\ \dots\ 0\ \dots\ 0\ ),$$
$$\langle 1 \mid = (\ 0\ 1\ \dots\ 0\ \dots\ 0\ ),$$
$$\vdots$$
$$\langle i \mid = (\ 0\ 0\ \dots\ 1\ \dots\ 0\ ),$$
$$\vdots$$
$$\langle n - 1 \mid = (\ 0\ 0\ \dots\ 0\ \dots\ 1\ ).$$

An n-dimensional `ket` vector $\mid \psi \rangle$ can be expressed in this basis as a linear combination of the orthonormal `ket` vectors of the basis

$$\mid \psi \rangle = \alpha_0 \mid 0 \rangle + \alpha_1 \mid 1 \rangle + \dots + \alpha_i \mid i \rangle + \dots + \alpha_{n-1} \mid n - 1 \rangle$$

where $\alpha_0, \alpha_1, \dots, \alpha_i, \dots, \alpha_{n-1}$ are complex numbers.

For each `ket` vector $\mid \psi \rangle$ there is a *dual*, the `bra` vector denoted by $\langle \psi \mid$. The `bra` and `ket` vectors are related by Hermitian conjugation

$$\mid \psi \rangle = (\langle \psi \mid)^\dagger \quad \text{and} \quad \langle \psi \mid = (\mid \psi \rangle)^\dagger.$$

The `bra` vector $\langle \psi \mid$, the *dual* of the `ket` vector $\mid \psi \rangle$, is expressed as a linear combination of the orthonormal `bra` vectors of the basis

$$\langle \psi \mid = \alpha_0^* \langle 0 \mid + \alpha_1^* \langle 1 \mid + \dots + \alpha_i^* \langle i \mid + \dots + \alpha_{n-1}^* \langle n - 1 \mid$$

where $\alpha_0^*, \alpha_1^*, \dots, \alpha_i^*, \dots, \alpha_{n-1}^*$ are the complex conjugates of $\alpha_0, \alpha_1, \dots, \alpha_i, \dots, \alpha_{n-1}$.

The `ket` vector $\mid \psi \rangle$ is expressed as the column vector

$$\mid \psi \rangle = \begin{pmatrix} \alpha_0 \\ \alpha_1 \\ \vdots \\ \alpha_i \\ \vdots \\ \alpha_{n-1} \end{pmatrix}$$

and the dual `bra` vector is expressed as the row vector

$$\langle \psi \mid = (\ \alpha_0^*\ \alpha_1^*\ \dots\ \alpha_i^*\ \dots\ \alpha_{n-1}^*\ ).$$

2.6 THE INNER PRODUCT IN AN n-DIMENSIONAL HILBERT SPACE

The inner product $\langle \psi_a \mid \psi_b \rangle$ of two vectors $\mid \psi_a \rangle, \mid \psi_b \rangle \in \mathcal{H}_n$ is a complex number. The inner product in a Hilbert space has the following properties:

1. The inner product of a vector with itself is a non-negative real number

$$\langle \psi \mid \psi \rangle \in \mathbb{R}$$

$$\langle \psi \mid \psi \rangle = \begin{cases} = 0 & \text{if} \mid \psi \rangle = \mid 0 \rangle \\ > 0 & \text{otherwise} \end{cases}.$$

2. Linearity: If $\mid \psi_a \rangle, \mid \psi_b \rangle, \mid \psi_c \rangle \in \mathcal{H}_n$ and $a, b, c \in \mathbb{C}$

$$\langle \psi_a \mid (c \mid \psi_b \rangle) = c \langle \psi_a \mid \psi_b \rangle;$$

$$(a \langle \psi_a \mid + b \langle \psi_b \mid) \mid \psi_c \rangle = a \langle \psi_a \mid \psi_c \rangle + b \langle \psi_b \mid \psi_c \rangle;$$

$$\langle \psi_c \mid (a \mid \psi_a \rangle + b \mid \psi_b \rangle) = a \langle \psi_c \mid \psi_a \rangle + b \langle \psi_c \mid \psi_b \rangle.$$

3. Skew symmetry:

$$\langle \psi_a \mid \psi_b \rangle = \langle \psi_b \mid \psi_a \rangle^*.$$

Let us observe that the skew symmetry implies a skew linearity in the second factor

$$\begin{aligned} &\langle \psi_a \mid (b \mid \psi_b \rangle + c \mid \psi_c \rangle) \\ &= [(b \langle \psi_b \mid + c \langle \psi_c \mid) \mid \psi_a \rangle]^* \\ &= [b \langle \psi_b \mid \psi_a \rangle + c \langle \psi_c \mid \psi_a \rangle]^* \\ &= b^* (\langle \psi_b \mid \psi_a \rangle)^* + c^* (\langle \psi_c \mid \psi_a \rangle)^* \\ &= b^* \langle \psi_a \mid \psi_b \rangle + c^* \langle \psi_a \mid \psi_c \rangle. \end{aligned}$$

The inner product maps an ordered pair of vectors in $\mathcal{H}_n$ to a complex number in $\mathbb{C}$. If $\mid \psi_a \rangle, \mid \psi_b \rangle \in \mathcal{H}_4$, and

$$\mid \psi_a \rangle = \alpha_0 \mid 0 \rangle + \alpha_1 \mid 1 \rangle + \alpha_2 \mid 2 \rangle + \alpha_3 \mid 3 \rangle$$

$$\mid \psi_b \rangle = \beta_0 \mid 0 \rangle + \beta_1 \mid 1 \rangle + \beta_2 \mid 2 \rangle + \beta_3 \mid 3 \rangle$$

then

$$\langle \psi_a \mid \psi_b \rangle = \begin{pmatrix} \alpha_0^* & \alpha_1^* & \alpha_2^* & \alpha_3^* \end{pmatrix} \begin{pmatrix} \beta_0 \\ \beta_1 \\ \beta_2 \\ \beta_3 \end{pmatrix} = \alpha_0^* \beta_0 + \alpha_1^* \beta_1 + \alpha_2^* \beta_2 + \alpha_3^* \beta_3.$$

For example, if

$$\mid \psi_a \rangle = (1 + i) \mid 0 \rangle + (2 - 3i) \mid 1 \rangle, \quad \mid \psi_b \rangle = (1 - 2i) \mid 0 \rangle + (3 + 2i) \mid 1 \rangle$$

then

$$\begin{aligned} \langle \psi_a \mid \psi_b \rangle &= (1 + i)^* (1 - 2i) + (2 - 3i)^* (3 + 2i) \\ &= (1 - i)(1 - 2i) + (2 + 3i)(3 + 2i) \\ &= -1 + 10i. \end{aligned}$$

Recall that the inner product of a vector with itself is a real number. Indeed

$$\langle \psi_a \mid \psi_a \rangle = \begin{pmatrix} \alpha_0^* & \alpha_1^* & \alpha_2^* & \alpha_3^* \end{pmatrix} \begin{pmatrix} \alpha_0 \\ \alpha_1 \\ \alpha_2 \\ \alpha_3 \end{pmatrix} = \alpha_0^*\alpha_0 + \alpha_1^*\alpha_1 + \alpha_2^*\alpha_2 + \alpha_3^*\alpha_3$$

$$= \mid \alpha_0 \mid^2 + \mid \alpha_1 \mid^2 + \mid \alpha_2 \mid^2 + \mid \alpha_3 \mid^2$$

with $\mid \alpha_i \mid^2$, the square of the modulus of the complex number α_i being equal to the sum of the squares of its real and imaginary components

$$\mid \alpha_i \mid^2 = [\Re(\alpha_i)]^2 + [\Im(\alpha_i)]^2.$$

For example, if

$$\mid \psi_a \rangle = (1 + 2i) \mid 0 \rangle + (4 - 3i) \mid 1 \rangle$$

then

$$\langle \psi_a \mid \psi_a \rangle = \mid 1 + 2i \mid^2 + \mid 4 - 3i \mid^2 = (1^2 + 2^2) + (4^2 + 3^2) = 5 + 25 = 30.$$

A few comments regarding the notations: $\langle \psi_a \mid \psi_b \rangle$ is an abbreviated notation for $\langle \psi_a \| \psi_b \rangle$; a complete bracket expression $\langle \mid \rangle$ denotes either a real or a complex number while an incomplete bracket expression such as $\langle \mid$, or $\mid \rangle$, denotes a row or a column vector, respectively. The notation $P_1 \Longrightarrow P_2$ means "proposition P_1 implies proposition P_2." The notation $\mid \alpha_i \mid$ denotes the modulus of the complex number α_i, while $\| \mid \alpha \rangle \|$ denotes the norm of the vector $\mid \alpha \rangle$.

Two vectors $\mid \psi_a \rangle$ and $\mid \psi_b \rangle$ in $\mathcal{H}_n$ are *orthogonal*, $\mid \psi_a \rangle \perp \mid \psi_b \rangle$, if their inner product is zero

$$\mid \psi_a \rangle \perp \mid \psi_b \rangle \quad \Longrightarrow \quad \langle \psi_a \mid \psi_b \rangle = 0.$$

The skew symmetry implies that orthogonality is a symmetric relation

$$\mid \psi_a \rangle \perp \mid \psi_b \rangle \quad \Longrightarrow \quad \mid \psi_b \rangle \perp \mid \psi_a \rangle.$$

A *normal unitary basis* of an n-dimensional space is a set of n vectors

$$\{\mid \psi_1 \rangle, \mid \psi_2 \rangle, \ldots, \mid \psi_i \rangle, \ldots, \mid \psi_n \rangle\}$$

where each vector has the norm (or "length") equal to 1 and any two vectors are orthogonal

$$\| \mid \psi_i \rangle \| = 1 \quad \text{and} \quad \langle \psi_i \mid \psi_j \rangle = 0 \quad \text{if } i \neq j.$$

The *unit vectors* $\langle 0 \mid = (1, 0, \ldots, 0), \ldots, \langle n - 1 \mid = (0, 0, \ldots, 1)$ have unit length and are mutually orthogonal; they form a normal unitary basis. It can be proven that *any set of $m < n$ mutually orthogonal vectors of length one of a unitary space forms part of a normal unitary basis of the space.*

The inner product satisfies the Schwartz inequality

$$\langle \psi_a \mid \psi_a \rangle \langle \psi_b \mid \psi_b \rangle \geq |\langle \psi_a \mid \psi_b \rangle|^2.$$

A non-zero vector $\mid a\rangle$ is an *eigenvector* and the scalar a is an *eigenvalue* of the linear operator $\mathbf{A}^7$ if the following equation is satisfied:

$$A \mid a\rangle = a \mid a\rangle.$$

The eigenvalues of the operator **A** are the solutions of the characteristic equation

$$c(\lambda) = 0$$

where the characteristic polynomial is

$$c(\lambda) \equiv \det(A - \lambda I).$$

According to the fundamental theorem of algebra a polynomial over $\mathbb{C}$ has at least one complex root. Thus every operator **A** in a finite dimensional vector space over $\mathbb{C}$ has at least one eigenvalue (actually it has n eigenvalues if n is the dimension of the vector space.)

The *eigenspace* corresponding to an eigenvalue a of operator **A** is the set of vectors $\{\mid a_i\rangle\}$ which have eigenvalue a. It is a subspace of the vector space that **A** operates on.

Any operator **A** can be written as

$$A = \sum_i a_i \mid b_i\rangle\langle c_i \mid.$$

A diagonal representation of the operator **A** is a representation of the form

$$A = \sum_i a_i \mid b_i\rangle\langle b_i \mid$$

where the vectors $\mid b_i\rangle$ form an orthogonal set of eigenvectors of **A** corresponding to the eigenvalues a_i. When such representation exists, it is unique.

It is easy to prove that a normal matrix is Hermitian if and only if it has real eigenvalues. The proof is left as an exercise to the reader, and here we only show that if a matrix in $\mathcal{H}_2$ is Hermitian it has real eigenvalues. Let

$$A = \begin{pmatrix} a_{11} & a_{12} \\ a_{21} & a_{22} \end{pmatrix}$$

with $a_{11}, a_{12}, a_{21}, a_{22} \neq 0$. Then

$$A^\dagger = \begin{pmatrix} a_{11}^* & a_{21}^* \\ a_{12}^* & a_{22}^* \end{pmatrix}.$$

The eigenvalues of A are the solutions of the equation

$$\det(A - \lambda I_2) = 0$$

---

[7] A is the matrix associated with the linear operator **A**.

or

$$(a_{11} - \lambda)(a_{22} - \lambda) - a_{12}a_{22} = 0.$$

The quadratic equation

$$\lambda^2 - (a_{11} + a_{22})\lambda + (a_{11}a_{22} - a_{12}a_{21}) = 0$$

has real roots if

$$(a_{11} + a_{22})^2 - 4(a_{11}a_{22} - a_{12}a_{21}) = (a_{11} - a_{22})^2 + 4a_{12}a_{21} \geqslant 0.$$

The fact that A is Hermitian ($A = A^{\dagger}$) implies that

$$a_{11} = a_{11}^*, \quad a_{22} = a_{22}^*, \quad a_{12} = a_{21}^*, \quad \text{and} \quad a_{21} = a_{12}^*.$$

Thus, indeed a Hermitian matrix in $\mathcal{H}_2$ has real eigenvalues.

2.7 TENSOR AND OUTER PRODUCTS

If $\mathcal{A}(F)$ and $\mathcal{B}(F)$ are vector spaces over the field F, then the set $\mathcal{F}(\mathcal{A}, \mathcal{B})$ of all bilinear functions $A(\alpha, \beta)$ with $\alpha \in \mathcal{A}$ and $\beta \in \mathcal{B}$ is also a vector space over the field F.

An $n \times m$ matrix A is regarded as a linear operator from an n-dimensional Hilbert space, $\mathcal{H}_n$, to an m-dimensional Hilbert space, $\mathcal{H}_m$, namely $A : \mathcal{H}_n \mapsto \mathcal{H}_m$.

The *tensor product* $\mathcal{A} \otimes \mathcal{B}$ of two vector spaces $\mathcal{A}$ and $\mathcal{B}$ over the same field F is the dual of the space $\mathcal{F}(\mathcal{A}, \mathcal{B})$ of bilinear functions from $\mathcal{A}$ and $\mathcal{B}$ to F.

The tensor product of two finite dimensional Hilbert spaces, $\mathcal{H}_n$ and $\mathcal{H}_m$, is $\mathcal{H}_n \otimes \mathcal{H}_m = \mathcal{H}_{nm}$.

Similarly, one defines the *tensor product of two linear operators*, in particular the tensor product of two matrices. For example, if A is an $m \times n$ matrix and B is a $p \times q$ matrix then using the *Kronecker product* representation for the tensor product we have

$$A \otimes B = \begin{pmatrix} a_{11}B & a_{12}B & \dots & a_{1n}B \\ a_{21}B & a_{22}B & \dots & a_{2n}B \\ a_{31}B & a_{32}B & \dots & a_{3n}B \\ \vdots & \vdots & & \vdots \\ a_{m1}B & a_{m2}B & \dots & a_{mn}B \end{pmatrix}$$

with

$$A = \begin{pmatrix} a_{11} & a_{12} & \dots & a_{1n} \\ a_{21} & a_{22} & \dots & a_{2n} \\ a_{31} & a_{32} & \dots & a_{3n} \\ \vdots & \vdots & & \vdots \\ a_{m1} & a_{m2} & \dots & a_{mn} \end{pmatrix} \qquad B = \begin{pmatrix} b_{11} & b_{12} & \dots & b_{1q} \\ b_{21} & b_{22} & \dots & b_{2q} \\ b_{31} & b_{32} & \dots & b_{3n} \\ \vdots & \vdots & & \vdots \\ b_{p1} & b_{p2} & \dots & b_{pq} \end{pmatrix}.$$

Here $a_{ij}B, 1 \leqslant i \leqslant n, 1 \leqslant j \leqslant m$, is a sub-matrix whose entries are the products of elements of matrix B and a_{ij}.

The tensor product of an $m \times n$ matrix and a $p \times q$ matrix is an $mp \times nq$ matrix. For example, the tensor product of vectors (a, b) and (c, d) is the vector

$$(a, b)^T \otimes (c, d)^T = \begin{pmatrix} a \\ b \end{pmatrix} \otimes \begin{pmatrix} c \\ d \end{pmatrix} = \begin{pmatrix} ac \\ ad \\ bc \\ bd \end{pmatrix}.$$

The tensor product of vectors $(\mid 0\rangle, \mid 1\rangle) \in \mathcal{H}_2$ is

$$\mid 0\rangle \otimes \mid 1\rangle = \begin{pmatrix} 1 \\ 0 \end{pmatrix} \otimes \begin{pmatrix} 0 \\ 1 \end{pmatrix} = \begin{pmatrix} 0 \\ 1 \\ 0 \\ 0 \end{pmatrix}.$$

The tensor product of two vectors, one in $\mathcal{H}_p$, and the other in $\mathcal{H}_q$, is a vector in $\mathcal{H}_{pq}$. In the previous example the tensor product of two vectors in $\mathcal{H}_2$ is a vector in $\mathcal{H}_4$.

The $m \times n$ matrix obtained as a product of an $m \times 1$ matrix and a $1 \times n$ matrix is a special case of a tensor product. Such a product of a row and a column vector is sometimes called the *outer product*. For example, in $\mathcal{H}_4$ the outer product of a ket vector and a bra vector, $\mid \psi_a\rangle\langle\psi_b \mid$, is

$$\mid \psi_a\rangle\langle\psi_b \mid = \begin{pmatrix} \alpha_0 \\ \alpha_1 \\ \alpha_2 \\ \alpha_3 \end{pmatrix} \begin{pmatrix} \beta_0^* & \beta_1^* & \beta_2^* & \beta_3^* \end{pmatrix} = \begin{pmatrix} \alpha_0\beta_0^* & \alpha_0\beta_1^* & \alpha_0\beta_2^* & \alpha_0\beta_3^* \\ \alpha_1\beta_0^* & \alpha_1\beta_1^* & \alpha_1\beta_2^* & \alpha_1\beta_3^* \\ \alpha_2\beta_0^* & \alpha_2\beta_1^* & \alpha_2\beta_2^* & \alpha_2\beta_3^* \\ \alpha_3\beta_0^* & \alpha_3\beta_1^* & \alpha_3\beta_2^* & \alpha_3\beta_3^* \end{pmatrix}.$$

2.8 QUANTUM STATES

A state is a complete description of a physical system. In quantum mechanics, *a state is represented by a vector of length (norm) equal to 1 in the Hilbert space* $\mathcal{H}_n$. The traditional notation for a state in quantum mechanics is

$$\mid \Psi_a\rangle = \alpha_0 \mid 0\rangle + \alpha_1 \mid 1\rangle \ldots + \alpha_i \mid i\rangle \ldots + \alpha_{n-1} \mid n - 1\rangle$$

and we shall follow this notation rather than $\mid \psi\rangle$ for the rest of this chapter. Two vectors $\mid \Psi_a\rangle$ and $\mid \Psi_{a'}\rangle$ such that $\mid \Psi_a\rangle = c \mid \Psi_{a'}\rangle$ with c a complex number of modulus equal to one, represent the same state. A zero ket vector does not correspond to a state of a quantum system.

The *length* or *norm* of a ket vector $\mid \Psi_a\rangle$ or of the corresponding bra vector $\langle\Psi_a \mid$ is defined as the square root of the positive number $\langle\Psi_a \mid \Psi_a\rangle$. By convention, state vectors are assumed to be normalized, that is, $\langle\Psi_a \mid \Psi_a\rangle = 1$. Therefore,

$$\sum_{i=0}^{n-1} \mid \alpha_i \mid^2 = 1.$$

The `ket` or the `bra` vectors characterizing a state are defined only as direction and their length is determined up to a factor. In fact, a quantum state is best described as a *ray* in a Hilbert space. A ray is a mathematical abstraction that exhibits only direction and can be represented as a straight line emanating from the origin of the coordinate system. While a vector has a well defined magnitude and direction, a ray has only a relative direction. A ray is a representative element of an equivalence class of vectors that differ by multiplication by a nonzero complex scalar. For any non-vanishing vector we can choose the element of this class with unit norm

$$\langle \Psi_a \mid \Psi_a \rangle = 1.$$

For such a normalized vector we can say that $\mid \Psi_a \rangle$ and $e^{i\gamma} \mid \Psi_a \rangle$, where $\mid e^{i\gamma} \mid = 1$, describe the same physical state. In this expression γ represents the *relative phase.*

The inner product of two state vectors $\mid \Psi_a \rangle$ and $\mid \Psi_b \rangle$ represents the "generalized angle" between the two states and gives an estimate of the *overlap* between the states $\mid \Psi_a \rangle$ and $\mid \Psi_b \rangle$. The interpretation of $\langle \Psi_a \mid \Psi_b \rangle = 0$ as representing orthogonal states and the implication of $\langle \Psi_a \mid \Psi_b \rangle = 1$ that $\mid \Psi_a \rangle$ and $\mid \Psi_b \rangle$ represent one and the same state are immediately evident. The inner product of two state vectors $\langle \Psi_a \mid \Psi_b \rangle$ is a complex number and its modulus, $\| \langle \Psi_a \mid \Psi_b \rangle \|$, can be considered a quantitative measure of the "relative orthogonality" between the states $\mid \Psi_a \rangle$ and $\mid \Psi_b \rangle$.

If the state of a dynamical system is the result of a *superposition* of other states, its corresponding `ket` vector can be expressed as a linear combination of the `ket` vectors representing the states entering the superposition. The states involved in a superposition are said to be *dependent.* The "superposition principle" can be formulated as: every *ray* in $\mathcal{H}_n$ corresponds to a possible state, so that given two states $\mid \Psi_a \rangle$ and $\mid \Psi_b \rangle$, we can form another state as $a \mid \Psi_a \rangle + b \mid \Psi_b \rangle$, a superposition of the original two states. The *relative phase* in such a superposition $a \mid \Psi_a \rangle + b \mid \Psi_b \rangle$ is associated to $e^{i\gamma}(a \mid \Psi_a \rangle + b \mid \Psi_b \rangle)$ and it is physically significant.

There is a fundamental difference between a quantum superposition and a classical superposition. For example, a superposition of a classical membrane vibration state with itself results in a different state with twice the magnitude of the original oscillation. There is no physical characterization of a quantum state which corresponds to the magnitude of a classical oscillation. One more significant difference: consider a classical state with amplitude of oscillation zero everywhere corresponding to a membrane at rest. No equivalent state exists for a quantum system due to the fact that a zero `ket` vector does not represent a valid state.

The set of unit vectors $\{\mid 0 \rangle, \mid 1 \rangle, \ldots, \mid i \rangle, \ldots, \mid n-1 \rangle\}$ (see Section 2.5) forms a *normal unitary basis* in the n-dimensional state vector space. For example, in $\mathcal{H}_4$ a state can be represented by the vector

$$\mid \Psi_a \rangle \mapsto \alpha_0 \mid 0 \rangle + \alpha_1 \mid 1 \rangle + \alpha_2 \mid 2 \rangle + \alpha_3 \mid 3 \rangle$$

where

$$| 0\rangle = \begin{pmatrix} 1 \\ 0 \\ 0 \\ 0 \end{pmatrix} \quad | 1\rangle = \begin{pmatrix} 0 \\ 1 \\ 0 \\ 0 \end{pmatrix} \quad | 2\rangle = \begin{pmatrix} 0 \\ 0 \\ 1 \\ 0 \end{pmatrix} \quad | 3\rangle = \begin{pmatrix} 0 \\ 0 \\ 0 \\ 1 \end{pmatrix}.$$

We can work with `bra` vectors instead of `ket` vectors. Then

$$\langle \Psi_a | \longrightarrow \alpha_0^* \langle 0 | + \alpha_1^* \langle 1 | + \alpha_2^* \langle 2 | + \alpha_3^* | 3\rangle.$$

The unit vectors satisfy the relation

$$\langle i | j\rangle = \delta_{i,j}$$

where $\delta_{i,j}$ is the *Kronecker delta* function previously defined.

The same physical state $| \Psi_a\rangle$ can be expressed in different bases. For example, the same state $| \Psi_a\rangle \in \mathcal{H}_2$ can be expressed in two different bases

$$\{| 0\rangle, | 1\rangle\}$$

as

$$| \Psi_a\rangle = \alpha_0 | 0\rangle + \alpha_1 | 1\rangle$$

and in

$$\{| x\rangle = \tfrac{1}{\sqrt{2}}(| 0\rangle + | 1\rangle), | y\rangle = \tfrac{1}{\sqrt{2}}(| 0\rangle - | 1\rangle)\}$$

as

$$| \Psi_a\rangle = \tfrac{1}{\sqrt{2}}(\alpha_0 + \alpha_1) | x\rangle + \tfrac{1}{\sqrt{2}}(\alpha_0 - \alpha_1) | y\rangle.$$

2.9 QUANTUM OBSERVABLES—QUANTUM OPERATORS

An observable is a property of a physical system that, in principle, can be measured. The formalism of quantum mechanics associates an observable with a *self-adjoint*, or Hermitian, operator. A measured value of an observable is an eigenvalue of the operator.

An operator $\mathbf{U}$ in a Hilbert space $\mathcal{H}_n$ is

Hermitian(self-adjoint)	if	$\mathbf{U} = \mathbf{U}^\dagger$
Unitary	if	$\mathbf{U}\mathbf{U}^\dagger = \mathbf{U}^\dagger\mathbf{U} = \mathbf{I}$
Normal	if	$[\mathbf{U}, \mathbf{U}^\dagger] = \mathbf{U}\mathbf{U}^\dagger - \mathbf{U}^\dagger\mathbf{U} = 0$

with $\mathbf{U}^\dagger$ the adjoint of $\mathbf{U}$ and $\mathbf{I}$ the identity operator. Clearly, a unitary Hermitian operator is normal. A unitary operator $\mathbf{U}$ in a Hilbert space preserves the inner product, thus it preserves the distance.

A matrix $U = [u_{ij}]$ with complex elements, $u_{ij} \in \mathbb{C}$ is said to be unitary if $U^\dagger U = 1$. Here $U^\dagger$ is the *adjoint* of U, a matrix obtained from U by first constructing U^T, the *transpose* of U, and then taking the complex conjugate of each element (or by first taking the complex conjugate of each element and then transposing the matrix). The determinant of a unitary matrix is equal to 1.

A unitary operator $\mathbf{U}$ maps state vectors to state vectors in $\mathcal{H}_n$. If we take a quantum system in state $| \Psi_a\rangle$ and apply to it a transformation described by the unitary operator $\mathbf{U}$, we get a different state $| \Psi_b\rangle$. This transformation, regardless of whether it is the operation of a rotation, the operation "waiting for some time Δt," or the operation of performing a measurement, is described mathematically as

$$| \Psi_b\rangle = \mathbf{U} \mid \Psi_a\rangle.$$

An operator $\mathbf{U}$ acts from the *left* on a `ket` state vector and acts from the *right* on a `bra` state vector

$$| \Psi_a\rangle \rightarrow \mathbf{U} \mid \Psi_a \rangle \quad \text{and} \quad \langle \Psi_a \mid \rightarrow \langle \Psi_a \mid \mathbf{U}.$$

Quantum states are represented by vectors having unit length therefore, we are only concerned with unitary linear transformations and with unitary linear operators. Thus, for any two state vectors, $| \Psi_a\rangle, | \Psi_b\rangle \in \mathcal{H}_n$, and any complex numbers $a, b \in \mathbb{C}$

$$\mathbf{U}(a \mid \Psi_a\rangle + b \mid \Psi_b\rangle) = a\mathbf{U} \mid \Psi_a\rangle + b\mathbf{U} \mid \Psi_b\rangle.$$

The state vector $| \Psi_a\rangle$ can be expressed as a linear combination of the *basis states*

$$\{| 0\rangle, | 1\rangle, \ldots, | i\rangle, \ldots, | n - 1\rangle\}$$

as follows

$$| \Psi_a\rangle = \sum_{i=0}^{n-1} \alpha_i \mid i\rangle$$

with $\alpha_i, 0 \leqslant i \leqslant n - 1$, complex numbers representing state *amplitudes*. It is easy to show that

$$\alpha_j = \langle j \mid \Psi_a\rangle.$$

Indeed, if we multiply the expression giving the expansion of $| \Psi_a\rangle$ in terms of the basis vectors with the `bra` state vector $\langle j |$ we obtain

$$\langle j \mid \Psi_a\rangle = \langle j \mid \sum_{i=0}^{n-1} \alpha_i \mid i\rangle = \sum_{i=0}^{n-1} \alpha_i \langle j \mid i\rangle = \alpha_j.$$

Recall that the unit vectors satisfy the relation

$$\langle j \mid i\rangle = \begin{cases} 0 & \text{if } j \neq i \\ 1 & \text{if } j = i. \end{cases}$$

Thus, the amplitudes $\langle j \mid \Psi_a\rangle = \alpha_j, 0 \leq j \leq n - 1$, give the amount of each basis state $| j\rangle$ we find in $| \Psi_a\rangle$. Now let us consider $| \Psi_b\rangle = \mathbf{U} \mid \Psi_a\rangle$ as an algebraic equation and multiply it by $\langle j |$ to the left

$$\langle j \mid \Psi_b\rangle = \langle j \mid \mathbf{U} \mid \Psi_a\rangle.$$

If we express the state vector $| \Psi_a\rangle$ in terms of the basis vectors and substitute the value of α_i we obtain

$$\langle j \mid \Psi_b\rangle = \langle j \mid \mathbf{U} \sum_{i=0}^{n-1} \alpha_i \mid i\rangle = \sum_{i=0}^{n-1} \langle j \mid \mathbf{U}\alpha_i \mid i\rangle = \sum_{i=0}^{n-1} \langle j \mid \mathbf{U}\langle i \mid \Psi_a\rangle \mid i\rangle.$$

We reverse the order of the last two elements, the complex number $\langle i \mid \Psi_a\rangle$ with the `ket` vector $| i\rangle$ and obtain

$$\langle j \mid \Psi_b\rangle = \sum_{i=0}^{n-1} \langle j \mid \mathbf{U} \mid i\rangle\langle i \mid \Psi_a\rangle.$$

where the states $| j\rangle$ are from the same set as $| i\rangle$. In this algebraic equation the complex numbers $\langle j \mid \mathbf{U} \mid i\rangle$, are the inner products of the `bra` state vector $\langle j |$ with the `ket` state vector $\mathbf{U} \mid i\rangle$ and they express how much of each amplitude $\langle i \mid \Psi_a\rangle$ goes into each term. The operator $\mathbf{U}$ has an associated matrix $U = [u_{ji}]$ with the complex elements

$$u_{ji} = \langle j \mid \mathbf{U} \mid i\rangle.$$

This matrix can be transformed from one vector basis to another (i.e., from one *representation* to another). When we want to obtain the results of a particular transformation in a specific basis we have to give the components of the state vector and we have to identify the operator $\mathbf{U}$ by its matrix $U = [u_{ij}]$ in terms of the same set of basis states. Once we know a matrix for one particular set of basis states, we can calculate the corresponding matrix for another basis.

In general, we do not have to specify a particular set of basis states, any basis will do. Thus, the original equation where the basis states are not specified, $| \Psi_b\rangle = \mathbf{U} \mid \Psi_a\rangle$, is convenient to use.

We now discuss the procedure to construct an operator given its desired function. Consider a two-dimensional Hilbert space $\mathcal{H}_2$ with basis $\{| 0\rangle, | 1\rangle\}$. We introduce an operator $\mathbf{U}$ whose function is to interchange the projection of a state vector on $| 0\rangle$ with the one on $| 1\rangle$, that is

$$\alpha_0 \mid 0\rangle + \alpha_1 \mid 1\rangle \mapsto \alpha_1 \mid 0\rangle + \alpha_0 \mid 1\rangle.$$

Thus $\mathbf{U}$ is defined as

$$\mathbf{U} = | 0\rangle\langle 1 | + | 1\rangle\langle 0 | .$$

The matrix expression of this operator is

$$U = \begin{pmatrix} 1 \\ 0 \end{pmatrix}\begin{pmatrix} 0 & 1 \end{pmatrix} + \begin{pmatrix} 0 \\ 1 \end{pmatrix}\begin{pmatrix} 1 & 0 \end{pmatrix} = \begin{pmatrix} 0 & 1 \\ 0 & 0 \end{pmatrix} + \begin{pmatrix} 0 & 0 \\ 1 & 0 \end{pmatrix} = \begin{pmatrix} 0 & 1 \\ 1 & 0 \end{pmatrix}.$$

When we apply the operator $\mathbf{U}$ to a `ket` state vector $| \Psi_a\rangle = \alpha_0 \mid 0\rangle + \alpha_1 \mid 1\rangle$ we obtain another state vector $| \Psi_b\rangle$

$$| \Psi_b\rangle = \mathbf{U} \mid \Psi_a\rangle = (| 0\rangle\langle 1 | + | 1\rangle\langle 0 |)(\alpha_0 \mid 0\rangle + \alpha_1 \mid 1\rangle)$$

or

$$| \Psi_b\rangle = \alpha_0[(| 0\rangle\langle 1 | + | 1\rangle\langle 0 |) \mid 0\rangle] + \alpha_1[(| 0\rangle\langle 1 | + | 1\rangle\langle 0 |) \mid 1\rangle].$$

Knowing that $\langle 0 \mid 1\rangle = \langle 1 \mid 0\rangle = 0$ and that $\langle 0 \mid 0\rangle = \langle 1 \mid 1\rangle = 1$ it is easy to show that

$$(\mid 0\rangle\langle 1 \mid) \mid 0\rangle = \mid 0\rangle(\langle 1 \mid 0\rangle) = \begin{pmatrix} 0 \\ 0 \end{pmatrix}$$

$$(\mid 1\rangle\langle 0 \mid) \mid 0\rangle = \mid 1\rangle(\langle 0 \mid 0\rangle) = \mid 1\rangle = \begin{pmatrix} 0 \\ 1 \end{pmatrix}$$

$$(\mid 0\rangle\langle 1 \mid) \mid 1\rangle = \mid 0\rangle(\langle 1 \mid 1\rangle) = \mid 0\rangle = \begin{pmatrix} 1 \\ 0 \end{pmatrix}$$

$$(\mid 1\rangle\langle 0 \mid) \mid 1\rangle = \mid 1\rangle(\langle 0 \mid 1\rangle) = \begin{pmatrix} 0 \\ 0 \end{pmatrix}.$$

Thus,

$$\mid \Psi_b\rangle = \mathbf{U} \mid \Psi_a\rangle = \alpha_0 \mid 1\rangle + \alpha_1 \mid 0\rangle.$$

Similarly, for a `bra` state vector we obtain

$$\langle \Psi_b \mid = \langle \Psi_a \mid \mathbf{U} = (\alpha_0^* \langle 0 \mid + \alpha_1^* \langle 1 \mid)(\mid 0\rangle\langle 1 \mid + \mid 1\rangle\langle 0 \mid) = \alpha_0^* \langle 1 \mid + \alpha_1^* \langle 0 \mid .$$

The same results could be obtained using the matrix formalism, with U the matrix corresponding to the linear operator $\mathbf{U}$:

$$\mid \Psi_b\rangle = U \mid \Psi_a\rangle \Longrightarrow \begin{pmatrix} 0 & 1 \\ 1 & 0 \end{pmatrix}\begin{pmatrix} \alpha_0 \\ \alpha_1 \end{pmatrix} = \begin{pmatrix} \alpha_1 \\ \alpha_0 \end{pmatrix} \Longrightarrow \mid \Psi_b\rangle = \alpha_1 \mid 0\rangle + \alpha_0 \mid 1\rangle$$

and

$$\langle \Psi_b \mid = \langle \Psi_a \mid U \Longrightarrow \begin{pmatrix} \alpha_0^* & \alpha_1^* \end{pmatrix}\begin{pmatrix} 0 & 1 \\ 1 & 0 \end{pmatrix} = \begin{pmatrix} \alpha_1^* & \alpha_0^* \end{pmatrix} \Longrightarrow \langle \Psi_b \mid = \alpha_1^*\langle 0 \mid + \alpha_0^* \langle 1 \mid .$$

Let us now turn our attention to another type of operator that plays an important role in the measurement process, the *projection operator* also called *projector*. Recall that any state vector has unitary length. The outer product of any state vector with itself is a *projection operator*

$$\mid \Psi_a\rangle\langle \Psi_a \mid = \mathbf{P}_{\Psi_a}.$$

A projection operator has the following property

$$(\mathbf{P}_{\Psi_a})^2 = \mid \Psi_a\rangle\langle \Psi_a \mid \Psi_a\rangle\langle \Psi_a \mid = \mid \Psi_a\rangle\langle \Psi_a \mid = \mathbf{P}_{\Psi_a}.$$

Two projectors $\mathbf{P}_i$, $\mathbf{P}_j$ are *orthogonal* if, for every state in the Hilbert space, $\forall \mid \Psi_a\rangle \in \mathcal{H}_n$

$$\mathbf{P}_i\mathbf{P}_j \mid \Psi_a\rangle = 0 \quad \text{with} \quad 0 \le i, j \le n - 1.$$

This condition is often written as

$$\mathbf{P}_i\mathbf{P}_j = 0.$$

We now give several examples of quantum mechanics operators. In these examples x, y, and z represent indexing variables, rather than the coordinates of a three-dimensional Hilbert space.

1. The *rotation operator* $\mathbf{R}_y(\theta)$ takes a state $| \Psi_a \rangle$ into a new state, in fact the old state as seen in a rotated coordinate system.
2. The *inversion* (*parity*) *operator* $\mathbf{P}$ creates a new state by reversing all coordinates.
3. The spin one-half operators given by Pauli matrices σ_x, σ_y, and σ_z.
4. The *displacement operator* $\mathbf{D}_x(L)$ along the x axis, by distance L. If we cause a small displacement Δx along the x axis, the state $| \Psi_a \rangle$ is transformed into another state $| \Psi_b \rangle$

$$| \Psi_b \rangle = \mathbf{D}_x(\Delta x) \, | \Psi_a \rangle.$$

If Δx goes to zero, then $| \Psi_b \rangle$ should be the initial state $| \Psi_a \rangle$. Thus

$$\mathbf{D}_x(0) = I.$$

For infinitesimally small displacement distances, δx, the change of $\mathbf{D}_x$ should be proportional to δx. The coefficient of proportionality is a product of a constant, $(i/\hbar)$, with the *momentum operator* for the x component, $\mathbf{p}_x$. Thus

$$\mathbf{D}_x(\delta x) = \left(1 + \frac{i}{\hbar} \mathbf{p}_x \, \delta x\right).$$

This expression serves as the definition of the momentum operator $\mathbf{p}_x$.

Several examples of rotation matrices are now presented:

(a) *Rotation matrices for spin one-half about the* z *axis*, $\mathrm{R}_z(\phi)$. There are two basis states, $| + \rangle$, spin "up" and $| - \rangle$, spin "down." The *spin numbers* of these states are respectively, $s = +\frac{1}{2}$ and $s = -\frac{1}{2}$.

$\mathrm{R}_z(\phi)$	$\| + \rangle$	$\| - \rangle$
$\langle + \|$	$e^{+i\frac{\phi}{2}}$	0
$\langle - \|$	0	$e^{-i\frac{\phi}{2}}$

(b) *Rotation matrices for spin one about the* z *and* y *axis*, $\mathrm{R}_z(\phi)$ and $\mathrm{R}_y(\theta)$. There are three basis states $| + \rangle$, $| 0 \rangle$ and $| - \rangle$. The *spin numbers* of these states are, respectively, $s = +1$, $s = 0$, and $s = -1$.

$\mathrm{R}_z(\phi)$	$\| + \rangle$	$\| 0 \rangle$	$\| - \rangle$
$\langle + \|$	$e^{+i\phi}$	0	0
$\langle 0 \|$	0	1	0
$\langle - \|$	0	0	$e^{-i\phi}$

The $R_y(\theta)$ matrix depends upon the particular choice of phase

$R_y(\theta)$	$\| +\rangle$	$\| 0\rangle$	$\| -\rangle$
$\langle + \|$	$\frac{1}{2}(1 + \cos\theta)$	$+\frac{1}{\sqrt{2}}\sin\theta$	$\frac{1}{2}(1 - \cos\theta)$
$\langle 0 \|$	$-\frac{1}{\sqrt{2}}\sin\theta$	$\cos\theta$	$+\frac{1}{\sqrt{2}}\sin\theta$
$\langle - \|$	$\frac{1}{2}(1 - \cos\theta)$	$-\frac{1}{\sqrt{2}}\sin\theta$	$\frac{1}{2}(1 + \cos\theta)$

(c) *Rotation matrices for photons of circular polarization in the* xy *plane,* $R_z(\phi)$. The xy plane is perpendicular to the flight path of the photons (the z direction). The photons could have Right-Hand Circular (RHC) or Left-Hand Circular (LHC) polarization. There are two basis states

$$| R\rangle = \frac{1}{\sqrt{2}}(| x\rangle + i | y\rangle), \quad m = +1 \text{ (RHC polarized)}$$

$$| L\rangle = \frac{1}{\sqrt{2}}(| x\rangle - i | y\rangle), \quad m = -1 \text{ (LHC polarized)}$$

$R_z(\phi)$	$\| R\rangle$	$\| L\rangle$
$\langle R \|$	$e^{+i\phi}$	0
$\langle L \|$	0	$e^{-i\phi}$

Another important operator discussed in Section 2.16 in conjecture with Schrödinger's Equation is the Hamiltonian. The Hamiltonian corresponds to the energy observable of a quantum system.

2.10 SPECTRAL DECOMPOSITION OF A QUANTUM OPERATOR

The *spectral decomposition* of a normal operator represents the operator as a linear combination of a set of projectors. The eigenvalues of the linear operator are the coefficients of the projectors in the linear combination.

Let $| \Psi\rangle$ be a vector in $\mathcal{H}_n$ and let $\mathbf{U}$ be a normal operator. $| \Psi\rangle$ is called an *eigenvector* and λ the corresponding *eigenvalue* of $\mathbf{U}$ if

$$\mathbf{U} | \Psi\rangle = \lambda | \Psi\rangle.$$

The identity operator $\mathbf{I}$ has the unity (1) as an eigenvalue

$$\mathbf{I} | \Psi\rangle = 1 | \Psi\rangle.$$

The equation defining the eigenvector $| \Psi\rangle$ and the eigenvalue λ of $\mathbf{U}$ can be rewritten as

$$\mathbf{U} | \Psi\rangle = \lambda\mathbf{I} | \Psi\rangle$$

or

$$(\mathbf{U} - \lambda \mathbf{I}) \mid \Psi\rangle = 0.$$

Let $\{| e_0\rangle, | e_1\rangle, \ldots, | e_j\rangle \ldots, | e_{n-1}\rangle\}$ be an orthonormal basis in $\mathcal{H}_n$ and express the state vector as

$$| \Psi\rangle = \sum_{i=0}^{n-1} \gamma_i \mid e_i\rangle.$$

Then the previous equation becomes

$$(\mathbf{U} - \lambda \mathbf{I}) \sum_{i=0}^{n-1} \gamma_i \mid e_i\rangle = 0.$$

Both operators **U** and **I** can be expressed as matrices, $U = [U_{ij}]$ and $I = [\delta_{ij}]$. Then the previous equation involving **U** and **I** is transformed into a matrix equation

$$\sum_{i=0}^{n-1} (u_{ij} - \lambda \delta_{ij}) \gamma_i = 0$$

which has a non-trivial solution if and only if

$$\det(U - \lambda I) = 0.$$

The eigenvector of the operator **U** corresponding to the eigenvalue λ can be found by solving the system of n linear equation above.

By definition, an *observable* is any Hermitian operator whose eigenvectors form a basis. The following statements are true but will not be proved here:

1. The eigenvalues of a Hermitian operator are real numbers.
2. Eigenvectors corresponding to different eigenvalues are mutually orthogonal.
3. If two Hermitian operators commute they have a common basis of orthonormal eigenvectors, a so called *eigenbasis*. If they do not commute, then no common eigenbasis exists.
4. A complete set of commuting observables is the minimal set of Hermitian operators with a unique common eigenbasis.

In a finite-dimensional vector space, every normal operator has a complete set of orthonormal eigenvectors. Thus in $\mathcal{H}_n$ every normal operator **N** has n eigenvectors, $| n_i\rangle, 0 \leqslant i \leqslant n - 1$, each corresponding to an eigenvalue λ_i

$$\mathbf{N} \mid n_i\rangle = \lambda_i \mid n_i\rangle.$$

Every state vector $| \Psi_a\rangle \in \mathcal{H}_n$ can be expressed on the basis formed by the n eigenvectors of the normal operator **N**

$$| \Psi_a\rangle = \sum_{i=0}^{n-1} \alpha_i \mid n_i\rangle.$$

The normality of **N** implies that the moduli of the amplitudes sum to unity

$$\sum_{i=0}^{n-1} | \alpha_i |^2 = 1.$$

Now the transformation of the state vector performed by the normal operator **N** can be expressed as

$$\mathbf{N} \mid \Psi_a\rangle = \mathbf{N} \sum_i \alpha_i \mid n_i\rangle = \sum_i \alpha_i \mathbf{N} \mid n_i\rangle = \sum_i \alpha_i \lambda_i \mid n_i\rangle.$$

Recall that the outer product of any state vector with itself is a *projection operator*

$$\mid \Psi_a\rangle\langle\Psi_a \mid = \mathbf{P}_{\Psi_a}.$$

Let us construct the projection operators of the basis vectors $\mid n_i\rangle$

$$\mathbf{P}_i = \mid n_i\rangle\langle n_i \mid .$$

Now we apply the projector operator to the state vector $\mid \Psi_a\rangle$

$$\mathbf{P}_i \mid \Psi_a\rangle = \mid n_i\rangle\langle n_i \mid \sum_{j=0}^{n-1} \alpha_j \mid n_j\rangle = \sum_{j=0}^{n-1} \alpha_j \mid n_i\rangle(\langle n_i \mid n_j\rangle).$$

But $\langle n_i \mid n_j\rangle = \delta_{ij}$, and so

$$\mathbf{P}_i \mid \Psi_a\rangle = \alpha_i \mid n_i\rangle.$$

We substitute $\mathbf{P}_i \mid \Psi_a\rangle$ for $\alpha_i \mid n_i\rangle$ in the expression of $\mathbf{N} \mid \Psi_a\rangle$

$$\mathbf{N} \mid \Psi_a\rangle = \sum_{i=0}^{n-1} \alpha_i \lambda_i \mid n_i\rangle = \sum_{i=0}^{n-1} \lambda_i \mathbf{P}_i \mid \Psi_a\rangle.$$

From this expression we conclude that

$$\mathbf{N} = \sum_{i=0}^{n-1} \lambda_i \mathbf{P}_i$$

and this means that this *spectral decomposition of the normal operator* **N** *is true for any state and it is independent of the basis*. An observable specifies an exhaustive measurement through its spectral decomposition.

Finally, we give an example of spectral decomposition of an operator in $\mathcal{H}_2$. Consider an operator **N** with two eigenvalues λ_a and λ_b and with the corresponding orthonormal eigenstates characterized by the eigenvectors $\mid a\rangle$ and $\mid b\rangle$

$$\mid a\rangle = \alpha_0 \mid 0\rangle + \alpha_1 \mid 1\rangle \mapsto \begin{pmatrix} \alpha_0 \\ \alpha_1 \end{pmatrix}$$

$$\mid b\rangle = \beta_0 \mid 0\rangle + \beta_1 \mid 1\rangle \mapsto \begin{pmatrix} \beta_0 \\ \beta_1 \end{pmatrix}.$$

The projection operators corresponding to these eigenvectors can be written, respectively, as

$$\mathbf{P}_a = | a\rangle\langle a | = \begin{pmatrix} \alpha_0 \\ \alpha_1 \end{pmatrix} \begin{pmatrix} \alpha_0^* & \alpha_1^* \end{pmatrix} = \begin{pmatrix} | \alpha_0 |^2 & \alpha_0\alpha_1^* \\ \alpha_1\alpha_0^* & | \alpha_1 |^2 \end{pmatrix}$$

$$\mathbf{P}_b = | b\rangle\langle b | = \begin{pmatrix} \beta_0 \\ \beta_1 \end{pmatrix} \begin{pmatrix} \beta_0^* & \beta_1^* \end{pmatrix} = \begin{pmatrix} | \beta_0 |^2 & \beta_0\beta_1^* \\ \beta_1\beta_0^* & | \beta_1 |^2 \end{pmatrix}.$$

We can use the spectral decomposition to write the operator using the matrix notation

$$\mathbf{U} \leftrightarrow \mathrm{U} = \lambda_a \begin{pmatrix} | \alpha_0 |^2 & \alpha_0\alpha_1^* \\ \alpha_1\alpha_0^* & | \alpha_1 |^2 \end{pmatrix} + \lambda_b \begin{pmatrix} | \beta_0 |^2 & \beta_0\beta_1^* \\ \beta_1\beta_0^* & | \beta_1 |^2 \end{pmatrix}.$$

2.11 THE MEASUREMENT OF OBSERVABLES

So far we have learned that the numerical outcome of a measurement of an observable $\mathcal{U}$ is an eigenvalue, λ_i, of the operator $\mathbf{U}$. Immediately after the measurement, the quantum state is the eigenstate $| u_i\rangle$ of $\mathbf{U}$ with the measured eigenvalue λ_i. Quantum mechanics postulates that: *mutually exclusive measurement-outcomes correspond to orthogonal projection operators (projectors)* $\{\mathbf{P}_0, \mathbf{P}_1, \ldots\}$.

Given the Hilbert space $\mathcal{H}_n$, a *complete set of orthogonal projectors* is a set $\{\mathbf{P}_0, \mathbf{P}_1, \ldots, \mathbf{P}_i, \ldots, \mathbf{P}_{m-1}\}$ such that

$$\sum_{i=0}^{m-1} \mathbf{P}_i = 1.$$

It follows that the number of projectors in a complete orthogonal set must be less than, or equal to the dimension of the Hilbert space

$$m \leqslant n.$$

The postulate can be reformulated as: *A complete set of orthogonal projectors specifies an exhaustive measurement.* If we take a closer look at the spectral decomposition of an operator we conclude that an observable specifies an exhaustive measurement. Whenever we say that "we measure an observable q" we mean that we measure the set of projectors $\mathbf{P}_i^{(q)}$ and we associate the eigenvalue $\lambda_i^{(q)}$ with the occurrence of the i-th outcome of observable q.[8]

We now compute the probability to obtain the value λ_i as the outcome of the measurement, when the quantum state just prior to the measurement is $| \Psi_a\rangle$:

$$\begin{aligned} \mathrm{Prob}(\lambda_i) &= | \mathbf{P}_i | \Psi_a\rangle |^2 \\ &= (\mathbf{P}_i | \Psi_a\rangle)^\dagger \mathbf{P}_i | \Psi_a\rangle \end{aligned}$$

[8]In $\mathbf{P}_i^{(q)}$ (q) denotes the index of the observable being measured, while in $\mathbf{P}_i^2$ the projector is raised to the power of 2.

$$= \langle \Psi_a \mid \mathbf{P}_i^\dagger \mathbf{P}_i \mid \Psi_a \rangle$$
$$= \langle \Psi_a \mid (\mathbf{P}_i)^2 \mid \Psi_a \rangle$$
$$= \langle \Psi_a \mid \mathbf{P}_i \mid \Psi_a \rangle.$$

In this derivation we used the fact that

$$\mathbf{P}_i^2 = (\mid n_i\rangle\langle n_i \mid)(\mid n_i\rangle\langle n_i \mid) = \mid n_i\rangle(\langle n_i \mid n_i\rangle)\langle n_i \mid = \mid n_i\rangle\langle n_i \mid = \mathbf{P_i}.$$

The total probability for all possible outcomes of a measurement is

$$\sum_{i=0}^{m-1} \text{Prob}(\lambda_i) = 1$$

due to the completeness of the set of projectors. The post measurement normalized pure quantum state, after the outcome λ_i, is

$$\frac{\mathbf{P}_i \mid \Psi_a \rangle}{\sqrt{\langle \Psi_a \mid \mathbf{P}_i \mid \Psi_a \rangle}}.$$

As an example, consider the case of a two-dimensional Hilbert space with an orthonormal basis formed by the vectors $\mid x\rangle$ and $\mid y\rangle$. Consider two state vectors $\mid \Psi_a\rangle, \mid \Psi_b\rangle \in \mathcal{H}_2$

$$\mid \Psi_a\rangle = \alpha_x \mid x\rangle + \alpha_y \mid y\rangle \quad \text{and} \quad \mid \Psi_b\rangle = \beta_x \mid x\rangle + \beta_y \mid y\rangle.$$

We start with a number of systems in the same state and we perform a large number M of measurements corresponding to the projectors

$$\mathbf{P}_x = \mid x\rangle\langle x \mid \quad \text{and} \quad \mathbf{P}_y = \mid y\rangle\langle y \mid .$$

We assume that before the measurement a system is a *mixed ensemble* of quantum states, it could be in state $\mid \Psi_a\rangle$ with probability p or in state $\mid \Psi_b\rangle$ with probability $1 - p$

$$\text{Prob}(\mid \Psi_a\rangle) = p \quad \text{and} \quad \text{Prob}(\mid \Psi_b\rangle) = 1 - p.$$

The probability that a system in state $\mid \Psi_a\rangle$ before the measurement will produce the outcome λ_x is

$$\text{Prob}(\lambda_x \mid \Psi_a\rangle) = \langle \Psi_a \mid \mathbf{P}_x \mid \Psi_a \rangle = \mid \alpha_x \mid^2$$

while the probability that a system in state $\mid \Psi_b\rangle$ before the measurement will produce the outcome λ_x is

$$\text{Prob}(\lambda_x \mid \Psi_b\rangle) = \langle \Psi_b \mid \mathbf{P}_x \mid \Psi_b \rangle = \mid \beta_x \mid^2 .$$

We try to predict the number of times m_x, out of the total number M of measurements, in which we expect to obtain the measurement outcome corresponding to the basis vector $| x\rangle$.

Given a system in state $| \Psi\rangle$, the average value of a measurement A with outcomes a is denoted as

$$\langle A\rangle = \langle \Psi | A | \Psi\rangle.$$

The standard deviation associated to observations of A is

$$\Delta A = \langle (A - \langle A\rangle)^2\rangle = \langle A^2\rangle - \langle A\rangle^2.$$

According to the probability theory

$$\begin{aligned} m_x &= M[\text{Prob}(| \Psi_a\rangle)\,\text{Prob}(\lambda_x | \Psi_a\rangle) + \text{Prob}(| \Psi_b\rangle)\,\text{Prob}(\lambda_x | \Psi_b\rangle)] \\ &= M[p\langle \Psi_a | \mathbf{P}_x | \Psi_a\rangle + (1 - p)\langle \Psi_b | \mathbf{P}_x | \Psi_b\rangle] \\ &= M[p | \alpha_x |^2 + (1 - p) | \beta_x |^2]. \end{aligned}$$

Since $0 \leq p \leq 1$, the quantity m_x/M is bounded from below by the smaller of $| \alpha_x |^2$ and $| \beta_x |^2$

$$\frac{m_x}{M} = p | \alpha_x |^2 + (1 - p) | \beta_x |^2 .$$

Now we consider that the system is in a *coherent superposition* of the states $| \Psi_a\rangle$ and $| \Psi_b\rangle$ with amplitudes γ_a and γ_b. The state $| \Psi(\gamma_a, \gamma_b)\rangle$ is a *pure* state

$$\begin{aligned} | \Psi(\gamma_a, \gamma_b)\rangle &= \gamma_a | \Psi_a\rangle + \gamma_b | \Psi_b\rangle \\ &= (\gamma_a\alpha_x + \gamma_b\beta_x) | x\rangle + (\gamma_a\alpha_y + \gamma_b\beta_y) | y\rangle. \end{aligned}$$

Here, γ_a and γ_b are chosen such that the state $| \Psi(\gamma_a, \gamma_b)\rangle$ is normalized. In this case, the probability of a measurement outcome corresponding to basis vector $| x\rangle$ is

$$\begin{aligned} p_x &= \langle \Psi(\gamma_a, \gamma_b) | \mathbf{P}_x | \Psi(\gamma_a, \gamma_b)\rangle \\ &= | \gamma_a a_x + \gamma_b b_x |^2 . \end{aligned}$$

In certain cases it is possible to choose γ_a and γ_b such that

$$\gamma_a\alpha_x = -\gamma_b\beta_x \neq 0.$$

Then $p_x = 0$ due to the phenomenon of *destructive interference* even though

$$| \alpha_x |^2 > 0 \quad \text{and} \quad | \beta_x |^2 > 0.$$

This reflects an important distinction between coherent superpositions and incoherent admixtures. A coherent superposition produces a single *pure state* in which case destructive interference is possible; an incoherent admixture produces a *mixed ensemble of quantum states* and destructive interference is not possible in that case.

2.12 MORE ABOUT MEASUREMENTS—THE DENSITY OPERATOR

Let us assume that we have N independent versions of a quantum system, each in a quantum state $\mid \psi_i \rangle$, with $1 \leqslant i \leqslant N$. We measure the same observable A on all N versions of the system and we want to compute the ensemble average of A, and call it $\langle A \rangle$ [1].

Let $\{\mid \varphi_a \rangle\}$ denote the set of vectors of an orthonormal basis and assume that these orthonormal vectors are the eigenstates of a Hermitian operator.

$\langle A \rangle$ is an ensemble average:

$$\langle A \rangle = \frac{1}{N} \sum_{i=1}^{N} \langle \psi_i \mid A \mid \psi_i \rangle.$$

But

$$\langle \psi_i \mid A \mid \psi_i \rangle = \sum_a \langle \psi_i \mid \varphi_a \rangle \langle \varphi_a \mid A \mid \psi_i \rangle = \sum_a \langle \varphi_a \mid A \mid \psi_i \rangle \langle \psi_i \mid \varphi_a \rangle$$

and

$$\langle A \rangle = \frac{1}{N} \sum_{i=1}^{N} \sum_a \langle \varphi_a \mid A \mid \psi_i \rangle \langle \psi_i \mid \varphi_a \rangle = \sum_a \langle \varphi_a \mid A \mid \varphi_a \rangle \frac{1}{N} \sum_{i=1}^{N} \mid \psi_i \rangle \langle \psi_i \mid .$$

If the *density matrix* ρ is

$$\rho = \frac{1}{N} \sum_{i=1}^{N} \mid \psi_i \rangle \langle \psi_i \mid$$

then the ensemble average $\langle A \rangle$ can be written as

$$\langle A \rangle = \sum_a \langle \varphi_a \mid \rho A \mid \varphi_a \rangle = \text{Tr}(\rho A).$$

If $\mathbf{P}_a = \mid \varphi_a \rangle \langle \varphi_a \mid$ is the projector operator onto the eigenstate $\mid \varphi_a \rangle$ of the Hermitian operator then the probability to find the system in a particular state $\mid \varphi_a \rangle$ is

$$\text{Prob}(\varphi_a) = \frac{1}{N} \sum_{i=1}^{N} \mid \langle \psi_i \mid \varphi_a \rangle \mid^2 = \text{Tr}(\rho \mathbf{P}_a).$$

When the set of state vectors $\{\mid \varphi_a \rangle\}$ form an orthonormal basis, the density matrix $\rho = [\rho_{ba}]$ is Hermitian. Indeed, its elements are

$$\rho_{ba} = \frac{1}{N} \sum_{i=1}^{N} \langle \psi_i \mid \varphi_a \rangle \langle \psi_i \mid \varphi_b \rangle^*.$$

The trace of ρ is the expected value of the identity operator

$$\text{Tr}(\rho) = 1.$$

If all N independent systems are in the same state $| \psi_i \rangle = | \psi \rangle$ then

$$\text{Tr}(\rho^2) = \text{Tr}(\rho) = 1.$$

Indeed,

$$\rho_{ba} = \frac{1}{N} \sum_{i=1}^{N} \langle \psi \mid \varphi_a \rangle \langle \psi \mid \varphi_b \rangle^*.$$

The eigenvalue λ_a corresponding to the eigenvector $| \varphi_a \rangle$ of ρ defined by the relation

$$\rho \mid \varphi_a \rangle = \lambda_a \mid \varphi_a \rangle$$

is non-negative

$$\lambda_a = \rho_{aa} = \frac{1}{N} \sum_{i=1}^{N} \langle \psi \mid \varphi_a \rangle \langle \psi \mid \varphi_a \rangle^* = \frac{1}{N} \sum_{i=1}^{N} | \langle \psi \mid \varphi_a \rangle |^2 \geqslant 0.$$

It is also easy to see that

$$0 \leqslant \text{Tr}(\rho^2) \leqslant 1.$$

Indeed,

$$\text{Tr}(\rho^2) = \sum_a \rho_{aa}^2 = \sum \lambda_a^2 \leqslant \left[\sum_a \lambda_a \right]^2.$$

Multiple ensembles of states may have the same density matrix. For example, a quantum system may be in state

$$| a \rangle = \frac{\sqrt{3}}{2} | 0 \rangle + \frac{1}{2} | 1 \rangle$$

with probability $p_a = 1/2$, or in state

$$| b \rangle = \frac{\sqrt{3}}{2} | 0 \rangle - \frac{1}{2} | 1 \rangle$$

with probability $p_b = 1/2$. In this case the density matrix is

$$\rho = \frac{3}{4} | 0 \rangle \langle 0 | + \frac{1}{4} | 1 \rangle \langle 1 | = \frac{1}{4} \begin{pmatrix} 3 & 0 \\ 0 & 1 \end{pmatrix}.$$

But the following states and probabilities lead to the same density matrix

$$| a' \rangle = | 0 \rangle \quad p_{a'} = \frac{3}{4}$$

and

$$| b' \rangle = | 1 \rangle \quad p_{b'} = \frac{1}{4}.$$

2.13 DOUBLE-SLIT EXPERIMENTS

To understand the difference between corpuscular and wave behavior we consider classical systems first and briefly discuss two double-slit experiments, one involving bullets and the other water waves.

In the first experiment, a gun shoots bullets at random at a barrier with two slits a and b. The bullets go through the two slits, some of them may bounce off the edges of the slits, but all are collected by a backstop. We repeat the experiment three times, first with slit a open, then with slit b open, and finally with both slits open. For each experiment we place a mobile detector at the lower edge of the backstop $x = x_{min}$ and fire n bullets and count the number of bullets reaching the detector. Then we move the detector to a new position, fire the same number of bullets, count them, and repeat the process until we reach the upper edge of the backstop, $x = x_{max}$.

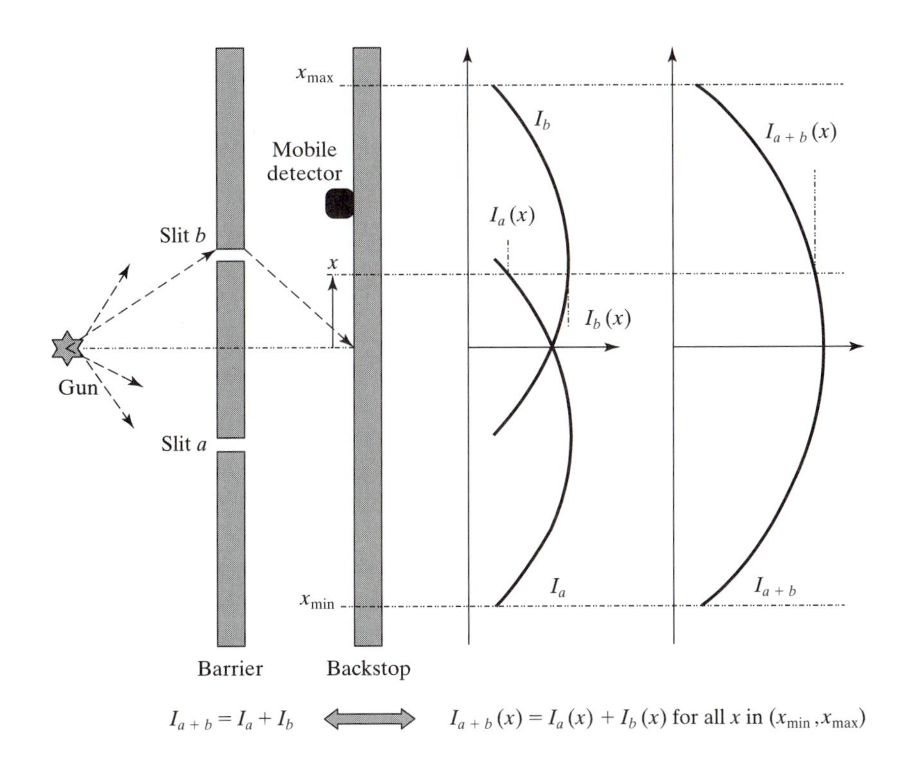

FIGURE 2.2: The double slit experiment with bullets. If $I_a(x)$, $I_b(x)$, and $I_{a+b}(x)$ are the number of bullets per unit time recorded at position x when only slit a is open, only slit b is open, and both slits are open, respectively, then $I_{a+b}(x) = I_a(x) + I_b(x)$, $\forall x \in (x_{min}, x_{max})$.

Call $I_a(x)$, $I_b(x)$, and $I_{a+b}(x)$ the number of bullets per unit time recorded at position x when only slit a is open, only slit b is open, and both slits are open, respectively. We observe that

$$I_{a+b}(x) = I_a(x) + I_b(x) \quad \forall x \in (x_{min}, x_{max}).$$

The three functions $I_a(x)$, $I_b(x)$, and $I_{a+b}(x)$ are plotted in Figure 2.2. All three curves are smooth; we do not observe any particular pattern in I_{a+b}. Now, we decide to divide the number of bullets per unit time recorded at each position x by the total number of bullets shot at the double-slit screen. Instead of the curves $I_a(x)$, $I_b(x)$, and $I_{a+b}(x)$ we obtain similar, smooth curves representing the probabilities of detecting bullets at each position x when only slit a is open, $P_a(x)$, when only slit b is open, $P_b(x)$, and when both slits are open, $P_{a+b}(x)$. We observe again that

$$P_{a+b}(x) = P_a(x) + P_b(x) \quad \forall x \in (x_{\text{min}}, x_{\text{max}})$$

and we conclude that the probability of recording bullets when both slits are open is the sum of the probabilities of recording bullets when each slit is open alone, that is, *the probabilities sum up*.

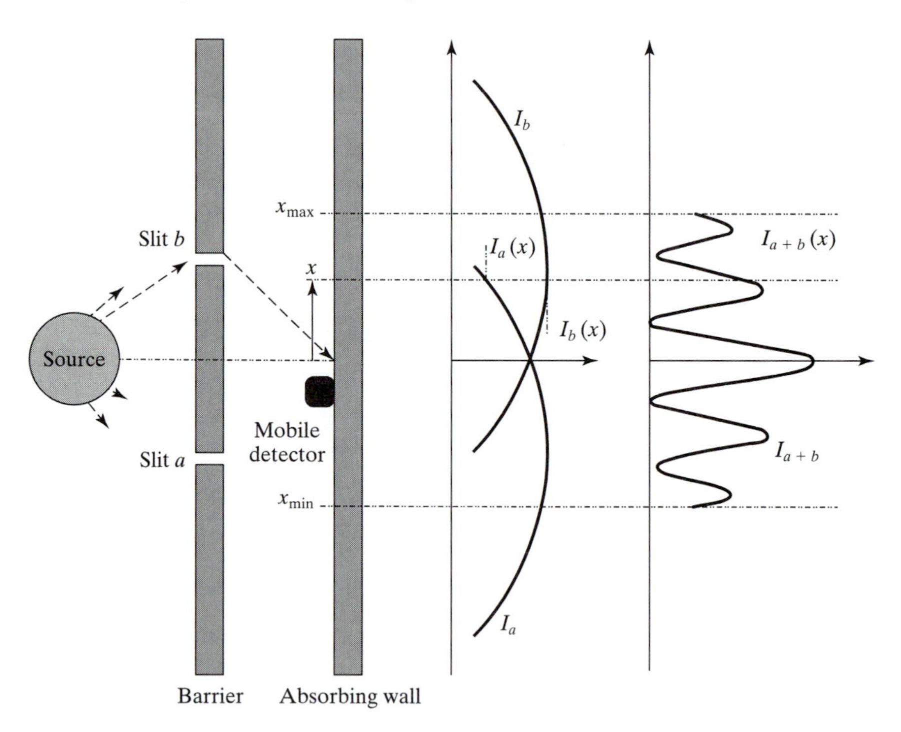

FIGURE 2.3: The double slit experiment with water waves. The wave coming through slit a is $\beta_a e^{i\omega t}$ and its intensity is $I_a = |\beta_a|^2$; the wave coming through slit b is $\beta_b e^{i\omega t}$ and its intensity is $I_b = |\beta_b|^2$. When both slits are open the height of the wave reaching the detector is the sum of the individual waves coming through slit a and slit b (i.e., $(\beta_a + \beta_b)e^{i\omega t}$ and its intensity is $I_{a+b} = |\beta_a + \beta_b|^2$). In this case $I_{a+b} = I_a + I_b + 2\sqrt{I_a I_b}\cos\delta \neq I_a + I_b$.

A similar setup is provided for a second experiment shown in Figure 2.3. In this case, water waves produced by a wave source arrive at the barrier, penetrate through the slits and reach the mobile detector. The detector is mounted on the absorber wall behind the barrier to avoid reflection of the waves. The instantaneous height of the water wave reaching the detector is given by the real part of the expression

$$w(t) = \beta e^{i\omega t} = \beta(\cos \omega t + i \sin \omega t)$$

where β represents the amplitude, a function of position x, sometimes given as a complex number, and ω is the angular frequency of the wave.

The detector is a device which measures the instantaneous height of the wave and converts it into the intensity of the wave. The intensity of the wave is estimated as the mean squared height, or in the case of complex numbers the absolute value squared $I = | \beta |^2$. When both slits are open, the result of the measurement is a curve I_{a+b} with maxima and minima. The instantaneous height and, respectively, the intensity are recorded as a function of the detector position.

As before, we repeat the experiment three times, once with only slit a open, then with slit b open, and finally with both slits open and obtain three curves, I_a, I_b, and I_{a+b}. Now we observe that I_a and I_b have a maximum centered at the respective slit position, as in the case of the experiment with bullets. The curve representing the intensity in the case both slits are open presents a certain pattern. The maxima and minima of I_{a+b} are the result of constructive and, respectively, destructive *interference* of the waves. We also see that $I_{a+b} \neq I_a + I_b$.

We assume that the wave coming through slit a has the height $\beta_a e^{i\omega t}$ and the intensity $I_a = | \beta_a |^2$, while the wave coming through slit b has the height $\beta_b e^{i\omega t}$ and the intensity $I_b = | \beta_b |^2$. When both slits are open the height of the wave reaching the detector is the sum of the individual waves coming through slit a and slit b (i.e., $(\beta_a + \beta_b)e^{i\omega t}$ and its intensity is $I_{a+b} = | \beta_a + \beta_b |^2$). Here,

$$|\beta_a + \beta_b|^2 = |\beta_a|^2 + |\beta_b|^2 + 2|\beta_a|\,|\beta_b| \cos \delta$$

where δ is the phase difference between β_a and β_b. Henceforth

$$I_{a+b} = I_a + I_b + 2\sqrt{I_a I_b} \cos \delta$$

and

$$I_{a+b} \neq I_a + I_b.$$

In this experiment we observe the phenomenon of interference among the waves diffracted by slit a and those diffracted by slit b.

Let us now consider a third experiment, this time involving electrons, see Figure 2.4, where we expect to notice quantum effects. The electrons produced by an electron gun pass through one of the slits or through both and then are recorded by a Geiger counter. The electrons arrive at the Geiger counter individually, very much like the bullets of the first experiment. Indeed, the Geiger counter emits a sound, a "click," whose intensity is proportional to the electric charge carried by each electron. The fact that we hear distinct clicks of the same intensity indicates that the electrons are indistinguishable and that they arrive individually, rather than in packets of more than one, or as fractional entities.

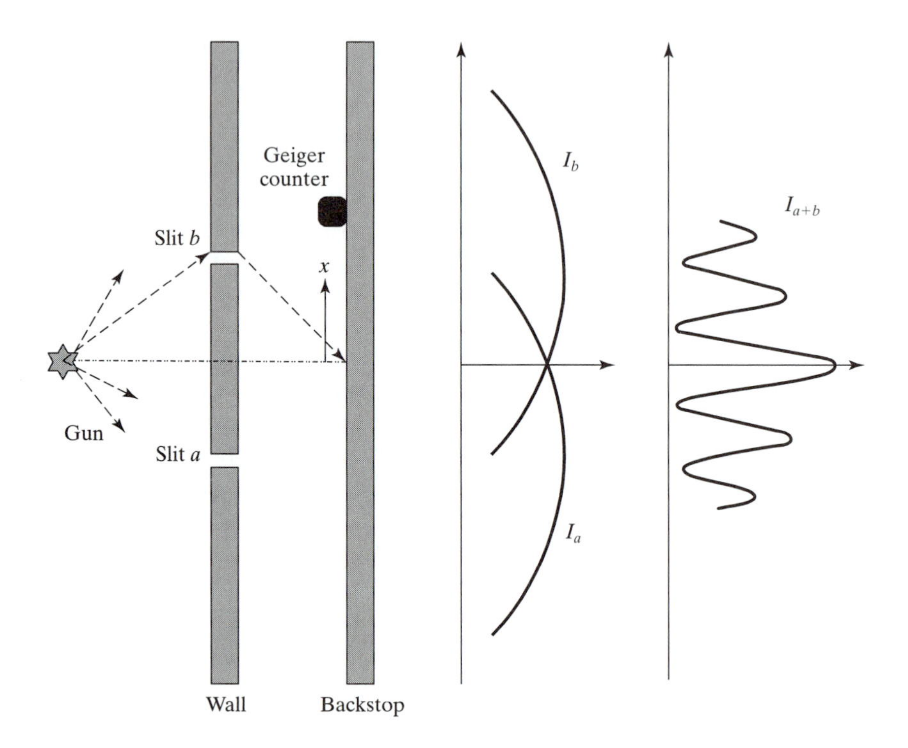

FIGURE 2.4: The double slit experiment showing electron interference. We observe experimentally that $I_{a+b} \neq I_a + I_b$, the number of electrons that arrives at a particular point when both slits are open is *not equal* to the number of electrons coming only through slit a plus the number of electrons coming through slit b.

First, we measure electrons coming through one slit at a time, either slit a or slit b. The number of electrons measured at a given spot gives the current intensity which is proportional to the rate of clicks. The intensities for electrons coming through either slit a or slit b are I_a and I_b, respectively. The result of the measurement when both slits are open, the curve I_{a+b}, shows an interference pattern, similar to the experiment with waves. At the same time the clicks emitted by the Geiger counter indicate that the electrons are in fact recorded as individual entities, like the bullets in the first experiment.

We observe experimentally that $I_{a+b} \neq I_a + I_b$, the number of electrons that arrives at a particular point when both slits are open is *not equal* to the number of electrons coming through slit a plus the number of electrons coming through slit b. At certain positions corresponding to the minima of the I_{a+b} curve the number of electrons detected by the Geiger counter is clearly smaller than the number expected to come through slits a and b.

The only possible explanation is that electrons, being quantum entities, arrive like particles and the probability of arrival of these particles is distributed like the intensity of a wave, as if each electron were travelling through *both* slits at the same time and then interfering with itself upon arrival at the detector. The intensities can be converted into probabilities by dividing the number of electrons arriving at each position of the detector with the total number of electrons recorded for each of the three stages of the experiment. We obtain the relation

$$P_{a+b} \neq P_a + P_b$$

where, P_a, P_b, P_{a+b} are the probabilities of recording electrons when only slit a is open, when only slit b is open, and, respectively, when both slits are open.

The probability P of an event in a quantum experiment is given by the square of the absolute value of a complex number α which is called the *probability amplitude*:

$$P = |\alpha|^2 .$$

When an event can occur in several alternative ways, the probability amplitude for the event is the sum of the probability amplitudes for each way considered separately (*the superposition probability rule*).

Assume that $I_a = |\alpha_1|^2$ and $I_b = |\alpha_2|^2$, where α_1 and α_2 are complex numbers. The combined effect of the two slits is $I_{a+b} = |\alpha_1 + \alpha_2|^2$ and the resultant curve is similar to that obtained in the waves case.

Richard Feynman discusses a variation of the last experiment in Chapter 3 of Volume 3 of the *Lecture Notes on Physics* [53]. The aim of the experiment is to identify the slit each electron passes through. A strong light source is placed behind the wall between the two slits. The electrons have electric charges and scatter the photons of the light beam. In this experiment we hear the click of the Geiger counter signaling an incoming electron and, at the same time, we see a flash of light (scattered photons) indicating if the electron has passed through slit a or through slit b.

Let us call $I'_a(x)$, $I'_b(x)$, and $I'_{a+b}(x)$ the values corresponding to $I_a(x)$, $I_b(x)$, and $I_{a+b}(x)$ defined for the previous experiment. We notice that $I'_a(x)$, and $I'_b(x)$ like $I_a(x)$ and $I_b(x)$ are smooth curves with a maximum at the position of the respective slit. Moreover, whether the slits are open one at a time or both at the same time, the electrons we see coming through when both slits a and b are open are distributed in the same way as those coming through when only slit a or only slit b is open. In this case

$$I'_{a+b} = I'_a + I'_b = I_a + I_b.$$

When we identify which slit the electrons are coming through, I'_{a+b} does not look like I_{a+b}; it shows no interference effect. The photons colliding with the electrons change the momentum and the trajectory of the electrons, not very much, but enough to "smear" the total probability distribution.

We can change the intensity or the wavelength (frequency) of the light. When we reduce the light intensity, we reduce the number of photons flying into the paths of electrons; we can reduce it so far that some electrons coming either through slit a or slit b are not seen before being detected (their presence being marked by a click) because the probability of an electron-photon collision decreases significantly. Assume we record the electrons which have been detected and seen as coming either through slit a or slit b, as well as the electrons which are coming through slit a and slit b and are detected without being seen.

The distributions for slit *a* and slit *b* are similar to I_a and I_b, as before, but I'_{a+b} is similar to I_{a+b} showing an interference pattern.

When we reduce the frequency of the light (thus increase its wavelength) we do not notice any change; we record a monotonous curve I'_{a+b} until we reach values of the wavelength larger than the distance between the two slits. At that moment we begin to see big fuzzy flashes of light, we can no longer distinguish which slit the electron went through, and we notice the appearance of some interference effect. As the wavelength continues to increase, the change in the electron momentum becomes small enough to observe a curve similar to I_{a+b}. We can say that as long as we just record the arrival of the electrons and do not try to identify the slit they are coming from (the path taken), the electrons behave like waves and show an interference pattern. As soon as we try to identify the path taken, the electrons behave like particles (like bullets).

In summary, *when a quantum event may occur in several alternative ways, the probability amplitude of the event, α, is the sum of the probability amplitudes for each way considered separately*. In this case we observe the interference between the separate paths. For example, when there are two possible ways for an event to occur, the probability amplitude and the actual probability of the occurrence of the event are respectively

$$\alpha = \alpha_1 + \alpha_2$$

$$P = |\alpha_1 + \alpha_2|^2 \neq P_1 + P_2$$

with $P_1 = |\alpha_1|^2$ and $P_2 = |\alpha_2|^2$ the probability of the occurrence of each path.

But, after we have determined that one of the alternatives was actually taken, the probability of the event is the sum of the probabilities for each alternative. The interference is lost. In the two-path case we have

$$P = P_1 + P_2.$$

Werner Heisenberg had suggested that the laws of quantum mechanics could be consistent *only* if there were some basic limitations, previously not recognized, on our experimental capabilities. Heisenberg's Uncertainty Principle sets a lower limit, *h*, Planck's constant, on the product of the position uncertainty and the momentum uncertainty. In fact, the entire theory of quantum mechanics depends on the correctness of the Uncertainty Principle.

In terms of the last experiment presented above, Feynman proposes to state Heisenberg's Uncertainty Principle in the following way: "It is *impossible* to design an apparatus to determine which slit the electron passes through, that will not at the same time disturb the electrons enough to destroy the interference pattern."

The correct interpretation of this experiment would be: if one has an experimental setup capable of determining whether the electrons go through slit *a* or slit *b*, then one *can* say that they go either through slit *a* or slit *b*. But, when one does not try to tell which way the electrons go and there is nothing in the experiment to disturb the electrons, then one *may not* say that an electron goes either through slit *a* or through slit *b*.

The motion of all forms of matter (even bullets) must be described in terms of waves. In the experiment with bullets no interference patterns could be observed because the wavelengths of the bullets are extremely short and the finite size of the detector would not allow us to distinguish the individual maxima and minima. The result was an average over all the rapid oscillations of the total probability, hence, the classical curve.

2.14 STERN-GERLACH TYPE EXPERIMENTS

An experiment first performed in 1922 by Otto Stern and Walther Gerlach to measure the magnetic moment of an atom played a crucial role in signaling the existence of a new intrinsic property of atoms and particles, the *spin*. At that time the physicists understood that the electrons orbiting around the nucleus represented an electric current. The existence of this orbital electric current meant that the atom had a magnetic field, a *magnetic dipole moment*.

Therefore, atoms placed in a magnetic field would behave like little bar magnets and they should be deflected by the applied field. This is precisely what Stern and Gerlach expected to see during their experiment.

The Stern-Gerlach experimental setup uses a magnet with asymmetric polar caps to create a non-uniform magnetic field. One polar cap is quasi-planar and the other one is wedge-shaped. This configuration results in a non-uniform magnetic field whose z axis component is normal to the planar cap. The components of the magnetic field along x and y axis are negligible. Neutral silver atoms evaporated from an oven are collimated with the aid of a slit in a screen and then are "beamed" perpendicular to the gradient of the magnetic field, see Figure 2.5.

The deflection of each atom depends on the atom's magnetic moment and the magnetic field between the polar caps of the two magnets (the component along z axis). The atoms emitted by the oven are expected to have their magnetic moments oriented randomly in every direction. Thus, according to the classical view, the atoms were expected to be deflected by the magnetic field at all angles, resulting in a continuous distribution of the number of atoms versus the deflection angles. Instead, the atoms were deflected at a discrete set of angles.

The experiment was repeated in 1927 with hydrogen atoms which have only one electron orbiting around the nucleus. What was very surprising was the number of peaks in the electron distribution seen in that experiment. The hydrogen atoms were expected to exit the magnetic field as an undeflected beam. Instead, two beams were observed, one deflected upwards and the other deflected downwards by the magnetic field (see Figure 2.6).

The result was difficult to explain unless it was postulated that the electron in the hydrogen atom had associated with it a quantity called *spin*, representing an internal rotation of the electron itself. Thus, the magnetic dipole moment of the hydrogen atom has two components, one due to the rotational motion of the electron, and the second one due to its spin.

The *spin* is the *intrinsic angular momentum* of the electron and it is in no way associated with the rotation of the electron around the nucleus. The spin (the

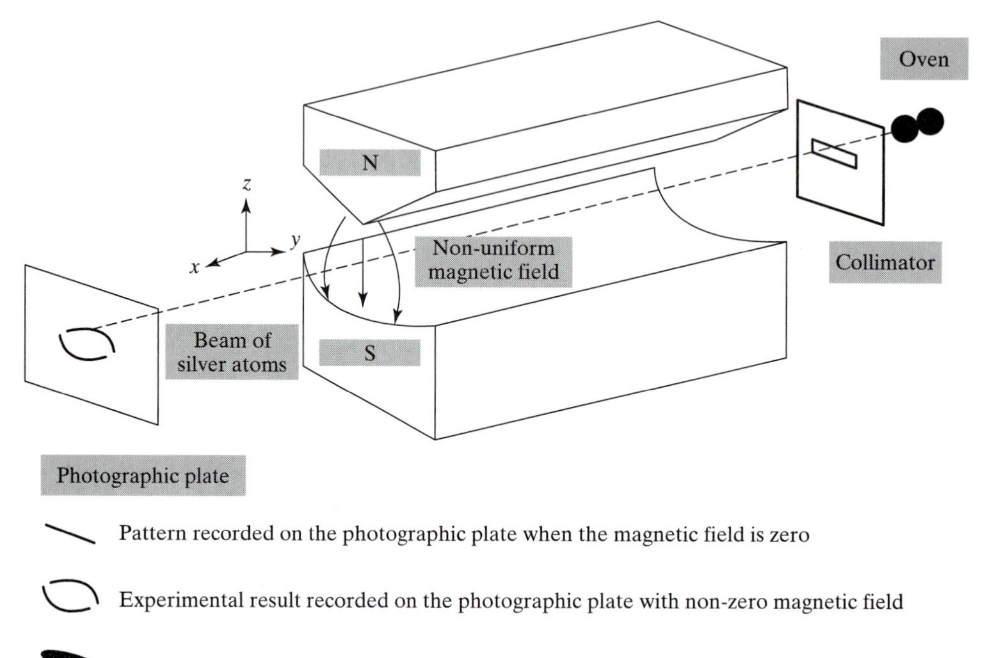

FIGURE 2.5: The Stern-Gerlach experiment with silver atoms. The image recorded on the photographic plate when a non-zero magnetic field is present differs from the classical expectation. The only possible explanation is that the atoms have spins which interact with the non-uniform magnetic field. In the absence of the magnetic field, the image observed on the photographic plate is a darkened image of the slit.

intrinsic angular momentum) of the electron has an intrinsic magnetic moment associated with it; this magnetic moment is proportional to the value of the spin.

It can be shown that a non-uniform magnetic field acts upon a magnetic moment with a force aligned with the direction of the gradient of the magnetic field. The value of the force is proportional to the gradient of the field and to the component of the magnetic moment (which is proportional to the spin) in the direction of the field gradient. The value of the spin can be estimated from the distance the beam is deflected on the screen.

In the experimental setup the gradient of the magnetic field was vertical and the initial direction of the electron beam was horizontal. Thus, the electrons had to be deflected upwards or downwards according to the component of their spin in the vertical direction. The electrons whose vertical spin component was "up" (i.e., positive) were deflected upward and those whose vertical spin component was "down" (i.e., negative) were deflected downwards.

2.15 THE SPIN AS AN INTRINSIC PROPERTY

At the time of its discovery, the spin represented a new physical quantity. What is the *spin* of a particle, for example, an electron? When Pauli concluded that

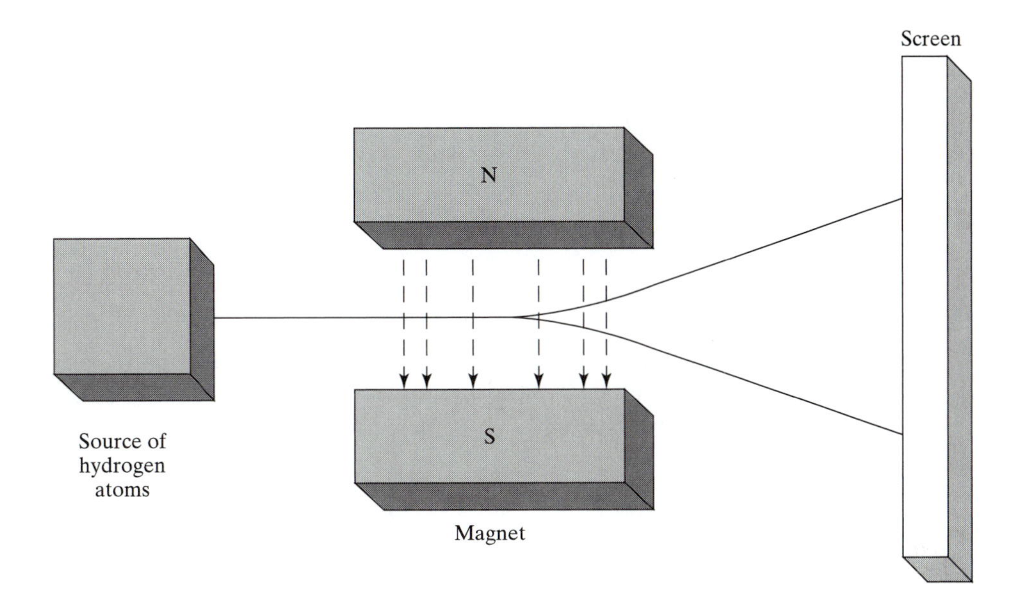

FIGURE 2.6: The Stern-Gerlach experiment with hydrogen atoms. Hydrogen atoms are expected to exit the magnetic field as an undeflected beam. In this experiment two beams emerge from the magnetic field, one deflected upwards and the other deflected downwards by the magnetic field.

a series of unexpected experimental observations of electrons' behavior could be explained by accepting the existence of a new quantum number (i.e., a new degree of freedom), Uhlenbeck and Goudsmit had the intuition to associate that new degree of freedom with an internal motion, a spinning, of the electron, different from the motion described by the dynamical variables position and momentum. The introduction of this two-valued quantum number for the electron also led Pauli to postulate his *Exclusion Principle*. According to Pauli's *Exclusion Principle*, no more than two electrons can occupy the same orbit and those two electrons must have anti-parallel spins. The non-relativistic electron was described by a two-component wave function, where the two components represented the states of spin $\pm\frac{1}{2}$.

The spin of an electron was predicted as an intrinsic property by Paul Dirac. Dirac was trying to develop a relativistic quantum mechanical wave equation for a relativistic free particle.[9] He proposed a differential equation of first order in the space and time derivatives, satisfying all the requirements of special relativity and quantum mechanics. For a non-relativistic ($v \ll c$) electron in a magnetic field this equation reduces to Schrödinger's Equation with a correction term. The correction term takes into account the interaction of the intrinsic magnetic moment of the electron with the external magnetic field and predicts the electron spin as an intrinsic property.

[9]A relativistic particle moves at a speed v close to the speed of light c ($v \approx c$).

Protons and neutrons are also characterized by spin one-half as an intrinsic property. The spin of the atoms is related to the spins of the electrons orbiting around its nucleus and to the spin of the nucleus itself.

The spin of the electron is associated with the physical rotation of an electrical charge. This motion of the electrical charge generates a magnetic field parallel to the axis of rotation. The spin is randomly distributed in space. The electron is characterized by a *spin magnetic moment*

$$\mu_z = -\frac{e}{2m_e} g\mathbf{S}$$

where e and m_e are the electron charge and mass, respectively, g is the spin gyromagnetic factor, and $\mathbf{S}$ is the spin angular momentum. The z component of the spin angular momentum is

$$S_z = \pm\frac{1}{2}\hbar$$

where + or − specify the orientation "up" or "down." The value of the z component of the magnetic moment can be written as

$$\mu_z = \pm\frac{e}{2m_e} gS_z = \pm\frac{e}{2m_e} g\frac{1}{2}\hbar = \pm\frac{1}{2} g\mu_B$$

where the universal constant $\mu_B = e\hbar/2m_e = 9.2740154 \times 10^{-24}$ J/T is called the *Bohr magneton.* In classical electromagnetism, a spinning sphere of electrical charge can produce a magnetic moment, but the magnitude of the electron spin magnetic moment calculated above cannot be reasonably modeled by considering the electron as a spinning sphere. According to high energy electron-electron scattering, the diameter of the electron is smaller than 10^{-18} m. A sphere of this size would have to spin at a preposterously high rate of some 10^{32} radian/s to match the observed angular momentum.

In the Stern-Gerlach experiment, the direction of the measurement axis (i.e., of the spin) was determined by the selection of the magnets and the orientation of the applied magnetic field. The particles used in the experiment were neutral silver atoms. Since the magnetic moments of the electrons in a silver atom cancel out in pairs except for one electron, the resultant magnetic moment is equal to that of the unpaired electron. When only spin is considered, each silver atom behaves like an electron.

The (neutral) atoms are separated into two beams, one deflected upward, the other downward, recorded as two distinct spots on the photographic plate being used as detector, with nothing in between. If the atoms were behaving as classical particles, the image would have been smeared into a single big spot. These two beams correspond to the only two states of spin possible along the measurement axis.

In quantum mechanics the intrinsic angular moment, the spin, is quantized and the values it may take are multiples of the rationalized Planck's constant $\hbar$ ($\hbar = h/2\pi$). The spin of an atom or of a particle is characterized by the *spin quantum number* s, which may assume integer and half-integer values. For a given value of s, the projection of the spin on any axis may assume $2s + 1$ values ranging from $-s$ to $+s$ by unit steps, in other words the spin is quantized. The electron has spin $s = \frac{1}{2}$ and the spin projection can assume the values $+\frac{1}{2}$, referred to as *spin up*, and $-\frac{1}{2}$, referred to as *spin down*.

The electron spin is one of the possible physical implementations of the quantum bit, the *qubit*. The spin is a measurable quantity, an observable; the spin states for which the probability of obtaining a particular result is unity play a central role in defining the states of a qubit.

2.16 SCHRÖDINGER'S WAVE EQUATION

To study the properties of a quantum system we have to know its state, as well as the evolution in time of its state. Schrödinger's Equation allows us to describe the state of a stationary system (a system whose state is independent of time) and the evolution in time of a non-stationary system.

Erwin Schrödinger proposed the following equation for the wave function (state vector) $\Psi_n(q)$ of a *stationary state* of energy E_n of an atom

$$H\left(q, \frac{h}{2\pi i}\frac{\partial}{\partial q}\right)\Psi_n(q) = E_n\Psi_n(q).$$

H is the *Hamiltonian operator*, Ψ_n is the *wave function* (state vector) associated with the atom in a stationary state of *energy* E_n. The energy E_n is the eigenvalue associated with the eigenvector $| \Psi_n \rangle$. The Hamilton function $H(q, p)$ represents the classical energy of the atom when expressed in terms of position q and momentum p coordinates. Schrödinger replaced the momentum variable p by the differential operator $(h/2\pi i)(\partial/\partial q)$.

Schrödinger assumed that the time evolution, the dynamics of the wave function, is governed by another partial differential equation

$$i\frac{h}{2\pi}\frac{\partial}{\partial t}\Psi(q, t) = H\left(q, \frac{h}{2\pi i}\frac{\partial}{\partial q}\right)\Psi(q, t).$$

Two important observations are in order:

1. The presence of the complex number i in this equation implies that the wave function is complex,
2. In general, a solution to this equation, $\Psi_n(q, t)$, is a function of time and represents the time evolution of the stationary wave function $\Psi_n(q)$

$$\Psi_n(q, t) = e^{-i\frac{2\pi}{h}E_n t}\Psi_n(q).$$

Schrödinger realized that the wave function of a many-electron atom was not defined in the ordinary, physical, three-dimensional space, as de Broglie had used, but in a more abstract *configuration space*, the space described by the coordinates of the positions of all the electrons. The wave associated with this system is different from an electromagnetic wave; the wave function exists in a Hilbert space and its values are complex. At the beginning, the physical interpretation of such a complex function posed serious challenges.

By the end of the same year, 1926, Max Born suggested the probabilistic interpretation of the wave function.[10] According to Born, the *probability density* that a particle can be found at a given location and time is equal to the *square of the amplitude of the wave function* at that location

$$\text{Prob} = \mid \Psi(q, t) \mid^2 .$$

Born made an analogy between the scattering of an electron colliding with an atom and the diffraction of X-rays and concluded that the electron can be anywhere in the space where the wave function is different from zero, but there is no way to pinpoint its position because this is a random event.

This formula is extremely important and represents the essence of what the quantum theory can give us. In the world of classical physics the position and the speed of an object can be measured or predicted with 100% certainty (in principle), while in the world of quantum physics our predictions are only *statistical* in nature. Schrödinger's Equation allows us to make probabilistic predictions; we can determine where a particle will be (if the position observable is considered) only in terms of *probabilities of different outcomes*, or, equivalently, what fraction of a large number of particles will be found at a specific location. But, it does not give any indication as to one *specific* outcome.

The *superposition principle* is an essential element of the quantum theory brought forth by Schrödinger's Equation. The equation is linear, therefore, when two function Ψ_1 and Ψ_2 are among its solutions, their sum $\Psi_1 + \Psi_2$ is another solution. The corresponding probability is proportional to $\mid \Psi_1 + \Psi_2 \mid^2$ and it can show interference effects. For example, an electron can be found in a state that is a superposition of two other states. A solution of Schrödinger's Equation can be any linear combination of wave functions $\Psi_n(q, t)$ given above.

The superposition can be extended to a single particle, a particle can be superposed with itself. In Young's experiment when the light is reduced to one photon emitted at a time, we still find an interference pattern on the screen (after enough photons have been detected). The explanation is that a single photon goes through both slits and then it interferes with itself, as two waves do by superposition. Similar experiments were later performed with electrons, neutrons, atoms, and even bucky balls.[11] All these particles behave like waves and create interference patterns.

Schrödinger himself realized that if a quantum system contains more than one particle, the superposition principle gives rise to the phenomenon of *entanglement*, the interference of the system with itself. The result is an *entangled*

---

[10]Einstein had the intuition of the probabilistic nature of the wave function at about the same time, but had reservations about the possible randomness of the physical world.

[11]A *bucky ball* is a molecule (also called *fullerene*) of sixty or seventy carbon atoms arranged in a structure resembling the geodesic domes built by the architect Buckminster Fuller.

system. The term *entanglement* was first used by Schrödinger in 1935 [114] in his discussion of the Einstein, Podolsky, and Rosen (EPR) paper [50].

2.17 HEISENBERG'S UNCERTAINTY PRINCIPLE

The concept of probability in quantum mechanics is very different from the one in classical physics. Classical probabilities only express some lack of information about the fine details of a given situation. The randomness is due to uncontrolled causes that are recognized to exist and that, if known better, would make the predictions better. For example, we can predict very accurately the arrival time of a transatlantic flight if we know precisely the speed of the airplane and of the tail or head wind throughout the entire journey. But if the atmospheric conditions change during the flight, so does the accuracy of the predicted arrival time.

On the other hand, the quantum probabilities assume, as a matter of principle, that a more precise knowledge at the quantum level is impossible. This limitation cannot be avoided as shown by many experiments performed over the years. Einstein, who doubted that "God is playing dice" questioned the truth of such an indeterminacy.

The *Uncertainty Principle*,[12] discovered by Werner Heisenberg in 1927, is at the heart of the special nature of the probabilistic aspect of the quantum theory. We use the term *observable* for a measurable physical property of a quantum system. The mathematical equivalent to measuring a quantum system is to apply a measurement operator to the vector describing the state of the system.

Consider two such observables, X the position of a particle and P_X the momentum of the same particle at position X. We assume that X and P_X are three-dimensional vectors. The operators corresponding to these observables, $\mathcal{X}$ and $\mathcal{P}_X$ are non-commutative. This means that if operator $\mathcal{X}$ is applied first to the state vector and the operator $\mathcal{P}_X$ is second (the state vector is projected first to a basis corresponding to property X and then to a basis corresponding to property $\mathcal{P}_X$) the results are not the same as when the operator $\mathcal{P}_X$ is applied first and the operator $\mathcal{X}$ is second (now, the state vector is projected first to a basis corresponding to property P_X and then to a basis corresponding to property X).

The *Uncertainty Principle* states that ΔX, the uncertainty in determining the position, and ΔP_X, the uncertainty in determining the momentum at position X, are constrained by the inequality

$$\Delta X \, \Delta P_X \geq \frac{\hbar}{2}$$

where $\hbar = h/2\pi$ is a modified form of Planck's constant.

The uncertainty is an intrinsic property of quantum systems. The accuracy of measured values of basic physical properties, such as position and momentum along the axis used to measure the position is limited, the precise knowledge of both the position and the momentum is simply forbidden in a quantum system.

[12] Heisenberg called it the Indeterminacy Principle, but in this book we use the term commonly found in the literature, the Uncertainty Principle.

A relatively simple derivation of this famous result follows. Consider a quantum state $| \psi \rangle$ and two Hermitian operators, **A** and **B**. Recall from Section 2.4 that **[A,B]** is the commutator of the two operators and $\{\mathbf{A}, \mathbf{B}\}$ is the anti-commutator of the two operators. Recall also that $\langle \psi \mid \psi \rangle$ represents the probability of obtaining the value λ_a when we use **A** as a projector and the system is in state $| \psi \rangle$ before the measurement.

Let us assume that the probability of getting the outcome λ_{AB} in a measurement of the system in state $| \psi \rangle$ is

$$\langle \psi \mid \mathbf{A}\,\mathbf{B} \mid \psi \rangle = a + ib.$$

with $a, b \in \mathbb{R}$. Then it is easy to see that

$$\langle \psi \mid [\mathbf{A}\,\mathbf{B}] \mid \psi \rangle = (a + ib) - (a - ib) = 2ib$$

and

$$\langle \psi \mid \{\mathbf{A}\,\mathbf{B}\} \mid \psi \rangle = (a + ib) + (a - ib) = 2a.$$

Now

$$| \langle \psi \mid [\mathbf{A}\,\mathbf{B}] \mid \psi \rangle |^2 = 4b^2.$$

$$| \langle \psi \mid \{\mathbf{A}\,\mathbf{B}\} \mid \psi \rangle |^2 = 4a^2$$

and

$$| \langle \psi \mid \mathbf{A}\,\mathbf{B} \mid \psi \rangle |^2 = a^2 + b^2.$$

It follows that

$$| \langle \psi \mid [\mathbf{A}\,\mathbf{B}] \mid \psi \rangle |^2 + | \langle \psi \mid \{\mathbf{A}\,\mathbf{B}\} \mid \psi \rangle |^2 = 4 \,| \langle \psi \mid \mathbf{A}\,\mathbf{B} \mid \psi \rangle |^2 .$$

Recall the Schwartz inequality involving the inner product of two vectors, **x** and **y** and their norms in an n-dimensional Euclidean space

$$(\mathbf{x}, \mathbf{y}) \leq (\mathbf{x}, \mathbf{x})(\mathbf{y}, \mathbf{y}).$$

Applying this inequality we obtain

$$| \langle \psi \mid \mathbf{A}\,\mathbf{B} \mid \psi \rangle |^2 \leq \langle \psi \mid \mathbf{A}^2 \mid \psi \rangle \langle \psi \mid \mathbf{B}^2 \mid \psi \rangle.$$

If we combine this inequality with the previous result and drop a non-negative term we obtain

$$| \langle \psi \mid [\mathbf{A}\,\mathbf{B}] \mid \psi \rangle |^2 \leq 4 \langle \psi \mid \mathbf{A}^2 \mid \psi \rangle \langle \psi \mid \mathbf{B}^2 \mid \psi \rangle.$$

Consider now two observables O_1 and O_2. Let us construct the corresponding matrices

$$A = O_1 - \langle O_1 \rangle$$

and

$$B = O_2 - \langle O_2 \rangle.$$

We substitute A and B with these values to obtain

$$\Delta O_1 \Delta O_2 \geqslant \frac{|\langle \psi | [O_1, O_2] | \psi \rangle|}{2}$$

When O_1 is the position X and O_2 is the momentum, P_X, then the commutator of the two operators is $[X, P_X] = i\hbar I$. Thus, the inequality becomes

$$\Delta X \, \Delta P_X \geqslant \frac{\hbar}{2}.$$

2.18 A BRIEF HISTORY OF QUANTUM IDEAS

The ideas behind quantum mechanics appeared at the very beginning of the twentieth century. During the following three decades they evolved rapidly into a new mathematical model which provides a description of the physical world, consistent with the experimental evidence collected over the years. For the past twenty years, or so, quantum mechanics, once considered a domain of interest only to physicists, has revealed that it may have a profound effect on how we manipulate information. Thus, it seems appropriate to discuss the evolution of quantum ideas before introducing the basic concepts of quantum computing.

Quantum mechanics is the description of the behavior of matter and light, in particular, of what is happening on the atomic and sub-atomic scale. *Quantum physics* would be a more appropriate name because it provides a general framework for the whole of physics rather than dealing with only the field of mechanics.

Things on a very small scale do not behave like waves, nor do they behave like particles; they behave like none of the objects or phenomena we encounter in our surrounding, macroscopic world.

Isaac Newton thought that light was made up of particles, though he was aware of the diffraction and interference phenomena which were indications that it behaved like waves. Sometime at the beginning of the nineteenth century, Thomas Young (1773–1829), a British physician and physicist, conducted the now famous "double-slit experiment" on light, demonstrating the wave-theory effect of interference. In his experiment Young used a light source, a barrier with two slits in it, and a screen behind the barrier. He shone light from the source on the barrier with two slits and obtained an *interference* pattern on the screen. An *interference* pattern is the characteristic signature of *wave* behavior. Particles, as we know them from daily experience, do not interfere with each other.

In physics the notion of *quanta* as elementary units of energy was introduced in 1900, when Max Planck presented his theory of *blackbody radiation*. The blackbody radiation is the electromagnetic radiation emitted by a body in thermodynamical equilibrium, or the radiation contained in a cavity when its walls are at a uniform temperature. The radiation is allowed to escape through a small aperture so that its frequency spectrum and energy density can be measured. Classical thermodynamics predicted that the intensity of the radiation emitted in a small frequency interval $\Delta\nu$ should be proportional to the square of the frequency, ν^2; when integrated over all frequencies that gives an infinite total intensity. The theoretical predictions were in contradiction with the experimental results at higher frequencies, where the measured intensity rather than increasing like ν^2 was decreasing exponentially.

Max Planck assumed, as a new postulate of physics, that the energy of the emitted radiation does not vary continuously, but by small amounts which are multiples of a basic *quantum*. Planck denoted this quantum of energy by $h\nu$, where ν is the frequency of the radiation (which is a wave) and h is a fundamental constant, now known as Planck's constant. (The value of this constant is $h = 6.6262 \times 10^{-34} Joule \times second$ and represents the product $Energy \times Time$). Planck proposed the following formula for the energy levels of the blackbody radiation (approximated as a Maxwell-Hertz oscillator)

$$E = 0, h\nu, 2h\nu, 3h\nu, 4h\nu, \ldots, nh\nu,$$

where n is a non-negative integer.

In 1905, Albert Einstein showed that the empirical properties of the photoelectric effect[13] could be explained by assuming that light consists of "particles," each one of them having an energy $h\nu$ and moving at the speed of light. Einstein's light particle became known as a *photon*.[14] According to Einstein, the existence of quanta was not due to the emission process but was a property of the light itself. The best evidence of this idea came later, when Arthur Compton found that the scattering of gamma rays[15] by electrons has the kinematical characteristics of the collision between two particles (or billiard balls).

Based on results produced by alpha particles scattering experiments, Ernest Rutherford established in 1911 that an atom consists of electrons surrounding a positively charged heavy nucleus. A planetary model of the atom was proposed at that time.

In 1913, Niels Bohr modified the planetary model of the atom based on the 1908 discovery of Walter Ritz that all the frequencies in the spectrum of a given atom can be obtained with a simple formula $\nu = \nu_n - \nu_m$, where the frequencies ν_i $(i = 1, 2, 3, \ldots)$ characterize the atom. Bohr noted that the angular momentum of the orbiting electron in his model of the hydrogen atom[16] has the same dimensions as Planck's constant h.

Bohr postulated that the angular momentum of the orbiting electron must be a multiple of the Planck's constant divided by 2π, that is

$$mvr = \frac{h}{2\pi}, 2\frac{h}{2\pi}, 3\frac{h}{2\pi}, \ldots$$

---

[13] Einstein received the Nobel prize for his theory of the photoelectric effect and not for his special and general relativity theories for which he is so well known.

[14] Photon is derived from the Greek word "photos" meaning light.

[15] Different regions of the energy spectrum of the electromagnetic radiation have different names such as gamma rays, X-rays, and (visible) light.

[16] The hydrogen atom has one electron orbiting the nucleus which contains one proton.

where mvr is the classical definition of the angular momentum, m is the electron mass, v is the electron velocity, and r is the radius of the electron orbit.

The quantization of the angular momentum led Bohr to postulate the quantization of the atom energy. Bohr also postulated that when the hydrogen atom made a transition from one energy state ("level") to a lower one, the difference between its initial and final energies was emitted in the form of a *quantum of energy*

$$E_a - E_b = h\nu_{ab}.$$

Here, E_a is the initial energy level of the atom, E_b is the final energy level after the transition from its prior state, h is Planck's constant and ν_{ab} is the frequency of the "light" quantum (gamma ray) emitted during the transition from the energy level E_a to energy level E_b.

The quantum mechanics developments in the 1920s and the 1930s reinforced the view that light behaves as a collection of particles as well as waves. In its interaction with matter, light exhibits phenomena that are characteristic of waves, such as interference and diffraction, as well as phenomena that are characteristic of particles, as it happens in the case of the photoelectric effect.

The next significant step was made by the French physicist Louis (Prince) de Broglie in 1923. Drawing an analogy with light and its dual character as both a wave and a particle, de Broglie proposed that a wave should be associated with every kind of particle, in particular with the electron. De Broglie also proposed an equation that linked the momentum, p, of a particle to the wavelength, λ, of the wave associated with it and Planck's constant

$$p = \frac{h}{\lambda}.$$

This assumption was fundamentally new[17]; it was no longer a correction, or a new chapter of classical physics imposed by quantum constraints. The experimental confirmation of de Broglie's assumption came in 1928 when Clinton Davisson and Lester Germer observed the diffraction of electrons by crystals which proved the wavelike characteristics of electrons.

The equation describing the evolution of the wave associated with a particle was proposed in 1926 by Erwin Schrödinger, who, thus, gave a precise formulation to de Broglie's wave hypothesis. In 1926, Werner Heisenberg developed a theory of the quantum mechanics based on matrices, equivalent to Schrödinger's theory which was based on the wave equation. In Heisenberg's more abstract approach, infinite matrices represent properties of observable entities and the mathematics used is that of matrix algebra. The matrix multiplication is non-commutative and that has important consequences in quantum mechanics. Heisenberg's leading idea was that physics should use only *observable* quantities, quantities that could be measured. In his opinion, such quantities as the classical "orbits" should not even be mentioned, since no experiment has ever shown their existence.

[17]Louis de Broglie received the Nobel Prize for his contributions to quantum theory in 1929.

Max Born, Pascual Jordan, and Paul Dirac were the first to realize that the use of non-commutative quantities to replace position and momentum was an essential feature of Heisenberg's theory and they recognized the rules of matrix calculus in their own theories. They concentrated upon a new kind of mechanics where the dynamical variables are not ordinary numbers, but new *non-commutative mathematical objects*.

In 1925, Born and Jordan developed a complete formulation of this new mechanics using infinite matrices to represent the basic physical quantities. The same year, 1925, Dirac introduced abstract mathematical objects, such as Q_j for the position coordinates and P_j for the momentum coordinates, without trying to specify them; the multiplication rules were those of the new mechanics. He postulated a general form for the commutator between two quantum dynamical variables using the Poisson brackets. Dirac's abstract quantities Q and P can be identified with some *operators* acting upon the wave functions. Heisenberg's matrices representing position and momentum can be obtained from the wave functions.

In 1926, Schrödinger and Dirac showed the equivalence of Heisenberg's matrix formulation and Dirac's algebraic one with Schrödinger's wave function. In 1926, Dirac and, independently, Born, Heisenberg, and Jordan, obtained a complete formulation of quantum dynamics that could be applied to any physical system; the first calculations were done for the hydrogen atom.

In 1926, John von Neumann introduced the concept of *Hilbert space* to quantum mechanics. The idea came to him while attending a lecture of Heisenberg who was presenting his matrix mechanics and the difference between his model and the one based upon Schrödinger's wave equation. During the presentation, David Hilbert, considered the greatest mathematician of the time, asked for clarifications. Then, John von Neumann decided to formulate the quantum theory in terms familiar to the great mathematician. He wrote a note for Hilbert, explaining Heisenberg's version of quantum theory in terms of what would later be known as *Hilbert spaces*. A Hilbert space is a vector space with a measure of the distance, called the norm, and the property of completeness. John von Neumann expanded this explanation into a book *The Mathematical Foundations of Quantum Mechanics* published in 1932.

John von Neumann demonstrated that the geometry of vectors over the complex plane has the same formal properties as the states of a quantum mechanical system. The states of the quantum systems, the wave functions, are represented as vectors in a Hilbert space and the operations associated with the position and the momentum act like matrices upon these vectors. He also derived a theorem, using some assumptions about the physical world, which proved that there are no "hidden variables." The inclusion of such hidden variables would completely eliminate the uncertainty inherent in the quantum mechanical description of physical systems. In 1966, John Bell (successfully) challenged von Neumann's assumptions and proved his own theorem establishing that indeed hidden variables could not exist.

In 1928, Paul Dirac developed the *relativistic quantum mechanics* that combined quantum mechanics with relativity. He introduced corrections for relativistic effects to the quantum mechanics equations for particles moving at close to the speed of light. This allowed the properties of *spin* to be obtained in a natural way from the relativistic Schrödinger's Equation. In January 1925, Wolfgang Pauli had suggested the existence of a fourth quantum number of the electron with values $(+\frac{1}{2}, -\frac{1}{2})$ and formulated the "exclusion principle" bearing his name. At that time, George Uhlenbeck and Samuel Goudsmit had interpreted the new quantum number as the *spin* of the electron, associated with the intrinsic rotation (spin) of the electron. In 1930, Dirac predicted the existence of *anti-electrons*, particles with an opposite charge to that of the electron.

In 1931, Carl Anderson discovered the *positron*, the anti-electron, in cosmic radiation. In 1949, Madame Chien-Shiung Wu and Irving Shaknov of Columbia University produced *positronium*, an artificial association between an electron and a positron circling each other. This element lives for a fraction of a second, then the electron and the positron spiral toward each other, causing mutual annihilation. Two photons of gamma radiation, each with an energy of 0.511 MeV, are emitted in opposite directions in this process.

This experiment verified an assumption made by John Wheeler in 1946 that the two photons produced when an electron and a positron annihilate each other, have opposite polarizations; if one is vertically polarized, the other must be horizontally polarized. What is remarkable about this experiment is the fact that it is the first one in history to produce the so called *entangled photons*. This important fact was recognized only eight years later, in 1957, by David Bohm and Yakir Aharonov.

2.19 SUMMARY AND FURTHER READINGS

Phenomena, such as interference and diffraction of light, can only be explained on the basis of a wave theory, while other phenomena including the photo-electric emission, show that light is composed of small particles called photons. Classical mechanics could not accept this duality and a new model of the physical world was necessary.

Quantum theory is such a mathematical model developed by Werner Heisenberg and Erwin Schrödinger in the mid-1920s. The cornerstones of quantum mechanics are Schrödinger's Equation and Heisenberg's Uncertainty Principle. The quantum mechanical model is characterized by the way it represents the states of a physical system, the observables of the system, the measurements of these observables, and the dynamics of the system.

The stage for quantum mechanics is an n-dimensional Hilbert space. An n-dimensional Hilbert space $\mathcal{H}_n$ is a *vector space* over the field of complex numbers with an inner product that helps define the norm, which in turn gives the "length" of a vector. The elements of $\mathcal{H}_n$ are n-dimensional vectors. Traditionally, quantum mechanics texts use Dirac notations for vectors. In addition to inner products of two vectors, quantum mechanics uses the *tensor product* of two vectors or matrices and the *outer product* of two vectors.

The state vector $| \Psi_a \rangle \in \mathcal{H}_n$ can be expressed as a linear combination of the *basis states*, $| 0\rangle, | 1\rangle \dots | i\rangle, \dots | n-1\rangle$ as $| \Psi_a\rangle = \sum_{i=0}^{n-1} \alpha_i | i\rangle$, with

$\alpha_i, 0 \leqslant i \leqslant n - 1$, complex numbers representing the *amplitudes*, with the property that $\alpha_j = \langle j \mid \Psi_a \rangle$. The "superposition principle" tells us that given two states $\mid \Psi_a \rangle$ and $\mid \Psi_b \rangle$, we can form another state as $(a \mid \Psi_a \rangle + b \mid \Psi_b \rangle)$, a superposition of the original two states.

Transformations of states in $\mathcal{H}_n$ are described by Hermitian operators. An operator $\mathbf{O}$ maps state vectors to state vectors in $\mathcal{H}_n$. If we take a quantum system in state $\mid \Psi_a \rangle$ and apply to it a transformation described by the operator $\mathbf{O}$, we get a different state $\mid \Psi_b \rangle = \mathbf{O} \mid \Psi_a \rangle$. Each operator $\mathbf{O}$ has an associated matrix $O = [O_{ij}]$ with elements given by $O_{ij} = \langle i \mid \mathbf{O} \mid j \rangle$, where $\mid i \rangle$ and $\mid j \rangle$ are unit vectors and $\langle i \mid j \rangle = \delta_{ij}$.

A *projection operator* is given by the outer product of any state vector $\mid \Psi_a \rangle$ with itself $\mathbf{P}_{\Psi_a} = \mid \Psi_a \rangle \langle \Psi_a \mid$. A complete set of orthogonal projectors in $\mathcal{H}_n$ is a set $\{\mathbf{P}_0, \mathbf{P}_1, \ldots, \mathbf{P}_i, \ldots, \mathbf{P}_{m-1}\}$ such that $\sum_{i=0}^{m-1} \mathbf{P}_i = 1$. The number of projectors in a complete orthogonal set must be less than or equal to the dimension of the Hilbert space $m \leqslant n$.

In $\mathcal{H}_n$ every normal operator $\mathbf{N}$ has n eigenvectors $\mid n_i \rangle$ and, correspondingly, n eigenvalues λ_i. If $\mathbf{P}_i$ are the projectors corresponding to these eigenvectors, $\mathbf{P}_i = \mid n_i \rangle \langle n_i \mid$, then the operator $\mathbf{N}$ has the spectral decomposition $\mathbf{N} = \sum_i \lambda_i \mathbf{P}_i$.

An *observable* $\mathbb{O}$ is a property of a physical system that, in principle, can be measured. The observable $\mathbb{O}$ has an associated Hermitian operator $\mathbf{O}$. The outcome of the measurement of the observable $\mathbb{O}$ of a system in state $\mid \Psi_a \rangle$ is an eigenvalue λ_i corresponding to the eigenvector $\mid o_i \rangle$ of the operator $\mathbf{O}$. Immediately after the measurement the state of the system becomes $\mid \Psi_b \rangle = \mathbf{O} \mid \Psi_a \rangle = \lambda_i \mid o_i \rangle$. The probability to observe outcome λ_i is equal to $\langle \Psi_a \mid \mathbf{P}_i \mid \Psi_a \rangle$, with $\mathbf{P}_i$ the projector corresponding to the eigenvector $\mid o_i \rangle$.

The density matrix contains all the information regarding the results of measurements of an ensemble of N independent versions of a quantum system and gives the expected value of any observable of the system. The density matrix does not uniquely determine the states of individual particles. The density matrix is Hermitian and its eigenvalues are nonnegative.

There are extremely well-written books on quantum mechanics and it is very difficult to single out only a few of them. The first three chapters of Paul Dirac's book *The Principles of Quantum Mechanics* [45] are accessible to a large audience while another classical book, *Mathematical Foundations of Quantum Mechanics* [142] by John von Neumann seems accessible only to the most mathematically sophisticated students. The third volume of Richard Feynman's *Lectures on Physics* [53] provides a comprehensive and clear exposition of quantum mechanics, vintage 1960s. David Bohm's *Quantum Theory* [24] is also an excellent reference.

Among the more modern texts we recommend, *The Interpretation of Quantum Mechanics* by Roland Omnes [100], a very readable account of "classical" as well as more "modern" ideas and concepts of quantum mechanics, such as the density matrix.

An excellent reference is E. Mertzbacher's book *Quantum Mechanics* [92]. A very accessible and modern text is also *Quantum Mechanics* by E. S. Abers [1]. The book of M. A. Nielsen and I. L. Chuang *Quantum Computing and*

Quantum Information [98] provides a concise review of the basic concepts of quantum mechanics for quantum computing.

John Preskill's notes [110] summarize the quantum concepts directly related to quantum computing.

A very clear and concise book on linear algebra is *Lectures on Linear Algebra* by I. M. Gelfand [62]. An excellent reference on functions and functional analysis is A. N. Kolmogorov and S. V. Fomin's book *Elements of the Theory of Functions and Functional Analysis* [85].

2.20 EXERCISES AND PROBLEMS

2.1. Show that any vector $x = (x_1, x_2, \dots, x_n) \in \mathbb{C}^n$ is a linear combination of n unit vectors:

$$\epsilon_1 = (1, 0, 0, \dots, 0), \epsilon_2 = (0, 1, 0, \dots, 0), \dots \epsilon_n = (0, 0, 0, \dots, 1).$$

2.2. Let n vectors span a vector space $\mathcal{A}$ containing r linearly independent vectors. Prove that $n \geq r$.

2.3. Show that all bases of any finite-dimensional vector space $\mathcal{A}$ have the same finite number of dimensions.

2.4. Show that if a vector space $\mathcal{A}$ has dimension n, then (i) any set of $(n + 1)$ elements of $\mathcal{A}$ is linearly dependent, and (ii) no set of $(n - 1)$ elements can span $\mathcal{A}$.

2.5. Show that row-equivalent matrices have the same row space.

2.6. Show the three properties of the trace of a matrix from Section 2.3.

2.7. Consider two vectors over $\mathbb{C}_2$:

$$| \psi\rangle = \alpha_0 \mid 0\rangle + \alpha_1 \mid 1\rangle$$
$$| \phi\rangle = \beta_0 \mid 0\rangle + \beta_1 \mid 1\rangle.$$

Show that:

$$| \psi\rangle \otimes | \phi\rangle = \alpha_0\beta_0 \mid 00\rangle + \alpha_0\beta_1 \mid 01\rangle + \alpha_1\beta_0 \mid 10\rangle + \alpha_1\beta_1 \mid 11\rangle.$$

2.8. For an $m \times m$ matrix A show that

$$(A^\dagger)^\dagger = A.$$

2.9. Show that a normal matrix is Hermitian if and only if it has real eigenvalues.

2.10. Consider vectors ψ, ϕ, ξ and scalars m, n. Show that

$$m\psi \otimes n\phi = mn(\psi \otimes \phi)$$
$$\xi \otimes (\psi + \phi) = \xi \otimes \psi + \xi \otimes \phi$$
$$(\psi + \phi) \otimes \xi = \psi \otimes \xi + \phi \otimes \xi.$$

2.11. Consider matrices V, W, X, Y, Z and vectors ψ, φ having an appropriate number of rows and columns. Show that

$$(V \otimes W)(X \otimes Y) = VX \otimes WY$$

$$(V \otimes W)(\psi \otimes \varphi) = (V\psi) \otimes (W\varphi)$$

$$\begin{pmatrix} V & W \\ X & Y \end{pmatrix} \otimes Z = \begin{pmatrix} V \otimes Z & W \otimes Z \\ X \otimes Z & Y \otimes Z \end{pmatrix}.$$

2.12. Consider n unitary matrices $V_i, 1 \leqslant i \leqslant n$. Let $W = V_1 \otimes V_2 \otimes \ldots \otimes V_n$. Show that W is unitary.

2.13. Show that in $\mathcal{H}_2$

$$(| 0\rangle\langle 1 |) | 0\rangle = | 0\rangle(\langle 1 | 0\rangle)$$

$$(| 1\rangle\langle 0 |) | 0\rangle = | 1\rangle(\langle 0 | 0\rangle) = | 1\rangle$$

$$(| 0\rangle\langle 1 |) | 1\rangle = | 0\rangle(\langle 1 | 1\rangle) = | 0\rangle$$

$$(| 1\rangle\langle 0 |) | 1\rangle = | 1\rangle(\langle 0 | 1\rangle).$$

2.14. Given the operator

$$\mathbf{O} = | 0\rangle\langle 1 | + | 1\rangle\langle 0 |$$

and the vector

$$| \Psi_a | = \alpha_0 | 0\rangle + \alpha_1 | 1\rangle,$$

show that for $\forall | \Psi_a\rangle \in \mathcal{H}_2$ we have

$$\langle \Psi_a | \mathbf{O} = (\alpha_0^* \langle 0 | + \alpha_1^* \langle 1 |)(| 0\rangle\langle 1 | + | 1\rangle\langle 0 |) = \alpha_0^* \langle 1 | + \alpha_1^* \langle 0 | .$$

2.15. Show that $| +\rangle = 1/\sqrt{2}(| 0\rangle + | 1\rangle)$ and $| -\rangle = 1/\sqrt{2}(| 0\rangle - | 1\rangle)$ form an orthonormal basis in $\mathcal{H}_2$. Construct the projectors $\mathbf{P}_+$ and $\mathbf{P}_-$ corresponding to these basis vectors.

2.16. Show that $| 0\rangle, | 1\rangle, | 2\rangle, | 3\rangle$ form an orthonormal basis in $\mathcal{H}_4$. Construct the operator Π_4 that permutes circularly the basis vectors in $\mathcal{H}_4$ as follows $| 0\rangle \rightarrow | 1\rangle$, $| 1\rangle \rightarrow | 2\rangle$, $| 2\rangle \rightarrow | 3\rangle$ and $| 3\rangle \rightarrow | 0\rangle$. Construct its matrix representation. Calculate $\Pi_4 | \psi\rangle$ and $\langle \psi | \Pi_4$ with $| \psi\rangle = \alpha_0 | 0\rangle + \alpha_1 | 1\rangle + \alpha_2 | 2\rangle + \alpha_3 | 3\rangle$.

2.17. Consider an operator $\mathbf{O}$ in $\mathcal{H}_3$ with three eigenvalues λ_0, λ_1, and λ_2 and with the corresponding orthonormal eigenstates $| o_0\rangle, | o_1\rangle$, and $| o_2\rangle$

$$| o_0\rangle = \alpha_0 | 0\rangle + \alpha_1 | 1\rangle + \alpha_2 | 2\rangle$$

$$| o_1\rangle = \beta_0 | 0\rangle + \beta_1 | 1\rangle + \beta_2 | 2\rangle$$

$$| o_2\rangle = \gamma_0 | 0\rangle + \gamma_1 | 1\rangle + \gamma_2 | 2\rangle.$$

Construct the three projectors corresponding to these eigenvectors. Express the matrix O corresponding to $\mathbf{O}$ in terms of the eigenvalues and the projectors.

CHAPTER 3

Qubits and Their Physical Realization

A classical bit is an abstraction of a physical system capable of being in one of two states, 0 or 1. To perform any computation we should be able to: (a) store information, (b) retrieve the information from the physical system, and (c) change the state of the physical system.

To perform a meaningful computation we need a fair amount of bits and the ability to access and modify them very fast. This is why a computer using electro-mechanical relays is more useful than the traditional abacus, computers using electronic circuits are more capable than the ones based upon electro-mechanical relays, and so on.

From solid state studies we already know that the "smaller" and "simpler" the physical systems used to implement a bit, the less energy is necessary to store and process information and the faster is the resulting computing engine. By pushing the limits of the physical systems used to implement a bit we inevitably end at the level of atomic, or even sub-atomic particles. At this level we enter the realm of quantum systems governed by quantum mechanics.

3.1 ONE QUBIT—A VERY SMALL BIT

A *quantum bit* or *qubit* is an elementary quantum object used to store information. Since it is difficult at this stage to explain what a quantum object is, for the time being we view a qubit as a mathematical abstraction and we hint to possible physical implementations of this abstract object. We review the basic facts revealed by the simple model discussed in Chapter 1 and reinforced by the elements of quantum mechanics presented in Chapter 2.

A qubit $\mid \psi \rangle$ is a vector in a two-dimensional complex vector space. In this space a vector has two components and the projections of the vector on a basis of the vector space are complex numbers. We use Dirac's notation to represent a vector $\mid \psi \rangle$ as

$$\mid \psi \rangle = \alpha_0 \mid 0 \rangle \ + \ \alpha_1 \mid 1 \rangle$$

with α_0 and α_1 complex numbers and with $\mid 0 \rangle$ and $\mid 1 \rangle$ the vectors forming an orthonormal basis for this two-dimensional vector space [46].

A classical bit can be in one of two states, 0 or 1. Thus, we can represent the state of a bit as $b = a_1 0 + a_2 1$, which has exactly two forms: either $a_1 = 1$ and $a_2 = 0$ and the value of the bit is $b = 0$, or $a_1 = 0$ and $a_2 = 1$ and the value of the bit is $b = 1$.

In contrast, the state of a qubit can be represented by $\mid \psi \rangle = \alpha_1 \mid 0 \rangle + \alpha_2 \mid 1 \rangle$ where $\mid 0 \rangle$ and $\mid 1 \rangle$ is an orthonormal pair of basis vectors called *computational basis states*. The only restrictions on the coefficients are that (1) α_0 and α_1 are complex numbers and (2) $\mid \alpha_1 \mid^2 + \mid \alpha_2 \mid^2 = 1$. Such a state is called a *superposition* of the basis vectors.

When we observe or measure a classical bit we determine its state with a probability of 1; the bit is either in state 0 or in state 1 and the result of a measurement is strictly deterministic. On the other hand, when we observe/measure a qubit we get the result

$$\mid 0 \rangle \text{ with probability } \mid \alpha_0 \mid^2$$

$$\mid 1 \rangle \text{ with probability } \mid \alpha_1 \mid^2 .$$

For these statements to be true we need the vector length, or the *norm of the vector* to be equal to 1, otherwise the probabilities do not sum to unity. This means that

$$\mid \alpha_0 \mid^2 + \mid \alpha_1 \mid^2 = \alpha_0^* \alpha_0 \ + \ \alpha_1^* \alpha_1 = 1.$$

We say that a qubit is in a continuum of states between $\mid 0 \rangle$ and $\mid 1 \rangle$ until we measure it. For example, a qubit can be in state

$$\frac{1}{2} \mid 0 \rangle \ + \ \frac{\sqrt{3}}{2} \mid 1 \rangle$$

and a measurement of the qubit yields the result $\mid 0 \rangle$ with probability 1/4 and $\mid 1 \rangle$ with probability 3/4.

The superposition and the effect of the measurement of a quantum state (the state of the qubit) really mean that there is *hidden information* that is preserved in a *closed quantum system* until it is forced to reveal itself to an external observer. We say that the system is closed until it interacts with the outside world (e.g., until we perform an observation of the system). An important question is how to use this hidden information.

So far we have used the vectors $| 0\rangle$ and $| 1\rangle$ to represent a qubit. But this is one of many choices; we can choose a different set of vectors as an orthonormal basis. For example, we can choose as the basis vectors

$$| +\rangle \equiv \frac{| 0\rangle + | 1\rangle}{\sqrt{2}}$$

and

$$| -\rangle \equiv \frac{| 0\rangle - | 1\rangle}{\sqrt{2}}.$$

In this case, a qubit state can be represented as

$$| \psi\rangle = \alpha_0 \, | 0\rangle \; + \; \alpha_1 \, | 1\rangle = \alpha_0 \frac{| +\rangle + | -\rangle}{\sqrt{2}} \; + \; \alpha_1 \frac{| +\rangle - | -\rangle}{\sqrt{2}},$$

$$| \psi\rangle = \frac{\alpha_0 \; + \; \alpha_1}{\sqrt{2}} | +\rangle \; + \; \frac{\alpha_0 \; - \; \alpha_1}{\sqrt{2}} | -\rangle.$$

Let us discuss briefly the transformations of a qubit by means of operators[1] represented by the Pauli matrices. The traditional notations for the Pauli matrices in quantum mechanics are: σ_x (also denoted as σ_1 or X); σ_y (also denoted as σ_2 or Y); and σ_z (also denoted as σ_3 or Z). The notation σ_0 for I, the identity matrix, is sometimes encountered in the literature.

Let us now reinforce our understanding of the formalism of the Dirac `ket` and `bra` notations, by deriving the Pauli matrices from the descriptions of the transformations they perform. Throughout this derivation we use the outer product of vectors defined in Section 2.7.

We start with the basics and show the basis vectors and their duals in matrix notation

$$| 0\rangle = \begin{pmatrix} 1 \\ 0 \end{pmatrix} \quad \text{and} \quad \langle 0 | = \begin{pmatrix} 1 & 0 \end{pmatrix};$$

$$| 1\rangle = \begin{pmatrix} 0 \\ 1 \end{pmatrix} \quad \text{and} \quad \langle 1 | = \begin{pmatrix} 0 & 1 \end{pmatrix}.$$

The identity matrix, I, performs an identity transformation of the two basis vectors, $| 0\rangle \mapsto | 0\rangle$ and $| 1\rangle \mapsto | 1\rangle$. This can be written as:

$$I = | 0\rangle\langle 0 | + | 1\rangle\langle 1 | = \begin{pmatrix} 1 \\ 0 \end{pmatrix} \begin{pmatrix} 1 & 0 \end{pmatrix} + \begin{pmatrix} 0 \\ 1 \end{pmatrix} \begin{pmatrix} 0 & 1 \end{pmatrix}.$$

Thus,

$$I = \begin{pmatrix} 1 & 0 \\ 0 & 0 \end{pmatrix} + \begin{pmatrix} 0 & 0 \\ 0 & 1 \end{pmatrix} = \begin{pmatrix} 1 & 0 \\ 0 & 1 \end{pmatrix}.$$

[1] The terms operator and matrix are used interchangeably throughout this chapter and the rest of the book.

The X matrix transforms a qubit as follows: it flips, or negates it, $| 0\rangle \mapsto | 1\rangle$ and $| 1\rangle \mapsto | 0\rangle$. This can be written as:

$$X = | 0\rangle\langle 1 | + | 1\rangle\langle 0 | = \begin{pmatrix} 1 \\ 0 \end{pmatrix}\begin{pmatrix} 0 & 1 \end{pmatrix} + \begin{pmatrix} 0 \\ 1 \end{pmatrix}\begin{pmatrix} 1 & 0 \end{pmatrix}.$$

Thus,

$$X = \begin{pmatrix} 0 & 1 \\ 0 & 0 \end{pmatrix} + \begin{pmatrix} 0 & 0 \\ 1 & 0 \end{pmatrix} = \begin{pmatrix} 0 & 1 \\ 1 & 0 \end{pmatrix}.$$

The Y matrix transforms a qubit as follows: it multiplies it by i and then flips or negates it, $| 0\rangle \mapsto i\,| 1\rangle$ and $| 1\rangle \mapsto -i\,| 0\rangle$. This can be written as:

$$Y = -i\,| 0\rangle\langle 1 | + i\,| 1\rangle\langle 0 | = -i\begin{pmatrix} 1 \\ 0 \end{pmatrix}\begin{pmatrix} 0 & 1 \end{pmatrix} + i\begin{pmatrix} 0 \\ 1 \end{pmatrix}\begin{pmatrix} 1 & 0 \end{pmatrix}.$$

Thus,

$$Y = -i\begin{pmatrix} 0 & 1 \\ 0 & 0 \end{pmatrix} + i\begin{pmatrix} 0 & 0 \\ 1 & 0 \end{pmatrix} = \begin{pmatrix} 0 & -i \\ i & 0 \end{pmatrix}.$$

The Z transformation requires a phase shift operation, $| 0\rangle \mapsto | 0\rangle$ and $| 1\rangle \mapsto -\,| 1\rangle$. This can be written as:

$$Z = | 0\rangle\langle 0 | - | 1\rangle\langle 1 | = \begin{pmatrix} 1 \\ 0 \end{pmatrix}\begin{pmatrix} 1 & 0 \end{pmatrix} - \begin{pmatrix} 0 \\ 1 \end{pmatrix}\begin{pmatrix} 0 & 1 \end{pmatrix}.$$

Thus,

$$Z = \begin{pmatrix} 1 & 0 \\ 0 & 0 \end{pmatrix} - \begin{pmatrix} 0 & 0 \\ 0 & 1 \end{pmatrix} = \begin{pmatrix} 1 & 0 \\ 0 & -1 \end{pmatrix}.$$

The Hadamard matrix H is also used frequently:

$$H = \frac{1}{\sqrt{2}}\begin{pmatrix} 1 & 1 \\ 1 & -1 \end{pmatrix}.$$

We present now the I matrix, the Pauli matrices, and the result of the transformation of the original state of the qubit, $| \psi\rangle = \alpha_0\,| 0\rangle + \alpha_1\,| 1\rangle$, into a new state , $| \varphi\rangle = \sigma\,| \psi\rangle$ under their action.

$$\sigma_0 = I = \begin{pmatrix} 1 & 0 \\ 0 & 1 \end{pmatrix}, \quad | \varphi\rangle = \sigma_0\,| \psi\rangle = \begin{pmatrix} 1 & 0 \\ 0 & 1 \end{pmatrix}\begin{pmatrix} \alpha_0 \\ \alpha_1 \end{pmatrix} = \begin{pmatrix} \alpha_0 \\ \alpha_1 \end{pmatrix},$$
$$| \varphi\rangle = \alpha_0\,| 0\rangle + \alpha_1\,| 1\rangle;$$

$$\sigma_x = X = \begin{pmatrix} 0 & 1 \\ 1 & 0 \end{pmatrix}, \quad | \varphi\rangle = \sigma_x\,| \psi\rangle = \begin{pmatrix} 0 & 1 \\ 1 & 0 \end{pmatrix}\begin{pmatrix} \alpha_0 \\ \alpha_1 \end{pmatrix} = \begin{pmatrix} \alpha_1 \\ \alpha_0 \end{pmatrix},$$
$$| \varphi\rangle = \alpha_1\,| 0\rangle + \alpha_0\,| 1\rangle;$$

$$\sigma_y = Y = \begin{pmatrix} 0 & -i \\ i & 0 \end{pmatrix}, \quad | \varphi\rangle = \sigma_y\,| \psi\rangle = \begin{pmatrix} 0 & -i \\ i & 0 \end{pmatrix}\begin{pmatrix} \alpha_0 \\ \alpha_1 \end{pmatrix} = i\begin{pmatrix} -\alpha_1 \\ \alpha_0 \end{pmatrix},$$
$$| \varphi\rangle = -i\alpha_1\,| 0\rangle + i\alpha_0\,| 1\rangle;$$

$$\sigma_z = Z = \begin{pmatrix} 1 & 0 \\ 0 & -1 \end{pmatrix}, \quad | \varphi\rangle = \sigma_z\,| \psi\rangle = \begin{pmatrix} 1 & 0 \\ 0 & -1 \end{pmatrix}\begin{pmatrix} \alpha_0 \\ \alpha_1 \end{pmatrix} = i\begin{pmatrix} \alpha_0 \\ -\alpha_1 \end{pmatrix},$$
$$| \varphi\rangle = \alpha_0\,| 0\rangle - \alpha_1\,| 1\rangle.$$

It is easy to determine if any two of these operators commute[2] under multiplication. For example, we see that X and Y do not commute. Indeed, the commutator of X and Y is non-zero

$$[X, Y] = XY - YX$$
$$= \begin{pmatrix} 0 & 1 \\ 1 & 0 \end{pmatrix} \begin{pmatrix} 0 & -i \\ i & 0 \end{pmatrix} - \begin{pmatrix} 0 & -i \\ i & 0 \end{pmatrix} \begin{pmatrix} 0 & 1 \\ 1 & 0 \end{pmatrix} = 2i \begin{pmatrix} 1 & 0 \\ 0 & -1 \end{pmatrix} = 2iZ.$$

It is equally trivial to show that

$$[Y, Z] = 2iX, \quad [Z, X] = 2iY.$$

Now let us briefly mention several physical realizations of a qubit. A qubit can be realized as the polarization of a photon as we have seen in the example discussed in Section 1.1. A laser and a polarizing lens form a source of polarized photons. Another possible physical realization of a qubit is the spin of the electron with states "up" or "down."

The energy "ground" state and an "excited" state of a bound electron may be another physical embodiment for a qubit. In this case, assuming that an electron or an atom is in the"ground" state one can provide enough energy (e.g., by shining laser light) to move the electron from this "ground" state to an "excited" state; we change the qubit state in this way. The physical methods of constructing a qubit are discussed in more detail later in this chapter.

3.2 THE BLOCH SPHERE REPRESENTATION OF ONE QUBIT

It is always useful to have a graphic representation of an abstract concept and this is what we are going to attempt next. First, we express the state of a qubit using three real numbers θ, φ, γ, as

$$| \psi \rangle = e^{i\gamma} \left[\cos \frac{\theta}{2} \mid 0 \rangle + e^{i\varphi} \sin \frac{\theta}{2} \mid 1 \rangle \right] = \alpha_0 \mid 0 \rangle + \alpha_1 \mid 1 \rangle$$

with

$$\alpha_0 = e^{i\gamma} \cos \frac{\theta}{2} \qquad \alpha_1 = e^{i\gamma} e^{i\varphi} \sin \frac{\theta}{2}.$$

In this representation $e^{i\gamma}$ is an overall phase factor that is not observable; it is generally ignored in our calculations. The normalization condition

$$| \alpha_0 |^2 + | \alpha_1 |^2 = 1.$$

can be easily verified when we substitute the expressions for α_0 and α_1 given above

$$| e^{i\gamma} \cos \tfrac{\theta}{2} |^2 + | e^{i\gamma} e^{i\varphi} \sin \tfrac{\theta}{2} |^2 = | e^{i\gamma} |^2 \cos^2 \tfrac{\theta}{2} + | e^{i\gamma} |^2 | e^{i\varphi} |^2 \sin^2 \tfrac{\theta}{2}$$
$$= \cos^2 \tfrac{\theta}{2} + \sin^2 \tfrac{\theta}{2} = 1.$$

[2]The concept of commutativity of two operators was introduced in Section 2.4. Recall that in the general case matrix multiplication is non-commutative. Yet, the product of any pair of square diagonal matrices with real or complex elements and with the same number of rows/columns does commute.

Here, we take into account that

$$| e^{i\gamma} |^2 = | e^{i\varphi} |^2 = 1.$$

This representation lends itself to an interesting geometrical interpretation: the state of a qubit $| \varphi \rangle$ is a vector $\mathbf{r}$ from the origin to a point on the three-dimensional sphere with a radius of one, called the *Bloch sphere*, see Figure 3.1. In this representation, the position of a point is defined by θ, the angle between the vector $\mathbf{r}$ and the z axis, and φ, the angle between the projection of the vector in the xy plane and the x axis.

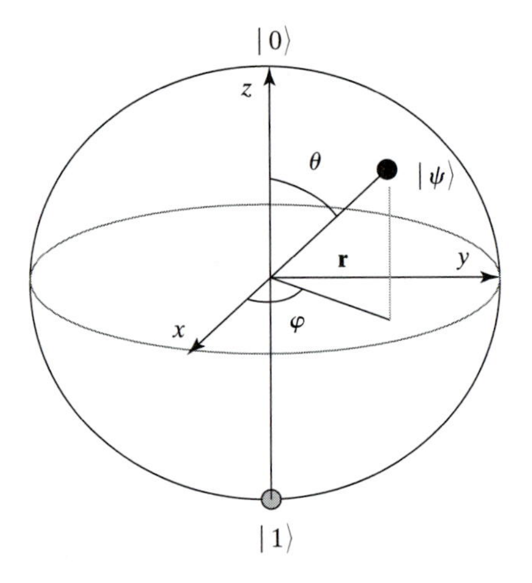

FIGURE 3.1: A qubit $| \psi \rangle$ is represented as a vector $\mathbf{r}$ from the origin to a point on the sphere with a radius of one, the Bloch sphere. θ is the angle of the vector $\mathbf{r}$ with the z axis and φ is the angle of the projection of the vector in the xy plane with the x axis; γ has no observable effect.

Figure 3.2 shows the Bloch sphere representation of one qubit in the superposition state $| \psi \rangle = (| 0 \rangle + | 1 \rangle)/\sqrt{2}$. In this case $\alpha_0 = \alpha_1 = 1/\sqrt{2}$. Then we see that

$$\alpha_0 = \cos\frac{\theta}{2} = \frac{1}{\sqrt{2}} \Longrightarrow \frac{\theta}{2} = 45 \quad \text{thus} \quad \theta = 90^\circ$$

and

$$\alpha_1 = e^{i\varphi}\sin\frac{\theta}{2} = \frac{1}{\sqrt{2}} \Longrightarrow e^{i\varphi} = 1 \quad \text{thus} \quad \varphi = 0^\circ .$$

A classical bit can be in one of two states, 0 or 1, see Figure 3.3(a), while a qubit can be in a continuum of states represented as points on the Bloch sphere, Figure 3.3(b). The state space of a qubit contains the two "basis" or "logical", states, $| 0 \rangle$ and $| 1 \rangle$. The initial state of a qubit is always assumed to be one of the basis states.

The classical information revealed by one qubit is the label of one of the two basis states, while the contents of the quantum information encoded to a qubit is considerably richer.

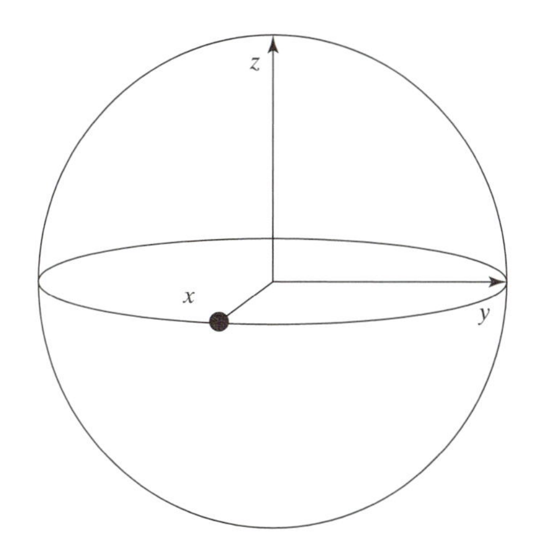

FIGURE 3.2: A qubit in a superposition state $| \psi\rangle = (| 0\rangle + | 1\rangle)/\sqrt{2}$. In this case $\theta = 90^\circ$ and $\varphi = 0^\circ$.

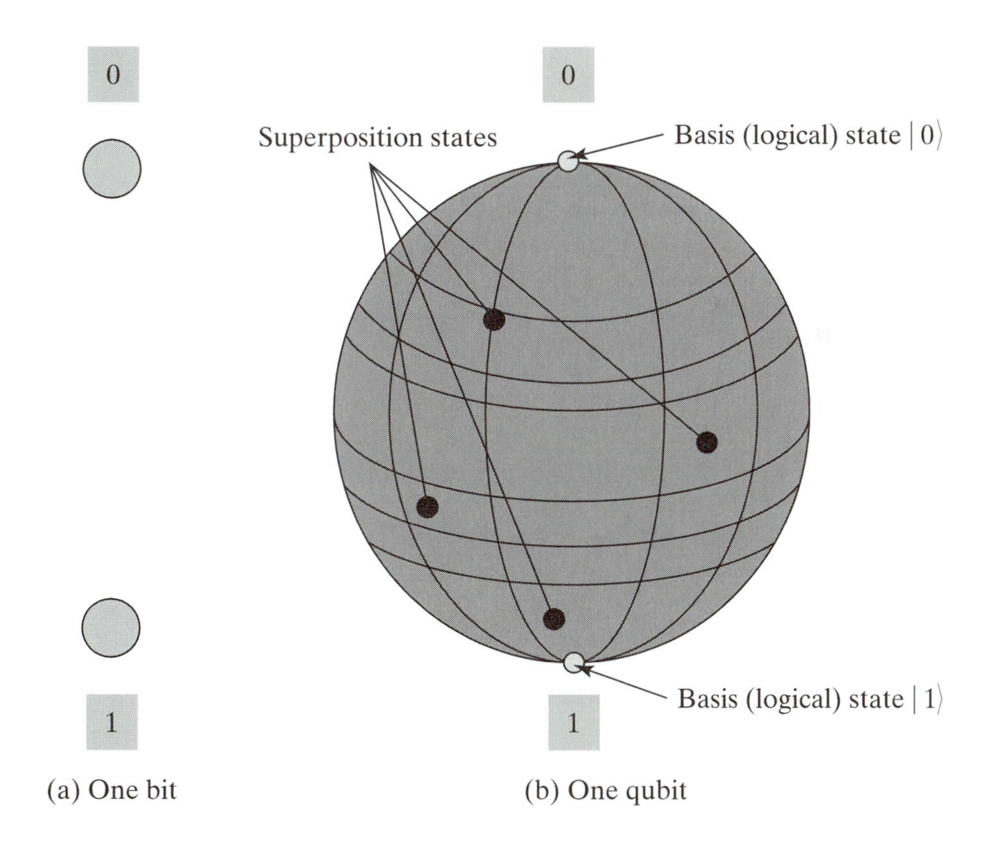

FIGURE 3.3: A bit can be in one of two states, 0 or 1. The state $| \psi\rangle$ of a qubit is represented as a vector from the origin to a point on the Bloch sphere. A qubit can be in a *basis state*, $| 0\rangle$ or $| 1\rangle$, or in a *superposition state* $| \psi\rangle = \alpha_0 | 0\rangle + \alpha_1 | 1\rangle$ with $| \alpha_0 |^2 + | \alpha_1 |^2 = 1$.

3.3 ROTATION OPERATIONS ON THE BLOCH SPHERE

A transformation of the state of one qubit performed by one of the quantum gates discussed in Chapter 4 corresponds to a rotation of the qubit. This motivates our interest in the rotation operators.

Single qubit operations are defined as rotations on the Bloch sphere. When a qubit is represented as a spin one-half particle (discussed in more detail in Section 3.9), the Pauli spin matrices, $\sigma = \{\sigma_x, \sigma_y, \sigma_z\}$, introduced in the preceding section, describe spin rotations on the Bloch sphere. Before we derive the rotation matrices we first discuss matrix exponentiation. If β is a real number and if matrix A is such that $A^2 = I$ then

$$e^{i\beta A} = \cos(\beta)I + i\sin(\beta)A.$$

Recall the following Taylor series expansions for $x \in \mathbb{R}$ [65]

$$e^x = 1 + x + \frac{x^2}{2!} + \ldots = \sum_{k=0}^{\infty} \frac{x^k}{k!}$$

$$\sin x = x - \frac{x^3}{3!} + \frac{x^5}{5!} + \ldots = \sum_{k=0}^{\infty} (-1)^k \frac{x^{2k+1}}{(2k+1)!}$$

$$\cos x = 1 - \frac{x^2}{2!} + \frac{x^4}{4!} + \ldots = \sum_{k=0}^{\infty} (-1)^k \frac{x^{2k}}{(2k)!}.$$

Given a square matrix A we can expand e^A as

$$e^A = I + A + \frac{A^2}{2!} + \frac{A^3}{3!} + \ldots + \frac{A^k}{k!} + \ldots.$$

Therefore,

$$e^{i\beta A} = I + (i\beta A) + \frac{(i\beta A)^2}{2!} + \frac{(i\beta A)^3}{3!} + \ldots + \frac{(i\beta A)^k}{k!} + \ldots.$$

But $A^2 = I$ and $i = \sqrt{-1}$ and we can group the terms of the sum as follows

$$\left(1 - \frac{\beta^2}{2!} + \frac{\beta^4}{4!} + \ldots + (-1)^k \frac{(\beta)^{2k}}{(2k)!} + \ldots\right) I + i\left(\beta - \frac{\beta^3}{3!} + \frac{\beta^5}{5!} + \ldots (-1)^k \frac{(\beta)^{2k+1}}{(2k+1)!} + \ldots\right) A.$$

Thus,

$$e^{i\beta A} = \cos(\beta)I + i\sin(\beta)A.$$

If $\boldsymbol{\sigma}$ is a Pauli matrix, then a finite rotation through an angle β about a given vector $\mathbf{n}$ on the Bloch sphere is defined as a linear operator with matrix representation

$$\mathcal{R}_{\mathbf{n}}(\beta) = \exp(-i(\beta/2)\mathbf{n} \cdot \boldsymbol{\sigma}) = \cos(\beta/2)I - i\sin(\beta/2)\mathbf{n} \cdot \boldsymbol{\sigma}$$

with $\mathbf{n} \cdot \boldsymbol{\sigma}$ the scalar product of the two vectors. In this expression $\mathbf{n} = (n_x, n_y, n_z)$ and $\boldsymbol{\sigma} = (\sigma_x, \sigma_y, \sigma_z)$ with

$$\sigma_x = \begin{pmatrix} 0 & 1 \\ 1 & 0 \end{pmatrix} \quad \sigma_y = \begin{pmatrix} 0 & -i \\ i & 0 \end{pmatrix} \quad \sigma_z = \begin{pmatrix} 1 & 0 \\ 0 & -1 \end{pmatrix}.$$

The rotations about the x, y, and z axes with the same angle β, denoted as $\mathcal{R}_x(\beta)$, $\mathcal{R}_y(\beta)$, and $\mathcal{R}_z(\beta)$, respectively, are computed observing that $\mathbf{n} \cdot \boldsymbol{\sigma}$ becomes $[(1,0,0) \cdot (\sigma_x, 0, 0)]$, $[(0,1,0) \cdot (0, \sigma_y, 0)]$, and $[(0,0,1) \cdot (0,0,\sigma_z)]$ respectively.

$$\begin{aligned} \mathcal{R}_x(\beta) &= \cos(\beta/2)I \;-\; i\sin(\beta/2)\sigma_x \\ &= \begin{pmatrix} \cos(\beta/2) & 0 \\ 0 & \cos(\beta/2) \end{pmatrix} + \begin{pmatrix} 0 & -i\sin(\beta/2) \\ -i\sin(\beta/2) & 0 \end{pmatrix}. \end{aligned}$$

Thus,

$$\mathcal{R}_x(\beta) = \begin{pmatrix} \cos(\beta/2) & -i\sin(\beta/2) \\ -i\sin(\beta/2) & \cos(\beta/2) \end{pmatrix}.$$

Now

$$\begin{aligned} \mathcal{R}_y(\beta) &= \cos(\beta/2)I \;-\; i\sin(\beta/2)\sigma_y \\ &= \begin{pmatrix} \cos(\beta/2) & 0 \\ 0 & \cos(\beta/2) \end{pmatrix} + \begin{pmatrix} 0 & -\sin(\beta/2) \\ \sin(\beta/2) & 0 \end{pmatrix}. \end{aligned}$$

Thus,

$$\mathcal{R}_y(\beta) = \begin{pmatrix} \cos(\beta/2) & -\sin(\beta/2) \\ \sin(\beta/2) & \cos(\beta/2) \end{pmatrix}.$$

Finally,

$$\begin{aligned} \mathcal{R}_z(\beta) &= \cos(\beta/2)I \;-\; i\sin(\beta/2)\sigma_z \\ &= \begin{pmatrix} \cos(\beta/2) & 0 \\ 0 & \cos(\beta/2) \end{pmatrix} + \begin{pmatrix} -i\sin(\beta/2) & 0 \\ 0 & i\sin(\beta/2) \end{pmatrix}. \end{aligned}$$

Thus,

$$\mathcal{R}_z(\beta) = \begin{pmatrix} \cos(\beta/2) - i\sin(\beta/2) & 0 \\ 0 & \cos(\beta/2) + i\sin(\beta/2) \end{pmatrix} = \begin{pmatrix} e^{-i\beta/2} & 0 \\ 0 & e^{i\beta/2} \end{pmatrix}.$$

The composition of two rotations with angles δ and β is

$$\mathcal{R}_z(\delta)\mathcal{R}_x(\beta) = \begin{pmatrix} e^{-i\delta/2}\cos(\beta/2) & -ie^{-i\delta/2}\sin(\beta/2) \\ -ie^{i\delta/2}\sin(\beta/2) & e^{i\delta/2}\cos(\beta/2) \end{pmatrix}.$$

Any rotation on the Bloch sphere can be reduced to the previous expression for some angles δ and β.

It is easy to see that the rotation operations are invertible; a rotation about the x, y, or z axis with an angle β followed by a rotation with an angle $-\beta$ about the same axis is equivalent to an identity transformation, it leaves the qubit in the original state

$$\mathcal{R}_x(\beta)\mathcal{R}_x(-\beta) = I \qquad \mathcal{R}_y(\beta)\mathcal{R}_y(-\beta) = I \qquad \mathcal{R}_z(\beta)\mathcal{R}_z(-\beta) = I.$$

Recall that $\cos\beta$ is an even function, while $\sin\beta$ is an odd function[3]. We only show that the first equality (the rotation about the x axis) holds:

$$\mathcal{R}_x(\beta)\mathcal{R}_x(-\beta) = \begin{pmatrix} \cos(\beta/2) & -i\sin(\beta/2) \\ -i\sin(\beta/2) & \cos(\beta/2) \end{pmatrix} \begin{pmatrix} \cos(\beta/2) & i\sin(\beta/2) \\ i\sin(\beta/2) & \cos(\beta/2) \end{pmatrix}$$

$$= \begin{pmatrix} \cos^2(\beta/2) + \sin^2(\beta/2) & \begin{matrix} i\cos(\beta/2)\sin(\beta/2) \\ -i\cos(\beta/2)\sin(\beta/2) \end{matrix} \\ \begin{matrix} -i\cos(\beta/2)\sin(\beta/2) \\ +i\cos(\beta/2)\sin(\beta/2) \end{matrix} & \cos^2(\beta/2) + \sin^2(\beta/2) \end{pmatrix}.$$

Thus,

$$\mathcal{R}_x(\beta)\mathcal{R}_x(-\beta) = \begin{pmatrix} 1 & 0 \\ 0 & 1 \end{pmatrix} = I.$$

An alternative method to prove the reversibility of rotation operations[4] is to apply to the qubit in state $|\psi\rangle$

$$|\psi\rangle = \alpha_0 \,|\, 0\rangle + \alpha_1 \,|\, 1\rangle = \cos\theta/2 \,|\, 0\rangle + e^{i\varphi}\sin\theta/2 \,|\, 1\rangle$$

a rotation with an angle β around the x axis, $\mathcal{R}_x(\beta)$, followed by a rotation with an angle $-\beta$ around the x axis, $\mathcal{R}_x(-\beta)$. The result of composing these two transformations should be the original vector $|\psi\rangle$.

$$|\psi^\beta\rangle = \mathcal{R}_x(\beta)\,|\,\psi\rangle = \begin{pmatrix} \cos(\beta/2) & -i\sin(\beta/2) \\ -i\sin(\beta/2) & \cos(\beta/2) \end{pmatrix} \begin{pmatrix} \cos(\theta/2) \\ e^{i\varphi}\sin(\theta/2) \end{pmatrix}.$$

It follows that:

$$|\psi^\beta\rangle = \begin{pmatrix} \cos(\beta/2)\cos(\theta/2) - ie^{i\varphi}\sin(\beta/2)\sin(\theta/2) \\ -i\sin(\beta/2)\cos(\theta/2) + e^{i\varphi}\cos(\beta/2)\sin(\theta/2) \end{pmatrix}.$$

Now

$$|\,(\psi^\beta)^{-\beta}\rangle = \mathcal{R}_x(-\beta)\,|\,\psi^\beta\rangle$$

[3]Indeed, $\cos\beta = \cos(-\beta)$ and $\sin\beta = -\sin(-\beta)$.

[4]An operator is invertible and a transformation is reversible. A transformation of a quantum system has an operator associated with it and we use the terms invertible and reversible interchangeably throughout the book.

$$| (\psi^{\beta})^{-\beta} \rangle = \begin{pmatrix} \cos(\beta/2) & i\sin(\beta/2) \\ i\sin(\beta/2) & \cos(\beta/2) \end{pmatrix}$$

$$\begin{pmatrix} \cos(\frac{\beta}{2})\cos(\theta/2) - ie^{i\varphi}\sin(\beta/2)\sin(\theta/2) \\ -i\sin(\beta/2)\cos(\theta/2) + e^{i\varphi}\cos(\beta/2)\sin(\theta/2) \end{pmatrix}$$

$$= \begin{pmatrix} \cos^2(\beta/2)\cos(\theta/2) - ie^{i\varphi}\sin(\beta/2)\sin(\theta/2) \\ + \sin^2(\beta/2)\cos(\theta/2) + e^{i\varphi}\sin^2(\beta/2)\sin(\theta/2) \\ \cdots\cdots\cdots\cdots\cdots\cdots\cdots\cdots\cdots\cdots\cdots\cdots \\ i\sin(\beta/2)\cos(\theta/2) + ie^{i\varphi}\cos(\beta/2)\sin(\beta/2)\sin(\theta/2) \\ -i\sin(\beta/2)\cos(\beta/2)\cos(\theta/2) + e^{i\varphi}\cos^2(\beta/2)\sin(\theta/2) \end{pmatrix}.$$

Thus,

$$| (\psi^{\beta})^{-\beta} \rangle = \begin{pmatrix} \cos(\theta/2) \\ e^{i\varphi}\sin(\theta/2) \end{pmatrix} = | \psi \rangle.$$

The composition of two rotations with angles β_1 and β_2 is a rotation with angle $\beta_1 + \beta_2$ about the same axis

$$\mathcal{R}_x(\beta_1)\mathcal{R}_x(\beta_2) = \mathcal{R}_x(\beta_1 + \beta_2),$$
$$\mathcal{R}_y(\beta_1)\mathcal{R}_y(\beta_2) = \mathcal{R}_y(\beta_1 + \beta_2),$$
$$\mathcal{R}_z(\beta_1)\mathcal{R}_z(\beta_2) = \mathcal{R}_z(\beta_1 + \beta_2).$$

We only discuss the second identity and leave the derivation of the others as an exercise. Recall that [65]

$$\sin(\beta_1 \pm \beta_2) = \sin(\beta_1)\cos(\beta_2) \pm \cos(\beta_1)\sin(\beta_2)$$
$$\cos(\beta_1 \pm \beta_2) = \cos(\beta_1)\cos(\beta_2) \mp \sin(\beta_1)\sin(\beta_2).$$

We have

$$\mathcal{R}_y(\beta_1)\mathcal{R}_y(\beta_2) = \begin{pmatrix} \cos(\beta_1/2) & -\sin(\beta_1/2) \\ \sin(\beta_1/2) & \cos(\beta_1/2) \end{pmatrix}\begin{pmatrix} \cos(\beta_2/2) & -\sin(\beta_2/2) \\ \sin(\beta_2/2) & \cos(\beta_2/2) \end{pmatrix}$$
$$= \begin{pmatrix} \cos[(\beta_1 + \beta_2)/2] & -\sin[(\beta_1 + \beta_2)/2] \\ \sin[(\beta_1 + \beta_2)/2] & \cos[(\beta_1 + \beta_2)/2] \end{pmatrix}$$
$$= \mathcal{R}_y(\beta_1 + \beta_2).$$

3.4 THE MEASUREMENT OF A SINGLE QUBIT

Quantum measurements of one qubit in state

$$| \psi \rangle = \alpha_0 | 0 \rangle + \alpha_1 | 1 \rangle$$

are characterized by a set of linear operators with the associated matrices, $\{\mathcal{M}_k\}$. The probability that the outcome with index k occurs as a result of the measurement is

$$p(k) = \langle \psi | \mathcal{M}_k^{\dagger}\mathcal{M}_k | \psi \rangle.$$

The sum of the probabilities of all possible outcomes of the measurement must be equal to 1

$$\sum_k p(k) = \sum_k \langle \psi \mid \mathcal{M}_k^\dagger \mathcal{M}_k \mid \psi \rangle = 1.$$

The measurement causes the qubit to change its state. If the state of the qubit immediately prior to the measurement is $\mid \psi \rangle$, then, after the measurement $\mathcal{M}_k$ it becomes $\mid \varphi_k \rangle$

$$\mid \psi \rangle \mapsto \mid \varphi_k \rangle = \frac{\mathcal{M}_k \mid \psi \rangle}{\sqrt{\langle \psi \mid \mathcal{M}_k^\dagger \mathcal{M}_k \mid \psi \rangle}}.$$

There are only two possible outcomes of a measurement in the $\{\mid 0\rangle, \mid 1\rangle\}$ basis; we can only observe the basis state $\mid 0\rangle$ or $\mid 1\rangle$, see Figure 3.4. The corresponding matrices of the measurement operators are:

$$\mathcal{M}_0 = \mid 0\rangle\langle 0 \mid = \begin{pmatrix} 1 \\ 0 \end{pmatrix} (1\ 0) = \begin{pmatrix} 1 & 0 \\ 0 & 0 \end{pmatrix} \quad \text{and}$$

$$\mathcal{M}_1 = \mid 1\rangle\langle 1 \mid = \begin{pmatrix} 0 \\ 1 \end{pmatrix} (0\ 1) = \begin{pmatrix} 0 & 0 \\ 0 & 1 \end{pmatrix}.$$

It is easy to see that these two operators are Hermitian and that $\mathcal{M}_0^\dagger \mathcal{M}_0 = \mathcal{M}_0$ and $\mathcal{M}_1^\dagger \mathcal{M}_1 = \mathcal{M}_1$,

$$\mathcal{M}_0^\dagger = \begin{pmatrix} 1 & 0 \\ 0 & 0 \end{pmatrix} = \mathcal{M}_0, \quad \mathcal{M}_0^\dagger \mathcal{M}_0 = \mathcal{M}_0^2 = \begin{pmatrix} 1 & 0 \\ 0 & 0 \end{pmatrix} \begin{pmatrix} 1 & 0 \\ 0 & 0 \end{pmatrix} = \begin{pmatrix} 1 & 0 \\ 0 & 0 \end{pmatrix} = \mathcal{M}_0$$

and

$$\mathcal{M}_1^\dagger = \begin{pmatrix} 0 & 0 \\ 0 & 1 \end{pmatrix} = \mathcal{M}_1, \quad \mathcal{M}_1^\dagger \mathcal{M}_1 = \mathcal{M}_1^2 = \begin{pmatrix} 0 & 0 \\ 0 & 1 \end{pmatrix} \begin{pmatrix} 0 & 0 \\ 0 & 1 \end{pmatrix} = \begin{pmatrix} 0 & 0 \\ 0 & 1 \end{pmatrix} = \mathcal{M}_1.$$

The probability of the outcome corresponding to $\mid 0\rangle$ is:

$$p_0 = \langle \psi \mid \mathcal{M}_0^\dagger \mathcal{M}_0 \mid \psi \rangle = \langle \psi \mid \mathcal{M}_0 \mid \psi \rangle.$$

We calculate first the effect of applying the operator $\mathcal{M}_0$ to the qubit state $\mid \psi \rangle$:

$$\mathcal{M}_0 \mid \psi \rangle = \begin{pmatrix} 1 & 0 \\ 0 & 0 \end{pmatrix} \begin{pmatrix} \alpha_0 \\ \alpha_1 \end{pmatrix} = \begin{pmatrix} \alpha_0 \\ 0 \end{pmatrix}.$$

The bra vector $\langle \psi \mid = (\alpha_0^* \quad \alpha_1^*)$ is the dual of the original vector, $\mid \psi \rangle$, thus

$$p_0 = \langle \psi \mid (\mathcal{M}_0 \mid \psi \rangle) = (\alpha_0^* \quad \alpha_1^*) \begin{pmatrix} \alpha_0 \\ 0 \end{pmatrix} = \mid \alpha_0 \mid^2.$$

But

$$\mathcal{M}_0 \mid \psi \rangle = \begin{pmatrix} \alpha_0 \\ 0 \end{pmatrix} = \alpha_0 \begin{pmatrix} 1 \\ 0 \end{pmatrix} = \alpha_0 \mid 0 \rangle.$$

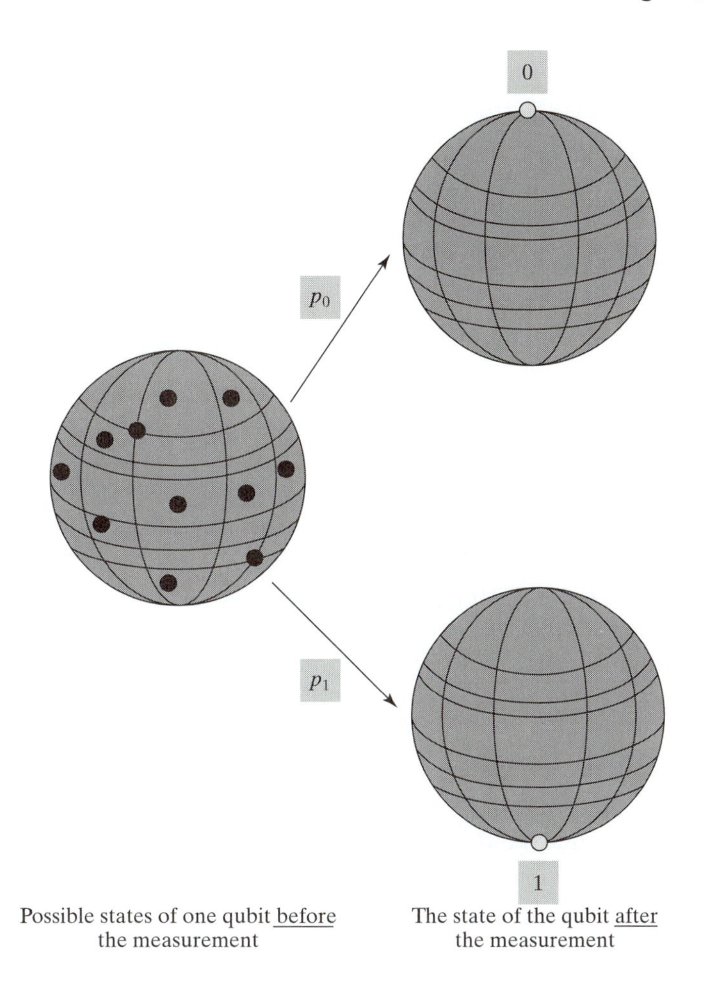

FIGURE 3.4: The effect of a measurement upon the state of one qubit. A measurement forces a qubit in a superposition state to one of the two basis states. A qubit in the superposition state $| \psi\rangle = \alpha_0 | 0\rangle + \alpha_1 | 1\rangle$ behaves like $| 0\rangle$ with probability $p_0 =| \alpha_0 |^2$ and like $| 1\rangle$ with probability $p_1 =| \alpha_1 |^2$.

Thus, the state of the qubit after applying the measurement operator $\mathcal{M}_0$ is:

$$| \varphi_0\rangle = \frac{\mathcal{M}_0 | \psi\rangle}{\sqrt{\langle\psi | \mathcal{M}_0^\dagger \mathcal{M}_0 | \psi\rangle}} = \frac{\alpha_0 | 0\rangle}{| \alpha_0 |} = \frac{\alpha_0}{| \alpha_0 |} | 0\rangle.$$

Similarly, the probability of the outcome corresponding to $| 1\rangle$ is:

$$p_1 = \langle\psi | \mathcal{M}_1^\dagger \mathcal{M}_1 | \psi\rangle = \langle\psi | \mathcal{M}_1 | \psi\rangle.$$

The effect of applying the operator $\mathcal{M}_1$ to the state $| \psi\rangle$ is:

$$\mathcal{M}_1 | \psi\rangle = \begin{pmatrix} 0 & 0 \\ 0 & 1 \end{pmatrix} \begin{pmatrix} \alpha_0 \\ \alpha_1 \end{pmatrix} = \begin{pmatrix} 0 \\ \alpha_1 \end{pmatrix}$$

or

$$\mathcal{M}_1 \mid \psi\rangle = \alpha_1 \mid 1\rangle.$$

Then

$$p_1 = \langle\psi \mid (\mathcal{M}_1 \mid \psi\rangle) = \begin{pmatrix} \alpha_0^* & \alpha_1^* \end{pmatrix} \begin{pmatrix} 0 \\ \alpha_1 \end{pmatrix} = \mid \alpha_1 \mid^2 .$$

The state of the qubit after applying the measurement operator $\mathcal{M}_1$ is

$$\mid \varphi_1\rangle = \frac{\mathcal{M}_1 \mid \psi\rangle}{\sqrt{\langle\psi \mid \mathcal{M}_1^\dagger \mathcal{M}_1 \mid \psi\rangle}} = \frac{\alpha_1 \mid 1\rangle}{\mid \alpha_1 \mid} = \frac{\alpha_1}{\mid \alpha_1 \mid} \mid 1\rangle.$$

The *completeness condition* is satisfied, since

$$p(0) + p(1) = \mid \alpha_0 \mid^2 + \mid \alpha_1 \mid^2 = 1.$$

3.5 PURE AND IMPURE STATES OF A QUBIT

We can define the Bloch sphere more rigourously using the density matrix of a single qubit. The elements of the density matrix represent the probabilities of the possible outcomes of a measurement performed on a quantum system. The *density operator* is

$$\rho = \sum_\mu p_\mu \mid \psi_\mu\rangle$$

with p_μ the probability of state $\mid \psi_\mu\rangle$. The individual states in the expression of the density matrix are

$$\mid \psi_\mu\rangle = \sum_i \alpha_{\mu i} \mid \varphi_i\rangle$$

with $\mid \varphi_i\rangle$ the basis vectors for some observable operator with matrix representation M.

The density operator is Hermitian, $\rho = \rho^\dagger$. Indeed, for a pure state $\mid \psi\rangle$

$$\rho = \mid \psi\rangle\langle\psi \mid$$

and for a mixed, or superposition state, ρ is the sum of Hermitian operators, thus it is Hermitian. The trace of the density matrix is $\mathrm{Tr}(\rho) = 1$. The trace of a matrix is the sum of its main diagonal elements. Indeed, the trace is

$$\mathrm{Tr}(\rho) = \sum_i p_i = 1.$$

Using these constraints, it is relatively easy to prove that the most general form of the density matrix for a single qubit is a 2×2 matrix obtained as the linear combination of Pauli matrices $(\sigma_x, \sigma_y, \sigma_z)$ and the identity matrix (I) with real coefficients $(\beta_x, \beta_y, \beta_z)$

$$\rho = \frac{1}{2}(I + \beta_x\sigma_x + \beta_y\sigma_y + \beta_z\sigma_z) = \frac{1}{2}\begin{pmatrix} 1 + \beta_z & \beta_x - i\beta_y \\ \beta_x + i\beta_y & 1 - \beta_z \end{pmatrix}.$$

The density matrix must have a non-negative determinant:

$$\begin{vmatrix} 1 + \beta_z & \beta_x - i\beta_y \\ \beta_x + i\beta_y & 1 - \beta_z \end{vmatrix} \geqslant 0.$$

This means that:

$$(1 + \beta_z)(1 - \beta_z) - (\beta_x - i\beta_y)(\beta_x + i\beta_y) = 1 - \beta_x^2 - \beta_y^2 - \beta_z^2 \geqslant 0.$$

In turn, this inequality can be written as

$$\beta^2 = \beta_x^2 + \beta_y^2 + \beta_z^2 \leqslant 1.$$

The fact the determinant of the density matrix is equal to $1 - \beta^2$ implies that there is a one-to-one correspondence between the possible density matrices of a single qubit and the points on the sphere $\beta \leqslant 1$, the Bloch sphere. At the same time we get a criteria for distinguishing the *pure* states of a single qubit from the *impure* or *mixed states*. Recall that a quantum system whose state $|\psi\rangle$ is not known precisely is said to be in a mixed state. A mixed state is a superposition of different pure states.

Let us first compute ρ^2:

$$\rho^2 = \frac{1}{2}\begin{pmatrix} 1 + \beta_z & \beta_x - i\beta_y \\ \beta_x + i\beta_y & 1 - \beta_z \end{pmatrix}\frac{1}{2}\begin{pmatrix} 1 + \beta_z & \beta_x - i\beta_y \\ \beta_x + i\beta_y & 1 - \beta_z \end{pmatrix}.$$

Then

$$\rho^2 = \frac{1}{4}\begin{pmatrix} 1 + \beta_x^2 + \beta_y^2 + \beta_z^2 + 2\beta_z & 2(\beta_x - i\beta_y) \\ 2(\beta_x + i\beta_y) & 1 + \beta_x^2 + \beta_y^2 + \beta_z^2 - 2\beta_z \end{pmatrix}.$$

Now

$$\begin{aligned} \mathrm{Tr}(\rho^2) &= \frac{1}{4}[(1 + \beta_x^2 + \beta_y^2 + \beta_z^2 + 2\beta_z) + (1 + \beta_x^2 + \beta_y^2 + \beta_z^2 - 2\beta_z)] \\ &= \frac{1}{2}(1 + \beta^2). \end{aligned}$$

Pure states are represented by points on the Bloch sphere and in that case $\beta^2 = 1$ *and this implies that* $\mathrm{Tr}(\rho^2) = 1$. *Impure states are represented by points inside the sphere,* $\beta^2 < 1$ *and this implies that* $\mathrm{Tr}(\rho^2) < 1$.

3.6 A PAIR OF QUBITS—ENTANGLEMENT

We consider now a system of two qubits in a four-dimensional complex vector space $\mathcal{H}_4$. In this Hilbert space of dimension four we choose as a basis the four vectors $|00\rangle$, $|01\rangle$, $|10\rangle$, and $|11\rangle$. A vector describing the state $|\psi\rangle$ of a two-qubit system is a linear combination of the basis vectors with complex coefficients $\alpha_{00}, \alpha_{01}, \alpha_{10}$, and α_{11}

$$|\psi\rangle = \alpha_{00}\,|00\rangle + \alpha_{01}\,|01\rangle + \alpha_{10}\,|10\rangle + \alpha_{11}\,|11\rangle.$$

When we measure (observe) a pair of qubits we project the state $| \psi \rangle$ of the system to one of the four basis states $| 00 \rangle$, $| 01 \rangle$, $| 10 \rangle$ and $| 11 \rangle$ with corresponding probabilities $| \alpha_{00} |^2$, $| \alpha_{01} |^2$, $| \alpha_{10} |^2$, or $| \alpha_{11} |^2$, respectively. The normalization condition reflects the fact that the norm of the vector $| \psi \rangle$ is one, thus the sum of probabilities must also be one:

$$| \alpha_{00} |^2 + | \alpha_{01} |^2 + | \alpha_{10} |^2 + | \alpha_{11} |^2 = 1.$$

Note that before the measurement, the state of the two qubits is uncertain. After the measurement the state is certain, it is $| 00 \rangle$, $| 01 \rangle$, $| 10 \rangle$, or $| 11 \rangle$, similar to the case of a classical two-bit system, which takes one of the four values, 00, 01, 10, or 11.

But nobody forces us to observe both qubits. What happens if we observe only the first qubit, what conclusions can we draw? Intuitively, we expect the system to be left in an uncertain state, because we did not measure the second qubit which can still be in a continuum of states. The first qubit can be 0 with probability

$$p_0^I = | \alpha_{00} |^2 + | \alpha_{01} |^2$$

or 1 with probability

$$p_1^I = | \alpha_{10} |^2 + | \alpha_{11} |^2 .$$

The normalization condition is satisfied, the sum of the two probabilities is unity

$$p_0^I + p_1^I = 1.$$

Call $| \psi_0^I \rangle$ the post-measurement state when the first qubit is measured to be 0 and $| \psi_1^I \rangle$ when the first qubit is measured to be 1. The two post-measurement states are

$$| \psi_0^I \rangle = \frac{\alpha_{00} | 00 \rangle + \alpha_{01} | 01 \rangle}{\sqrt{| \alpha_{00} |^2 + | \alpha_{01} |^2}}$$

and

$$| \psi_1^I \rangle = \frac{\alpha_{10} | 10 \rangle + \alpha_{11} | 11 \rangle}{\sqrt{| \alpha_{10} |^2 + | \alpha_{11} |^2}}.$$

Now let us measure the second qubit only. The second qubit can be 0 with probability

$$p_0^{II} = | \alpha_{00} |^2 + | \alpha_{10} |^2$$

or 1 with probability

$$p_1^{II} = | \alpha_{01} |^2 + | \alpha_{11} |^2 .$$

Again, the normalization condition requires that

$$p_0^{II} + p_1^{II} = 1.$$

The two post-measurement states are:

$$| \psi_0^{II} \rangle = \frac{\alpha_{00} | 00 \rangle + \alpha_{10} | 10 \rangle}{\sqrt{| \alpha_{00} |^2 + | \alpha_{10} |^2}}$$

and

$$|\psi_1^{II}\rangle = \frac{\alpha_{01} \mid 01\rangle \;+\; \alpha_{11} \mid 11\rangle}{\sqrt{\mid \alpha_{01} \mid^2 + \mid \alpha_{11} \mid^2}}.$$

Let us now consider a special state of a two-qubit system when

$$\alpha_{00} = \alpha_{11} = 1/\sqrt{2}$$

and

$$\alpha_{01} = \alpha_{10} = 0.$$

This state is called a *Bell state* and the pair of qubits is called an *EPR pair.* [5] If the two-qubit system is in this state, when we measure the first qubit the two possible outcomes are 0 with probability 1/2 and 1 with probability 1/2. The corresponding post-measurement states are:

$$|\psi_0^{I}\rangle = |00\rangle$$

and

$$|\psi_1^{I}\rangle = |11\rangle.$$

When we measure the second qubit, the two possible outcomes are 0 with probability 1/2 and 1 with probability 1/2. The corresponding post-measurement states are:

$$|\psi_0^{II}\rangle = |00\rangle$$

and

$$|\psi_1^{II}\rangle = |11\rangle.$$

This is quite an amazing result! The two measurements are correlated, once we measure the first qubit we get exactly the same result as when we measure the second one. The two qubits need not be physically constrained to be at the same location and yet, because of the strong coupling between them, measurements performed on the second one allow us to determine the state of the first.

There are four special states called *Bell states* and they form an orthonormal basis:

1. $|\beta_{00}\rangle = \frac{|00\rangle + |11\rangle}{\sqrt{2}}$,

2. $|\beta_{01}\rangle = \frac{|01\rangle + |10\rangle}{\sqrt{2}}$,

3. $|\beta_{10}\rangle = \frac{|00\rangle - |11\rangle}{\sqrt{2}}$, and

4. $|\beta_{11}\rangle = \frac{|01\rangle - |10\rangle}{\sqrt{2}}$.

---

[5] Einstein, Podolsky, and Rosen (thus, the abbreviation EPR) were the first ones to consider the strange behavior of states like Bell states.

The Bell states can be distinguished from one another. All four states are called *maximally entangled* states. The last one is called *anti-correlated* state. An EPR pair is in one of the four Bell states.

This strange behavior hints of possible applications of quantum information that are well beyond what we could possibly envision in a classical universe. This is the basis of a phenomenon called *teleportation* discussed later in Section 6.3.

3.7 THE FRAGILITY OF QUANTUM INFORMATION—SCHRÖDINGER'S CAT

By now we should be convinced that classical information can be encoded into a quantum system. Having a potentially infinite number of states a qubit offers dazzling possibilities. A two-qubit system is truly amazing, as we have seen in the case of EPR pairs. This seems too good to be true—is there a catch? Yes, the catch is that quantum information is encoded into very fragile non-local correlations of different parts of a physical system. Why do we say that these interactions are "fragile?" Because a quantum system interacts continually with its environment and these interactions with the environment destroy the correlations encoded into the quantum system. In time, the internal correlations of the quantum system are transferred into correlations between the quantum system and the environment and, thus, the information encoded into the quantum system is lost.

Schrödinger gave an extreme example of the apparent ambiguity of the quantum information and of the superposition states. Consider the quantum description of a physical entity we are familiar with—a cat:

$$| \text{cat}\rangle = \frac{1}{\sqrt{2}}(| \text{dead}\rangle + | \text{alive}\rangle).$$

Since all real cats are either dead or alive, a layperson would view Schrödinger's cat as a definitive proof of foolishness; even Schrödinger himself considered this example as a blemish on his theory [110]. Today we are more sophisticated and realize that in fact the state $| \text{cat} \rangle$ though possible, is practically inaccessible. Assuming that a $| \text{cat} \rangle$ could be constructed as a superposition of two possible states $| \text{dead} \rangle$ and $| \text{alive} \rangle$, the $| \text{cat} \rangle$ would be immediately transformed due to correlations between the $| \text{cat} \rangle$, which is not an isolated system, and the environment. The $| \text{cat} \rangle$ as a superposition of two possible states would become inaccessible. All the cats we have ever seen are in fact projections of a $| \text{cat} \rangle$ generated by the environment into either the $| \text{dead} \rangle$ or the $| \text{alive} \rangle$ state.

What can possibly go wrong with quantum information? A very serious problem is that we disturb the state of the quantum system when we attempt to measure it, as we discussed previously. Another concern is that quantum information cannot be copied with fidelity. What about errors? What type of errors should we expect?

With classical bits we observe *bit-flip errors*: a 0 becomes a 1 and a 1 becomes a 0. We expect the same to happen to qubits:

$$| 0\rangle \mapsto | 1\rangle \quad \text{and} \quad | 1\rangle \mapsto | 0\rangle.$$

In addition, a qubit may experience a phase error; the phase may flip and then

$$| 0\rangle \mapsto | 0\rangle \quad \text{and} \quad | 1\rangle \mapsto -| 1\rangle.$$

The quantum information is continuous. If the state of a qubit is

$$| \psi\rangle = \alpha_0 | 0\rangle + \alpha_1 | 1\rangle$$

then either α_0 or α_1, or both, may change by an infinitesimal quantity, ϵ, and then we experience a qubit error of a new type, one we cannot encounter when dealing with classical bits.

3.8 QUBITS—FROM HILBERT SPACES TO PHYSICAL IMPLEMENTATION

Let us now summarize what we have learned so far about qubits. Recall that an n-dimensional Hilbert space is a vector space over the field of complex numbers. If $| \psi\rangle, | \varphi\rangle \in \mathcal{H}_n$ the scalar product is denoted by $\langle\varphi | \psi\rangle$. The norm of the vector $| \varphi\rangle$ is defined by

$$\| \, | \varphi\rangle \, \| = \sqrt{\langle\varphi | \varphi\rangle}.$$

We use Dirac's notation for the inner product between two n-dimensional vectors $| \varphi\rangle$ and $| \psi\rangle$ [46]

$$\langle\varphi | \psi\rangle = \sum_{i=0}^{n-1} \alpha_i^* \beta_i$$

where the asterisk, (*), denotes complex conjugation. We can think of this as the product of the row vector

$$\langle\varphi | = (\alpha_0^* \; \alpha_1^* \; \dots \; \alpha_{n-1}^*)$$

and the column vector

$$| \psi\rangle = \begin{pmatrix} \beta_0 \\ \beta_1 \\ \vdots \\ \beta_{n-1} \end{pmatrix}.$$

A qubit is a mathematical model of a microscopic physical system (e.g., the spin of an electron or the polarization of a photon) and may exist in a continuum of "intermediate states", or "superpositions." We have already seen that individual states of a quantum system consisting of one qubit can be represented as unit vectors in a two-dimensional complex vector space denoted as $\mathcal{H}_2$.

We can choose different basis vectors associated with "basis states" to describe the "intermediate states" of a single qubit. For example, the polarization of a photon can be described using as basis vectors the pair $| 0\rangle$ and $| 1\rangle$, or the pair $\sqrt{1/2}(| 0\rangle + | 1\rangle)$ and $\sqrt{1/2}(| 0\rangle - | 1\rangle)$,[6] or any other pair of vectors in $\mathcal{H}_2$.

[6]In the example in Section 1.8, we used the more intuitive notation for the basis vectors, $| \updownarrow\rangle$ and $| \leftrightarrow\rangle$ instead of $| 0\rangle$ and $| 1\rangle$ and $| \nearrow\rangle$ and $| \searrow\rangle$ instead of $\sqrt{1/2}(| 0\rangle + | 1\rangle)$ and $\sqrt{1/2}(| 0\rangle - | 1\rangle)$.

Our ability to distinguish between the states of a single qubit is limited [16]. We have already seen that the superposition $| \psi \rangle = \alpha_0 | 0 \rangle + \alpha_1 | 1 \rangle$ behaves like $| 0 \rangle$ with probability $| \alpha_0 |^2$ and like $| 1 \rangle$ with probability $| \alpha_1 |^2$.

This gets even more complicated: *two quantum states of one qubit can be distinguished if and only if they are represented in the same basis.* We can distinguish $| 0 \rangle$ from $| 1 \rangle$ but we cannot distinguish $| 0 \rangle$ from $\sqrt{1/2}(| 0 \rangle - | 1 \rangle)$. One last observation: quantum states form what we call *equivalence classes*[7] (i.e., multiplication with a factor such as $e^{i\gamma}$ does not alter a unit vector). Therefore, the quantum state of one qubit can be represented by a *ray* in $\mathcal{H}_2$. A ray is the equivalence class of a vector under multiplication by a complex constant.

We expect more complications for quantum systems with more than one qubit. We have seen in Section 3.6 that a two-qubit system may be in one of four basis states $| 00 \rangle$, $| 01 \rangle$, $| 10 \rangle$, and $| 11 \rangle$ or in a superposition of them, in a Hilbert space with four dimensions, $\mathcal{H}_4$.

Now the two-qubit system can either be in a superposition state, when each qubit has a well-defined state or in an entangled state when neither qubit has a defined state, though the pair does have a well-defined state. For example, consider the case when the first qubit is in the state $| 1 \rangle$, while the second is the state $\sqrt{1/2}(| 0 \rangle + | 1 \rangle)$. Then the state of the two-qubit system can be represented as the tensor product of the individual states of the two qubits:

$$\frac{1}{\sqrt{2}} | 1 \rangle (| 0 \rangle + | 1 \rangle) = \frac{1}{\sqrt{2}} (| 10 \rangle + | 11 \rangle).$$

An entangled state such as

$$\frac{1}{\sqrt{2}} (| 00 \rangle + | 11 \rangle)$$

cannot be written as a product of the states of the individual qubits.

In general, a system of n qubits is represented by a complex unit vector in a 2^n-dimensional Hilbert space, $\mathcal{H}_{2^n}$ defined as a tensor product of n two-dimensional Hilbert spaces

$$\mathcal{H}_{2^n} = (\mathcal{H}_2)^n.$$

Classical systems consisting of many components can be described by providing the state of each individual component separately. It follows that the complexity of the description of a classical system, henceforth the number of states, grows linearly with the number of components. For a quantum system the previous equation shows that the dimensionality of the state space grows exponentially with the number of components or qubits. The majority of states of a quantum system are entangled and they are responsible for both the immense power of a quantum computer and the difficulties of building quantum computers.

The evolution of a quantum system in isolation is unitary; it is linear and conserves the inner product in a Hilbert space. This means that the evolution

[7]See Appendix A for a definition of the terms equivalence relation and equivalence class.

of the system preserves superposition and distinguishability of the system states.

A superposition of the input states of a quantum system consisting of a number $n > 1$ of qubits evolves into a corresponding superposition of output states as we see in the next chapter devoted to quantum gates and quantum circuits. Most gates map unentangled initial states of the quantum system to entangled output states. Conventional computations and communication destroy the entanglement, while quantum operations can create entanglement, preserve and use the entanglement to speed up computations and to transmit information over quantum channels.

Suppose that we have a qubit in such a general normalized (unknown) state

$$\mid \psi\rangle = \alpha_0 \mid 0\rangle \ + \alpha_1 \mid 1\rangle.$$

If we want to determine the computational value of this qubit, we have to perform a measurement of the qubit or, more exactly, of that particular property (observable) of the quantum system through which the qubit is implemented. The choice of the observable determines the frame of reference for the measurement; it specifies a basis, such as $\{\mid 0\rangle, \mid 1\rangle\}$, in the Hilbert space of the system. The basis vectors are eigenvectors of the observable operator. Mathematically, the physical measurement represents a projection of the state of the qubit on this vector basis.

The system randomly changes from the initial state characterized by state vector $\mid \psi\rangle$ to either the state with state vector $\mid 0\rangle$ with probability $\mid \alpha_0 \mid^2$, or to the state with state vector $\mid 1\rangle$ with probability $\mid \alpha_1 \mid^2$. The computational value of the qubit obtained as a result of the measurement is a real number, it is one of the eigenvalues λ_0 and λ_1, associated with the eigenvectors $\mid 0\rangle$ and $\mid 1\rangle$, respectively, of the observable operator.

As a result of the measurement we are not able to determine the initial (unprepared) state of the qubit. The coefficients α_0 and α_1 cannot be determined with this single measurement. What we do is preparing the qubit in a *known* state which, in general, is different from the initial one. The state vector of this known state is now one of the eigenvectors ($\mid 0\rangle$ or $\mid 1\rangle$) of the measured observable. A qubit in a known state is what we need before we do any manipulation of it (i.e., before we apply any of the quantum gates).

Two physical systems leading to the simplest possible embodiments of a qubit are:

1. the *electron* with two independent *spin* values, $\pm 1/2$, and
2. the *photon*, with two independent *polarizations*, say *horizontal* and *vertical* (in case of linear polarization), or *right hand* and *left hand* (in case of circular polarization).

The *spin* is an intrinsic angular momentum[8] of a quantum particle, related to an intrinsic rotation about an arbitrary direction.

[8]The intrinsic angular momentum of a quantum particle should be distinguished from its orbital angular momentum.

There are two classes of quantum particles, those with spin odd multiple of half-integer, called *fermions*, and those with integer spin, called *bosons*. The spin quantum number of fermions can be $s = +1/2$, $s = -1/2$, or an odd multiple of $s = \pm 1/2$. The spin quantum number of bosons can be $s = +1$, $s = -1$, $s = 0$, or a multiple of ± 1. The spin of a quantum particle can be observed as an interaction of the intrinsic angular momentum of the particle with an external magnetic field **B**.

3.9 QUBITS AS SPIN HALF-INTEGER PARTICLES

One embodiment of a qubit is the spin state of a particle with spin half-integer, such as the electron.[9] The concept of "spin" does not have a correspondent in any property in classical mechanics. Classical mechanics operates with the concept of an "angular momentum" arising from a rotation around a well-defined axis of a body. A quantum particle such as the electron is not a "body" in the classical sense and does not have a defined axis of rotation. The electron is characterized by a charge which has a non-stationary spatial distribution. The variation in time of this charge distribution can be associated with an intrinsic rotation of the electron about directions randomly oriented in space.

The observable associated with the electron intrinsic rotation is the intrinsic angular momentum, also called the *spin angular momentum* of the electron. The "spin" is the quantum number characterizing the intrinsic angular momentum of the electron. The electron spin is found to have either the value $+1/2$ or $-1/2$ along the measurement axis, regardless of what that axis is, see Figure 3.5(a).

The qubit states $| 0\rangle$ and $| 1\rangle$ correspond to the spin up $| \uparrow \rangle$ and spin down $| \downarrow \rangle$ states along a chosen axis such as the z axis. It is convenient to represent the spin states as orthogonal unit vectors

$$| 0\rangle = | \uparrow \rangle = \begin{pmatrix} 1 \\ 0 \end{pmatrix} \quad \text{and} \quad | 1\rangle = | \downarrow \rangle = \begin{pmatrix} 0 \\ 1 \end{pmatrix}.$$

The theory of the spin is based on the assumption that the components of the spin angular momentum are connected with the rotation operators σ_x, σ_y and σ_z [46] introduced in Section 3.1. The self-adjoint operators $\mathbf{S}_x$, $\mathbf{S}_y$, and $\mathbf{S}_z$ associated with the components of the intrinsic angular momentum of the electron can be expressed in a two-dimensional representation as

$$\mathbf{S}_k = \frac{1}{2}\sigma_k, \quad k \in \{x, y, z\}$$

where σ_k are the Pauli matrices σ_x, σ_y, and σ_z and represent the components of the rotation operator along the x, y, and z axes in real space. The eigenvalues of $\mathbf{S}_k$ are $\pm 1/2$ in units $\hbar = 1$. Rotations about distinct axes do not commute (i.e., a series of rotations if performed in different sequences will not have the same result). The operators associated with rotations about distinct axes are

[9] The protons and the neutrons are other particles with spin 1/2.

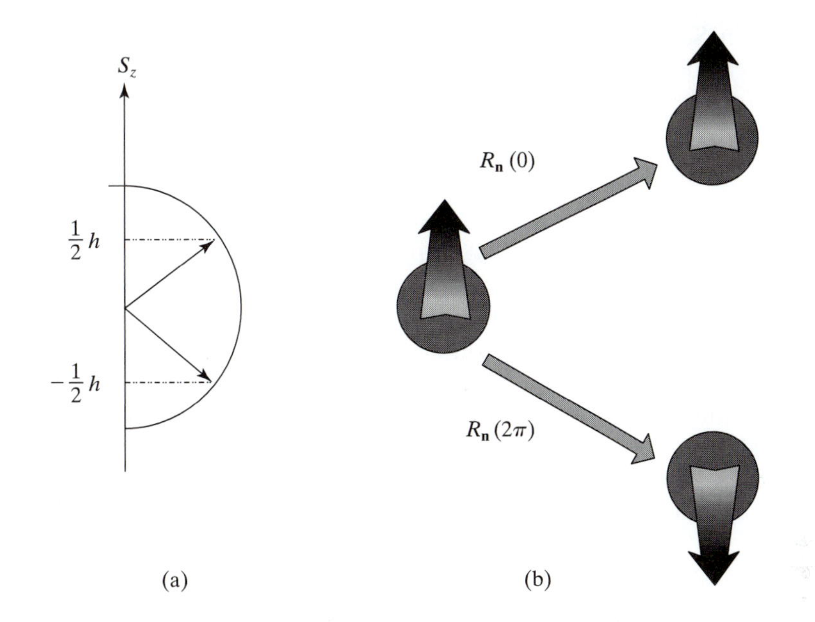

FIGURE 3.5: The spin of the electron. (a) Electrons and other particles (fermions) have intrinsic angular momentum characterized by the spin quantum number $\frac{1}{2}$. (b) There are two rotations operators—the first keeps the spin unchanged and the second flips the spin to an orthogonal state.

non-commutative. The commutators of the Pauli matrices are:

$$[\sigma_x, \sigma_y] = \sigma_x\sigma_y - \sigma_y\sigma_x = 2i\sigma_z,$$
$$[\sigma_y, \sigma_z] = \sigma_y\sigma_z - \sigma_z\sigma_y = 2i\sigma_x,$$
$$[\sigma_z, \sigma_x] = \sigma_z\sigma_x - \sigma_x\sigma_z = 2i\sigma_y.$$

The squares of the Pauli matrices are equal to I

$$\sigma_x^2 = \sigma_y^2 = \sigma_z^2 = I.$$

Indeed, $\sigma_k^\dagger = \sigma_k$ and

$$\sigma_k^2 = \sigma_k^\dagger\sigma_k = \sigma_k^2 = I, \quad k \in \{x, y, z\}.$$

These equations can be reduced to a simpler form

$$\sigma_y\sigma_z = -\sigma_z\sigma_y = i\sigma_x,$$
$$\sigma_z\sigma_x = -\sigma_x\sigma_z = i\sigma_y,$$
$$\sigma_x\sigma_y = -\sigma_y\sigma_x = i\sigma_z,$$

and

$$\sigma_x\sigma_y\sigma_z = iI.$$

In Section 3.3, we saw that finite rotations about a given vector $\mathbf{n}$ in the Bloch sphere can be represented as

$$\mathcal{R}_{\mathbf{n}}(\theta) = \exp\left(-i\frac{\theta}{2}\mathbf{n}\cdot\boldsymbol{\sigma}\right) = \cos\left(\frac{\theta}{2}\right)I - i\sin\left(\frac{\theta}{2}\right)\mathbf{n}\cdot\boldsymbol{\sigma},$$

where $\boldsymbol{\sigma} = (\sigma_x, \sigma_y, \sigma_z)$. It is easy to show that the matrices $\mathcal{R}_x(\theta), \mathcal{R}_y(\theta), \mathcal{R}_z(\theta)$ representing rotations about the individual axes x, y, z are unitary and have determinant 1. Moreover, any general 2×2 unitary matrix with determinant 1 can be expressed in this form.

When a qubit is represented as the spin state of an electron, the only arbitrary unitary transformations possible in acting on the state are, in fact, rotations of the spin. In $\mathcal{H}_2$ a rotation operator has two eigenvalues corresponding to the angles $\theta = 0$ and $\theta = 2\pi$ and thus there are two possible rotations of the spin, one by an angle $\theta = 0$ which keeps the spin unchanged, and one by $\theta = 2\pi$ flips the spin to an orthogonal state, as in Figure 3.5(b)

$$\begin{aligned} \theta = 0 \quad &\Longrightarrow \mathcal{R}_{\mathbf{n}}(0) = \cos(0)I - i\sin(0)\mathbf{n}\cdot\boldsymbol{\sigma} = I, \\ \theta = 2\pi &\Longrightarrow \mathcal{R}_{\mathbf{n}}(2\pi) = \cos(\pi)I - i\sin(\pi)\mathbf{n}\cdot\boldsymbol{\sigma} = -I. \end{aligned}$$

3.10 THE MEASUREMENT OF THE SPIN

Formally, measuring the spin along an axis $\mathbf{n}$ is equivalent to first applying a rotation transformation that rotates the $\mathbf{n}$ axis to the z axis and then measuring along z. In the act of taking a measurement along a defined direction, chosen by the observer in the laboratory frame of reference, the electron assumes a definite axis of rotation.

How do we measure the spin? We use the intrinsic magnetic moment and let it interact with an external magnetic field. In a Stern-Gerlach type of experiment we can measure the spin component along the vertical axis by observing which way the spin half-integer particle is deflected and by how much, as seen in Figure 3.6.

The applied magnetic field interacts with the spin magnetic moment of the electrons, defines their axes of rotation and separates the electrons with states of given spin. At the same time the magnetic field exerts a force on the electrons; the force acts upward on the electron if the spin points up and downward if the spin points down.

At any instant before a measurement an electron's spin state $| \psi\rangle$ can be represented by a linear combination of those two possible observable states. The choice of a measurement direction is equivalent to choosing a *basis* for expressing the spin components of the state $| \psi\rangle$.

In the case of a spin half-integer, there are two such states that are generally denoted by $| \uparrow \rangle$ and $| \downarrow \rangle$, according to the sign of the corresponding value 1/2 and $-1/2$, respectively. These two states are vectors in a Hilbert space and have all the properties associated with such a space. They form an orthonormal basis in a two-dimensional Hilbert space. A general normalized spin state, for which positive and negative coefficients are possible, can be constructed as a linear

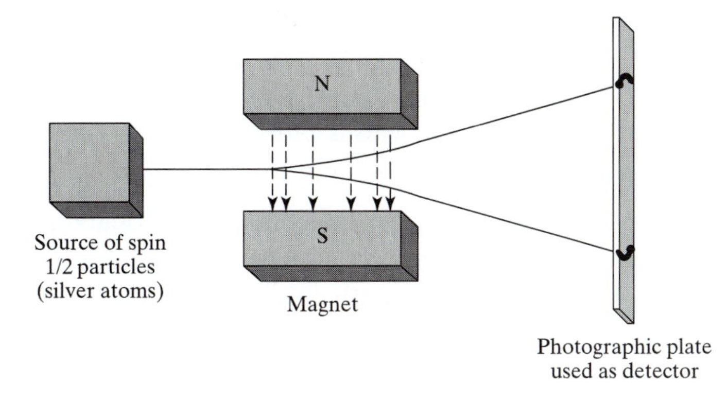

FIGURE 3.6: The measurement of the spin in a Stern-Gerlach type setup. The nonhomogeneous magnetic field exerts a force on the electrons while the electron is traversing the magnet. The force acts upward on the electron if the spin is up and downward if the spin is down. Once the electrons exit the magnet they continue on a rectilinear trajectory.

superposition of these two states:

$$| \psi\rangle = c_+ | \uparrow \rangle \ + c_- | \downarrow \rangle.$$

A *qubit* as a state vector in a two-dimensional Hilbert space can take any value of the form mentioned previously, where the spin states $| \uparrow \rangle$ and $| \downarrow \rangle$ would correspond to the qubit states $| 0\rangle$ and $| 1\rangle$, respectively. The coefficients c_+ and c_- are arbitrary complex numbers, subject to the *normalization condition*

$$| c_+ |^2 + | c_- |^2 = 1$$

and they contain *all the information we can ever obtain* about a quantum state of spin one-half.

If this maximum information is available, the system is said to be in a *pure state*. In such a state, the probabilities for the two possible results of a measurement similar to the Stern-Gerlach experiment are

$$P_+ = | c_+ |^2 \quad \text{and} \quad P_- = | c_- |^2$$

and they correspond to the two components of the beam split by the magnetic field, one deflected upwards and the other downwards, respectively, as shown in Figure 3.6. We may have two detectors, one for the upwards deflected beam and the other for the downward deflected beam. Depending on which detector is clicking, we know that the initial general state of the system (electron or atom) is reduced to either spin state $| +\rangle$ or $| -\rangle$ through the measurement.

In general, the measuring apparatus consists of multiple components. Each of these components contributes to the overall definition of the measurement direction, and has its own vector basis. Therefore, the travel of the electron through the apparatus is associated with measurements in several bases. Each one of these measurements contributes to the overall probability of finding the electron in a particular final state.

A modified Stern-Gerlach apparatus containing more than one set of magnets can be used to *filter* an incident beam of atoms, so that after the last magnet we end up with a polarized beam consisting of atoms in a definite pure spin state.

In general, a beam of particles is called:

1. "completely polarized"—if all particles are in a pure state,
2. "unpolarized"—if particles have an equal probability to have any of the possible values of the spin, or
3. "partially polarized"—if particles have unequal probabilities to have any of the possible values of the spin.

Consider a basis consisting of three orthogonal axes x, y, and z. Then, the three principal components of the spin angular momentum of the particle, S_x, S_y, and S_z, corresponding to measurements along the three axes x, y, and z are

$$S_x = \frac{1}{2}\sigma_x = \frac{1}{2}\begin{pmatrix} 0 & 1 \\ 1 & 0 \end{pmatrix}, \quad S_y = \frac{1}{2}\sigma_y = \frac{1}{2}\begin{pmatrix} 0 & -i \\ i & 0 \end{pmatrix}, \quad S_z = \frac{1}{2}\sigma_z = \frac{1}{2}\begin{pmatrix} 1 & 0 \\ 0 & -1 \end{pmatrix}$$

with $\sigma_x, \sigma_y, \sigma_z$ the Pauli matrices. In the case of an electron in a magnetic field, the Pauli matrices have a special geometrical significance, but in the general case they can be used simply as matrices.[10]

In quantum mechanics, each possible measurement basis is associated with an operator whose eigenvalues represent the possible outcomes of the corresponding measurement. The eigenvalues of the operator associated with the measurement (the spin angular momentum operator, in this case) along one of the directions x, y, or z determine the probabilities of the possible outcomes ($+1/2$ or $-1/2$). The eigenvectors of the spin operators along the three principal directions S_x, S_y, S_z corresponding to the eigenvalues $+1/2$ or $-1/2$ are

	$-\frac{1}{2}$	$+\frac{1}{2}$
S_x	$\frac{1}{\sqrt{2}}\begin{pmatrix} -1 \\ 1 \end{pmatrix}$	$\frac{1}{\sqrt{2}}\begin{pmatrix} 1 \\ 1 \end{pmatrix}$
S_y	$\frac{1}{\sqrt{2}}\begin{pmatrix} 1 \\ -i \end{pmatrix}$	$\frac{1}{\sqrt{2}}\begin{pmatrix} 1 \\ i \end{pmatrix}$
S_z	$\begin{pmatrix} 1 \\ 0 \end{pmatrix}$	$\begin{pmatrix} 0 \\ 1 \end{pmatrix}$

[10] It is left as an exercise to the reader to prove that any 2×2 matrix A, such as the Hamiltonian of any two-state system, can be represented as a linear combination of Pauli matrices and the identity matrix I with complex coefficients $c_0, c_1, c_2, c_3 \in \mathbb{C}$

$$A = \begin{pmatrix} a_{11} & a_{12} \\ a_{21} & a_{22} \end{pmatrix} = c_0 I + c_1\sigma_x + c_2\sigma_y + c_3\sigma_z.$$

Each pair of eigenvectors constitutes a basis for the state space. The state vector of the spin half-integer particle can be expressed as a linear combination of the basis vectors for the desired measurement (along one of the principal axes) and the coefficients give the probability of that measurement yielding either the *spin up* or the *spin down* eigenvalues.

Let us consider the case where we measure the spin of an electron along an arbitrary axis determined by an applied magnetic field **B**, as in a Stern-Gerlach type experiment. As a result of the measurement, the spin points along the magnetic field direction. The magnetic field **B** is not along the z direction, but in a direction z' defined by the polar angle θ and φ, as in Figure 3.7. Here θ is the angle between the direction of the magnetic field and the positive z axis, and φ is the azimuthal angle between the projection of the magnetic field in the xy plane and the x axis, as in Figure 3.7.

Given that the spin points in the direction of the magnetic field **B**, we want to express the state of the electron in the vector basis corresponding to the z axis as

$$| \psi_{\theta,\varphi}\rangle = \alpha_0 \mid \uparrow_z\rangle \ + \ \alpha_1 \mid \downarrow_z\rangle$$

where $| \uparrow_z\rangle$ and $| \downarrow_z\rangle$ are the states spin *up*, and spin *down* along the z axis.

We want to determine the values of the coefficients α_0 and α_1 which give the probabilities to find the electron with its spin up or down, respectively, along the z axis, when we know that its spin is oriented along the axis defined by the angles θ and φ. Solving the Schrödinger equation for an atomic particle in a magnetic field[11] shows that

$$\alpha_0 = \cos\frac{\theta}{2}e^{-i\varphi/2}$$

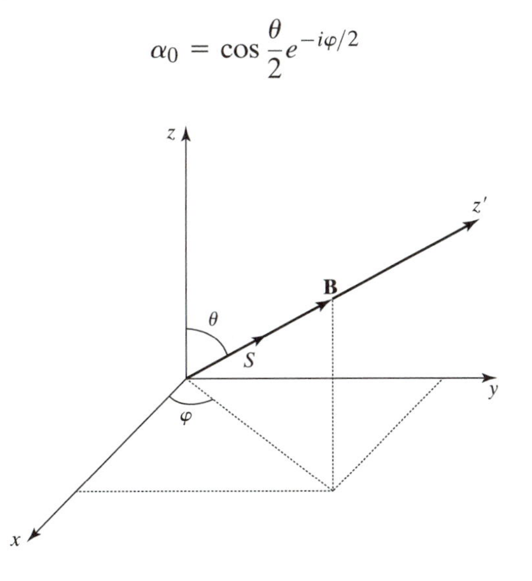

FIGURE 3.7: The spin S of an electron is oriented along an axis defined by the polar angle θ and φ. Here θ is the angle between the direction of the magnetic field **B** and the positive z axis, and φ is the azimuthal angle between the projection of the magnetic field in the xy plane and the x axis.

[11] A clear presentation of the calculations to solve the Schrödinger equation for an atomic particle in a magnetic field can be found in Section 10, Volume III of [53].

and

$$\alpha_1 = \sin\frac{\theta}{2}e^{+i\varphi/2}$$

up to an exponential factor with an imaginary exponent which does not count in the evaluation of the probabilities.

We notice that the value of the magnetic field **B** does not appear in these two expressions; hence, the result is the same for the case when $\mathbf{B} = 0$. This result is the solution of the more general case of an arbitrary particle of spin half-integer whose spin is pointing along an arbitrary axis. For example, assume that $|\uparrow_z\rangle$ represents a state with spin up along the z axis and $|\downarrow_z\rangle$ represents the spin down state along the z axis. Given another vector $|\uparrow_{z'}\rangle$ representing a state of "spin up" along a different z' axis which makes an angle θ with the z axis and φ is the angle between the x axis and the projection of the z' axis in the xy plane, we can write its projections on the z axis vector basis as

$$\langle\uparrow_z|\uparrow_{z'}\rangle = \cos\frac{\theta}{2}e^{-i\varphi/2},$$

$$\langle\downarrow_z|\uparrow_{z'}\rangle = \sin\frac{\theta}{2}e^{+i\varphi/2}.$$

Let us assume that we perform an experiment, similar to the Stern-Gerlach experiment, where the electrons are moving along the y axis and their spins are measured in the z direction; that means the z component of the electron spin vector is pinned down. If we filter out the electrons with "spin down" in the z direction, we are left with electrons with the "spin up" in the z direction. Assume that now we perform a new spin measurement on the remaining "spin up" electrons along a direction in the xz plane at an angle θ with the positive z axis. The measurement spin operator in this direction, S_θ, will be given by its projections on the S_x and S_z operators (the probabilities to obtain the spin values along the basis axes are interpreted as the projections of the electron state vector onto those axes):

$$\begin{aligned} S_\theta &= \sin(\theta)S_x + \cos(\theta)S_z \\ &= \frac{1}{2}\sin(\theta)\begin{pmatrix} 0 & 1 \\ 1 & 0 \end{pmatrix} + \frac{1}{2}\cos(\theta)\begin{pmatrix} 1 & 0 \\ 0 & -1 \end{pmatrix} \\ &= \frac{1}{2}\begin{pmatrix} \cos(\theta) & \sin(\theta) \\ \sin(\theta) & -\cos(\theta) \end{pmatrix}. \end{aligned}$$

The eigenvalues of this operator are $+1/2$ and $-1/2$ in units of $\hbar = 1$ and the corresponding eigenvectors are:

$$\begin{pmatrix} \cos(\theta/2) \\ \sin(\theta/2) \end{pmatrix} \quad \begin{pmatrix} -\sin(\theta/2) \\ \cos(\theta/2) \end{pmatrix}.$$

Each electron, after being measured in the first stage of the experiment, has now the state vector as "spin up" in the z direction. This state vector is the initial state vector for the next measurement along the axis in the xz plane and it can be expressed as a linear combination of these new basis vectors

$$\begin{pmatrix} 1 \\ 0 \end{pmatrix} = c_1 \begin{pmatrix} \cos(\theta/2) \\ \sin(\theta/2) \end{pmatrix} + c_2 \begin{pmatrix} -\sin(\theta/2) \\ \cos(\theta/2) \end{pmatrix}.$$

The coefficients are estimated to be $c_1 = \cos(\theta/2)$ and $c_2 = -\sin(\theta/2)$. The probabilities of *spin up* and *spin down* for the measurement of such an electron along the direction rotated through angle θ are $|\cos(\theta/2)|^2$ and $|\sin(\theta/2)|^2$, respectively, where θ is the angle between the two measurement directions. In fact, θ is the relative angle between the initial and the final states of the electron. In quantum mechanics, the relative orientation (angle) of two state vectors is significant, not their absolute orientations. In this context, we understand why quantum states can be represented by rays in $\mathcal{H}_n$.

3.11 THE QUBIT AS A POLARIZED PHOTON

A *photon* is another important two-state quantum system used to embody a qubit. A photon can have two independent polarizations and systems using the polarization of a photon to encode binary information have been used in real-life experiments.

Photons differ from the spin one-half electrons in two ways: (1) they are massless and (2) they have spin one. A photon is characterized by its vector momentum (the vector momentum determines the frequency) and its polarization. In the classical theory light is described as an electromagnetic radiation whose electric field component oscillates. The electric field can oscillate vertically, in a plane perpendicular to the direction of propagation, the z axis, and then we say that the light is *x polarized*, as in Figure 3.8(a). The electric field can oscillate horizontally in a plane perpendicular to the direction of propagation, and then we say the light is *y polarized* as shown in Figure 3.8(b).

If the electric field has an arbitrary orientation in the xy plane, then it will have x and y components. If these components are out of phase by $90°$, the electric field rotates and the light is *elliptically polarized.* When the x and y components are equal and out of phase by $90°$ the light is *circularly polarized.* Circularly polarized light can be *right-hand* polarized or *left-hand* polarized, depending on which way it propagates along the z direction.

If we look at the individual photons participating in the "light," we can not talk about an electric field associated with a single photon, but a single photon must have a property as the analog of the classical phenomena of polarization.

From the point of view of polarization, a photon can be described as a two-state system; a photon can be in state $|h\rangle$ or in state $|v\rangle$. All photons in a classically y polarized beam of light are said to be in polarization state $|h\rangle$ and, similarly, all photons in a classically x polarized beam of light are said to be in polarization state $|v\rangle$. The states $|h\rangle$ and $|v\rangle$ can be used as basis states to describe the polarization of a photon (in a linearly polarized beam of light) with given momentum oriented along the z direction.

Actually, light contains photons in these two states of polarization. If we use a polarization filter (or polarization analyzer) and set its axis to let y polarized

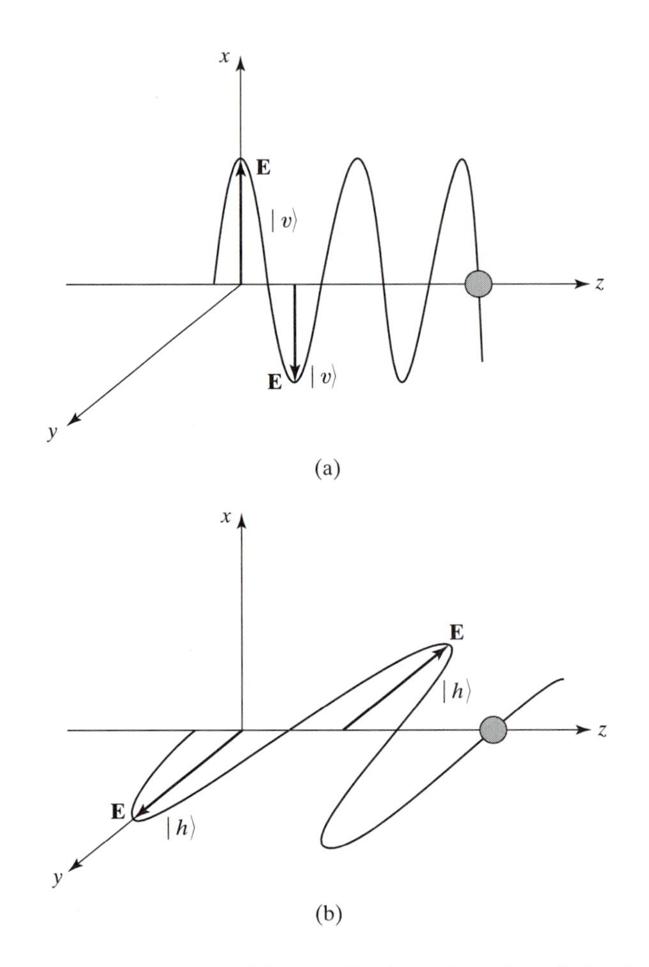

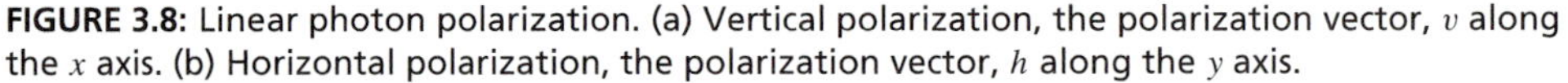
FIGURE 3.8: Linear photon polarization. (a) Vertical polarization, the polarization vector, v along the x axis. (b) Horizontal polarization, the polarization vector, h along the y axis.

light pass, then all photons in the state $| \, v\rangle$ will be absorbed in the filter and only the photons in state $| \, h\rangle$ will pass through. If the axis of the polarization filter is set to let x polarized light pass, then all photons in state $| \, h\rangle$ will be absorbed and only photons in state $| \, v\rangle$ will pass through.

A material called *calcite* has an inverse action on a beam of light: the light beam will split it into an $| \, h\rangle$ state beam and a $| \, v\rangle$ state beam. That is similar to the action of a Stern-Gerlach apparatus that separates the spin $s = +1/2$ states from spin $-1/2$ states. If the polarization filter is set in a new position and the light passing through is polarized in a y' direction which makes an angle θ with the y direction, as in Figure 3.9(a), then the photons coming out of the filter will be in a $| \, h'\rangle$ polarization state. This state can be expressed as a linear combination of the basis states $| \, h\rangle$ and $| \, v\rangle$ as

$$| \, h'\rangle = \cos\theta \, | \, h\rangle \; + \; \sin\theta \, | \, v\rangle.$$

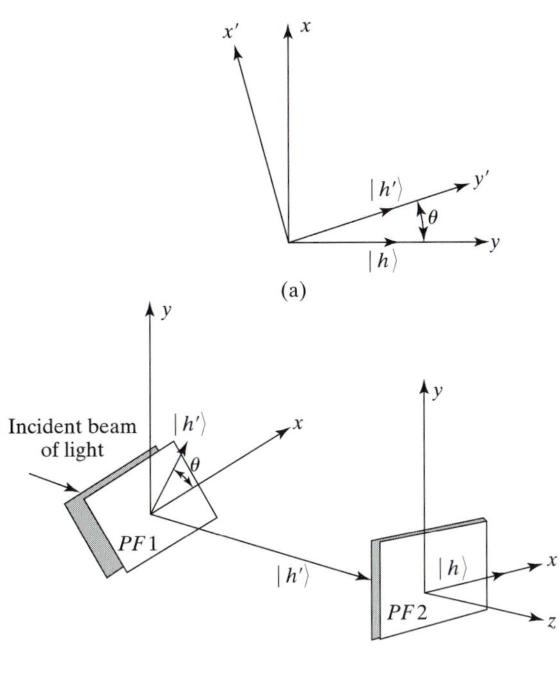

FIGURE 3.9: The effect of a polarization filter. (a) The polarization filter $PF1$ is set at angle θ with respect to the coordinate system of the incoming beam of light. The emerging photons are in a superposition state, $\mid h'\rangle = \alpha_0 \mid h\rangle \ + \ \alpha_1 \mid v\rangle$, with $\alpha_0 = \cos\theta$ and $\alpha_1 = \sin\theta$. (b) The polarization filter at angle θ, $PF1$, is followed by a second filter, $PF2$, which lets only h-polarized photons pass with a probability equal to $\cos\theta$.

Assume now that these photons in state $\mid h'\rangle$ (i.e., polarized along direction y') have to go through a second polarization analyzer set in position $\theta = 0$, as in Figure 3.9(b). This filter will let pass only photons in state $\mid h\rangle$. We have to estimate the probability that a photon in state $\mid h'\rangle$ is also in state $\mid h\rangle$ and in this case we have to calculate the projection

$$\langle h \mid h'\rangle = \cos\theta\langle h \mid h\rangle \ + \ \sin\theta\langle h \mid v\rangle$$

where $\mid h\rangle$ and $\mid v\rangle$ is an orthonormal basis, so that $\langle h \mid h\rangle = 1$ and $\langle h \mid v\rangle = 0$. Thus, the probability that a photon in state $\mid h'\rangle$ is also in state $\mid h\rangle$ is

$$\mid \langle h \mid h'\rangle \mid^2 = \cos^2\theta.$$

We have seen that under a rotation of angle θ about the axis of propagation the horizontal polarization state $\mid h\rangle$ is transformed into $\mid v'\rangle$

$$\mid h\rangle \mapsto \mid h'\rangle = \cos(\theta) \mid h\rangle \ + \ \sin(\theta) \mid v\rangle.$$

At the same time the vertical polarization state $\mid v\rangle$ is transformed into $\mid v'\rangle$

$$\mid v\rangle \mapsto \mid v'\rangle = -\sin(\theta) \mid h\rangle \ + \ \cos(\theta) \mid v\rangle.$$

The matrix of this transformation is

$$\begin{pmatrix} \cos\theta & \sin\theta \\ -\sin\theta & \cos\theta \end{pmatrix}$$

and it has the eigenstates

$$| R\rangle = \frac{1}{\sqrt{2}}\begin{pmatrix} 1 \\ i \end{pmatrix} \quad | L\rangle = \frac{1}{\sqrt{2}}\begin{pmatrix} i \\ 1 \end{pmatrix}.$$

It is easy to see that the eigenvalues are $e^{i\theta}$ and $e^{-i\theta}$, respectively, for these eigenstates which we call *right* and *left* polarization states.
These states are also eigenstates of the rotation operator

$$\sigma_y = \begin{pmatrix} 0 & -i \\ i & 0 \end{pmatrix}$$

with eigenvalues ± 1. Because the eigenvalues are ± 1 we say that the photon has spin integer.

Remember the experiment with polarized light (photons) presented in Section 3.2. We had a polarization filter that allowed only one of the two linear photon polarizations, say $| h\rangle$, to pass through. Only 1/2 of the photons could get through. We added a filter for polarization $| v\rangle$ and no photon could get through. Next, we interposed a 45° rotated polarizer between the h and v analyzers. As a result, an h-polarized photon coming out of the first analyzer had a probability equal to 1/2 of passing through the 45° polarizer and a 45° polarized photon coming out of the 45° polarizer had a probability equal to 1/2 of passing through the v-polarizing filter.

A device can be constructed that rotates the linear polarization of a photon. Such a device applies the first transformation mentioned previously to our qubit and sets the qubit in a mixed state. We can use another device that alters the relative phase ω of the two orthogonal linear polarization states and performs the following transformations

$$| h\rangle \mapsto e^{i\omega/2} \, | h\rangle$$

and

$$| v\rangle \mapsto e^{-i\omega/2} \, | v\rangle.$$

When the two devices are used together they apply a 2×2 unitary transformation to the photon polarization state.

3.12 ENTANGLEMENT

Assume that an experiment is set up to measure the spins of two electrons, emitted in opposite directions following the decay of a two-electron singlet state with zero total spin. A singlet electron state corresponds to the *anti-symmetric* superposition state of a pair of electrons with *anti-parallel* spins; [12] the electrons

[12] The three possible superposition states of a pair of electrons with *parallel* spins (i.e., with equal spin quantum numbers $s = +1/2$ and $s = +1/2$ or $s = -1/2$ and $s = -1/2$) is called a *triplet*. The total spin of a triplet state is equal to $+1, 0,$ or -1.

have different quantum numbers $s = +1/2$ and $s = -1/2$ and the total spin of the singlet state is zero.

For such a state, the conservation of the angular momentum requires that the spin vectors of the two particles are oriented in opposite directions. Hence, if we measure the spin of one of these two particles along a certain direction and find it in the state "spin up," then along that direction, the other particle must be in the state "spin down." By measuring the spin of one particle and thus reducing its state vector to one of the eigenvectors of the measurement basis, we automatically project the state vector of the other particle onto the same basis. Instead of a set of probabilistically possible states, we obtain one well-defined state.

Now, assume that we perform the following experiment with a pair of particles in a singlet state. First, we measure the spin of one particle (call it `particle I`) along a direction in the xz plane and find it in the "spin down" state; then the other particle (call it `particle II`) is in a pure "spin up" state along that direction. Next we measure the spin of `particle II` along a direction at an angle θ with the direction of first measurement. We find that `particle II` is in a combination of "spin up" and "spin down" states with probabilities $\cos^2(\theta/2)$ and $\sin^2(\theta/2)$, respectively. If the first measurement on `particle I` yields "spin up," then `particle II` is in a pure "spin down" state along that direction. A measurement of the spin of `particle II` along a new direction at an angle θ relative to the first measurement, finds `particle II` in a combination of "spin up" and "spin down" states with probabilities $\sin^2(\theta/2)$ and $\cos^2(\theta/2)$, respectively.

Now let us perform successive measurements of each of the two particles along directions at an angle θ relative to each other. The probability that the successive measurements give the same result (up-up or down-down) is $\sin^2(\theta/2)$ while the probability that they yield opposite results (up-down or down-up) is $\cos^2(\theta/2)$.

The two particles emitted from the singlet are said to be *entangled*; regardless of how far apart they travel before the spin measurements are made, the joint results will exhibit these *joint* probabilities.

These probabilities can be evaluated from the results of multiple measurements; such measurements have to be performed over several particle pairs, not just one pair. In principle a large number of particle pairs can be prepared in an identical way, in space-separated locations (in a string like formation) and the measurements can be performed independently on the pairs. According to quantum mechanics, the results are expected to satisfy the same correlations.

At this time, we can imagine a quantum computer as a collection of prepared pairs of particles whose entanglement (correlations) satisfy all the requirements for carrying out a parallel computation described by a parallel algorithm; alternatively we can consider an algorithm designed to take advantage of a specific entanglement. Two entangled particles in a singlet state have total spin 0 because the spins of the individual particles about any axis are always opposite; their spin will exist only as potentialities until an observer chooses a definite axis and performs the measurement.

The measurement axis defines the vector basis for both electrons and the instant `particle I` is measured, `particle II` acquires a definite spin along the chosen axis, instantaneously. If we let a vertical axis represent the binary value 1 and a horizontal axis represent 0, it is possible to pass 0s and 1s apparently "instantaneously" across large distances by selecting the appropriate axis of measurement. We use the word "instantaneously" in quotes to emphasize the fact that the transmission of information does not really happen with a speed greater than the speed of light. According to a postulate of the relativity theory [49] "light is always propagated in empty space with a definite velocity c which is independent of the state of the motion of the emitting body." The relativity theory was confirmed by numerous experiments.

At the receiving end of the communication channel the measurement must not only be performed after the initial measurement at the sending end, but it must also be carried out in a matching vector basis. Thus, the sender must use some classical means of communication to inform the receiver about the vector basis used for the measurement. The delay between the moments the two results are obtained prevents a truly instantaneous transmission of information.

Entanglement is an elegant, almost exact, translation of the German term *Verschränkung* used by Schrödinger who was the first to recognize this quantum effect. An *entangled pair* is a single quantum system in a superposition of equally possible states. The entangled state contains no information about the individual particles, only that they are in opposite states. The important property of an entangled pair is that the measurement of one particle influences the state of the other particle. Einstein called that phenomenon "*Spooky action at a distance*."

3.13 THE EXCHANGE OF INFORMATION USING ENTANGLED PARTICLES

The entanglement of a pair of qubits can be exploited to transmit information as discussed in depth in Chapter 6. Here we only sketch the basic idea of communication using qubits in antisymmetric states. In this case, when one of the qubits is forced into one state, then the second member of the pair is forced into an opposite state. The conceptual idea is strikingly simple, but poses immense practical challenges. We are able to create pairs of entangled particles, but separating them, while maintaining their entanglement is very challenging.

The method described in [14] can be used in conveying information in a secure way at a distance. Assume that we have prepared three particles. The first of them, `particle1` is located in New York and its initial state is $| \psi_1 \rangle$. The second and the third ones, called `particle2` and `particle3`, are entangled and then separated; `particle2` is sent to New York together with `particle1`, while `particle3` is sent to London. The entire process is depicted in Figure 3.10.

In the following discussion we use the abbreviated notation $| 01 \rangle$ to describe a state of a two-particle quantum system when `particle1` is in state $| 0 \rangle$ while `particle2` is in state $| 1 \rangle$. The essential point of the experiment is to perform specific measurements that project the ensemble consisting of `particle1` and `particle2` onto the entangled state $| \psi_{12} \rangle$

$$| \psi_{12} \rangle = | \beta_{11} \rangle = \frac{1}{\sqrt{2}}(| 01 \rangle - | 10 \rangle).$$

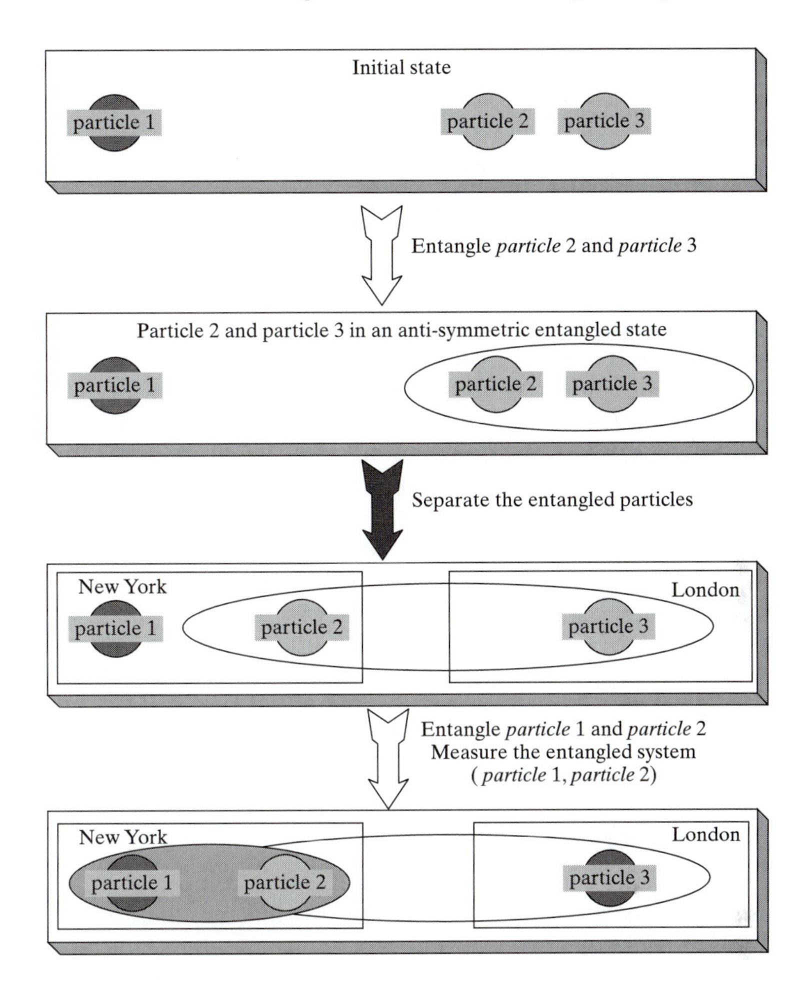

FIGURE 3.10: Communication with entangled particles. Given three particles, `particle1`, `particle2`, and `particle3` we wish to transmit the state of `particle1` in New York to `particle3` in London. We first entangle `particle2` and `particle3`. Then we separate the two entangled particles, `particle2` goes to New York and `particle3` to London. Once in New York we entangle `particle1` with `particle2`. Then we measure the entangled system by projecting it into the state $|\psi_{12}\rangle$. This measurement process forces `particle2` into an antisymmetric state to the one of `particle1`. But `particle2` is also in an antisymmetric entanglement relation with `particle3`, thus `particle3` is forced into a state opposite to `particle2`. This is precisely the state of `particle1`.

This is one of the four possible maximally entangled states of a pair of particles, as discussed in Section 3.6, the other three states being

$$|\beta_{00}\rangle = \frac{1}{\sqrt{2}}(|00\rangle + |11\rangle),$$

$$|\beta_{01}\rangle = \frac{1}{\sqrt{2}}(|01\rangle + |10\rangle),$$

and

$$| \beta_{10} \rangle = \frac{1}{\sqrt{2}}(| 00 \rangle - | 11 \rangle).$$

The $| \beta_{11} \rangle$ state is distinguished from the other three by the fact that it changes sign when `particle1` and `particle2` are interchanged. This anti-symmetric feature plays an important role in the experimental identification of this state.

Quantum physics predicts that once `particle1` and `particle2` are entangled and projected onto the entangled state $| \psi_{12} \rangle$, `particle3` is instantaneously projected into the initial state of `particle1`.

What is the explanation? Since we have entangled `particle1` and `particle2`, no matter what state `particle1` is in, `particle2` must be in the opposite state, a state which is orthogonal to the state of `particle1`. Initially, `particle2` and `particle3` were prepared in state $| \psi_{23} \rangle$ and that means that the state of `particle2` is also orthogonal to the state of `particle3`. That is only possible if `particle3` in London is in the same state as `particle1` was initially.

It is important to realize that as a result of this communication process we do not have two identical copies of `particle1`, one in New York and one in London. During the entanglement with `particle2`, `particle1` loses its identity, its state cannot be extricated from the state of the entangled system. The state $| \psi_1 \rangle$, where the information was "inscribed", is destroyed during the measurement process on the New York side, but the information has been "communicated" to `particle3` on the London side of the message exchange.

3.14 SUMMARY AND FURTHER READINGS

A *quantum bit* or *qubit* is a mathematical abstraction for an elementary quantum system used to store and to transform information. A qubit state $| \psi \rangle$ is a vector in $\mathcal{H}_2$, a two-dimensional Hilbert space.

The state space of one qubit contains the two "basis" or "logical" states

$$| 0 \rangle = \begin{pmatrix} 1 \\ 0 \end{pmatrix} \qquad | 1 \rangle = \begin{pmatrix} 0 \\ 1 \end{pmatrix}.$$

The initial state of a qubit is always one of the basis states. Using the transformation discussed in this chapter we can obtain states which are "superpositions" of the basis states, see Figure 3.3. Superpositions can be expressed as sums over the basis states with complex coefficients

$$| \psi \rangle = \alpha_0 \, | 0 \rangle \; + \; \alpha_1 \, | 1 \rangle.$$

In addition to the `ket` notation of Dirac displayed here, the state of a qubit can be expressed as a vector

$$| \psi \rangle = \begin{pmatrix} \alpha_0 \\ \alpha_1 \end{pmatrix}.$$

Throughout this book the two notations are used interchangeably.

A superposition is *a pure state* if the corresponding vector $| \psi \rangle$ has length 1, that is, if $| \alpha_0 |^2 + | \alpha_1 |^2 = 1$. Such a superposition, or vector, is said to be "normalized."

The amplitudes of a qubit in a superposition state can be expressed using three real numbers θ, φ, and γ such that

$$\alpha_0 = e^{i\gamma} \cos \frac{\theta}{2} \quad \text{and} \quad \alpha_1 = e^{i\gamma} e^{i\varphi} \sin \frac{\theta}{2}.$$

The elements of the density matrix give the probabilities of the possible outcomes of a measurement performed on a qubit. The most general form of the density matrix for a single qubit is

$$\rho = \frac{1}{2}(I + \beta_x \sigma_x + \beta_y \sigma_y + \beta_z \sigma_z) = \begin{pmatrix} 1 + \beta_z & \beta_x - i\beta_y \\ \beta_x + i\beta_y & 1 - \beta_z \end{pmatrix}$$

where I is the 2×2 identity matrix and β_x, β_y, and β_z are real numbers. There is a one-to-one correspondence between the possible density matrices of a single qubit and the points on the Bloch sphere $\beta^2 = 1$ with $\beta^2 = \beta_x^2 + \beta_y^2 + \beta_z^2$. *Pure states* are represented by points on the Bloch sphere, while *impure states* are represented by points inside the sphere.

Single qubit operations are defined as rotations on the Bloch sphere and described by the Pauli spin matrices $\{\sigma_x, \sigma_y, \sigma_z\}$.

Our ability to distinguish between the states of a single qubit is limited. The superposition states of one qubit cannot be reliably distinguished from the basis states. The superposition $| \psi \rangle = \alpha_0 | 0 \rangle + \alpha_1 | 1 \rangle$ behaves like $| 0 \rangle$ with probability $| \alpha_0 |^2$ and like $| 1 \rangle$ with probability $| \alpha_1 |^2$. A qubit in a superposition state is measured to one of the two basis states as shown in Figure 3.4.

Two quantum states of two qubits can be distinguished if and only if their vector representations are orthogonal. Quantum states form equivalence classes, meaning that multiplication with $e^{i\gamma}$ does not change the length of a unit vector. Therefore, the quantum state of one qubit is a "ray" in $\mathcal{H}_2$.

The state of a system of two qubits is a linear combination of the basis vectors with complex coefficients $\alpha_{00}, \alpha_{01}, \alpha_{10}, \alpha_{11}$:

$$| \psi \rangle = \alpha_{00} | 00 \rangle + \alpha_{01} | 01 \rangle + \alpha_{10} | 10 \rangle + \alpha_{11} | 11 \rangle.$$

When we measure a pair of qubits we determine that the system is in one of four basis states $| 00 \rangle$, $| 01 \rangle$, $| 10 \rangle$, or $| 11 \rangle$, with probabilities $| \alpha_{00} |^2$, $| \alpha_{01} |^2$, $| \alpha_{10} |^2$, and $| \alpha_{11} |^2$ respectively. The sum of probabilities must be one

$$| \alpha_{00} |^2 + | \alpha_{01} |^2 + | \alpha_{10} |^2 + \alpha_{11} |^2 = 1.$$

Before the measurement the state is unknown. The post measurement state of the qubit is $| 00 \rangle$, $| 01 \rangle$, $| 10 \rangle$, or $| 11 \rangle$.

When $\alpha_{00} = \alpha_{11} = 1/\sqrt{2}$ and $\alpha_{01} = \alpha_{10} = 0$ we have a *Bell state* and the pair of qubits is called an *EPR pair*. In this case, when we measure only the first qubit, we get the same results as when we measure only the second qubit.

A system of n qubits is represented by a complex unit vector in a 2^n-dimensional Hilbert space, $\mathcal{H}_{2^n}$ defined as a tensor product of n two-dimensional Hilbert spaces

$$\mathcal{H}_{2^n} = \mathcal{H}_2 \otimes \mathcal{H}_2 \otimes \cdots \otimes \mathcal{H}_2 = (\mathcal{H}_2)^{\otimes n}.$$

We usually abbreviate this as

$$\mathcal{H}_{2^n} = (\mathcal{H}_2)^n.$$

This equation shows that the dimensionality of the state space grows exponentially with the number of components or qubits. For a classical system the number of states grows linearly with the number of components.

The evolution of a quantum system in isolation is unitary; it is linear and conserves the inner product in a Hilbert space. The evolution of the system preserves superposition and distinguishability of the system states. A superposition of the input states of a quantum system consisting of a number $n > 1$ of qubits evolves into a corresponding superposition of output states. Conventional computations and communication destroy the entanglement, while quantum operations can create, preserve, and use the entanglement to speed up computations and to transmit information over quantum channels.

The physical systems leading to the simplest possible embodiments of a qubit are: the *electron* with two independent *spin* values, $\pm 1/2$, and the *photon*, with two independent *polarizations*.

The *spin* is the quantum number characterizing the intrinsic angular momentum of the electron. The electron spin is found to have either the value $+1/2$ or $-1/2$ along the measurement axis, regardless of what that axis is.

A *photon* can have two independent polarizations. Photons differ from the spin half-integer electrons; they are massless and have spin one. A photon is characterized by its vector momentum (the vector momentum determines the frequency) and its polarization. In the classical theory, light is described as an electromagnetic radiation whose electric field component oscillates either vertically (the light is *x polarized*), or horizontally (the light is *y polarized*) in a plane perpendicular to the direction of propagation (the z axis).

A singlet electron state corresponds to a pair of electrons with *anti-parallel* spins in an *anti-symmetric* superposition state. Consider two particles of spin half-integer (electrons) emitted in opposite directions following the decay of a singlet state with zero total spin; in this case, the conservation of the angular momentum requires that the spin vectors of the two particles should be oriented in opposite directions. The two particles emitted from the singlet are said to be *entangled*.

If we measure the spin of one of these two particles along a certain direction and find it in the state "spin up," then, along that direction, the other particle must be in the state "spin down." By measuring the spin of one particle and, thus, reducing its state vector to one of the eigenvectors of the measurement basis, we automatically project the state vector, *collapse the wave function*, of the other particle onto the same basis. Instead of a set of probabilistically possible states we obtain one, well-defined state.

The entanglement of a pair of qubits can be exploited to transmit information as discussed in Section 3.13.

The paper of E. Rieffel and W. Polack [112] as well as M.A. Nielsen and I. L. Chuang's book [98] supplement the information provided in this chapter on qubits. A recent collection of articles on the physics of quantum information edited by D. Bouwmester, A. Ekert, and A. Zeilinger [26] contains several insightful papers on the subject of the physical realization of qubits.

3.15 EXERCISES AND PROBLEMS

3.1. Show that the Pauli matrices σ_k have the following property:

$$\sigma_k^\dagger = \sigma_k, \quad k \in \{x, y, z\}.$$

Show that: (i) they are Hermitian, (ii) $\sigma_x^2 = \sigma_y^2 = \sigma_z^2 = I$, their squares are equal to the identity matrix, (iii) $\sigma_x \sigma_y = i\sigma_z$ (this is also true for a cyclic permutation of indices), and (iv) they satisfy the relation $\sigma_x \sigma_y + \sigma_y \sigma_x = 0$ (this is also true for a cyclic permutation of indices).

3.2. Prove the following relations among the identity (I), the Hadamard (H), and Pauli matrices

$$\sigma_x \sigma_x = I; \quad H\sigma_x H = \sigma_z; \quad H\sigma_z H = \sigma_x; \quad H\sigma_y H = -\sigma_y.$$

3.3. Show that the composition of two rotations with angles θ_1 and θ_2 is a rotation with angle $\theta_1 + \theta_2$ along the same axis

$$R_{\mathbf{r}}(\theta_1) R_{\mathbf{r}}(\theta_2) = R_{\mathbf{r}}(\theta_1 + \theta_2).$$

3.4. Show that the rotation matrices $R_{\mathbf{r}}$ are unitary. Recall that a unitary operator preserves the distance. What is the implication of this property for the representation of a qubit using a Bloch sphere?

3.5. Prove that any 2×2 matrix A can be represented as a linear combination of Pauli matrices and the identity matrix I

$$A = \begin{pmatrix} a_{11} & a_{12} \\ a_{21} & a_{22} \end{pmatrix} = c_0 I + c_1 \sigma_x + c_2 \sigma_y + c_3 \sigma_z$$

with $c_0, c_1, c_2, c_3 \in \mathbb{C}$.

3.6. Prove that

$$[Y, Z] = 2iX \quad [Z, X] = 2iY$$

where X, Y, and Z are the Pauli matrices.

3.7. Write a Java applet with a graphics user interface (GUI) that displays a qubit on the Bloch sphere. The program requires as input the values of the complex coefficients α_0 and α_1 of the state vector, $\psi = \alpha_0 \mid 0\rangle + \alpha_1 \mid 1\rangle$.
(*Hint:* Once the user clicks on the link to the object stored on the Web server, a GUI form should be displayed allowing the user to enter the necessary data, verify that $\mid \alpha_0 \mid^2 + \mid \alpha_1 \mid^2 = 1$, carry out the necessary calculations, then download the applet.)

3.8. Augment the program in the previous assignment to compute the effect of a transformation performed on a qubit by: the three Pauli matrices, the Hadamard (H), the phase (S), the $\pi/8$ (T), the phase-shift with angle θ (P_θ and R_θ), and R_k matrices. The GUI should allow the user to specify: (i) the values of the complex coefficients α_0 and α_1 of the state vector, $| \psi \rangle = \alpha_0 | 0 \rangle + \alpha_1 | 1 \rangle$, (ii) the transformation, and (iii) the values of the elements of the transformation matrix for P_θ, R_θ, and R_k matrices. The program should display the original and the transformed qubits on the same Bloch sphere, and the matrix of the corresponding transformation.

CHAPTER 4

Quantum Gates and Quantum Circuits

In this chapter, we discuss quantum gates—the building blocks of a quantum computer. We review familiar gates and logic circuits used to transform the information in a classical computer.

To maintain consistency for those familiar with classical logic circuits we use the same formulation when talking about quantum gates; we say that a gate transforms its input to its output according to the rules hardwired into the truth table. The physical reality is that a quantum gate transforms the state of a quantum system into a new state. The state transformations performed by quantum gates and circuits are described by Hermitian operators. We call the matrix describing the state transformation a *transfer matrix*.

The matrices characterizing quantum gates are unitary[1] and the transformations performed by these gates are reversible. A unitary transformation corresponds to a length-preserving and information-preserving rotation in the

[1]A unitary matrix U with complex entries has an inverse equal to its Hermitian conjugate $U^{\dagger}$, $UU^{\dagger} = I$, and the length of $U\mathbf{x}$ is the same as the length of the vector $\mathbf{x}$.

vector space. The length-preserving requirement ensures that the total probability of the set of states is always equal to one.

A quantum gate has an equal number of qubits (one, two, or three) at the input and at the output. In each case, the input and the output states are described by vectors in a Hilbert space with the corresponding number of dimensions: $\mathcal{H}_2$ for a one-qubit gate, $\mathcal{H}_4$ for a two-qubit gate, and $\mathcal{H}_8$ for a three-qubit gate. The input/output state of an n-qubit quantum gate is computed as the tensor product of the corresponding state vectors of the individual input/output qubits. For example, the input state of a two-qubit gate is $| \varphi\rangle \otimes | \psi\rangle$ if the state vectors of the two input qubits are $| \varphi\rangle$ and $| \psi\rangle$.

The orthonormal basis used for the analysis of a quantum gate consists of:

1. two vectors in $\mathcal{H}_2$, $| 0\rangle$ and $| 1\rangle$
2. four vectors in $\mathcal{H}_4$, $| 00\rangle$, $| 01\rangle$, $| 10\rangle$, and $| 11\rangle$ and
3. eight vectors in $\mathcal{H}_8$, $| 000\rangle$, $| 001\rangle$, $| 010\rangle$, $| 011\rangle$, $| 100\rangle$, $| 101\rangle$, $| 110\rangle$, and $| 111\rangle$.

The eigenvalues corresponding to these eigenvectors represent the result of the measurement of the quantum system's state.

When discussing a quantum gate we start with an informal description of the transformation performed by the gate; from this informal description we construct the truth table and then the matrix corresponding to the quantum operator. We construct the eigenvectors in a specific orthonormal basis, then the projectors reflecting the transformation of the input performed by the quantum gate.

If $| V\rangle$ is the initial state of a quantum system applied as input to the gate performing a state transformation described by the operator **G** with a matrix representation $G = [g_{ij}]$, then the resulting state $| W\rangle$ of the quantum system is given by

$$| W\rangle = G | V\rangle.$$

There are classical equivalents of the quantum gates presented in this section.

We introduce single-qubit gates and two-qubit gates including an ample discussion of the `CNOT` gate. We move on to three-qubit gates and discuss the `Fredkin` and `Toffoli` gates and continue with the no-cloning theorem. Then we present quantum circuits. We discuss single- and multiple-qubit controlled operations and the Walsh-Hadamard Transform. We conclude the chapter with a section on mathematical models and quantum gate arrays and discuss errors in quantum computing.

4.1 CLASSICAL LOGIC GATES AND CIRCUITS

Boolean variables have values of either 0 or 1. The Boolean operations are: `NOT`, `AND`, `NAND`, `OR`, `NOR`, and `XOR`. Boolean algebra deals with Boolean variables and Boolean operations.

Logic gates are the active elements of a computer; they transform information using the laws of Boolean algebra, see Figure 4.1. Logic gates implement Boolean operations. Here we only describe standard logic gates with one or two inputs and one output. Gates with more than two inputs exist.

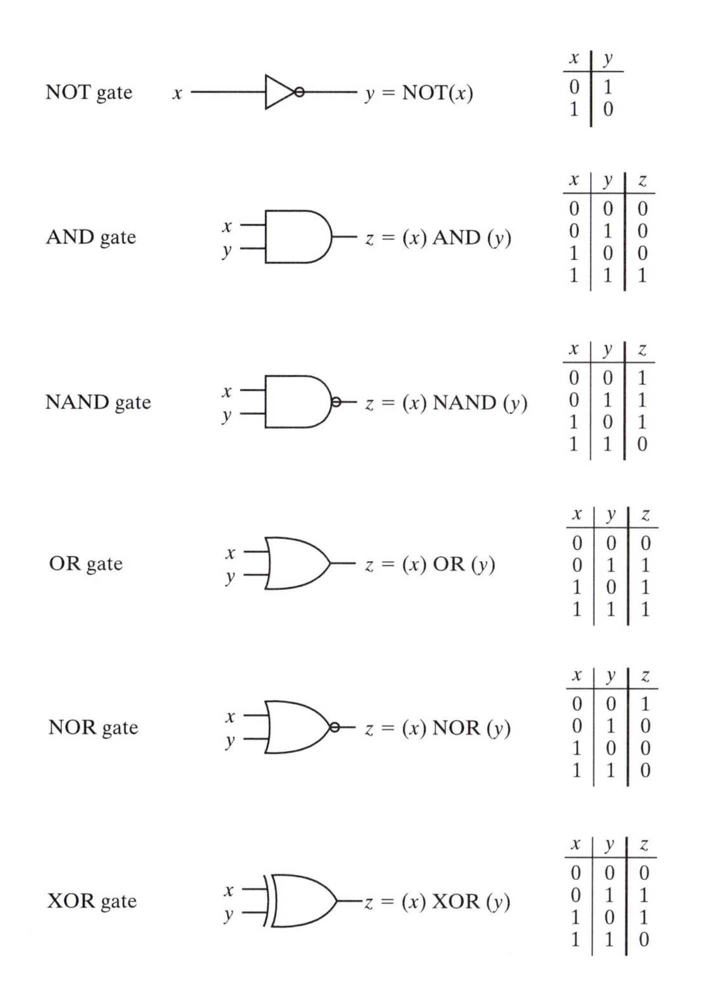

FIGURE 4.1: Classical logic gates. The truth table of each logic gate gives the output, function of the input(s) of the circuit.

Figure 4.1 presents six classical logic gates. For each gate we show the output as a Boolean function of the input. Each logic function is characterized by a *truth table* giving the output for different combinations of inputs.

We denote the *addition modulo 2* by $\oplus$; the output is 1 when the two inputs differ and is 0 if they are identical. If one of the inputs, say y, is equal to 0, then $x \oplus y = x$. The truth table of a modulo 2 adder is the same as the one of an XOR gate:

x	y	$x \oplus y$
0	0	0
0	1	1
1	0	1
1	1	0

It is not very difficult to prove that NAND gates are *universal*. This means that any logic function can be expressed using only NAND Boolean operations; thus, one can construct a logic circuit using only NAND gates. In contrast, XOR is not a universal gate; indeed, it does not change the parity[2] of its input. If the input has odd parity ($x = 0$ and $y = 1$ and the input string is 01, or $x = 1$ and $y = 0$ and then the input string is 10) so does the output, it is 1; if the input has even parity (00 or 11) so does the output, it is 0. Thus the class of Boolean functions constructed with XOR alone is limited.

All gates with two inputs in Figure 4.1 are *irreversible* or *non invertible*. This means that knowing the output we cannot determine the input for all possible combinations of input values. For example, knowing that the output of an AND gate is 0 we cannot identify the input combination; 0 can be produced by three possible combinations of inputs 00, 01 and 10. Clearly, the NOT gate is reversible and two cascaded NOT gates recover the input of the first one. The irreversibility of classical gates means that there is an irretrievable loss of information and this has very serious consequences regarding the energy consumption of classical gates, as we shall see in Chapter 6.

Now we give an example of a logic circuit, the one-bit full-adder, constructed with some of the logic gates presented in this section. This circuit has three inputs a, b, and *CarryIn* and two outputs *Sum* and *CarryOut*, see Figure 4.2(a).

The truth table of the one-bit full-adder is:

a	b	*CarryIn*	*Sum*	*CarryOut*
0	0	0	0	0
0	0	1	1	0
0	1	0	1	0
0	1	1	0	1
1	0	0	1	0
1	0	1	0	1
1	1	0	0	1
1	1	1	1	1

You may recall from an introductory course in computer architecture that one can derive the Boolean equations giving the outputs of a logic circuit function of the inputs from the truth table of the circuit as sums of products [102]. Here sum stands for the Boolean OR and product stands for the Boolean AND operation. In this section $\bar{a}$ denotes the negation of the Boolean variable a. Each term of the sum corresponds to an entry in the truth table where the output variable is 1; each term is the product of the corresponding input variables in that row negated if the value of the variable is 0, or without negation if the value is 1.

[2] A binary string has *odd parity* if it has an odd number of 1s, and has *even parity* if it has an even number of 1s.

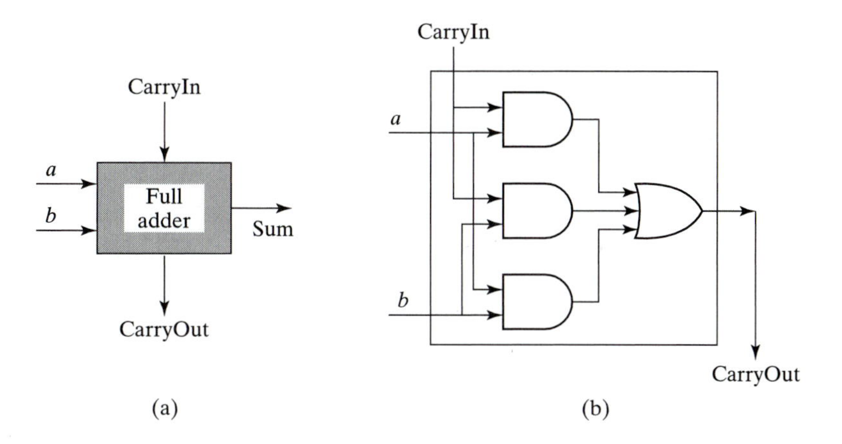

FIGURE 4.2: (a) A one-bit full-adder circuit. (b) The circuit to compute the *CarryOut* for the one-bit full-adder in (a).

From the truth table of the full adder it is easy to see that

$$Sum = \bar{a}\bar{b}CarryIn + \bar{a}b\overline{CarryIn} + a\bar{b}\overline{CarryIn} + abCarryIn$$

and

$$CarryOut = \bar{a}bCarryIn + a\bar{b}CarryIn + ab\overline{CarryIn} + abCarryIn.$$

A manipulation of the last Boolean expression shows that

$$CarryOut = ab + aCarryIn + bCarryIn.$$

We leave the analytical derivation of this equality as an exercise for the reader and use the truth table method to prove the equality of the two Boolean expressions. The truth table below gives the *CarryOut* computed as $ab + aCarryIn + bCarryIn$.

a	b	$CarryIn$	ab	$aCarryIn$	$bCarryIn$	$CarryOut$
0	0	0	0	0	0	0
0	0	1	0	0	0	0
0	1	0	0	0	0	0
0	1	1	0	0	1	1
1	0	0	0	0	0	0
1	0	1	0	1	0	1
1	1	0	1	0	0	1
1	1	1	1	1	1	1

If we compare the last columns of the truth tables of the full adder and the one presented above we see that indeed $CarryOut = ab + aCarryIn + bCarryIn$. Figure 4.2(b) shows the circuit implementing the last Boolean expression for the $CarryOut$.

4.2 ONE-QUBIT QUANTUM GATES

A one-qubit gate is a black box transforming an input qubit $\mid \psi\rangle = \alpha_0 \mid 0\rangle + \alpha_1 \mid 1\rangle$ into an output qubit $\mid \varphi\rangle = \alpha'_0 \mid 0\rangle + \alpha'_1 \mid 1\rangle$.

Mathematically, a gate G is represented by a 2×2 *transfer matrix* with complex entries g_{ij} with $i, j \in \{1, 2\}$:

$$G = \begin{pmatrix} g_{11} & g_{12} \\ g_{21} & g_{22} \end{pmatrix}.$$

Recall that the normalization condition requires that $\mid \alpha_0 \mid^2 + \mid \alpha_1 \mid^2 = 1$ and similarly $\mid \alpha'_0 \mid^2 + \mid \alpha'_1 \mid^2 = 1$. This requires that G must be a *unitary matrix*, in other words that $G^\dagger G = I$. Here $G^\dagger$ is the *adjoint* of G, a matrix obtained from G by first constructing G^T, the *transpose* of G and then taking the complex conjugate, g^*_{ij}, of each matrix element, or by first taking the complex conjugate of each element and then transposing the matrix. The transpose of a matrix has as rows the columns of the original matrix, thus:

$$G^T = \begin{pmatrix} g_{11} & g_{21} \\ g_{12} & g_{22} \end{pmatrix}.$$

It follows that:

$$G^\dagger = \begin{pmatrix} g^*_{11} & g^*_{21} \\ g^*_{12} & g^*_{22} \end{pmatrix}.$$

The condition for G to be unitary is:

$$G^\dagger G = \begin{pmatrix} g^*_{11}g_{11} + g^*_{21}g_{12} & g^*_{11}g_{21} + g^*_{21}g_{22} \\ g^*_{12}g_{11} + g^*_{22}g_{12} & g^*_{12}g_{21} + g^*_{22}g_{22} \end{pmatrix} = I$$

where I denotes the 2×2 identity matrix. Thus, G is invertible and its inverse, $G^\dagger$, is also unitary. This implies that a quantum gate with a unitary transfer matrix can always be inverted. This is extremely important since it shows that quantum gates are reversible, as opposed to classical gates which are irreversible.

Given the transfer matrix of a quantum gate, G, and the input and output qubits represented as column vectors, the transformation performed by the gate is given by the equation:

$$\mid \varphi\rangle = G \mid \psi\rangle.$$

For a single-qubit gate, this equation can be written as:

$$\begin{pmatrix} \alpha_0' \\ \alpha_1' \end{pmatrix} = \begin{pmatrix} g_{11} & g_{12} \\ g_{21} & g_{22} \end{pmatrix} \begin{pmatrix} \alpha_0 \\ \alpha_1 \end{pmatrix}.$$

Thus,

$$\alpha_0' = g_{11}\alpha_0 + g_{12}\alpha_1 \quad \text{and} \quad \alpha_1' = g_{21}\alpha_0 + g_{22}\alpha_1.$$

We examine a few important one-qubit gates and give their transfer matrices:

1. the I identity gate—it leaves a qubit unchanged $\Longrightarrow I = \begin{pmatrix} 1 & 0 \\ 0 & 1 \end{pmatrix}$.
2. the X or NOT gate—it transposes the components of a qubit $\Longrightarrow$ $X = \sigma_x = \begin{pmatrix} 0 & 1 \\ 1 & 0 \end{pmatrix}$.
3. the Y gate—it multiplies the input qubit by i and flips the two components of the qubit $\Longrightarrow Y = \sigma_y = \begin{pmatrix} 0 & -i \\ i & 0 \end{pmatrix}$.
4. the Z gate—it changes the phase (flips the sign) of a qubit $\Longrightarrow$ $Z = \sigma_z = \begin{pmatrix} 1 & 0 \\ 0 & -1 \end{pmatrix}$.
5. the Hadamard gate H $\Longrightarrow H = \frac{1}{\sqrt{2}} \begin{pmatrix} 1 & 1 \\ 1 & -1 \end{pmatrix}$.

The transfer matrices of the first four gates, I, X, Y, and Z are the identity matrix I and the *Pauli matrices* $\sigma_x, \sigma_y, \sigma_z$, respectively. The output vectors of these gates, $| \varphi\rangle$, for a given input $| \psi\rangle = \alpha_0 | 0\rangle + \alpha_1 | 1\rangle$ are listed here:

$$| \varphi\rangle = I | \psi\rangle = \begin{pmatrix} 1 & 0 \\ 0 & 1 \end{pmatrix} \begin{pmatrix} \alpha_0 \\ \alpha_1 \end{pmatrix} = \begin{pmatrix} \alpha_0 \\ \alpha_1 \end{pmatrix} \quad \text{or} \quad | \varphi\rangle = \alpha_0 | 0\rangle + \alpha_1 | 1\rangle.$$

$$| \varphi\rangle = \sigma_x | \psi\rangle = \begin{pmatrix} 0 & 1 \\ 1 & 0 \end{pmatrix} \begin{pmatrix} \alpha_0 \\ \alpha_1 \end{pmatrix} = \begin{pmatrix} \alpha_1 \\ \alpha_0 \end{pmatrix} \quad \text{or} \quad | \varphi\rangle = \alpha_1 | 0\rangle + \alpha_0 | 1\rangle.$$

$$| \varphi\rangle = \sigma_y | \psi\rangle = \begin{pmatrix} 0 & -i \\ i & 0 \end{pmatrix} \begin{pmatrix} \alpha_0 \\ \alpha_1 \end{pmatrix} = i \begin{pmatrix} -\alpha_1 \\ \alpha_0 \end{pmatrix} \quad \text{or} \quad | \varphi\rangle = -i\alpha_1 | 0\rangle + i\alpha_0 | 1\rangle.$$

$$| \varphi\rangle = \sigma_z | \psi\rangle = \begin{pmatrix} 1 & 0 \\ 0 & -1 \end{pmatrix} \begin{pmatrix} \alpha_0 \\ \alpha_1 \end{pmatrix} = \begin{pmatrix} \alpha_0 \\ -\alpha_1 \end{pmatrix} \quad \text{or} \quad | \varphi\rangle = \alpha_0 | 0\rangle - \alpha_1 | 1\rangle.$$

$$| \varphi\rangle = H | \psi\rangle = \frac{1}{\sqrt{2}} \begin{pmatrix} 1 & 1 \\ 1 & -1 \end{pmatrix} \begin{pmatrix} \alpha_0 \\ \alpha_1 \end{pmatrix} \quad \text{or}$$

$$| \varphi\rangle = \frac{\alpha_0}{\sqrt{2}}(| 0\rangle + | 1\rangle) + \frac{\alpha_1}{\sqrt{2}}(| 0\rangle - | 1\rangle).$$

The Hadamard gate, H, when applied to a pure state, $| 0\rangle$ or $| 1\rangle$, creates a superposition state,

$$H | 0\rangle \mapsto \frac{1}{\sqrt{2}}(| 0\rangle + | 1\rangle)$$

and

$$H \mid 1\rangle \quad \mapsto \quad \left(\frac{1}{\sqrt{2}}\right)(\mid 0\rangle - \mid 1\rangle).$$

It follows that the transformation of a qubit $\mid x\rangle$, with $x = 0$ or $x = 1$, carried out by a `Hadamard` gate can be expressed as

$$\mid x\rangle \quad \mapsto \quad \frac{1}{\sqrt{2}}\left(\mid 0\rangle + (-1)^x \mid 1\rangle\right).$$

In Section 3.1 we derived the Pauli matrices from the description of the transformation carried out by each one of them. Recall that single gate operations correspond to rotations and reflections on the Bloch sphere, as pointed out in Section 3.2.

4.3 THE HADAMARD GATE, BEAM SPLITTERS, AND INTERFEROMETERS

In Chapter 1, we discussed at length experiments involving a very simple device called a *beam splitter*. Let us note first that beam splitters have been constructed not only for photons, but also for other types of quantum particles.

We consider now a 50–50 beam splitter where an incident particle coming from above or from below has the same probability of emerging as an upwards or a downwards beam, as seen in Figure 4.3(a). It turns out that the transformation performed by the beam splitter is described by the Walsh-Hadamard Transform [26].

Indeed, let us call the input to a `Hadamard` gate $\mid \psi\rangle = \alpha_0 \mid 0\rangle + \alpha_1 \mid 1\rangle$ and call its output $\mid \varphi\rangle = H \mid \psi\rangle$. We have seen earlier that

$$\mid \varphi\rangle = \frac{1}{\sqrt{2}}\begin{pmatrix} \alpha_0 + \alpha_1 \\ \alpha_0 - \alpha_1 \end{pmatrix}$$

or

$$\mid \varphi\rangle = \frac{1}{\sqrt{2}}(\alpha_0 + \alpha_1) \mid 0\rangle + \frac{1}{\sqrt{2}}(\alpha_0 - \alpha_1) \mid 1\rangle = \frac{\alpha_0}{\sqrt{2}}(\mid 0\rangle + \mid 1\rangle) + \frac{\alpha_1}{\sqrt{2}}(\mid 0\rangle - \mid 1\rangle).$$

The probability amplitude for finding the particle in the outgoing beam directed upwards is $(1/\sqrt{2})(\alpha_0 + \alpha_1)$ and the probability amplitude for finding the particle in the outgoing beam directed downwards is $(1/\sqrt{2})(\alpha_0 - \alpha_1)$.

Let us now consider a system consisting of two cascaded beam splitters. The setup in Figure 4.3(b) reminds us of the experiment presented in Figure 1.7 in Section 1.9.

In this case the output $\mid \varphi\rangle$ is

$$\mid \varphi\rangle = H \times H \mid \psi\rangle.$$

The product of two matrices A and B is denoted as $A \times B$. It is easy to see that $H \times H = I$. Indeed,

$$H \times H = \frac{1}{\sqrt{2}}\begin{pmatrix} 1 & 1 \\ 1 & -1 \end{pmatrix} \times \frac{1}{\sqrt{2}}\begin{pmatrix} 1 & 1 \\ 1 & -1 \end{pmatrix} = \frac{1}{2}\begin{pmatrix} 2 & 0 \\ 0 & 2 \end{pmatrix} = I.$$

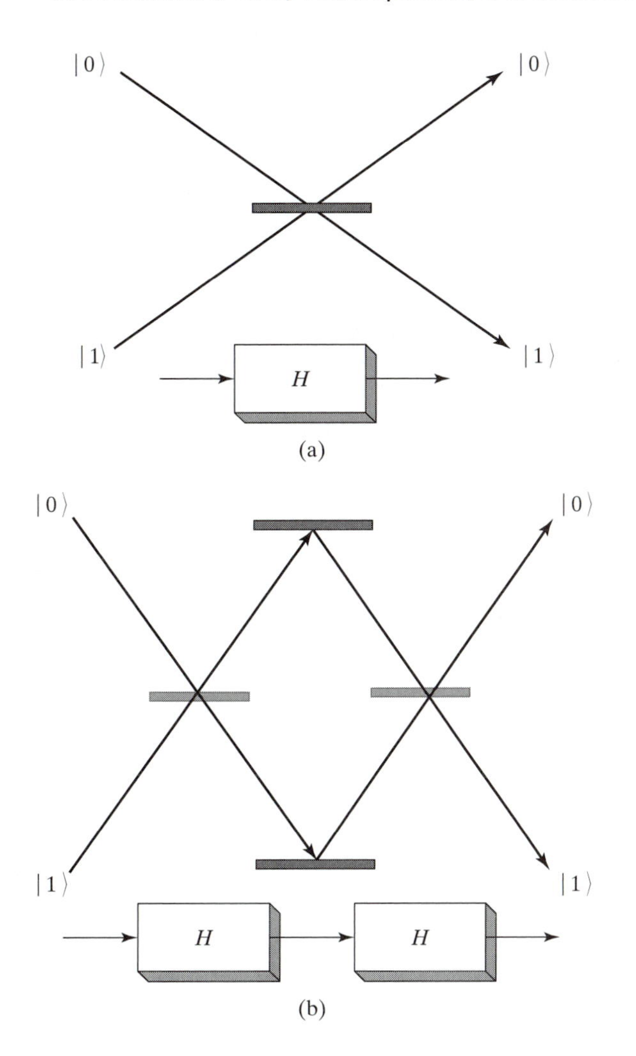

FIGURE 4.3: (a) A beam splitter performs a transformation described by a `Hadamard` gate. (b) An interferometer performs a transformation described by two cascaded `Hadamard` gates.

Thus,

$$| \varphi\rangle = H \times H \mid \psi\rangle = I \mid \psi\rangle = \mid \psi\rangle.$$

This result can be generalized. If we apply $n = 2k$ successive Walsh-Hadamard transformations to a qubit in state $| \psi\rangle$ the qubit ends up in the same state $| \varphi\rangle = | \psi\rangle$. Indeed,

$$| \varphi\rangle = H \times H \ldots \times H \mid \psi\rangle = H^{2k} \mid \psi\rangle = (H^2)^k \mid \psi\rangle = I^k \mid \psi\rangle = I \mid \psi\rangle = \mid \psi\rangle.$$

If $n = 2k + 1$, then n successive Walsh-Hadamard transformations of a qubit in state $| \psi\rangle$ leads to the same state as the one produced by a single

Hadamard gate

$$| \varphi \rangle = H^{2k+1} | \psi \rangle = (H^2)^k \times H | \psi \rangle = I^k \times H | \psi \rangle = I \times H | \psi \rangle = H | \psi \rangle.$$

4.4 TWO-QUBIT QUANTUM GATES—THE CNOT GATE

Now we describe a gate with two inputs and two outputs called CNOT, Controlled-NOT gate, see Figure 4.4. One of the inputs is called the *control* input, the other one is the *target* input. The first output is called the *control* and the second is called the *target*.

The classical equivalent of a quantum CNOT gate is the XOR gate: its output is the sum modulo two, $\oplus$, of its two inputs. For a classical CNOT gate the target output is equal to the target input if the control input is 0 and flipped[3] if the control input is 1.

A quantum CNOT gate has two inputs as well; the control input is a qubit in state $| \psi \rangle$ and the target input is a qubit in state $| \varphi \rangle$. The operation of the CNOT quantum gate is informally described as follows: the control input is transferred directly to the control output of the gate. The target output is equal to the target input if the control input is $| 0 \rangle$ and it is flipped if the control input is $| 1 \rangle$.

The input and the output qubits of a CNOT quantum gate can be represented as vectors in a four dimensional Hilbert space $\mathcal{H}_4$. Recall that the two qubits applied to the input of the CNOT gate in Figure 4.4 are a control qubit $| \psi \rangle$ and a target qubit $| \varphi \rangle$

$$| \psi \rangle = \alpha_0 | 0 \rangle + \alpha_1 | 1 \rangle, \quad | \varphi \rangle = \beta_0 | 0 \rangle + \beta_1 | 1 \rangle.$$

The input vector of the quantum CNOT gate is:

$$| V_{\text{CNOT}} \rangle = | \psi \rangle \otimes | \varphi \rangle = \begin{pmatrix} \alpha_0 \\ \alpha_1 \end{pmatrix} \otimes \begin{pmatrix} \beta_0 \\ \beta_1 \end{pmatrix} = \begin{pmatrix} \alpha_0\beta_0 \\ \alpha_0\beta_1 \\ \alpha_1\beta_0 \\ \alpha_1\beta_1 \end{pmatrix}.$$

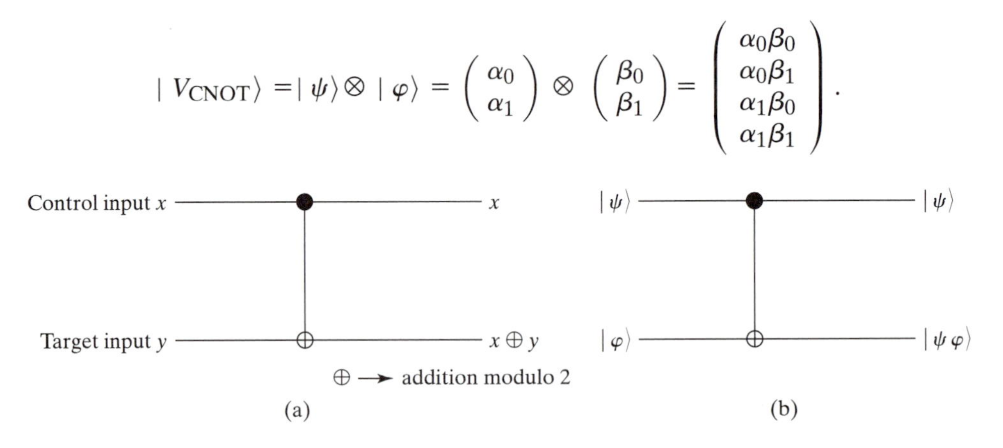

FIGURE 4.4: (a) A classical CNOT gate. The target output is equal to the target input if the control input is 0 and it is flipped if the control input is 1. (b) A quantum CNOT gate has two inputs as well; the control input is a qubit in state $| \psi \rangle$ and the target input is a qubit in state $| \varphi \rangle$.

[3] *Flipping a classical bit a* means complementing it, transforming it to $\bar{a}$: if $a = 0$, it becomes 1 and vice versa. *Flipping a qubit* $| \psi \rangle = \alpha_0 | 0 \rangle + \alpha_1 | 1 \rangle$ results in $| \varphi \rangle = \alpha_1 | 0 \rangle + \alpha_0 | 1 \rangle$, where the projections on the two basis vectors are swapped.

The components of the input vector are transformed by the CNOT quantum gate as follows

$$|\,00\rangle \mapsto |\,00\rangle \quad |\,01\rangle \mapsto |\,01\rangle \quad |\,10\rangle \mapsto |\,11\rangle \quad |\,11\rangle \mapsto |\,10\rangle.$$

The transfer matrix G_{CNOT} of the CNOT quantum gate can be written as a sum of the outer products of the components of the output and input vectors

$$G_{\text{CNOT}} = |\,00\rangle\langle 00\,| + |\,01\rangle\langle 01\,| + |\,11\rangle\langle 10\,| + |\,10\rangle\langle 11\,|$$

Let us start from the basics and construct the basis vectors $|\,00\rangle$, $|\,01\rangle$, $|\,10\rangle$, and $|\,11\rangle$ in the four-dimensional Hilbert space $\mathcal{H}_4$:

$$|\,00\rangle = |\,0\rangle \otimes |\,0\rangle = \begin{pmatrix} 1 \\ 0 \end{pmatrix} \otimes \begin{pmatrix} 1 \\ 0 \end{pmatrix} = \begin{pmatrix} 1 \\ 0 \\ 0 \\ 0 \end{pmatrix},$$

$$|\,01\rangle = |\,0\rangle \otimes |\,1\rangle = \begin{pmatrix} 1 \\ 0 \end{pmatrix} \otimes \begin{pmatrix} 0 \\ 1 \end{pmatrix} = \begin{pmatrix} 0 \\ 1 \\ 0 \\ 0 \end{pmatrix},$$

$$|\,10\rangle = |\,1\rangle \otimes |\,0\rangle = \begin{pmatrix} 0 \\ 1 \end{pmatrix} \otimes \begin{pmatrix} 1 \\ 0 \end{pmatrix} = \begin{pmatrix} 0 \\ 0 \\ 1 \\ 0 \end{pmatrix},$$

$$|\,11\rangle = |\,1\rangle \otimes |\,1\rangle = \begin{pmatrix} 0 \\ 1 \end{pmatrix} \otimes \begin{pmatrix} 0 \\ 1 \end{pmatrix} = \begin{pmatrix} 0 \\ 0 \\ 0 \\ 1 \end{pmatrix}.$$

The outer products of the basis vectors $|\,00\rangle$ and $|\,01\rangle$ with themselves as well as the outer products of $|\,10\rangle$ with $|\,11\rangle$ and $|\,11\rangle$ with $|\,10\rangle$ are:

$$|\,00\rangle\langle 00\,| = \begin{pmatrix} 1 \\ 0 \\ 0 \\ 0 \end{pmatrix} (1\,0\,0\,0) = \begin{pmatrix} 1 & 0 & 0 & 0 \\ 0 & 0 & 0 & 0 \\ 0 & 0 & 0 & 0 \\ 0 & 0 & 0 & 0 \end{pmatrix},$$

$$|\,01\rangle\langle 01\,| = \begin{pmatrix} 0 \\ 1 \\ 0 \\ 0 \end{pmatrix} (0\,1\,0\,0) = \begin{pmatrix} 0 & 0 & 0 & 0 \\ 0 & 1 & 0 & 0 \\ 0 & 0 & 0 & 0 \\ 0 & 0 & 0 & 0 \end{pmatrix},$$

$$| 10\rangle\langle 11 | = \begin{pmatrix} 0 \\ 0 \\ 1 \\ 0 \end{pmatrix} (0\,0\,0\,1) = \begin{pmatrix} 0&0&0&0 \\ 0&0&0&0 \\ 0&0&0&1 \\ 0&0&0&0 \end{pmatrix},$$

$$| 11\rangle\langle 10 | = \begin{pmatrix} 0 \\ 0 \\ 0 \\ 1 \end{pmatrix} (0\,0\,1\,0) = \begin{pmatrix} 0&0&0&0 \\ 0&0&0&0 \\ 0&0&0&0 \\ 0&0&1&0 \end{pmatrix}.$$

Therefore, the transition matrix of the circuit is

$$G_{\text{CNOT}} = \begin{pmatrix} 1&0&0&0 \\ 0&1&0&0 \\ 0&0&0&1 \\ 0&0&1&0 \end{pmatrix}.$$

It is easy to determine the output state vector $| W_{\text{CNOT}}\rangle$ given the input state vector $| V_{\text{CNOT}}\rangle$ and the transfer matrix of a CNOT gate:

$$| W_{\text{CNOT}}\rangle = G_{\text{CNOT}} \, | V_{\text{CNOT}}\rangle$$

$$| W_{\text{CNOT}}\rangle = \begin{pmatrix} 1&0&0&0 \\ 0&1&0&0 \\ 0&0&0&1 \\ 0&0&1&0 \end{pmatrix} \begin{pmatrix} \alpha_0\beta_0 \\ \alpha_0\beta_1 \\ \alpha_1\beta_0 \\ \alpha_1\beta_1 \end{pmatrix} = \begin{pmatrix} \alpha_0\beta_0 \\ \alpha_0\beta_1 \\ \alpha_1\beta_1 \\ \alpha_1\beta_0 \end{pmatrix}.$$

This result can be written as

$$| W_{\text{CNOT}}\rangle = \alpha_0\beta_0 \, | 00\rangle + \alpha_0\beta_1 \, | 01\rangle + \alpha_1\beta_1 \, | 10\rangle + \alpha_1\beta_0 \, | 11\rangle.$$

We see that the circuit in Figure 4.4 preserves the control qubit (the first and the second component of the input vector are replicated in the output vector) and flips the target qubit; the third and fourth component of the input vector $| V_{\text{CNOT}}\rangle$ become the fourth and, respectively, the third component of the output vector.

The output can also be written as

$$| W_{\text{CNOT}}\rangle = \alpha_0 \, | 0\rangle \, [\beta_0 \, | 0\rangle + \beta_1 \, | 1\rangle] + \alpha_1 \, | 1\rangle \, [\beta_1 \, | 0\rangle + \beta_0 \, | 1\rangle].$$

The CNOT gate is reversible. Indeed, the product $G_{\text{CNOT}}G_{\text{CNOT}} = I$:

$$\begin{pmatrix} 1&0&0&0 \\ 0&1&0&0 \\ 0&0&0&1 \\ 0&0&1&0 \end{pmatrix} \begin{pmatrix} 1&0&0&0 \\ 0&1&0&0 \\ 0&0&0&1 \\ 0&0&1&0 \end{pmatrix} = \begin{pmatrix} 1&0&0&0 \\ 0&1&0&0 \\ 0&0&1&0 \\ 0&0&0&1 \end{pmatrix}.$$

The control qubit, $| \psi\rangle$, is replicated at the output and thus, once we know it, we can reconstruct the target input qubit $| \varphi\rangle$ given the target output qubit of the CNOT gate, $| \psi\varphi\rangle$, see Figure 4.4.

Later, we show that CNOT is a universal quantum gate, any multiple-qubit gate can be constructed from single-qubit and CNOT gates.

4.5 CAN WE BUILD QUANTUM COPY MACHINES?

There are several ways to replicate an input signal using classical gates. In Figure 4.5(a), we see a classical CNOT gate, a binary circuit with two inputs, a control bit x and a target bit y. The circuit has two outputs. The control bit, x, is transferred directly to the output and the target bit, y, is transformed into $x \oplus y$. A classical CNOT gate can be implemented with an XOR gate. Indeed, the output of an XOR gate with x and y as input is $z = x \oplus y$.

If the target input of a classical CNOT gate is zero, $y = 0$, the circuit simply replicates input x on both output lines, as shown in Figure 4.5(b). Indeed, $x \oplus 0 = x$, thus a classical CNOT gate allows us to replicate an input bit.

Now that we have a CNOT gate and have figured out how to replicate a classical input bit, let us see if we can copy qubits. It is easy to show that the

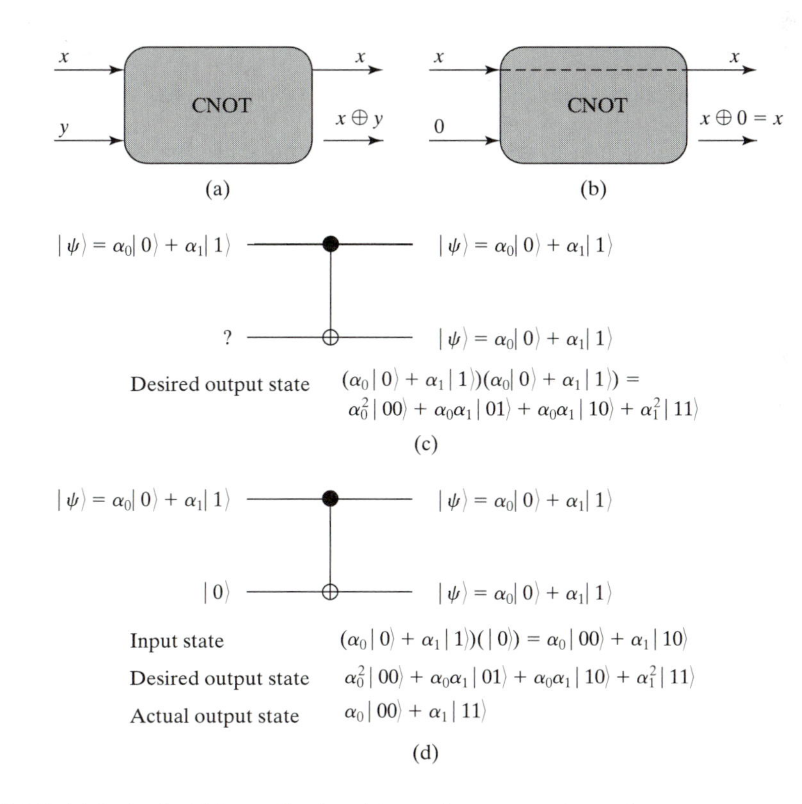

FIGURE 4.5: (a) A classical binary circuit with two inputs x and y and two outputs x and $x \oplus y$. (b) When $y = 0$ the circuit in (a) simply replicates input x on both output lines. (c) A quantum CNOT with an arbitrary input $|\psi\rangle = \alpha_0 |0\rangle + \alpha_1 |1\rangle$; we would like it to replicate $|\psi\rangle$ on its output lines. We know its desired output state but we do not know yet what the second input should be. (d) If we select the second input to be $|0\rangle$ then the output is $\alpha_0 |00\rangle + \alpha_1 |11\rangle$, not exactly what we wished for.

simple way to replicate a classical input discussed earlier does not extend to quantum gates.

In Figure 4.5(c), we show a CNOT quantum gate with an arbitrary control qubit $| \psi \rangle$ as input

$$| \psi \rangle = \alpha_0 \mid 0 \rangle \; + \; \alpha_1 \mid 1 \rangle .$$

We wish to replicate $| \psi \rangle$ on its output lines and try to determine the target qubit that will allow us to do so. To replicate the input implies that the output of this gate should be a vector $| W \rangle$ in $\mathcal{H}_4$

$$\begin{aligned} | W_{\text{desired}} \rangle &= | \psi \rangle \mid \psi \rangle = (\alpha_0 \mid 0 \rangle \; + \; \alpha_1 \mid 1 \rangle)(\alpha_0 \mid 0 \rangle \; + \; \alpha_1 \mid 1 \rangle) \\ &= {\alpha_0}^2 \mid 00 \rangle \; + \; \alpha_0 \alpha_1 \mid 01 \rangle \; + \; \alpha_0 \alpha_1 \mid 10 \rangle \; + \; {\alpha_1}^2 \mid 11 \rangle . \end{aligned}$$

Alternatively,

$$| W_{\text{desired}} \rangle = | \psi \rangle \mid \psi \rangle = | \psi \rangle \otimes \mid \psi \rangle = \begin{pmatrix} \alpha_0 \\ \alpha_1 \end{pmatrix} \otimes \begin{pmatrix} \alpha_0 \\ \alpha_1 \end{pmatrix} = \begin{pmatrix} \alpha_0^2 \\ \alpha_0 \alpha_1 \\ \alpha_1 \alpha_0 \\ \alpha_1^2 \end{pmatrix} .$$

We do not know yet what the input should be, but, based upon the analogy with the classical case, we suspect that $| 0 \rangle$ as the second input may do it. Let us try to determine the actual output state of the CNOT gate in Figure 4.5(d). First we determine the components of its input vector $| V \rangle$

$$| V \rangle = (| \psi \rangle)(| 0 \rangle) = (\alpha_0 \mid 0 \rangle \; + \; \alpha_1 \mid 1 \rangle)(| 0 \rangle) = \alpha_0 \mid 00 \rangle \; + \; \alpha_1 \mid 10 \rangle$$

or

$$| V \rangle = \begin{pmatrix} \alpha_0 \\ 0 \\ \alpha_1 \\ 0 \end{pmatrix} .$$

The actual output vector of the CNOT gate is

$$| W_{\text{actual}} \rangle = G_{\text{CNOT}} \mid V \rangle = \begin{pmatrix} 1 & 0 & 0 & 0 \\ 0 & 1 & 0 & 0 \\ 0 & 0 & 0 & 1 \\ 0 & 0 & 1 & 0 \end{pmatrix} \begin{pmatrix} \alpha_0 \\ 0 \\ \alpha_1 \\ 0 \end{pmatrix} = \begin{pmatrix} \alpha_0 \\ 0 \\ 0 \\ \alpha_1 \end{pmatrix} .$$

We notice that the actual output vector, $| W_{\text{actual}} \rangle$, is different from the desired one, $| W_{\text{desired}} \rangle$. The only conclusion we can draw from the exercise described previously is that the CNOT gate in Figure 4.5(d) cannot be used to copy qubits.

An informal explanation of the limitations of the circuit in Figure 4.5(d) is that once we measure one of the qubits associated with the output state $\alpha_0 \mid 00 \rangle + \alpha_1 \mid 11 \rangle$ we obtain the observed value 00 with probability $| \alpha_0 |^2$ or 11 with probability $|\alpha_1|^2$. Once we measure one qubit the other is completely determined.

In Section 4.9 we discuss the "No Cloning Theorem" which states that qubits cannot be cloned. Recall that a qubit is stored as the state of a binary quantum system, such as the spin of an electron, or the polarization of a photon. Replicating the state $|\psi\rangle$ of a quantum particle is clearly not possible because the state can only be known as a result of a measurement. At the time of a measurement the vector representing the state in a two-dimensional Hilbert space is projected on the two orthonormal basis vectors, $|0\rangle$ and $|1\rangle$ and we observe only one of the two possible eigenvalues, 0 or 1.

4.6 THREE-QUBIT QUANTUM GATES—THE FREDKIN GATE

Three-qubit quantum gates and their classical counterparts have three inputs, a, b, and c and three outputs, a', b', and c'. One or two of the inputs are referred to as *control* qubit(s) and are transferred directly to the output. The other input(s) are referred to as *target* qubit(s).

The gate in Figure 4.6 is called a `Fredkin` gate. There are quantum `Fredkin` gates which are reversible, as well as classical ones that perform similar functions, but are not reversible.

We use the classical version of this gate to construct the truth table and derive the Boolean expressions relating the inputs and the outputs. The classical `Fredkin` gate has two target inputs, a, b, and a control input c and three outputs, a', b' and c'. The input of the `Fredkin` gate determines the output as follows:

1. The control input c is transferred directly to the output, $c' = c$.
2. When $c = 0$, the two target inputs are transferred without modification to the output: $a' = a$ and $b' = b$, see Figure 4.6(a). The truth table for this case has four entries, one for each possible combination of values of a and b.
3. When $c = 1$, the two target inputs are swapped: $a' = b$ and $b' = a$, see Figure 4.6(b) for the circuit and its truth table.

The Boolean equations relating the output and the input of the `Fredkin` gate can be obtained from the truth table of the circuit. The full truth table of the `Fredkin` gate in Figure 4.6(c) is obtained by concatenating the truth tables for the two configurations in Figures 4.6(a) and (b).

The Boolean expressions for the output of the `Fredkin` gate are derived following the rules discussed for the full adder circuit presented at the beginning of this section. Each output is a logical sum consisting of four terms, one for every entry when the desired output is true. Each term is the logical product of the corresponding inputs taken either directly or negated.

$$a' = a\bar{b}\bar{c} + ab\bar{c} + \bar{a}bc + abc = a\bar{c}(\bar{b} + b) + bc(\bar{a} + a) = a\bar{c} + bc$$

$$b' = \bar{a}b\bar{c} + ab\bar{c} + a\bar{b}c + abc = b\bar{c}(\bar{a} + a) + ac(\bar{b} + b) = b\bar{c} + ac$$

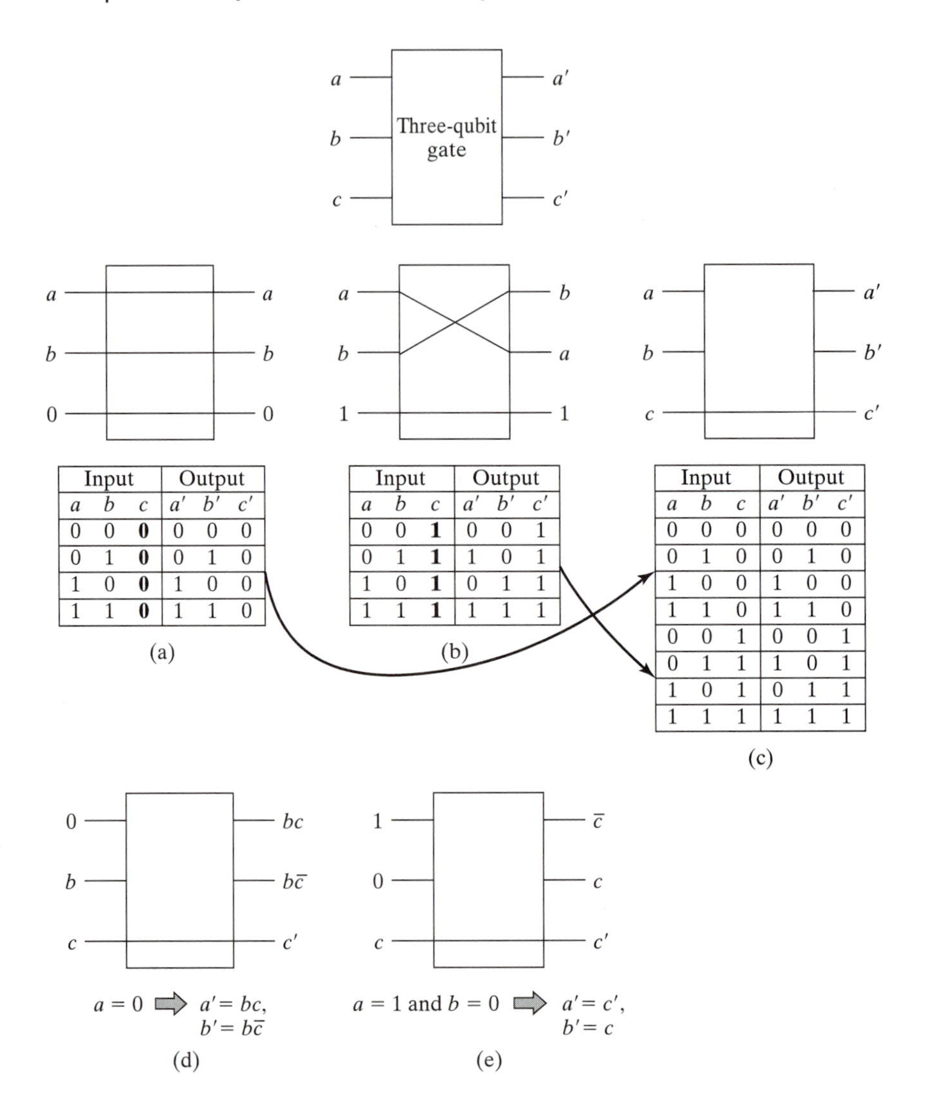

FIGURE 4.6: The `Fredkin` gate has three inputs, a, b, and control, or c; it also has three outputs, a', b', and $c' = c$. (a) When $c = 0$ the inputs appear at the output, $a' = a$ and $b' = b$. (b) When $c = 1$ the inputs are swapped, $a' = b$ and $b' = a$. (c) The truth table for the `Fredkin` gate. (d) When $a = 0$ then $a' = bc$ and $b' = b\bar{c}$, the `Fredkin` gate becomes an AND gate. (e) When $a = 1$ and $b = 0$ the `Fredkin` gate becomes a NOT gate.

$$c' = \bar{a}\bar{b}c + \bar{a}bc + a\bar{b}c + abc = \bar{a}c(\bar{b} + b) + ac(b + \bar{b}) = \bar{a}c + ac$$
$$= c(\bar{a} + a) = c.$$

The last equation confirms that the control input is transferred directly to the output.

Let us now turn our attention to the quantum `Fredkin` gate. A system of three qubits requires an eight-dimensional complex vector space with a basis consisting of eight vectors: $|000\rangle$, $|001\rangle$, $|010\rangle$, $|011\rangle$, $|100\rangle$, $|101\rangle$, $|110\rangle$, and $|111\rangle$.

We now compute the individual basis vectors. We see that the tensor product of the three single-qubit state vectors is:

$$\begin{pmatrix} a \\ b \end{pmatrix} \otimes \begin{pmatrix} c \\ d \end{pmatrix} \otimes \begin{pmatrix} e \\ f \end{pmatrix} = \begin{pmatrix} ac \\ ad \\ bc \\ bd \end{pmatrix} \otimes \begin{pmatrix} e \\ f \end{pmatrix} = \begin{pmatrix} ace \\ acf \\ cde \\ cdf \\ bce \\ bcf \\ bde \\ bdf \end{pmatrix}$$

where the basis vectors in $\mathcal{H}_2$ are:

$$|0\rangle = \begin{pmatrix} 1 \\ 0 \end{pmatrix} \qquad |1\rangle = \begin{pmatrix} 0 \\ 1 \end{pmatrix}.$$

Each of the basis vectors in $\mathcal{H}_8$ is a tensor product of three basis vectors in $\mathcal{H}_2$:

$$|000\rangle = \begin{pmatrix} 1 \\ 0 \\ 0 \\ 0 \\ 0 \\ 0 \\ 0 \\ 0 \end{pmatrix} \quad |001\rangle = \begin{pmatrix} 0 \\ 1 \\ 0 \\ 0 \\ 0 \\ 0 \\ 0 \\ 0 \end{pmatrix} \quad |010\rangle = \begin{pmatrix} 0 \\ 0 \\ 1 \\ 0 \\ 0 \\ 0 \\ 0 \\ 0 \end{pmatrix} \quad |011\rangle = \begin{pmatrix} 0 \\ 0 \\ 0 \\ 1 \\ 0 \\ 0 \\ 0 \\ 0 \end{pmatrix},$$

$$|100\rangle = \begin{pmatrix} 0 \\ 0 \\ 0 \\ 0 \\ 1 \\ 0 \\ 0 \\ 0 \end{pmatrix} \quad |101\rangle = \begin{pmatrix} 0 \\ 0 \\ 0 \\ 0 \\ 0 \\ 1 \\ 0 \\ 0 \end{pmatrix} \quad |110\rangle = \begin{pmatrix} 0 \\ 0 \\ 0 \\ 0 \\ 0 \\ 0 \\ 1 \\ 0 \end{pmatrix} \quad |111\rangle = \begin{pmatrix} 0 \\ 0 \\ 0 \\ 0 \\ 0 \\ 0 \\ 0 \\ 1 \end{pmatrix}.$$

We observe that each basis vector in $\mathcal{H}_8$ is a column vector with a single 1 in the row corresponding to the binary representation of the corresponding integer plus 1. Indeed, $|000\rangle$ has a single 1 in row 1 $(1 = 0 + 1)$, $|011\rangle$ has a single 1 in row 4 $(4 = 3 + 1)$, $|101\rangle$ has a single 1 in row 6 $(6 = 5 + 1)$, and so on. These basis vectors correspond to the respective columns of the 8×8 identity matrix

$$I_8 = \begin{pmatrix} 1&0&0&0&0&0&0&0 \\ 0&1&0&0&0&0&0&0 \\ 0&0&1&0&0&0&0&0 \\ 0&0&0&1&0&0&0&0 \\ 0&0&0&0&1&0&0&0 \\ 0&0&0&0&0&1&0&0 \\ 0&0&0&0&0&0&1&0 \\ 0&0&0&0&0&0&0&1 \end{pmatrix}.$$

Let us now construct the matrix representation, G_{Fredkin}, of the linear operator $\mathbf{G}_{\text{Fredkin}}$. The truth table in Figure 4.6(c) shows that the quantum circuit maps its input to the output as follows:

$$\begin{aligned} |000\rangle &\mapsto |000\rangle \\ |001\rangle &\mapsto |001\rangle \\ |010\rangle &\mapsto |010\rangle \\ |011\rangle &\mapsto |101\rangle \\ |100\rangle &\mapsto |100\rangle \\ |101\rangle &\mapsto |011\rangle \\ |110\rangle &\mapsto |110\rangle \\ |111\rangle &\mapsto |111\rangle. \end{aligned}$$

Therefore

$$\begin{aligned} G_{\text{Fredkin}} = {} & |000\rangle\langle 000| + |001\rangle\langle 001| + |010\rangle\langle 010| + |011\rangle\langle 101| \\ & + |100\rangle\langle 100| + |101\rangle\langle 011| + |110\rangle\langle 110| + |111\rangle\langle 111|. \end{aligned}$$

It is easy to compute the outer products in the expression above. We only compute two terms of this sum to illustrate the procedure:

$$|011\rangle\langle 101| = \begin{pmatrix} 0 \\ 0 \\ 0 \\ 1 \\ 0 \\ 0 \\ 0 \\ 0 \end{pmatrix} (0\,0\,0\,0\,0\,1\,0\,0) = \begin{pmatrix} 0&0&0&0&0&0&0&0 \\ 0&0&0&0&0&0&0&0 \\ 0&0&0&0&0&0&0&0 \\ 0&0&0&0&0&1&0&0 \\ 0&0&0&0&0&0&0&0 \\ 0&0&0&0&0&0&0&0 \\ 0&0&0&0&0&0&0&0 \\ 0&0&0&0&0&0&0&0 \end{pmatrix}$$

and

$$| 101\rangle\langle 011 | = \begin{pmatrix} 0 \\ 0 \\ 0 \\ 0 \\ 0 \\ 1 \\ 0 \\ 0 \end{pmatrix} (0\,0\,0\,1\,0\,0\,0\,0) = \begin{pmatrix} 0&0&0&0&0&0&0&0 \\ 0&0&0&0&0&0&0&0 \\ 0&0&0&0&0&0&0&0 \\ 0&0&0&0&0&0&0&0 \\ 0&0&0&0&0&0&0&0 \\ 0&0&0&1&0&0&0&0 \\ 0&0&0&0&0&0&0&0 \\ 0&0&0&0&0&0&0&0 \end{pmatrix}.$$

Finally, we add the eight terms and obtain

$$G_{\text{Fredkin}} = \begin{pmatrix} 1&0&0&0&0&0&0&0 \\ 0&1&0&0&0&0&0&0 \\ 0&0&1&0&0&0&0&0 \\ 0&0&0&0&0&1&0&0 \\ 0&0&0&0&1&0&0&0 \\ 0&0&0&1&0&0&0&0 \\ 0&0&0&0&0&0&1&0 \\ 0&0&0&0&0&0&0&1 \end{pmatrix}.$$

In $\mathcal{H}_8$, the input state of a `Fredkin` gate, $| V\rangle$, is a superposition of the basis vectors with complex coefficients $\alpha_{000}, \alpha_{001}, \alpha_{010}, \alpha_{011}, \alpha_{100}, \alpha_{101}, \alpha_{110}, \alpha_{111}$:

$$\begin{aligned} | V_{\text{Fredkin}}\rangle = {} & \alpha_{000} \,| 000\rangle + \alpha_{001} \,| 001\rangle + \alpha_{010} \,| 010\rangle + \alpha_{011} \,| 011\rangle \\ & + \alpha_{100} \,| 100\rangle + \alpha_{101} \,| 101\rangle + \alpha_{110} \,| 110\rangle + \alpha_{111} \,| 111\rangle. \end{aligned}$$

The input and output vectors of a quantum `Fredkin` gate are related by a set of linear equations

$$| W_{\text{Fredkin}}\rangle = G_{\text{Fredkin}} \,| V_{\text{Fredkin}}\rangle.$$

Now we discuss several properties of the `Fredkin` gate. First, we show that indeed the `Fredkin` gate is reversible: knowing a', b', and c' we can determine a, b, and c. This is easy to prove; for example, one can express a, b, and c as functions of a', b', and c'. Using the truth table in Figure 4.6(c) we obtain:

$$\begin{aligned} a &= a'\overline{b'}\,\overline{c'} + a'b'\overline{c'} + \overline{a'}b'c' + a''b''c' \\ &= a'\overline{c'}(\overline{b'} + b') + b'c'(\overline{a'} + a') = a'\overline{c'} + b'c', \\ b &= \overline{a'}b'\overline{c'} + a'b'\overline{c'} + a'\overline{b'}c' + a'b'c' \\ &= b'\overline{c'}(\overline{a'} + a'') + a'c'(\overline{b'} + b') = b'\overline{c'} + a'c', \\ c &= \overline{a'}\,\overline{b'}c' + a'\overline{b'}c' + \overline{a'}b'c + a'b'c' \\ &= \overline{b'}c''(\overline{a'} + a') + b'c'(\overline{a'} + a') = \overline{b'}c' + b'c' = c'. \end{aligned}$$

If we apply two consecutive Fredkin gates with (a, b, c) as input and (a', b', c') as output for the first, and (a', b', c') as input and (a'', b'', c'') as output for the second gate, then the output of the second gate is the same as the input of the first one

$$a'' = a, \quad b'' = b, \quad c'' = c.$$

We examine the truth table once again and observe that the Fredkin gate conserves the number of 1s at its input and for this reason is called a *conservative logic gate*. This property suggests analogies with other physics laws regarding the conservation of mass, energy, and momentum.

Last, but not least, we show that the Fredkin gate is universal, that is, it can simulate both AND and NOT gates. Consider the circuit in Figure 4.6(d) with $a = 0$. In this case,

$$a' = bc \quad \text{and} \quad b' = b\bar{c}.$$

Thus, the output of the Fredkin gate is identical to the output of an AND gate (recall that the product of b and c corresponds to the Boolean AND). Now we show that the Fredkin gate can simulate a NOT gate. The case when $a = 1$ and $b = 0$ is shown in Figure 4.6(e). We see that

$$a' = \bar{c} \quad b' = c.$$

Thus, the output of the Fredkin gate is identical to the output of a NOT gate. The configuration in Figure 4.6(e) generates two copies of the input c and can be used as a FANOUT gate. The Fredkin gate can perform a switching function called CROSSOVER when it swaps its two inputs as in Figure 4.6(b).

4.7 THE TOFFOLI GATE

While the Fredkin gate had only one control input, the Toffoli gate has two control inputs, a and b, and one target input c. The outputs are: $a' = a$, $b' = b$ and c', see Figure 4.7(a). The Toffoli gate is a universal gate and it is reversible.

There are both classical and quantum versions of a Toffoli gate. The truth table and the transfer matrix of classical and quantum Toffoli gates are identical. For the sake of clarity we describe first the function of a classical

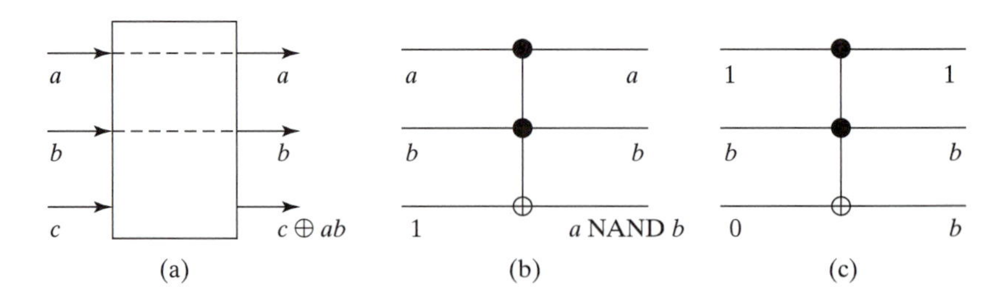

FIGURE 4.7: (a) The Toffoli gate has three inputs, two control inputs, a and b, and a target, c. The outputs are $a' = a$, $b' = b$, and c'. If both control bits or qubits are 1 then c is flipped, otherwise its state is unchanged. (b) A classical Toffoli gate can be used to implement a NAND gate; when $c = 1$ then $c' = 1 \oplus (a \text{ AND } b) = \text{NOT}(a \text{ AND } b)$. If $c = 0$ then $c' = a \text{ AND } b$. (c) A quantum Toffoli gate performs a FANOUT function.

`Toffoli` gate. If both control bits are 1 then c is flipped, otherwise its state is unchanged. The truth table of the classical `Toffoli` gate reflects the functional description, $c' = c$ for the first six entries and $c' = \overline{c}$ for the last two, when $a = b = 1$:

a	b	c	a'	b'	c'
0	0	0	0	0	0
0	0	1	0	0	1
0	1	0	0	1	0
0	1	1	0	1	1
1	0	0	1	0	0
1	0	1	1	0	1
1	1	0	1	1	1
1	1	1	1	1	0

From the truth table we see that $c' = c \oplus (a \texttt{ AND } b)$. From the definition of addition modulo two ($\oplus$) it follows immediately that for any binary variable a, $1 \oplus a = \overline{a}$ and $0 \oplus a = a$.

If $c = 1$ then $c' = a \texttt{ NAND } b$. If $c = 0$ then $c' = a \texttt{ AND } b$. The `Toffoli` gate implements both `NAND` and `AND` functions, see Figure 4.7(b). Figure 4.7(c) shows a `FANOUT` circuit; when $a = 1$ and $c = 0$, the target output $c' = c \oplus (a\texttt{AND}b) = 0 \oplus b = b$. The second control input bit is replicated.

Let us now examine the quantum `Toffoli` gate. Let the three input qubits corresponding to a, b, and c be

$$| \psi\rangle = \alpha_0 \mid 0\rangle + \alpha_1 \mid 1\rangle$$

$$| \varphi\rangle = \beta_0 \mid 0\rangle + \beta_1 \mid 1\rangle$$

$$| \zeta\rangle = \gamma_0 \mid 0\rangle + \gamma_1 \mid 1\rangle.$$

Then the input state is given by the tensor product of the three qubits, $| V_{\text{Toffoli}}\rangle = | \psi\rangle \otimes | \varphi\rangle \otimes | \zeta\rangle$. If we substitute the expressions for the three qubits and carry out the vector multiplication, we get

$$| V_{\text{Toffoli}}\rangle = \alpha_0\beta_0\gamma_0 \mid 000\rangle + \alpha_0\beta_0\gamma_1 \mid 001\rangle + \alpha_0\beta_1\gamma_0 \mid 010\rangle + \alpha_0\beta_1\gamma_1 \mid 011\rangle + \alpha_1\beta_0\gamma_0 \mid 100\rangle + \alpha_1\beta_0\gamma_1 \mid 101\rangle + \alpha_1\beta_1\gamma_0 \mid 110\rangle + \alpha_1\beta_1\gamma_1 \mid 111\rangle.$$

The vector describing the input state is:

$$| V_{\text{Toffoli}}\rangle = \begin{pmatrix} \alpha_0 \\ \alpha_1 \end{pmatrix} \otimes \begin{pmatrix} \beta_0 \\ \beta_1 \end{pmatrix} \otimes \begin{pmatrix} \gamma_0 \\ \gamma_1 \end{pmatrix} = \begin{pmatrix} \alpha_0\beta_0 \\ \alpha_0\beta_1 \\ \alpha_1\beta_0 \\ \alpha_1\beta_1 \end{pmatrix} \otimes \begin{pmatrix} \gamma_0 \\ \gamma_1 \end{pmatrix} = \begin{pmatrix} \alpha_0\beta_0\gamma_0 \\ \alpha_0\beta_0\gamma_1 \\ \alpha_0\beta_1\gamma_0 \\ \alpha_0\beta_1\gamma_1 \\ \alpha_1\beta_0\gamma_0 \\ \alpha_1\beta_0\gamma_1 \\ \alpha_1\beta_1\gamma_0 \\ \alpha_1\beta_1\gamma_1 \end{pmatrix}.$$

Let us now construct the matrix representation G_{Toffoli} of the linear operator $\mathbf{G}_{\text{Toffoli}}$. The truth table of the `Toffoli` gate shows that the quantum circuit maps its input to the output as follows

$$
\begin{aligned}
|\,000\rangle &\mapsto |\,000\rangle \\
|\,001\rangle &\mapsto |\,001\rangle \\
|\,010\rangle &\mapsto |\,010\rangle \\
|\,011\rangle &\mapsto |\,011\rangle \\
|\,100\rangle &\mapsto |\,100\rangle \\
|\,101\rangle &\mapsto |\,101\rangle \\
|\,110\rangle &\mapsto |\,111\rangle \\
|\,111\rangle &\mapsto |\,110\rangle.
\end{aligned}
$$

Therefore

$$
\begin{aligned}
G_{\text{Toffoli}} = {} & |\,000\rangle\langle 000\,| + |\,001\rangle\langle 001\,| + |\,010\rangle\langle 010\,| + |\,011\rangle\langle 011\,| \\
& + |\,100\rangle\langle 100\,| + |\,101\rangle\langle 101\,| + |\,110\rangle\langle 111\,| + |\,111\rangle\langle 110\,|\,.
\end{aligned}
$$

The transfer matrix of a `Toffoli` gate is

$$
G_{\text{Toffoli}} = \begin{pmatrix}
1&0&0&0&0&0&0&0\\
0&1&0&0&0&0&0&0\\
0&0&1&0&0&0&0&0\\
0&0&0&1&0&0&0&0\\
0&0&0&0&1&0&0&0\\
0&0&0&0&0&1&0&0\\
0&0&0&0&0&0&0&1\\
0&0&0&0&0&0&1&0
\end{pmatrix}.
$$

The output of the `Toffoli` gate is $|\,W_{\text{Toffoli}}\rangle = G_{\text{Toffoli}}\,|\,V_{\text{Toffoli}}\rangle$

$$
|\,W_{\text{Toffoli}}\rangle = \begin{pmatrix}
1&0&0&0&0&0&0&0\\
0&1&0&0&0&0&0&0\\
0&0&1&0&0&0&0&0\\
0&0&0&1&0&0&0&0\\
0&0&0&0&1&0&0&0\\
0&0&0&0&0&1&0&0\\
0&0&0&0&0&0&0&1\\
0&0&0&0&0&0&1&0
\end{pmatrix}
\begin{pmatrix}
\alpha_0\beta_0\gamma_0\\ \alpha_0\beta_0\gamma_1\\ \alpha_0\beta_1\gamma_0\\ \alpha_0\beta_1\gamma_1\\ \alpha_1\beta_0\gamma_0\\ \alpha_1\beta_0\gamma_1\\ \alpha_1\beta_1\gamma_0\\ \alpha_1\beta_1\gamma_1
\end{pmatrix}
=
\begin{pmatrix}
\alpha_0\beta_0\gamma_0\\ \alpha_0\beta_0\gamma_1\\ \alpha_0\beta_1\gamma_0\\ \alpha_0\beta_1\gamma_1\\ \alpha_1\beta_0\gamma_0\\ \alpha_1\beta_0\gamma_1\\ \alpha_1\beta_1\gamma_1\\ \alpha_1\beta_1\gamma_0
\end{pmatrix}.
$$

4.8 QUANTUM CIRCUITS

We learned in the previous chapters that quantum particles can be used to store information. The size of the state space of a classical system with n bits is 2^n, while the state of a quantum circuit with n qubits is a vector in $\mathcal{H}_{2^n}$. We could simulate very complex physical systems with an n-qubit quantum computer. Now we ask ourselves what a quantum computer might be and how the qubits can be transformed inside such a quantum computing device.

According to Andrew Steane [131], "the quantum computer is first and foremost a machine, which is a theoretical construct, like a thought experiment, whose purpose is to allow quantum information processing to be formally analyzed." In the same paper [131], Steane provides a more precise definition of a quantum computer put forward by David Deutsch "A quantum computer is a set of n qubits in which the following operations are feasible:

1. Each qubit can be prepared in some known state $| 0\rangle$.
2. Each qubit can be measured in the basis $\{| 0\rangle, | 1\rangle\}$.
3. A universal quantum gate (or set of gates) can be applied at will to any fixed-size subset of qubits.
4. The qubits do not evolve other than via the above transformations."

Quantum circuits are built by interconnecting quantum gates. Several limitations are imposed in the realization of quantum circuits. First, the circuits are acyclic. Feedback from one part of the circuit to another is not allowed; there are no loops. Second, we cannot copy qubits. These restrictions are necessary to ensure the reversibility of quantum circuits, a topic addressed in depth in Chapter 6.

4.9 THE NO CLONING THEOREM

The transformations carried out by quantum circuits are unitary. As a consequence of this fact, unknown quantum states cannot be copied or cloned. Several proofs of this theorem are available [98].

Here we present a proof by contradiction first given in [145]. Let us assume that a two input gate capable of cloning one of its inputs exists. By now we know that a gate performs a linear transformation of its input vector $| V\rangle$ into an output vector $| W\rangle = G | V\rangle$ with G the transfer matrix of the gate. We already know that transformations performed by matrices can be represented by linear operators. Let us call **G** the unitary transformation corresponding to the unitary matrix G of a two-input gate that allows us to replicate an input qubit.

Let $| \psi\rangle$ and $| \varphi\rangle$ be two orthogonal quantum states of two qubits. We apply each one of them at the first input of the gate independently, while the second input is $| 0\rangle$; assuming that the gate clones its input, it follows that

$$\mathbf{G}(| \psi\rangle \otimes | 0\rangle) = | \psi\rangle \otimes | \psi\rangle$$

or in a compact notation

$$\mathbf{G}(| \psi 0\rangle) = | \psi\psi\rangle.$$

Also,

$$\mathbf{G}(| \varphi\rangle \otimes | 0\rangle) = | \varphi\rangle \otimes | \varphi\rangle$$

or

$$\mathbf{G}(| \varphi 0\rangle) = | \varphi\varphi\rangle.$$

These two equations simply state that when the second input qubit is $| 0\rangle$, then the state of the quantum system at the output of the gate is composed of two replicas of the first qubit, namely $| \psi\rangle$ for the first case, and $| \varphi\rangle$ for the second case. Recall that the input and the output states are represented by vectors in $\mathcal{H}_4$ obtained by taking the tensor product of the two vectors describing the state of the two qubits at the input, respectively, the output of the gate.

Now consider another state $| \xi\rangle = (1/\sqrt{2})(| \psi\rangle + | \varphi\rangle)$. The operator $\mathbf{G}$ represents a linear transformation, thus

$$\mathbf{G}(| \xi 0\rangle) = \frac{1}{\sqrt{2}}[\mathbf{G}(| \psi 0\rangle) + \mathbf{G}(| \varphi 0\rangle)] = \frac{1}{\sqrt{2}}(| \psi\psi\rangle + | \varphi\varphi\rangle).$$

We could try to apply $| \xi\rangle$ at the input of our cloning gate. Then we expect $| \xi\rangle$ to be cloned

$$\mathbf{G}(| \xi 0\rangle) = | \xi\xi\rangle.$$

But

$$| \xi\xi\rangle = \frac{| \psi\rangle + | \varphi\rangle}{\sqrt{2}} \frac{| \psi\rangle + | \varphi\rangle}{\sqrt{2}} = \frac{1}{2}(| \psi\psi\rangle + | \psi\varphi\rangle + | \varphi\psi\rangle + | \varphi\varphi\rangle).$$

This contradicts the expression we found earlier when assuming linearity

$$\mathbf{G}(| \xi 0\rangle) = \frac{1}{\sqrt{2}}(| \psi\psi\rangle + | \varphi\varphi\rangle).$$

We have to conclude that there is no linear operation that reliably clones unknown quantum states. It is possible to clone known states, after a measurement has been performed. Before the measurement of a quantum system in an unknown state, the outcome of the measurement is uncertain. After the measurement, the outcome is the projection of the state on one of the basis vectors, thus it is well-determined.

In summary, non-orthogonal pure quantum states $\{| \psi_i\rangle\}$ *cannot be cloned. This means that no physical system able to carry out the transformation* $| \psi_i\rangle \longmapsto | \psi_i\rangle | \psi_i\rangle$ *may exist.* As Jozsa points out [79], "Although the impossibility of cloning in quantum theory can be attributed to the fact that such a process is non-unitary or non-linear, from an information point of view we can understand it by saying that two copies of a quantum system embody strictly more *information* than it is available in strictly one copy, so cloning must be impossible."

4.10 QUBIT SWAPPING AND FULL ADDER CIRCUITS

In this section, we give examples of quantum circuits. Figure 4.8 presents a two-bit adder circuit constructed with reversible gates. Figure 4.9 displays a circuit for swapping two qubits.

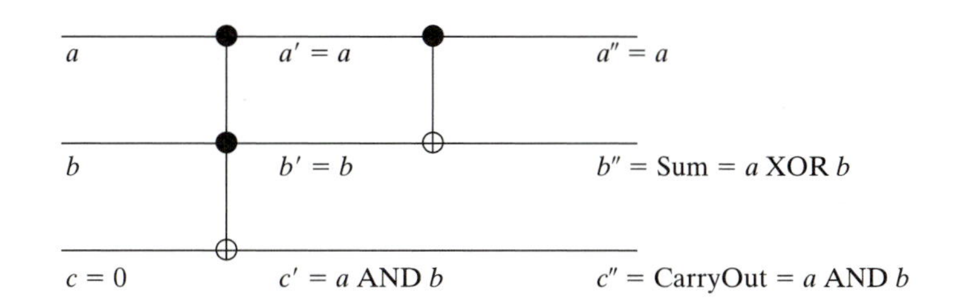

FIGURE 4.8: A two-bit adder made of reversible gates.

Let us first examine the circuit in Figure 4.8. The two-bit adder consists of a `Toffoli` gate followed by a `CNOT` gate. The two control inputs to the `Toffoli` gate are a and b and its target input is $c = 0$.

The target output of the `Toffoli` gate is (recall that $c = 0$)

$$c' = c \oplus (a \text{ AND } b) = a \text{ AND } b.$$

This state propagates to the output and

$$c' = a \text{ AND } b.$$

Then $a'' = a' = a$ as the control input of the `CNOT` propagates to its output unchanged. Finally, $b' = b$ and this becomes the target input of the `CNOT` gate. The target output of the `CNOT` gate is

$$b'' = a' \text{ XOR } b' = a \text{ XOR } b.$$

Thus, the two outputs of the circuit are the *Sum* and the *CarryOut* of the two inputs, a and b.

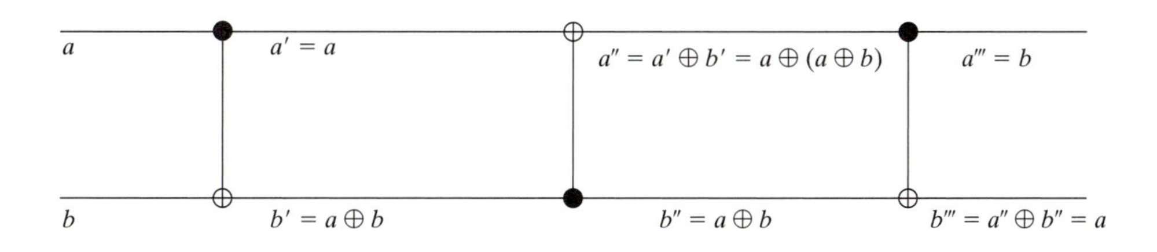

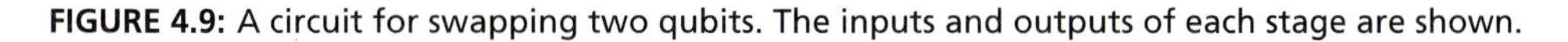

FIGURE 4.9: A circuit for swapping two qubits. The inputs and outputs of each stage are shown.

The circuit for swapping two qubits in Figure 4.9 consists of three `CNOT` gates. Given Boolean variables a and b we note that

$$a \oplus (a \oplus b) = (a \oplus a) \oplus b = b$$

and

$$b \oplus (a \oplus b) = (b \oplus b) \oplus a = a.$$

It is easy to see that $a \oplus a = 0$ and $0 \oplus b = b$.

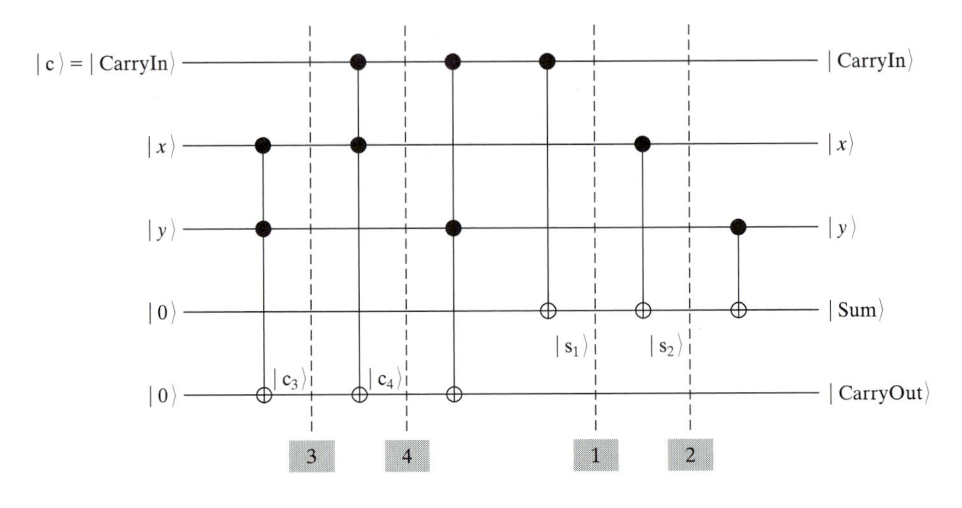

FIGURE 4.10: A full one-qubit adder constructed with `Toffoli` and `CNOT` gates.

In order to show that the two inputs are swapped we determine the output of each stage. The inputs to the second stage are $a' = a$ and $b' = a \oplus b$. The inputs to the third stage are $a'' = a \oplus (a \oplus b) = b$ and $b'' = a \oplus b$. Based upon the previous observations, we see that the outputs of the third stage are $a''' = b$ and $b''' = b \oplus (a \oplus b) = a$, thus the circuit swaps it two inputs.

Now we show that the circuit in Figure 4.10 constructed with `Toffoli` and `CNOT` gates implements a *full one-qubit adder*. First, we observe that the first three qubits from the top, namely $| c\rangle = | CarryIn\rangle$, $| x\rangle$, and $| y\rangle$ are the control qubits on all gates thus, they are transferred without any change to the output. We only have to compute expressions for $| Sum\rangle$ and $| CarryOut\rangle$. For simplicity, we drop the `ket` notations and denote qubit vector $| x\rangle$ simply as x.

If we call s_1 and s_2 the results of `CNOT` transformations at stages marked as 1 and 2 in Figure 4.10 we see that

$$\begin{aligned} s_1 &= c \oplus 0 = c, \\ s_2 &= s_1 \oplus x = c \oplus x, \\ Sum &= s_2 \oplus y = (c \oplus x) \oplus y. \end{aligned}$$

It is easy to show that

$$(c \oplus x) \oplus y = xyc + x\overline{yc} + y\overline{xc} + c\overline{xy}.$$

This is precisely the expression for the *Sum* derived in Section 4.1. To prove this equality we use the fact that $x \oplus y = x\overline{y} + \overline{x}y$ and also apply de Morgan's laws $\overline{x + y} = \overline{x}\overline{y}$ and $\overline{xy} = \overline{x} + \overline{y}$.

We call c_3 and c_4 the results of transformations performed by `Toffoli` gates at stages marked as 3 and 4 in Figure 4.10 we see that

$$\begin{aligned} c_3 &= 0 \oplus (xy) = xy \\ c_4 &= c_3 \oplus (cx) = (xy) \oplus (cx), \\ CarryOut &= c_4 \oplus (cy) = (xy) \oplus (cx) \oplus (cy). \end{aligned}$$

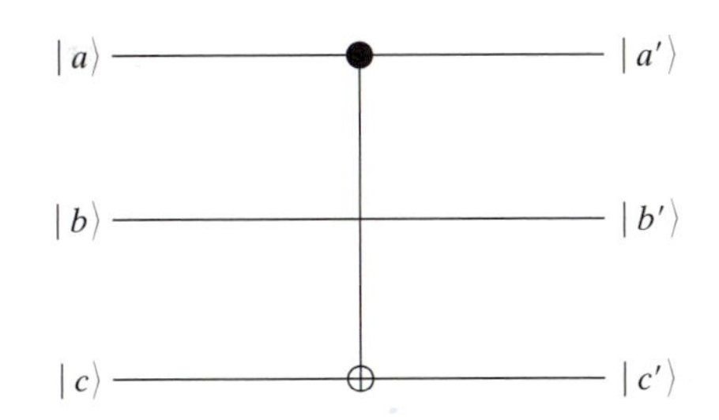

FIGURE 4.11: A three-qubit gate with a CNOT.

It is easy to show that

$$(xy) \oplus (cx) \oplus (cy) = xyc + \bar{x}yc + x\bar{y}c + xy\bar{c}.$$

We recognize the expression for the *CarryOut* derived in Section 4.1.

Let us now compute the transfer matrix of the circuit in Figure 4.11. We observe that the top most qubit $|\, a\rangle$ is the control qubit of a CNOT gate, thus it is transferred to the output unchanged, $|\, a'\rangle = |\, a\rangle$. The second qubit is not affected, $|\, b'\rangle = |\, b\rangle$. The third qubit is flipped when $|\, a\rangle = |\, 1\rangle$, and left unchanged if $|\, a\rangle = |\, 0\rangle$. Henceforth, the truth table of this circuit is:

a	b	c	a'	b'	c'
0	0	0	0	0	0
0	0	1	0	0	1
0	1	0	0	1	0
0	1	1	0	1	1
1	0	0	1	0	1
1	0	1	1	0	0
1	1	0	1	1	1
1	1	1	1	1	0

This means that the quantum circuit maps its input to the output as follows

$$|\, 000\rangle \mapsto |\, 000\rangle$$
$$|\, 001\rangle \mapsto |\, 001\rangle$$
$$|\, 010\rangle \mapsto |\, 010\rangle$$
$$|\, 011\rangle \mapsto |\, 011\rangle$$
$$|\, 100\rangle \mapsto |\, 101\rangle$$
$$|\, 101\rangle \mapsto |\, 100\rangle$$
$$|\, 110\rangle \mapsto |\, 111\rangle$$
$$|\, 111\rangle \mapsto |\, 110\rangle.$$

Therefore

$$G_Q = |\,000\rangle\langle 000\,| + |\,001\rangle\langle 001\,| + |\,010\rangle\langle 010\,| + |\,011\rangle\langle 011\,| \\ + |\,100\rangle\langle 101\,| + |\,101\rangle\langle 100\,| + |\,110\rangle\langle 111\,| + |\,111\rangle\langle 110\,|\,.$$

After we carry out the outer products we obtain

$$G_Q = \begin{pmatrix} 1 & 0 & 0 & 0 & 0 & 0 & 0 & 0 \\ 0 & 1 & 0 & 0 & 0 & 0 & 0 & 0 \\ 0 & 0 & 1 & 0 & 0 & 0 & 0 & 0 \\ 0 & 0 & 0 & 1 & 0 & 0 & 0 & 0 \\ 0 & 0 & 0 & 0 & 0 & 1 & 0 & 0 \\ 0 & 0 & 0 & 0 & 1 & 0 & 0 & 0 \\ 0 & 0 & 0 & 0 & 0 & 0 & 0 & 1 \\ 0 & 0 & 0 & 0 & 0 & 0 & 1 & 0 \end{pmatrix} = \begin{bmatrix} I_4 & 0 & \\ & \sigma_x & 0 \\ 0 & 0 & \sigma_x \end{bmatrix}$$

with I_4 the 4×4 identity matrix and σ_x the Pauli matrix, X.

4.11 MORE ABOUT UNITARY OPERATIONS AND ROTATION MATRICES

In the previous sections, we stressed the fact that single-qubit transformations must be unitary to preserve the norm. We also discussed the Pauli matrices associated with the rotation operators.

These rotation operators apply only to spin 1/2 particles, such as electrons with basis states denoted as $|\uparrow\rangle$ and $|\downarrow\rangle$. Their action can be summarized as follows

Operator	**Transformation**	**Comments**
σ_x	$\sigma_x \,|\uparrow\rangle = |\downarrow\rangle$ $\sigma_x \,|\downarrow\rangle = |\uparrow\rangle$	σ_x flips the components of a qubit
σ_y	$\sigma_y \,|\uparrow\rangle = i\,|\downarrow\rangle$ $\sigma_y \,|\downarrow\rangle = -i\,|\uparrow\rangle$	σ_y flips a qubit and multiplies it by i
σ_z	$\sigma_z \,|\uparrow\rangle = |\uparrow\rangle$ $\sigma_z \,|\downarrow\rangle = -\,|\downarrow\rangle$	σ_z changes the phase of a qubit
I	$I\,|\uparrow\rangle = |\uparrow\rangle$ $I\,|\downarrow\rangle = |\downarrow\rangle$	I leaves a qubit in the same state

The three matrices σ_x, σ_y, and σ_z can be constructed from the description of the transformations carried out by the corresponding operators. For example,

$$\sigma_y = -i\,|\,0\rangle\langle 1\,| + i\,|\,1\rangle\langle 0\,| = -i\begin{pmatrix} 1 \\ 0 \end{pmatrix}(0\,1) + i\begin{pmatrix} 0 \\ 1 \end{pmatrix}(1\,0) = -i\begin{pmatrix} 0 & 1 \\ 0 & 0 \end{pmatrix} + i\begin{pmatrix} 0 & 0 \\ 1 & 0 \end{pmatrix}.$$

Thus

$$\sigma_y = \begin{pmatrix} 0 & -i \\ i & 0 \end{pmatrix}.$$

We now introduce several other operators frequently used to transform a qubit and the transfer matrices of the corresponding gates: the Hadamard (H), the phase—(S), the $\pi/8$ (T), the phase-shift with angle θ, P_θ, R_θ, and R_k. The matrices corresponding to these operators are unitary:

$$H = \frac{1}{\sqrt{2}}\begin{pmatrix} 1 & 1 \\ 1 & -1 \end{pmatrix} \quad S = \begin{pmatrix} 1 & 0 \\ 0 & i \end{pmatrix} \quad T = \begin{pmatrix} 1 & 0 \\ 0 & e^{i\pi/4} \end{pmatrix},$$

$$P_\theta = \begin{pmatrix} e^{i\theta} & 0 \\ 0 & e^{i\theta} \end{pmatrix} \quad R_\theta = \begin{pmatrix} 1 & 0 \\ 0 & e^{i\theta} \end{pmatrix} \quad R_k = \begin{pmatrix} 1 & 0 \\ 0 & e^{2\pi i/2^k} \end{pmatrix}.$$

Recall from Section 3.2 that the state of a qubit can be represented as a unit vector on the Bloch sphere. When we apply a transformation to a qubit, the corresponding vector on the Bloch sphere is also transformed. In particular, we are interested in rotations with an angle θ about the x, y, and z axes. The three rotations are performed respectively by the following operators obtained by the exponentiation of the σ_x, σ_y, and σ_z Pauli matrices. For simplicity we drop the vector notations used when discussing rotation operations and the Bloch sphere in Section 3.3. We denote the rotation operators $\mathcal{R}_x(\theta)$, $\mathcal{R}_y(\theta)$, and $\mathcal{R}_z(\theta)$

$$\mathcal{R}_x(\theta) = \cos(\theta/2)I - i\sin(\theta/2)\sigma_x = \begin{pmatrix} \cos(\theta/2) & -i\sin(\theta/2) \\ -i\sin(\theta/2) & \cos(\theta/2) \end{pmatrix},$$

$$\mathcal{R}_y(\theta) = \cos(\theta/2)I - i\sin(\theta/2)\sigma_y = \begin{pmatrix} \cos(\theta/2) & -\sin(\theta/2) \\ \sin(\theta/2) & \cos(\theta/2) \end{pmatrix},$$

$$\mathcal{R}_z(\theta) = \cos(\theta/2)I - i\sin(\theta/2)\sigma_z = \begin{pmatrix} e^{-i\theta/2} & 0 \\ 0 & e^{i\theta/2} \end{pmatrix}.$$

We have seen that the rotation matrices satisfy the following properties:

$$\begin{aligned} \mathcal{R}_y(\theta_1)\mathcal{R}_y(\theta_2) &= \mathcal{R}_y(\theta_1 + \theta_2), \\ \mathcal{R}_z(\theta_1)\mathcal{R}_z(\theta_2) &= \mathcal{R}_z(\theta_1 + \theta_2), \\ \sigma_x\mathcal{R}_y(\theta)\sigma_x &= \mathcal{R}_y(-\theta), \\ \sigma_x\mathcal{R}_z(\theta)\sigma_x &= \mathcal{R}_z(-\theta). \end{aligned}$$

Every unitary 2×2 matrix M can be expressed as

$$M = \begin{pmatrix} e^{i\delta} & 0 \\ 0 & e^{i\delta} \end{pmatrix}\begin{pmatrix} e^{i\alpha/2} & 0 \\ 0 & e^{-i\alpha/2} \end{pmatrix}\begin{pmatrix} \cos(\theta/2) & \sin(\theta/2) \\ -\sin(\theta/2) & \cos(\theta/2) \end{pmatrix}\begin{pmatrix} e^{i\beta/2} & 0 \\ 0 & e^{-i\beta/2} \end{pmatrix}$$

with $\alpha, \beta, \delta, \theta \in \mathbb{R}$. Recall that a matrix is unitary if and only if its row and column vectors are orthonormal. The proof of this statement can be found in several textbooks including [62].

If U is a unitary 2×2 matrix then there exist unitary matrices A, B, and C such that $ABC = I$ and $U = A\sigma_x B\sigma_x C$ where σ_x is the Pauli matrix defined previously.

To show this we consider matrices A, B, and C defined by

$$\begin{aligned} A &= \mathcal{R}_z(\beta)\mathcal{R}_y(\gamma/2), \\ B &= \mathcal{R}_y(-\gamma/2)\mathcal{R}_z(-(\delta + \beta)/2), \\ C &= \mathcal{R}_z((\delta - \beta)/2). \end{aligned}$$

It is easy to see that $ABC = I$. Indeed,

$$ABC = \mathcal{R}_z(\beta)\mathcal{R}_y(\gamma/2)\mathcal{R}_y(-\gamma/2)\mathcal{R}_z(-(\delta + \beta)/2)\mathcal{R}_z((\delta - \beta)/2).$$

We use the properties of the rotation matrices

$$\mathcal{R}_y(\gamma/2)\mathcal{R}_y(-\gamma/2) = \mathcal{R}_y(\gamma/2 - \gamma/2) = \mathcal{R}_y(0) = I$$

and

$$\begin{aligned} \mathcal{R}_z(\beta)\mathcal{R}_z(-(\delta + \beta)/2)\mathcal{R}_z((\delta - \beta)/2) &= \mathcal{R}_z(\beta - (\delta + \beta)/2 + (\delta - \beta)/2) \\ &= \mathcal{R}_z(0) = I. \end{aligned}$$

To compute $A\sigma_x B\sigma_x C$ we recall that the Pauli matrix σ_x has the property $\sigma_x\sigma_x = I$, thus

$$A\sigma_x B\sigma_x C = A(\sigma_x\mathcal{R}_y(-\gamma/2)\sigma_x)(\sigma_x\mathcal{R}_z(-(\delta + \beta)/2)\sigma_x)C.$$

But we showed earlier that

$$\sigma_x\mathcal{R}_y(-\gamma/2)\sigma_x = \mathcal{R}_y(\gamma/2)$$

and

$$\sigma_x\mathcal{R}_z(-(\delta + \beta)/2)\sigma_x = \mathcal{R}_z((\delta + \beta)/2).$$

It follows that for our choices of matrices A, B, and C we have

$$A\sigma_x B\sigma_x C = \mathcal{R}_z(\beta)\mathcal{R}_y(\gamma)\mathcal{R}_z(\delta).$$

Sometimes the unitary matrix U is given as $U = e^{i\gamma}A\sigma_x B\sigma_x C$ with $\gamma \in \mathbb{R}$, an overall phase factor, as discussed in Section 3.2.

4.12 SINGLE-QUBIT CONTROLLED OPERATIONS

We address now the flow of control in quantum circuits, enabling us to implement "if ... then ... else ..." constructs. More precisely, we need quantum circuits able to execute a unitary transformation G depending upon the state of a single qubit. In this section, we discuss several transformations G: the Hadamard (H), the transformation described by the Pauli Z matrix, the R_k, and the P_α matrices.

The CNOT is one of the gates in this family of quantum circuits that allows us to control the target qubit depending upon the value of the control qubit. Throughout this chapter, we use the following convention: control qubits are represented as shaded circles, while the target qubits are open circles or boxes.

Figure 4.12 shows the diagram of a generic quantum circuit behaving as follows:

1. when the control qubit is $| c\rangle = | 0\rangle$, then the target qubit is transferred directly to the output;
2. when the control qubit is $| c\rangle = | 1\rangle$, then the transformation described by G is applied to the target qubit.

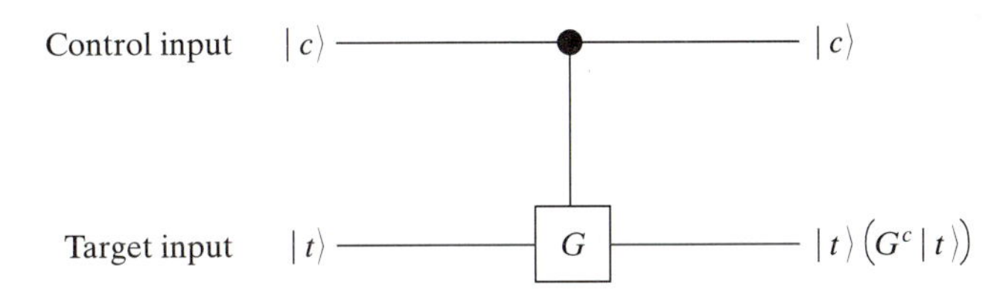

FIGURE 4.12: A controlled-G operation. When the control qubit is $| c\rangle = | 0\rangle$ then the target qubit is transferred directly to the output. When the control qubit is set, $| c\rangle = | 1\rangle$, then the transformation described by G is applied to the target qubit.

The Controlled-Z Operation. Figure 4.13(a) shows a quantum circuit for the controlled-Z operation. Recall that the Z gate flips the sign of the projection of a qubit on $| 1\rangle$. It maps

$$| 0\rangle \mapsto | 0\rangle \qquad | 1\rangle \mapsto -| 1\rangle.$$

Thus, the `controlled-Z` gate maps its input to the output as follows

$$| 00\rangle \mapsto | 00\rangle \qquad | 01\rangle \mapsto | 01\rangle \qquad | 10\rangle \mapsto | 10\rangle \qquad | 11\rangle \mapsto -| 11\rangle.$$

The transfer matrix of the `controlled-Z` circuit in Figure 4.13(a) is

$$G_{\text{controlled-Z}} = | 00\rangle\langle 00 | + | 01\rangle\langle 01 | + | 10\rangle\langle 10 | - | 11\rangle\langle 11 |$$

$$= \begin{pmatrix} 1 & 0 & \vdots & 0 & 0 \\ 0 & 1 & \vdots & 0 & 0 \\ \cdots & \cdots & \cdots & \cdots & \cdots \\ 0 & 0 & \vdots & 1 & 0 \\ 0 & 0 & \vdots & 0 & -1 \end{pmatrix}$$

or

$$G_{\text{controlled-Z}} = \begin{pmatrix} I_2 & 0 \\ 0 & \sigma_z \end{pmatrix}.$$

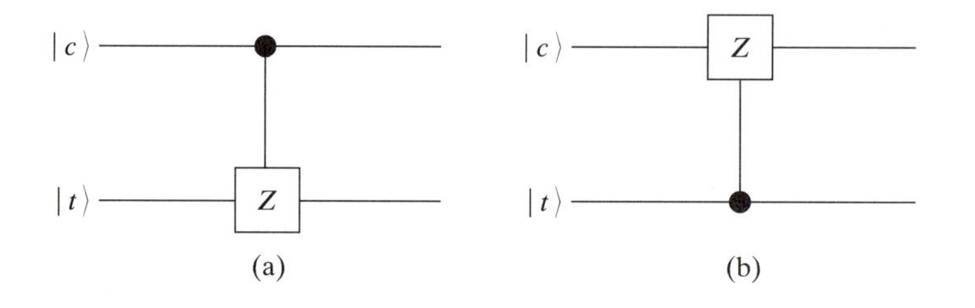

FIGURE 4.13: Quantum circuits for the controlled-Z operation with identical behavior. (a) Original `controlled-Z` circuit. If the control qubit $| c\rangle = | 0\rangle$, then the target qubit $| t\rangle$ is not changed. If $| c\rangle =| 1\rangle$, then the target qubit is flipped. (b) Circuit when the roles of the control and target qubits are reversed.

It is easy to see that the `controlled-Z` gate in Figure 4.13(b) maps its input to the output exactly like the one in Figure 4.13(a), therefore the two circuits are equivalent.

The roles of the control qubit and the target qubit can be reversed by an approximate change of basis as shown in Figure 4.14. We now dissect the circuit in Figure 4.14(a).

A quantum circuit could be viewed as consisting of several stages. For example, the circuit in Figure 4.14(a) consists of three stages. In principle, there are two ways to determine the output of a multi-stage quantum circuit. First, we may examine the transformations applied to each qubit, at each stage. This strategy works for the first qubit of the circuit in Figure 4.14(a). The first qubit

$$| \psi\rangle = \alpha_0 \mid 0\rangle + \alpha_1 \mid 1\rangle = \begin{pmatrix} \alpha_0 \\ \alpha_1 \end{pmatrix}$$

is applied at the input of a `Hadamard` gate and its state is transformed by the first stage of the circuit to

$$| \psi_1\rangle = H \mid \psi\rangle = \frac{1}{\sqrt{2}}\begin{pmatrix} 1 & 1 \\ 1 & -1 \end{pmatrix}\begin{pmatrix} \alpha_0 \\ \alpha_1 \end{pmatrix} = \frac{1}{\sqrt{2}}\begin{pmatrix} \alpha_0 + \alpha_1 \\ \alpha_0 - \alpha_1 \end{pmatrix}.$$

The second stage does not alter the state of the first qubit, as it acts as the control qubit of a `CNOT` gate. Thus,

$$| \psi_2\rangle = | \psi_1\rangle.$$

The third stage transforms the state of the first qubit as follows:

$$| \psi_3\rangle = H \mid \psi_2\rangle = \frac{1}{\sqrt{2}}\begin{pmatrix} 1 & 1 \\ 1 & -1 \end{pmatrix}\frac{1}{\sqrt{2}}\begin{pmatrix} \alpha_0 + \alpha_1 \\ \alpha_0 - \alpha_1 \end{pmatrix}$$

$$= \frac{1}{2}\begin{pmatrix} \alpha_0 + \alpha_1 + (\alpha_0 - \alpha_1) \\ \alpha_0 - \alpha_1 - (\alpha_0 - \alpha_1) \end{pmatrix} = \begin{pmatrix} \alpha_0 \\ \alpha_1 \end{pmatrix}.$$

Thus,

$$| \psi_3\rangle = | \psi\rangle.$$

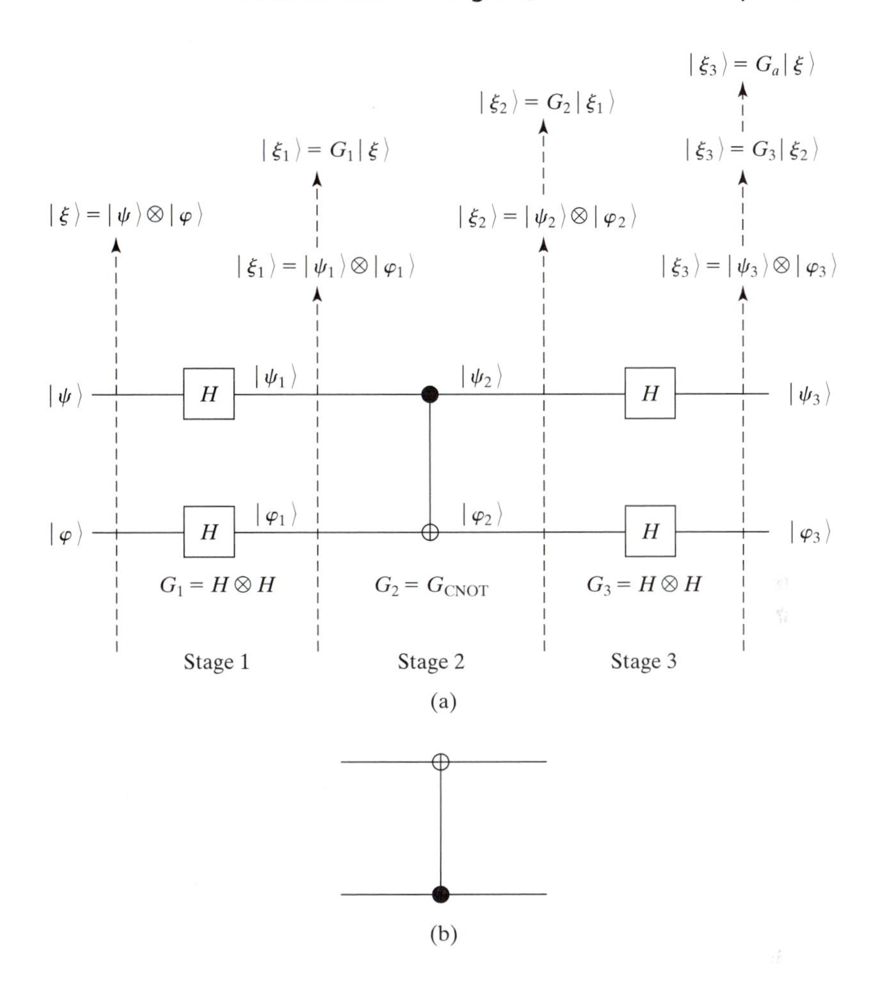

FIGURE 4.14: The roles of control and target qubits can be reversed by an appropriate change of basis. (a) A circuit where the basis of the control and that of the target qubit are changed by a pair of `Hadamard` gates. (b) A reversed `CNOT` gate equivalent to the circuit in (a).

This method cannot be easily applied when the state of one qubit is affected by the state of other qubits, as in the case of the second qubit

$$| \varphi\rangle = \beta_0 \mid 0\rangle + \beta_1 \mid 1\rangle = \begin{pmatrix} \beta_0 \\ \beta_1 \end{pmatrix}.$$

For the first stage we can still compute

$$| \varphi_1\rangle = H \mid \psi\rangle = \frac{1}{\sqrt{2}}\begin{pmatrix} 1 & 1 \\ 1 & -1 \end{pmatrix}\begin{pmatrix} \beta_0 \\ \beta_1 \end{pmatrix} = \frac{1}{\sqrt{2}}\begin{pmatrix} \beta_0 + \beta_1 \\ \beta_0 - \beta_1 \end{pmatrix}.$$

For the second stage, the state of the target qubit $| \varphi_2\rangle$ depends on the value of the control qubit. The more general method to compute the output state of a multi-stage quantum circuit is to construct the transfer matrices of individual stages and then to multiply them. For the first stage of the circuit in Figure 4.14(a) we first construct

$$G_1 = H \otimes H = \frac{1}{\sqrt{2}}\begin{pmatrix} 1 & 1 \\ 1 & -1 \end{pmatrix} \otimes \frac{1}{\sqrt{2}}\begin{pmatrix} 1 & 1 \\ 1 & -1 \end{pmatrix} = \frac{1}{2}\begin{pmatrix} 1 & 1 & 1 & 1 \\ 1 & -1 & 1 & -1 \\ 1 & 1 & -1 & -1 \\ 1 & -1 & -1 & 1 \end{pmatrix}.$$

The second stage is a CNOT gate thus

$$G_2 = G_{\text{CNOT}}.$$

Now, the transfer matrix of the first two stages is

$$G_2 G_1 = \begin{pmatrix} 1 & 0 & 0 & 0 \\ 0 & 1 & 0 & 0 \\ 0 & 0 & 0 & 1 \\ 0 & 0 & 1 & 0 \end{pmatrix} \frac{1}{2}\begin{pmatrix} 1 & 1 & 1 & 1 \\ 1 & -1 & 1 & -1 \\ 1 & 1 & -1 & -1 \\ 1 & -1 & -1 & 1 \end{pmatrix} = \frac{1}{2}\begin{pmatrix} 1 & 1 & 1 & 1 \\ 1 & -1 & 1 & -1 \\ 1 & -1 & -1 & 1 \\ 1 & 1 & -1 & -1 \end{pmatrix}.$$

For the third stage the transfer matrix is

$$G_3 = H \otimes H = \frac{1}{2}\begin{pmatrix} 1 & 1 & 1 & 1 \\ 1 & -1 & 1 & -1 \\ 1 & 1 & -1 & -1 \\ 1 & -1 & -1 & 1 \end{pmatrix}.$$

Finally, the transfer matrix of the entire circuit, G_a is

$$G_a = G_3 G_2 G_1 = \frac{1}{4}\begin{pmatrix} 1 & 1 & 1 & 1 \\ 1 & -1 & 1 & -1 \\ 1 & 1 & -1 & -1 \\ 1 & -1 & -1 & 1 \end{pmatrix}\begin{pmatrix} 1 & 1 & 1 & 1 \\ 1 & -1 & 1 & -1 \\ 1 & -1 & -1 & 1 \\ 1 & 1 & -1 & -1 \end{pmatrix} = \frac{1}{4}\begin{pmatrix} 4 & 0 & 0 & 0 \\ 0 & 0 & 0 & 4 \\ 0 & 0 & 4 & 0 \\ 0 & 4 & 0 & 0 \end{pmatrix}$$

or

$$G_a = \begin{pmatrix} 1 & 0 & 0 & 0 \\ 0 & 0 & 0 & 1 \\ 0 & 0 & 1 & 0 \\ 0 & 1 & 0 & 0 \end{pmatrix}.$$

We can examine the state of the entire circuit, described by vectors in $\mathcal{H}_4$. The input state is

$$| \xi \rangle = | \psi \rangle \otimes | \varphi \rangle = \begin{pmatrix} \alpha_0 \\ \alpha_1 \end{pmatrix} \otimes \begin{pmatrix} \beta_0 \\ \beta_1 \end{pmatrix} = \begin{pmatrix} \alpha_0\beta_0 \\ \alpha_0\beta_1 \\ \alpha_1\beta_0 \\ \alpha_1\beta_1 \end{pmatrix}.$$

The state of the two-qubit system after the first stage is

$$|\xi_1\rangle = |\psi_1\rangle \otimes |\varphi_1\rangle = \frac{1}{\sqrt{2}}\begin{pmatrix} \alpha_0 + \alpha_1 \\ \alpha_0 - \alpha_1 \end{pmatrix} \otimes \frac{1}{\sqrt{2}}\begin{pmatrix} \beta_0 + \beta_1 \\ \beta_0 - \beta_1 \end{pmatrix}$$

or

$$|\xi_1\rangle = \frac{1}{2}\begin{pmatrix} (\alpha_0 + \alpha_1)(\beta_0 + \beta_1) \\ (\alpha_0 + \alpha_1)(\beta_0 - \beta_1) \\ (\alpha_0 - \alpha_1)(\beta_0 + \beta_1) \\ (\alpha_0 - \alpha_1)(\beta_0 - \beta_1) \end{pmatrix}.$$

But

$$|\xi_1\rangle = G_1 |\xi\rangle = \frac{1}{2}\begin{pmatrix} 1 & 1 & 1 & 1 \\ 1 & -1 & 1 & -1 \\ 1 & 1 & -1 & -1 \\ 1 & -1 & -1 & 1 \end{pmatrix}\begin{pmatrix} \alpha_0\beta_0 \\ \alpha_0\beta_1 \\ \alpha_1\beta_0 \\ \alpha_1\beta_1 \end{pmatrix}$$

or

$$|\xi_1\rangle = \frac{1}{2}\begin{pmatrix} \alpha_0(\beta_0 + \beta_1) + \alpha_1(\beta_0 + \beta_1) \\ \alpha_0(\beta_0 - \beta_1) + \alpha_1(\beta_0 - \beta_1) \\ \alpha_0(\beta_0 + \beta_1) - \alpha_1(\beta_0 + \beta_1) \\ \alpha_0(\beta_0 - \beta_1) - \alpha_1(\beta_0 - \beta_1) \end{pmatrix}$$

which coincides with the expression for $|\xi_1\rangle$ we obtained earlier. Similarly, we can determine the state of the two-qubit system after the second and third stage

$$|\xi_2\rangle = G_2 |\xi_1\rangle,$$

$$|\xi_3\rangle = G_3 |\xi_2\rangle.$$

The state after the third stage can also be computed knowing the transfer matrix of the entire circuit, G_a, and the input state, $|\xi\rangle$, as

$$|\xi_3\rangle = G_a |\xi\rangle.$$

Let us now examine the circuit in Figure 4.14(b). This circuit performs the following mappings:

$$|00\rangle \mapsto |00\rangle \quad |01\rangle \mapsto |11\rangle \quad |10\rangle \mapsto |10\rangle \quad |11\rangle \mapsto |01\rangle.$$

Its transfer matrix is

$$G_b = |00\rangle\langle 00| + |01\rangle\langle 11| + |10\rangle\langle 10| + |11\rangle\langle 01|,$$

or

$$G_b = \begin{pmatrix} 1&0&0&0 \\ 0&0&0&0 \\ 0&0&0&0 \\ 0&0&0&0 \end{pmatrix} + \begin{pmatrix} 0&0&0&0 \\ 0&0&0&1 \\ 0&0&0&0 \\ 0&0&0&0 \end{pmatrix} + \begin{pmatrix} 0&0&0&0 \\ 0&0&0&0 \\ 0&0&1&0 \\ 0&0&0&0 \end{pmatrix} + \begin{pmatrix} 0&0&0&0 \\ 0&0&0&0 \\ 0&0&0&0 \\ 0&1&0&0 \end{pmatrix}.$$

Thus,

$$G_b = \begin{pmatrix} 1 & 0 & 0 & 0 \\ 0 & 0 & 0 & 1 \\ 0 & 0 & 1 & 0 \\ 0 & 1 & 0 & 0 \end{pmatrix}$$

and

$$G_a = G_b.$$

This completes the proof that the circuits in Figures 4.14(a) and (b) are equivalent.

The Controlled-H Operation. Figure 4.15 shows a quantum circuit for the controlled-H operation.

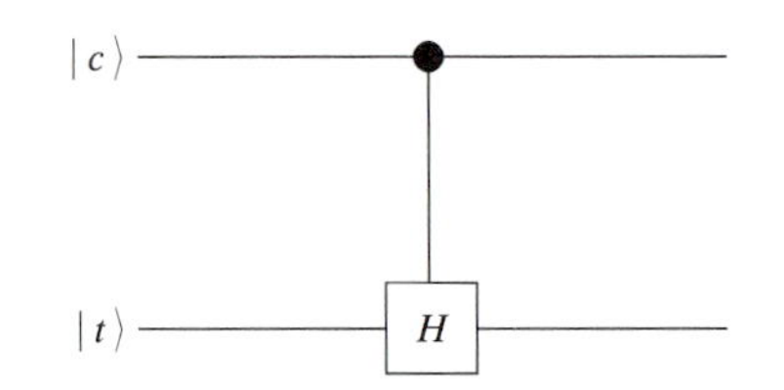

FIGURE 4.15: A quantum circuit for the controlled-H operation. If the control qubit $|c\rangle = |0\rangle$, then the target qubit $|t\rangle$ is not changed. If $|c\rangle = |1\rangle$, then the target qubit is set to a superposition state. If $|t\rangle = |0\rangle$ then it becomes $1/\sqrt{2}(|0\rangle + |1\rangle)$. If $|t\rangle = |1\rangle$ then it becomes $1/\sqrt{2}(|0\rangle - |1\rangle)$.

The `controlled-H` gate maps its input to the output as follows:

$$|00\rangle \mapsto |00\rangle \quad |01\rangle \mapsto |01\rangle,$$

$$|10\rangle \mapsto \frac{1}{\sqrt{2}}(|10\rangle + |11\rangle),$$

$$|11\rangle \mapsto \frac{1}{\sqrt{2}}(|10\rangle - |11\rangle).$$

The transfer matrix of the circuit in Figure 4.15 is

$$G_{\text{controlled-H}} = |00\rangle\langle 00| + |01\rangle\langle 01| + \frac{1}{\sqrt{2}}|10\rangle(\langle 10| + \langle 11|) + \frac{1}{\sqrt{2}}|11\rangle(\langle 10| - \langle 11|).$$

Thus,

$$G_{\text{controlled-H}} = \left(\begin{array}{cc:cc} 1 & 0 & 0 & 0 \\ 0 & 1 & 0 & 0 \\ \hdashline 0 & 0 & \frac{1}{\sqrt{2}} & \frac{1}{\sqrt{2}} \\ 0 & 0 & \frac{1}{\sqrt{2}} & -\frac{1}{\sqrt{2}} \end{array}\right) = \begin{pmatrix} I_2 & 0 \\ 0 & H \end{pmatrix}.$$

The Controlled-R_k Operation.

$$R_k = \begin{pmatrix} 1 & 0 \\ 0 & e^{2\pi i/2^k} \end{pmatrix}$$

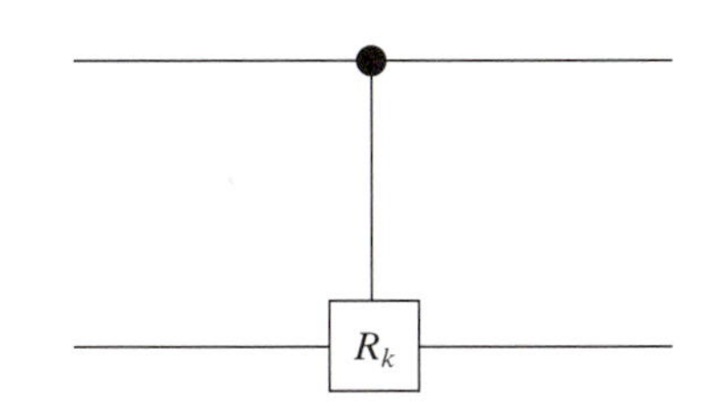

FIGURE 4.16: A quantum circuit for the controlled-R_k operation.

Figure 4.16 shows a quantum circuit for the controlled-R_k operation. The `controlled-`R_k gate maps its input to the output as follows:

$$|\,00\rangle \mapsto |\,00\rangle, \quad |\,01\rangle \mapsto |\,01\rangle, \quad |\,10\rangle \mapsto |\,10\rangle, \quad |\,11\rangle \ \mapsto e^{2\pi i/2^k}\,|\,11\rangle.$$

The transfer matrix of the circuit in Figure 4.16 is

$$G_{\text{controlled-R}_k} = |\,00\rangle\langle 00\,| + |\,01\rangle\langle 01\,| + |\,10\rangle\langle 10\,| + e^{2\pi i/2^k}\,|\,11\rangle\langle 11\,|\,.$$

Thus,

$$G_{\text{controlled-R}_k} = \begin{pmatrix} 1 & 0 & 0 & 0 \\ 0 & 1 & 0 & 0 \\ 0 & 0 & 1 & 0 \\ 0 & 0 & 0 & e^{2\pi i/2^k} \end{pmatrix}.$$

The Controlled-P Operation. Now we show that the controlled-P circuit in Figure 4.17(a) is equivalent to the phase shift circuit in Figure 4.17(b).

FIGURE 4.17: (a) A quantum circuit for the controlled-P operation. (b) An equivalent circuit.

P_α is a phase shift and T_α is a phase transformation

$$P_\alpha = \begin{pmatrix} e^{i\alpha} & 0 \\ 0 & e^{i\alpha} \end{pmatrix}, \qquad T_\alpha = \begin{pmatrix} 1 & 0 \\ 0 & e^{i\alpha} \end{pmatrix}.$$

It is easy to see that the quantum circuit in Figure 4.17(a), a controlled-P circuit, performs the following transformation of its input qubits

$$|\,00\rangle \mapsto |\,00\rangle, \quad |\,01\rangle \mapsto |\,01\rangle, \quad |\,10\rangle \ \mapsto e^{i\alpha}\,|\,10\rangle, \quad |\,11\rangle \ \mapsto e^{i\alpha}\,|\,11\rangle.$$

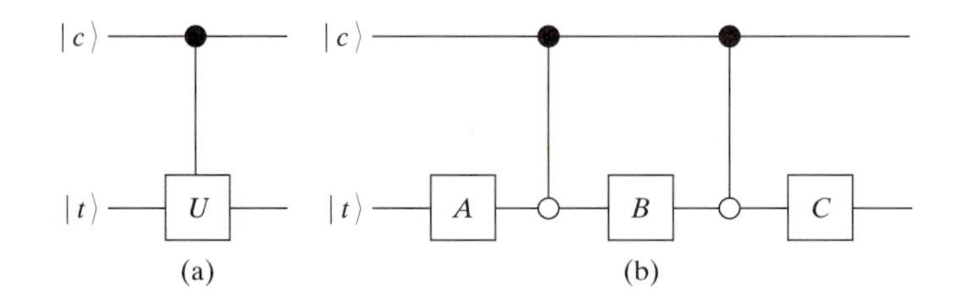

FIGURE 4.18: (a) A generic single qubit controlled gate with matrix $U = A\sigma_x B\sigma_x C$. (b) A circuit simulating the generic single qubit controlled gate in (a).

Thus, the transfer matrix of the controlled-P quantum circuit is

$$G_{\text{controlled-P}} = \mid 00\rangle\langle 00 \mid + \mid 01\rangle\langle 01 \mid + e^{i\alpha} \mid 10\rangle\langle 10 \mid + e^{i\alpha} \mid 11\rangle\langle 11 \mid$$
$$= \begin{pmatrix} 1 & 0 & 0 & 0 \\ 0 & 1 & 0 & 0 \\ 0 & 0 & e^{i\alpha} & 0 \\ 0 & 0 & 0 & e^{i\alpha} \end{pmatrix}.$$

On the other hand, the transfer matrix of the circuit in Figure 4.17(b) is

$$G_{T_\alpha} = T_\alpha \otimes I = \begin{pmatrix} 1 & 0 \\ 0 & e^{i\alpha} \end{pmatrix} \otimes \begin{pmatrix} 1 & 0 \\ 0 & 1 \end{pmatrix} = \begin{pmatrix} 1 & 0 & 0 & 0 \\ 0 & 1 & 0 & 0 \\ 0 & 0 & e^{i\alpha} & 0 \\ 0 & 0 & 0 & e^{i\alpha} \end{pmatrix},$$

the same as $G_{\text{controlled-P}}$, the transfer matrix of the circuit in 4.17(a); the two circuits are equivalent.

The quantum circuit in Figure 4.18(b) simulates the general one-qubit controlled gate in Figure 4.18(a), where A, B, and C are the rotation matrices defined in Section 4.11 . When the control qubit is $\mid c\rangle = \mid 0\rangle$, the target qubit becomes

$$\mid t\rangle \mapsto ABC \mid t\rangle = I \mid t\rangle = \mid t\rangle.$$

When $\mid c\rangle = \mid 1\rangle$, the target qubit becomes

$$\mid t\rangle \mapsto A\sigma_x B\sigma_x C \mid t\rangle = U \mid t\rangle$$

because $ABC = I$ and $U = A\sigma_x B\sigma_x C$ according to the last proposition of Section 4.11.

4.13 MULTIPLE-QUBIT CONTROLLED OPERATIONS

Consider a circuit performing a unitary transformation U with n control qubits

$$\mid c_0\rangle, \mid c_1\rangle, \ldots, \mid c_{n-1}\rangle$$

and k target qubits

$$\mid t_0\rangle, \mid t_1\rangle, \ldots, \mid t_{k-1}\rangle$$

at the input.

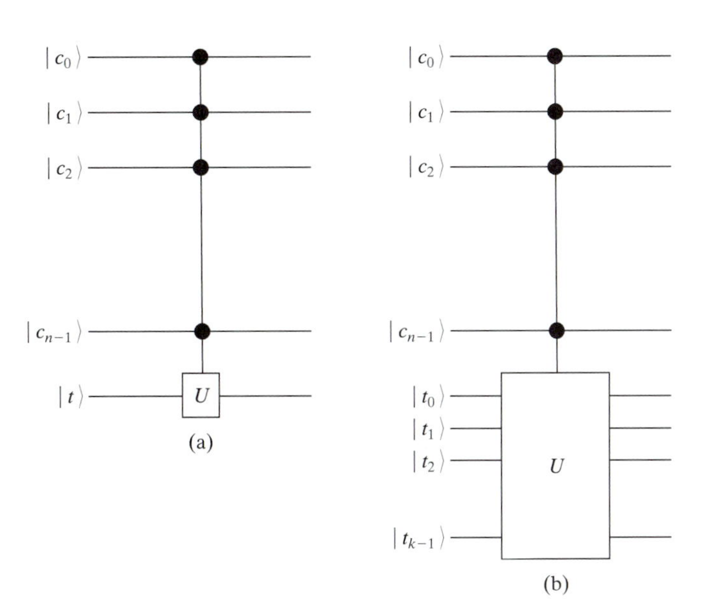

FIGURE 4.19: (a) An n-qubit controlled, single target qubit unitary function quantum circuit. (b) An n-qubit controlled, k-target qubit quantum circuit. The unitary transformation **U** is applied to the k target qubits.

In Figure 4.19(a) we present the case when $k = 1$. For $n = 2$ the unitary operator **U** has a matrix U given by

$$U = \begin{pmatrix} u_{00} & u_{01} \\ u_{10} & u_{11} \end{pmatrix}.$$

When all controlled qubits of the circuit in Figure 4.19(a) are set to 1 we apply the transformation **U** to the target qubit, $| t\rangle = \alpha_0 | 0\rangle + \alpha_1 | 1\rangle$. When the logical product, AND, of all control qubits is not 1 we transfer directly the qubits to the output of the circuit. Using Feynman's notation [57], we define the operator for this quantum circuit acting upon $n + 1$ qubits with n control qubits $| c_0\rangle, | c_1\rangle, \ldots, | c_{n-1}\rangle$ and one target qubit $| t\rangle$ as follows:

$$\wedge_n (U)(| c_0 c_1 \ldots c_{n-1} t\rangle)$$

$$= \begin{cases} | c_0 c_1 \ldots c_{n-1} t\rangle & \text{if } \wedge_{j=0}^{n-1} (c_j) = 0 \\ \alpha_0 | c_0 c_1 \ldots c_{n-2} 0\rangle + \alpha_1 | c_0 c_1 \ldots c_{n-1} 1\rangle & \text{if } \wedge_{j=0}^{n-1} (c_j) = 1. \end{cases}$$

The notation $\wedge_{j=0}^{n-1}(c_j) = 1$ means $c_0 AND c_1 AND \ldots c_{n-2} AND c_{n-1} = 1$. The transfer matrix of the circuit is

$$\begin{pmatrix} 1 & 0 & \ldots & 0 & 0 & 0 \\ 0 & 1 & \ldots & 0 & 0 & 0 \\ \ldots & \ldots & \ldots & \ldots & \ldots & \ldots \\ 0 & 0 & \ldots & 1 & 0 & 0 \\ 0 & 0 & \ldots & 0 & u_{00} & u_{01} \\ 0 & 0 & \ldots & 0 & u_{10} & u_{11} \end{pmatrix}.$$

Figure 4.19(b) presents a quantum circuit with $k > 1$ target qubits. An alternative notation for multiple-qubit controlled operation is $C^n(U_k)$ where n is the number of control qubits and k is the number of target qubits. The unitary transformation U_k is applied to the target qubits when the control condition is satisfied. Let $U^{c_0 c_1 \dots c_{n-1}} \mid t_1 t_2 \dots t_k\rangle$ denote that the transformation U is applied to the target qubits only if the logical product of $c_0 c_1 \dots c_{n-1}$ is equal to one. Then the transformation is

$$C^n(U_k) \mid c_0 c_1 \dots c_{n-1}\rangle \mid t_0 t_1 \dots t_{k-1}\rangle = \mid c_0 c_1 \dots c_{n-1}\rangle U^{c_0 c_1 \dots c_{n-1}} \mid t_0 t_1 \dots t_{k-1}\rangle.$$

Figure 4.20 shows a quantum circuit using `Toffoli` gates, with $n = 6$ control qubits labeled as $\mid c_1\rangle, \mid c_2\rangle, \mid c_3\rangle, \mid c_4\rangle, \mid c_5\rangle$, and $\mid c_6\rangle$ and one target qubit $\mid t\rangle$. The circuit is general and can be used for any number n of control qubits. The circuit has an additional $n - 1$ *work qubits* $\mid w_1\rangle, \mid w_2\rangle, \mid w_3\rangle, \mid w_4\rangle$, and $\mid w_5\rangle$ initially in state $\mid 0\rangle$. At the end of the computation the work qubits are returned to state $\mid 0\rangle$.

The circuit in Figure 4.20 consists of three stages. The first stage produces the logical product, `AND`, of all six control qubits in one of the work qubits, $\mid w_5\rangle$. The second stage performs a single qubit U-controlled transformation of the

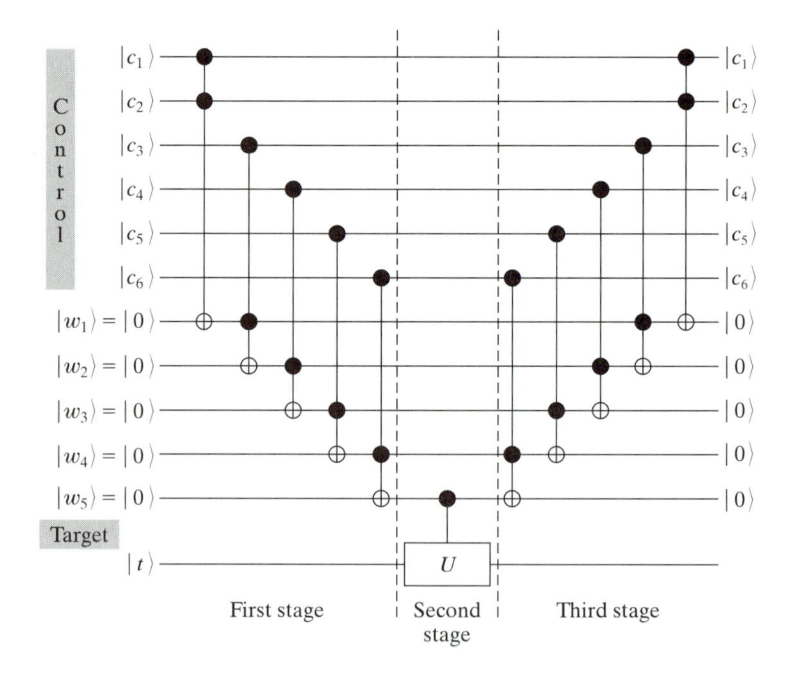

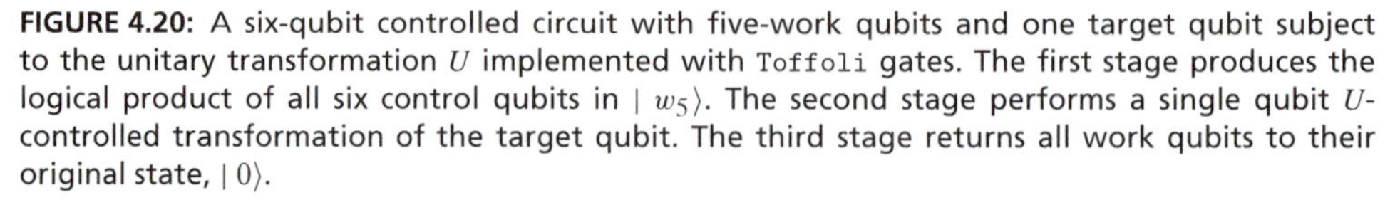

FIGURE 4.20: A six-qubit controlled circuit with five-work qubits and one target qubit subject to the unitary transformation U implemented with `Toffoli` gates. The first stage produces the logical product of all six control qubits in $\mid w_5\rangle$. The second stage performs a single qubit U-controlled transformation of the target qubit. The third stage returns all work qubits to their original state, $\mid 0\rangle$.

target qubit. The last stage returns the work qubits, $| w_1\rangle, | w_2\rangle, | w_3\rangle, | w_4\rangle$, and $| w_5\rangle$ to their original state, $| 0\rangle$.

We verify that the first stage operates as described. We observe that the first `Toffoli` gate from the left `AND`s $| c_1\rangle$ and $| c_2\rangle$ and changes the state of $| w_1\rangle$ from $| 0\rangle$ to $| c_1c_2\rangle$. The next `Toffoli` gate changes the state of $| w_2\rangle$ from $| 0\rangle$ to $| c_1c_2c_3\rangle$, and so on.

Recall that we are only interested in reversible quantum circuits, therefore we should return the work qubits to their original state. We only show that the last work qubit, $| w_5\rangle$, is set to $| 0\rangle$. Indeed, the first `Toffoli` gate of the third stage of the circuit in Figure 4.20 performs the following operation:

$$| w_5\rangle \texttt{ AND } [(| c_1\rangle \texttt{ AND } | c_2\rangle \ldots \texttt{ AND } | c_6\rangle) \texttt{ AND }$$
$$(| c_1\rangle \texttt{ AND } | c_2\rangle \ldots \texttt{ AND } | c_6\rangle)] = | w_5\rangle.$$

4.14 UNIVERSAL QUANTUM GATES

Several universal family of gates can be implemented. For example, the `Hadamard`, `phase`, `CNOT`, and $\pi/8$ gates are universal; any unitary operation can be approximated by a circuit using only these gates. The gates belonging to such a set of universal gates may be applied to any number of qubits to produce the desired unitary transformation.

First, we show that a `Toffoli` gate can be implemented using only `H`, `T`, and `S` one-qubit gates. Combined with the circuit presented in Figure 4.20, which can be generalized for any number n of control qubits, this result shows that we can construct an arbitrary n-qubit controlled circuit using only `Hadamard`, `H`, `phase`, `S`, `T`, and $\pi/8$ one-qubit gates where

$$H = \frac{1}{\sqrt{2}}\begin{pmatrix} 1 & 1 \\ 1 & -1 \end{pmatrix}, \quad S = \begin{pmatrix} 1 & 0 \\ 0 & i \end{pmatrix}, \quad T = \begin{pmatrix} 1 & 0 \\ 0 & e^{i\pi/4} \end{pmatrix}, \quad T^{\dagger} = \begin{pmatrix} 1 & 0 \\ 0 & e^{-i\pi/4} \end{pmatrix}.$$

The proof that the circuit in Figure 4.21 simulates a `Toffoli` gate is relatively straightforward. First, we write the transfer matrix of the `Toffoli` gate in a more compact form

$$G_{\text{Toffoli}} = \begin{pmatrix} 1&0&0&0&0&0&0&0 \\ 0&1&0&0&0&0&0&0 \\ 0&0&1&0&0&0&0&0 \\ 0&0&0&1&0&0&0&0 \\ 0&0&0&0&1&0&0&0 \\ 0&0&0&0&0&1&0&0 \\ 0&0&0&0&0&0&0&1 \\ 0&0&0&0&0&0&1&0 \end{pmatrix} = \begin{pmatrix} I & O & O & O \\ O & I & O & O \\ O & O & I & O \\ O & O & O & \sigma_x \end{pmatrix},$$

where

$$I = \begin{pmatrix} 1 & 0 \\ 0 & 1 \end{pmatrix}, \quad O = \begin{pmatrix} 0 & 0 \\ 0 & 0 \end{pmatrix}, \quad \sigma_x = \begin{pmatrix} 0 & 1 \\ 1 & 0 \end{pmatrix}.$$

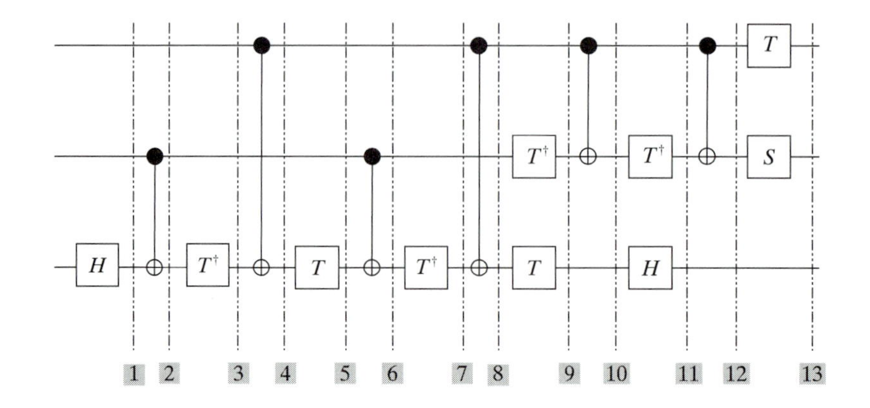

FIGURE 4.21: A quantum circuit using only `H`, `S` and `T` one-qubit gates to simulate a `Toffoli` gate. The $T^{\dagger}$ is the adjoint of the T matrix. The circuit has 13 stages. We calculate the transfer matrix of each stage.

These matrix labels are used to write in a more compact form the transfer matrices for all the 13 stages of the circuit in Figure 4.21. The transfer matrices for stages 1, 5, 9, 11, and 13 are

$$G^{(1)} = I \otimes I \otimes H = \begin{pmatrix} H & O & O & O \\ O & H & O & O \\ O & O & H & O \\ O & O & O & H \end{pmatrix},$$

$$G^{(5)} = I \otimes I \otimes T = \begin{pmatrix} T & O & O & O \\ O & T & O & O \\ O & O & T & O \\ O & O & O & T \end{pmatrix},$$

$$G^{(9)} = I \otimes T^{\dagger} \otimes T = \begin{pmatrix} T & O & O & O \\ O & e^{-i\pi/4}T & O & O \\ O & O & T & O \\ O & O & O & e^{-i\pi/4}T \end{pmatrix},$$

$$G^{(11)} = I \otimes T^{\dagger} \otimes H = \begin{pmatrix} H & O & O & O \\ O & e^{-i\pi/4}H & O & O \\ O & O & H & O \\ O & O & O & e^{-i\pi/4}H \end{pmatrix},$$

$$G^{(13)} = T \otimes S \otimes I = \begin{pmatrix} I & O & O & O \\ O & iI & O & O \\ O & O & e^{i\pi/4}I & O \\ O & O & O & ie^{i\pi/4}I \end{pmatrix}.$$

Several stages have identical transfer matrices

$$G^{(2)} = G^{(6)} = I \otimes G_{\text{CNOT}} = I \otimes \begin{pmatrix} I & O \\ O & \sigma_x \end{pmatrix} = \begin{pmatrix} I & O & O & O \\ O & \sigma_x & O & O \\ O & O & I & O \\ O & O & O & \sigma_x \end{pmatrix},$$

$$G^{(3)} = G^{(7)} = I \otimes I \otimes T^\dagger = \begin{pmatrix} T^\dagger & O & O & O \\ O & T^\dagger & O & O \\ O & O & T^\dagger & O \\ O & O & O & T^\dagger \end{pmatrix},$$

$$G^{(4)} = G^{(8)} = \begin{pmatrix} I & O & O & O \\ O & I & O & O \\ O & O & \sigma_x & O \\ O & O & O & \sigma_x \end{pmatrix},$$

$$G^{(10)} = G^{(12)} = G_{\text{CNOT}} \otimes I = \begin{pmatrix} I & O & O & O \\ O & I & O & O \\ O & O & O & I \\ O & O & I & O \end{pmatrix}.$$

Now we calculate several partial products

$$G^{(4)}G^{(3)}G^{(2)}G^{(1)} = \begin{pmatrix} T^\dagger H & O & O & O \\ O & T^\dagger \sigma_x H & O & O \\ O & O & \sigma_x T^\dagger H & O \\ O & O & O & \sigma_x T^\dagger \sigma_x H \end{pmatrix},$$

$$G^{(8)}G^{(7)}G^{(6)}G^{(5)} = \begin{pmatrix} IT^\dagger IT & O & O & O \\ O & IT^\dagger \sigma_x T & O & O \\ O & O & \sigma_x T^\dagger IT & O \\ O & O & O & \sigma_x T^\dagger \sigma_x T \end{pmatrix}$$

$$= \begin{pmatrix} I & O & O & O \\ O & T^\dagger \sigma_x T & O & O \\ O & O & \sigma_x & O \\ O & O & O & \sigma_x T^\dagger \sigma_x T \end{pmatrix},$$

$$G^{(12)}G^{(11)}G^{(10)}G^{(9)} = \begin{pmatrix} IHIT & O & O & O \\ O & e^{-i\pi/2} IHIT & O & O \\ O & O & e^{-i\pi/4} IHIT & O \\ O & O & O & e^{-i\pi/2} IHIT \end{pmatrix}$$

$$= \begin{pmatrix} HT & O & O & O \\ O & e^{-i\pi/2}HT & O & O \\ O & O & e^{-i\pi/4}HT & O \\ O & O & O & e^{-i\pi/4}HT \end{pmatrix}.$$

The transfer matrix of the circuit in Figure 4.21 is the product of the transfer matrices for the 13 stages; after elementary transformations of the exponential terms we obtain

$$G_{\text{oneQubitHST}} = G^{(13)}(G^{(12)}G^{(11)}G^{(10)}G^{(9)})(G^{(8)}G^{(7)}G^{(6)}G^{(5)}) \times (G^{(4)}G^{(3)}G^{(2)}G^{(1)}) = \begin{pmatrix} I & O & O & O \\ O & I & O & O \\ O & O & I & O \\ O & O & O & \sigma_x \end{pmatrix}.$$

Thus,

$$G_{\text{oneQubitHST}} = G_{\text{Toffoli}}.$$

The circuit in Figure 4.21 with one-qubit gates(H, S and T gates) has the same transfer matrix as a `Toffoli` gate, thus it simulates this three-qubit gate.

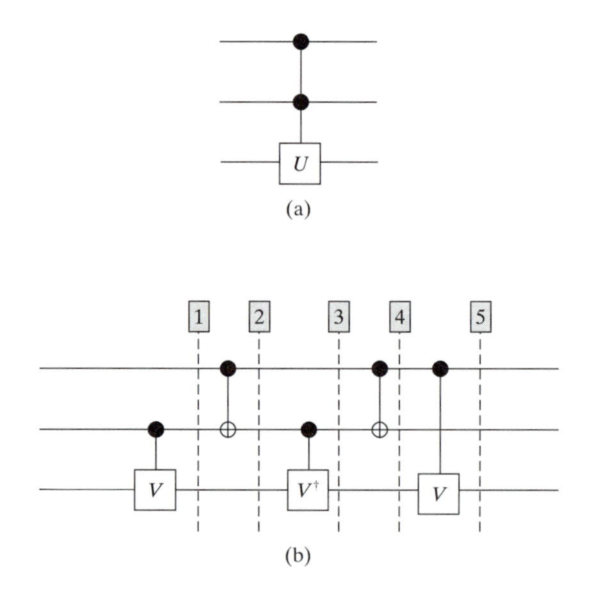

FIGURE 4.22: A three-qubit controlled operation can be implemented using a combination of two-qubit controlled gates. The transformations U and V implemented by the two-qubit controlled gates are related with each other by the expression $V^2 = U$. (a) A three-qubit controlled gate. (b) The three-qubit controlled gate in (a) implemented using the two-qubit controlled gates.

Now we show that a two-qubit controlled gate(see Figure 4.22(a)) can be implemented using a combination of one-qubit controlled gates, as shown in Figure 4.22.

We can construct the transfer matrix of the circuit in Figure 4.22(a) using our method of identifying the transformation of each basis vector and then constructing the outer products

$$G_{\text{twoQubitControlled}} = \begin{pmatrix} I & O & O & O \\ O & I & O & O \\ O & O & I & O \\ O & O & O & U \end{pmatrix}$$

with

$$U = \begin{pmatrix} u_{11} & u_{12} \\ u_{21} & u_{22} \end{pmatrix}.$$

The transfer matrices of the five stages of the circuit in Figure 4.22(b) are

$$G_{tC}^{(1)} = \begin{pmatrix} I & O & O & O \\ O & V & O & O \\ O & O & I & O \\ O & O & O & V \end{pmatrix}, \quad G_{tC}^{(2)} = G_{tC}^{(4)} = \begin{pmatrix} I & O & O & O \\ O & I & O & O \\ O & O & O & I \\ O & O & I & O \end{pmatrix},$$

$$G_{tC}^{(3)} = \begin{pmatrix} I & O & O & O \\ O & V^{\dagger} & O & O \\ O & O & I & O \\ O & O & O & V^{\dagger} \end{pmatrix}, \quad G_{tC}^{(5)} = \begin{pmatrix} I & O & O & O \\ O & I & O & O \\ O & O & V & O \\ O & O & O & V \end{pmatrix}.$$

The products of these matrices are

$$G_{tC}^{(2)} G_{tC}^{(1)} = \begin{pmatrix} I & O & O & O \\ O & V & O & O \\ O & O & O & V \\ O & O & I & O \end{pmatrix}, \quad G_{tC}^{(3)} G_{tC}^{(2)} G_{tC}^{(1)} = \begin{pmatrix} I & O & O & O \\ O & I & O & O \\ O & O & O & V \\ O & O & V^{\dagger} & O \end{pmatrix},$$

and

$$G_{tC}^{(4)} G_{tC}^{(3)} G_{tC}^{(2)} G_{tC}^{(1)} = \begin{pmatrix} I & O & O & O \\ O & I & O & O \\ O & O & V^{\dagger} & O \\ O & O & O & V \end{pmatrix},$$

$$G_{tC}^{(5)} G_{tC}^{(4)} G_{tC}^{(3)} G_{tC}^{(2)} G_{tC}^{(1)} = \begin{pmatrix} I & O & O & O \\ O & I & O & O \\ O & O & I & O \\ O & O & O & V^2 \end{pmatrix}.$$

But $V^2 = U$, thus the transfer matrix of the circuit in Figure 4.22(b) constructed with two-qubit gates is identical to the transfer matrix of two-qubit controlled gate in Figure 4.22(a). We conclude that the circuit in Figure 4.22(b) simulates the one in Figure 4.22(a).

4.15 A QUANTUM CIRCUIT FOR THE WALSH-HADAMARD TRANSFORM

Recall from Section 4.2 that the `Hadamard` gate describes a unitary quantum "fair coin flip" operation performed upon a single qubit

$$H = \frac{1}{\sqrt{2}} \begin{pmatrix} 1 & 1 \\ 1 & -1 \end{pmatrix}.$$

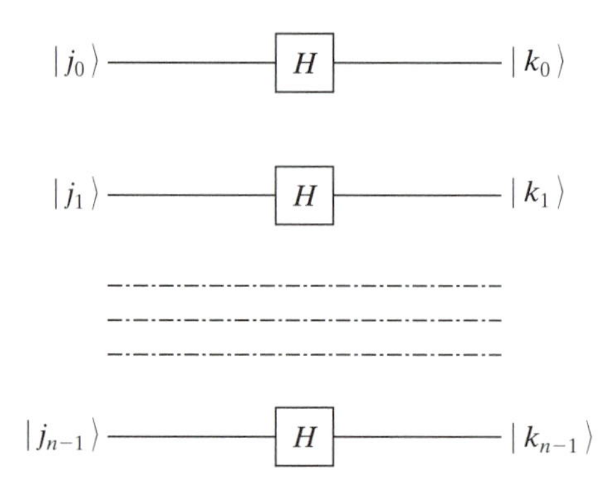

FIGURE 4.23: A quantum circuit for the Walsh-Hadamard transform.

The `Hadamard` gate transforms an input qubit in state $| 0\rangle$ into a superposition state $(| 0\rangle + | 1\rangle)/\sqrt{2}$. An input qubit in state $| 1\rangle$ is transformed into the superposition state $(| 0\rangle - | 1\rangle)/\sqrt{2}$. This transformation is its own inverse. Performing H twice on an initial state $| 0\rangle$ or $| 1\rangle$ gives $| 0\rangle$ or $| 1\rangle$, respectively, with probability 1.

The Walsh-Hadamard Transform is presented in more detail in Appendix C. This linear transformation is defined recursively as

$$W_1 = H,$$
$$W_{j+1} = H \otimes W_j.$$

When applied individually to n qubits in state $| 0\rangle$, the Walsh-Hadamard Transform creates a superposition of 2^n states

$$(H \otimes H \otimes H \ldots \otimes H) \,|\, 000\ldots 0\rangle$$
$$= \frac{1}{2^{n/2}}[(| 0\rangle + | 1\rangle) \otimes (| 0\rangle + | 1\rangle) \ldots \otimes (| 0\rangle + | 1\rangle)].$$

In the general case, the n-qubit Walsh-Hadamard transform implemented by the circuit in Figure 4.23 is described by a $2^n \times 2^n$ matrix G_{H^n}.

The entries of the matrix G_{H^n} are

$$g_{jk} = \left(\frac{1}{\sqrt{2}}\right)^n (-1)^{j \cdot k}$$

where $j \cdot k$ is a short hand notation for the number of 1s in the binary representation of integers j and k, or the scalar product of vectors $\mathbf{j}$ and $\mathbf{k}$

$$j = j_0 2^{n-1} + j_1 2^{n-2} + \ldots + j_{n-2} 2 + j_{n-1} 2^0, \quad \mathbf{j} = (j_0, j_1, \ldots, j_{n-2}, j_{n-1}),$$
$$k = k_0 2^{n-1} + k_1 2^{n-2} + \ldots + k_{n-2} 2 + k_{n-1} 2^0, \quad \mathbf{k} = (k_0, k_1, \ldots, k_{n-2}, k_{n-1}).$$

Let $| j\rangle$ be the input state vector and $| k\rangle$ the output state vector of the circuit in Figure 4.23

$$| j\rangle = | j_0\rangle + | j_1\rangle + \ldots + | j_{n-1}\rangle = j_0 \,| 000\ldots00\rangle + j_1 \,| 000\ldots01\rangle + \ldots + j_{n-1} \,| 111\ldots11\rangle,$$

$$| k\rangle = | k_1\rangle + | k_2\rangle + \ldots + | k_{n-1}\rangle = k_0 \,| 000\ldots00\rangle + k_1 \,| 000\ldots01\rangle + \ldots + k_{n-1} \,| 111\ldots11\rangle.$$

Then

$$| k\rangle = G_{H^n} \,| j\rangle = (H \otimes H \otimes H\ldots H)(| j_{n-1}\rangle \otimes | j_{n-2}\rangle \ldots \otimes | j_0\rangle,$$

or

$$G_{H^n} \,| j\rangle = \left(\frac{1}{\sqrt{2}}\right)^n [| 0\rangle + (-1)^{j_{n-1}} \,| 1\rangle]\ldots \otimes [| 0\rangle + (-1)^{j_0} \,| 1\rangle],$$

$$G_{H^n} \,| j\rangle = \left(\frac{1}{\sqrt{2}}\right)^n \sum_{k=0}^{2^n-1} (-1)^{j_{n-1}k_{n-1}} \,| j_{n-1}\rangle \otimes \ldots (-1)^{j_0 k_0} \,| j_0\rangle.$$

Finally,

$$G_{H^n} \,| j\rangle = \left(\frac{1}{\sqrt{2}}\right)^n \left(\sum_{k=0}^{2^n-1} (-1)^{j\cdot k}\right) | j\rangle.$$

The result of performing a sequence of n Hadamard transformations is a superposition of all 2^n possible binary strings of length n. The final probability amplitude of each string is $2^{-n/2}$. Simon points out that "as the transformations are applied in turn, the phase of the resulting state is changed" [127]. The output vector $| k\rangle$ for an input vector $| j\rangle$ is the superposition

$$2^{-\frac{n}{2}} \sum_k (-1)^{j\cdot k} \,| j\rangle.$$

4.16 THE STATE TRANSFORMATION PERFORMED BY QUANTUM CIRCUITS

Constructing the matrix associated with the linear operator describing the transformation performed by a quantum circuit demands some attention. Sometimes, we construct first the truth table of the circuit from the description of the functionality of the circuit as was the case with the `CNOT`, `Fredkin` and `Toffoli` gates, and with the circuit in Figure 4.11 and then we describe the transformations of the basis vectors. *The outer products used to define the transformation of the basis vectors have as the first term the new basis and as the second term the old basis*. For example, consider a two-qubit circuit performing the following transformation

$$| 00\rangle \mapsto | 01\rangle \quad | 01\rangle \mapsto | 11\rangle \quad | 11\rangle \mapsto | 10\rangle \quad | 10\rangle \mapsto | 00\rangle.$$

In this case,

$$G_{G2} = | 01\rangle\langle 00 | + | 11\rangle\langle 01 | + | 10\rangle\langle 11 | + | 10\rangle\langle 00 | .$$

It is easy to see that

$$| 01\rangle\langle 00 | = \begin{pmatrix} 0 \\ 1 \\ 0 \\ 0 \end{pmatrix} \begin{pmatrix} 1 & 0 & 0 & 0 \end{pmatrix} = \begin{pmatrix} 0 & 0 & 0 & 0 \\ 1 & 0 & 0 & 0 \\ 0 & 0 & 0 & 0 \\ 0 & 0 & 0 & 0 \end{pmatrix},$$

$$| 11\rangle\langle 01 | = \begin{pmatrix} 0 \\ 0 \\ 0 \\ 1 \end{pmatrix} \begin{pmatrix} 0 & 1 & 0 & 0 \end{pmatrix} = \begin{pmatrix} 0 & 0 & 0 & 0 \\ 0 & 0 & 0 & 0 \\ 0 & 0 & 0 & 0 \\ 0 & 1 & 0 & 0 \end{pmatrix},$$

$$| 10\rangle\langle 11 | = \begin{pmatrix} 0 \\ 0 \\ 1 \\ 0 \end{pmatrix} \begin{pmatrix} 0 & 0 & 0 & 1 \end{pmatrix} = \begin{pmatrix} 0 & 0 & 0 & 0 \\ 0 & 0 & 0 & 0 \\ 0 & 0 & 0 & 1 \\ 0 & 0 & 0 & 0 \end{pmatrix},$$

$$| 00\rangle\langle 10 | = \begin{pmatrix} 1 \\ 0 \\ 0 \\ 0 \end{pmatrix} \begin{pmatrix} 0 & 0 & 1 & 0 \end{pmatrix} = \begin{pmatrix} 0 & 0 & 1 & 0 \\ 0 & 0 & 0 & 0 \\ 0 & 0 & 0 & 0 \\ 0 & 0 & 0 & 0 \end{pmatrix},$$

and

$$G_{G2} = \begin{pmatrix} 0 & 0 & 1 & 0 \\ 1 & 0 & 0 & 0 \\ 0 & 0 & 0 & 1 \\ 0 & 1 & 0 & 0 \end{pmatrix}.$$

We can verify that indeed the circuit performs the required transformation. For example, that it maps $| 11\rangle \mapsto | 10\rangle$

$$G_{G2} \mid 11\rangle = \begin{pmatrix} 0 & 0 & 1 & 0 \\ 1 & 0 & 0 & 0 \\ 0 & 0 & 0 & 1 \\ 0 & 1 & 0 & 0 \end{pmatrix} \begin{pmatrix} 0 \\ 0 \\ 0 \\ 1 \end{pmatrix} = \begin{pmatrix} 0 \\ 0 \\ 1 \\ 0 \end{pmatrix} = | 10\rangle.$$

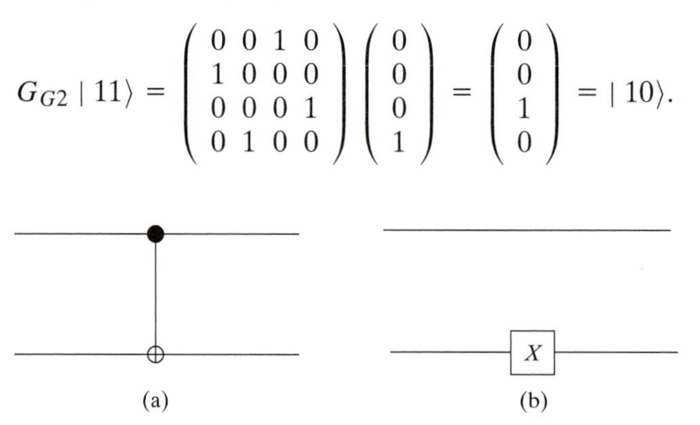

FIGURE 4.24: (a) A CNOT gate, (b) a two-qubit circuit with an X gate. The transfer matrices of the two circuits are different.

We have to be careful when there is coupling between the qubits. For example, Figure 4.24 shows two two-qubit gates: a CNOT and a circuit with an X gate. The functions of the two circuits are different thus, their transfer matrices are different. Indeed, in the case of the `CNOT` there is coupling between the two qubits, the target is flipped only when the control qubit is set to 1, while for the circuit on the right the second qubit is always flipped. The transfer matrix of the circuit in Figure 4.24(b) is:

$$G_{IX} = I \otimes X = \begin{pmatrix} 1 & 0 \\ 0 & 1 \end{pmatrix} \otimes \begin{pmatrix} 0 & 1 \\ 1 & 0 \end{pmatrix} = \begin{pmatrix} 0 & 1 & 0 & 0 \\ 1 & 0 & 0 & 0 \\ 0 & 0 & 0 & 1 \\ 0 & 0 & 1 & 0 \end{pmatrix}$$

while

$$G_{CNOT} = \begin{pmatrix} 1 & 0 & 0 & 0 \\ 0 & 1 & 0 & 0 \\ 0 & 0 & 0 & 1 \\ 0 & 0 & 1 & 0 \end{pmatrix}.$$

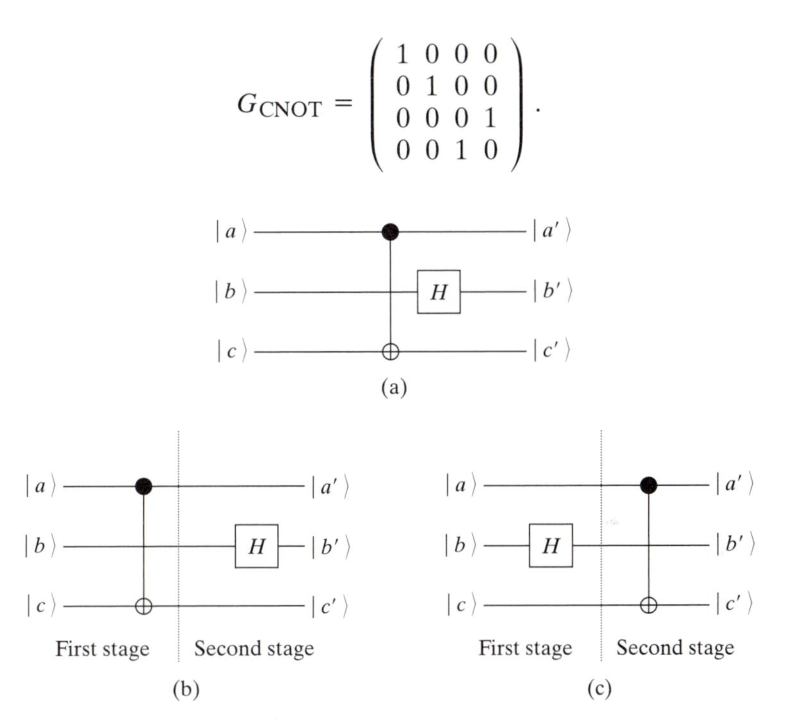

FIGURE 4.25: (a) A circuit with a `CNOT` gate and an `H` gate. (b) The first stage consists of a `CNOT` with the topmost qubit acting as control and the bottom one as the target qubit; the second stage consists of the `H` gate acting on the second qubit only. (c) The first stage consists of the `H` gate acting on the second qubit only; the second stage consists of a `CNOT` with the topmost qubit acting as control and the bottom one as the target qubit. The transfer matrices of the circuits in (b) and (c) are identical, confirming our intuition.

We have already seen that in general we decompose a quantum circuit in several stages and then determine the transfer matrix for each stage and then multiply them. An interesting case is shown in Figure 4.25(a). This circuit is related to the one in Figure 4.11 and we observe that there are two ways to define the two stages. In Figure 4.25(b) the first stage consists of a `CNOT` with the topmost qubit acting as control and the bottom one as the target qubit and the second stage consists of the `H` gate acting on the second qubit only. In Figure 4.25(c) the first stage consists of the H gate acting on the second qubit only; the second stage consists of a CNOT with the topmost qubit acting as control and the bottom one as the target qubit.

Our intuition tells us that the behavior of the quantum circuit is invariant to the manner we analyze it, thus the result of the analysis should be the same in the two cases.

It is easy to see that

$$G_{\mathrm{IHI}} = I \otimes H \otimes I = \frac{1}{\sqrt{2}} \begin{pmatrix} 1 & 0 & 0 & 0 & 0 & 0 & 0 & 0 \\ 0 & 1 & 0 & 1 & 0 & 0 & 0 & 0 \\ 1 & 0 & -1 & 1 & 0 & 0 & 0 & 0 \\ 0 & 1 & 0 & -1 & 0 & 0 & 0 & 0 \\ 0 & 0 & 0 & 0 & 1 & 0 & 0 & 0 \\ 0 & 0 & 0 & 0 & 0 & 1 & 0 & 1 \\ 0 & 0 & 0 & 0 & 1 & 0 & -1 & 1 \\ 0 & 0 & 0 & 0 & 0 & 1 & 0 & -1 \end{pmatrix}.$$

Now recall from Section 4.10 that for CNOT gates such as in Figure 4.25

$$G_{\mathrm{Q}} = \begin{pmatrix} 1 & 0 & 0 & 0 & 0 & 0 & 0 & 0 \\ 0 & 1 & 0 & 0 & 0 & 0 & 0 & 0 \\ 0 & 0 & 1 & 0 & 0 & 0 & 0 & 0 \\ 0 & 0 & 0 & 1 & 0 & 0 & 0 & 0 \\ 0 & 0 & 0 & 0 & 0 & 1 & 0 & 0 \\ 0 & 0 & 0 & 0 & 1 & 0 & 0 & 0 \\ 0 & 0 & 0 & 0 & 0 & 0 & 0 & 1 \\ 0 & 0 & 0 & 0 & 0 & 0 & 1 & 0 \end{pmatrix}.$$

The transfer matrices corresponding to Figures 4.25(b) and (c) are

$$G_{\mathrm{b}} = G_{\mathrm{IHI}} G_{\mathrm{Q}} = \begin{pmatrix} 1 & 0 & 1 & 0 & 0 & 0 & 0 & 0 \\ 0 & 1 & 0 & 1 & 0 & 0 & 0 & 0 \\ 1 & 0 & -1 & 0 & 0 & 0 & 0 & 0 \\ 0 & 1 & 0 & -1 & 0 & 0 & 0 & 0 \\ 0 & 0 & 0 & 0 & 0 & 1 & 0 & 1 \\ 0 & 0 & 0 & 0 & 1 & 0 & 1 & 0 \\ 0 & 0 & 0 & 0 & 0 & 1 & 0 & -1 \\ 0 & 0 & 0 & 0 & 1 & 0 & -1 & 0 \end{pmatrix},$$

$$G_{\mathrm{c}} = G_{\mathrm{Q}} G_{\mathrm{IHI}} = \begin{pmatrix} 1 & 0 & 1 & 0 & 0 & 0 & 0 & 0 \\ 0 & 1 & 0 & 1 & 0 & 0 & 0 & 0 \\ 1 & 0 & -1 & 0 & 0 & 0 & 0 & 0 \\ 0 & 1 & 0 & -1 & 0 & 0 & 0 & 0 \\ 0 & 0 & 0 & 0 & 0 & 1 & 0 & 1 \\ 0 & 0 & 0 & 0 & 1 & 0 & 1 & 0 \\ 0 & 0 & 0 & 0 & 0 & 1 & 0 & -1 \\ 0 & 0 & 0 & 0 & 1 & 0 & -1 & 0 \end{pmatrix}.$$

The two transfer matrices are identical and this result confirms our intuition.

4.17 MATHEMATICAL MODELS OF A QUANTUM COMPUTER

Several mathematical models for a quantum computer have been proposed, including the quantum Turing Machine model [20], and the quantum cellular automata model [41]. In a recent publication, Peter Shor provides a succinct description of the *quantum circuit model* [125].

A quantum computer consists of input and output "quantum wires" as conduits for qubits and quantum circuits, which in turn are made out of quantum gates. The number of input and output bits of a classical gate may differ, while a quantum gate maps q input qubits into precisely q output qubits. This is a necessary but not a sufficient condition for reversibility.

The mathematical representation of such a quantum gate is a unitary matrix with 2^q rows and 2^q columns, $G = [g_{ij}]\ 1 \leqslant i, j \leqslant 2^q$. If $\mid V \rangle$ is an input vector, then the output vector $\mid W \rangle$ is given by

$$\mid W \rangle = G \mid V \rangle.$$

The input vector $\mid V \rangle$ is the tensor product of q two-dimensional vectors, each one of them representing the state of one qubit. Each qubit can be in a superposition state

$$\mid \psi \rangle = \alpha_0 \mid u_0 \rangle + \alpha_1 \mid u_1 \rangle$$

where u_0 and u_1 are orthonormal base vectors in $\mathcal{H}_2$. The vectors $\mid 0 \rangle$ and $\mid 1 \rangle$ are often used as the basis vectors of $\mathcal{H}_2$.

The input vector of the gate represents a superposition state of a quantum system in a Hilbert space $\mathcal{H}_{2^q}$, see Section 3.8. Similarly, the output of the gate is a vector representing the superposition state of a quantum system in the Hilbert space $\mathcal{H}_{2^q}$ and can be regarded as the tensor product of q qubits.

Let us now consider an n-qubit quantum computer operating in the Hilbert space $\mathcal{H}_{2^n}$, the tensor product of the individual state spaces of the n qubits

$$\mathcal{H}_{2^n} = (\mathcal{H}_2)^n.$$

Let $b = b_0 b_1 b_2, \ldots, b_{n-1}$ be a binary string and let $\mid V_{b_0} \rangle, \mid V_{b_1} \rangle, \mid V_{b_2} \rangle, \ldots, \mid V_{b_{n-1}} \rangle$ be the vectors in $\mathcal{H}_2$ corresponding to the classical bits $b_0, b_1, b_2, \ldots, b_{n-1}$. The tensor product of these vectors is denoted as

$$\mid V_b \rangle = \mid V_{b_0 b_1 b_2, \ldots, b_{n-1}} \rangle = \mid V_0 \rangle \otimes \mid V_1 \rangle \otimes \mid V_2 \rangle \cdots \otimes \mid V_{n-1} \rangle.$$

An equivalent notation is

$$\mid V_b \rangle = \mid V_{b_0 b_1 b_2, \ldots, b_{n-1}} \rangle = \mid V_{b_0} V_{b_1} V_{b_2} \ldots V_{b_{n-1}} \rangle.$$

The basis vectors of the $\mathcal{H}_{2^n}$ Hilbert space are

$$u_0 = \mid 000 \ldots 000 \rangle, u_1 = \mid 000 \ldots 001 \rangle, u_2 = \mid 000 \ldots 010 \rangle, u_3 = \mid 000 \ldots 011 \rangle,$$
$$\ldots \quad \ldots \quad \ldots \quad \ldots \quad u_{2^n-2} = \mid 111 \ldots 110 \rangle, u_{2^n-1} = \mid 111 \ldots 111 \rangle.$$

The input to a quantum computer is classical information represented as a binary string of say $k \leqslant n$ bits. This string is mapped to an input string with the last $n - k$ bits set to zero

$$b' = b_0 b_1 \dots b_{k-1} 00 \dots 0.$$

Let $| V_{b'} \rangle$ be the vector in $\mathcal{H}_{2^n}$ corresponding to this string. Then the output of the n-qubit quantum computer is

$$| W \rangle = G \, | V_{b'} \rangle = \sum_{j=0}^{2^n - 1} \alpha_j \, | u_j \rangle.$$

The α_j are complex numbers called *probability amplitudes* and they must satisfy the condition

$$\sum_{j=0}^{2^n - 1} | \alpha_j |^2 = 1.$$

According to Heisenberg's Uncertainty Principle we cannot measure the complete state of a quantum system. If we observe the output of the quantum computer we decide that the result is j (represented by the binary string $b_0 b_1, b_2, \dots b_{n-1}$) with probability $| \alpha_j |^2$. For example, if $n = 3$, the output value $j = 6$ corresponds to the binary string 110 with $b_0 = 1, b_1 = 1, b_2 = 0$. If $\alpha_{110} = 0.3 + 0.4i$ then the probability of observing the value 6 of the output is $\sqrt{(0.3^2 + 0.4^2)} = 0.5$.

Each gate of a quantum computer applied to a subset of $q \leqslant n$ qubits induces a transformation of the state space $\mathcal{H}_{2^n}$ of the entire quantum computer. For example, the action of a two-qubit quantum gate with a 4×4 transfer matrix $G = [g_{kl}], k, l \in \{1, 4\}$ acting on two qubits, say j_1 and j_2, of an n-qubit quantum computer is represented by the tensor product of the matrix G with $n - 2$ identity matrices of size 2×2, each acting on one of the remaining qubits.

We discuss now the statement "*given any classical circuit capable of computing a function $f(x)$ we can build a reversible quantum circuit capable of computing $f(x)$*." Recall from Section 4.13 that a reversible circuit needs some *work qubits*, or *ancilla qubits*, to contain intermediary values required by the computation. For example, the circuit in Figure 4.20, needs five such qubits to store partial products of the control qubits. Ultimately, the product of all control qubits, stored in one of the work qubits, controls the gate that conditionally transforms the target qubit. These work qubits are initially set to $| 0 \rangle$ and the reversible circuit returns them to the same state before completing its function.

The `Fredkin` gate discussed in Section 4.6 has a control input and two target qubits. One of the target qubits can be thought of as a work qubit and the other as target qubit proper. The `Fredkin` gate is universal, it can simulate an `AND` and a `NOT` gate.

Given a function $f(x)$ we can construct a quantum circuit consisting of `Fredkin` gates only, capable of transforming two qubits $| x \rangle$ and $| y \rangle$ into $| x \rangle$ and $| y \oplus f(x) \rangle$, respectively. Thus, the output vector of the circuit in Figure 4.26

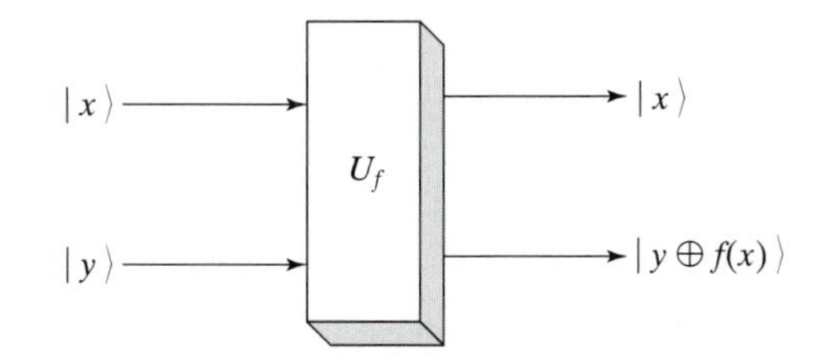

FIGURE 4.26: A two-qubit quantum gate mapping $U_f : | xy\rangle \mapsto | x(y \oplus f(x))\rangle$. When $| y\rangle = | 0\rangle$ then the transformation is $| x\, 0\rangle \mapsto | x\ f(x)\rangle$.

is $| x(y \oplus f(x))\rangle$. We stress the fact that the function $f(x)$ is hardwired into the circuit.

In the general case, we can have multiple control and work qubits and the transformation carried out by the quantum circuit is

$$| x\ y\rangle \mapsto | f(x)g(x)\rangle$$

with $f(x)$ the result and $g(x)$ some undesirable output that should be cleaned up. As before, $| x\ y\rangle$ means the tensor product of $| x\rangle$ and $| y\rangle$. If we have m control qubits then

$$| x\rangle = | x_0x_1 \dots x_j \dots x_{m-1}\rangle \quad \text{with} \quad x_j \in \{0, 1\} \text{ for } j = 0, 1, \dots, m - 1.$$

Therefore, x may take all possible values in the range 0 to $2^m - 1$. If we have k work qubits then

$$| y\rangle = | y_0y_1 \dots y_j \dots y_{k-1}\rangle \quad \text{with} \quad y_j \in \{0, 1\} \text{ for } j = 0, 1, \dots, k - 1.$$

The work qubits should start in state $| 0\rangle$. In this case, the transformation is

$$| x_0x_1 \dots x_{m-1}00 \dots 0\rangle \mapsto | f(x)g(x)\rangle$$

or in a compact form

$$| x\, 0\rangle \mapsto | f(x)g(x)\rangle.$$

We can gradually augment our circuit to carry out a reversible computation, see Figure 4.27. First, we create a copy of the input x using a CNOT gate. This copy will not be altered during the computation. Now, the transformation carried out by the circuit is

$$| x\, 0\, 0\rangle \mapsto | x f(x)g(x)\rangle.$$

Now we need to reverse the status of the work qubits to their original state. We construct a circuit with four input registers. The first contains the input $| x\rangle$. The second is originally $| 0\rangle$, and, at the end of the computation, will contain the result. The third contains the work qubits in state $| 0\rangle$, and, at some point in time, will contain the undesirable intermediate results, $| g(x)\rangle$. The fourth

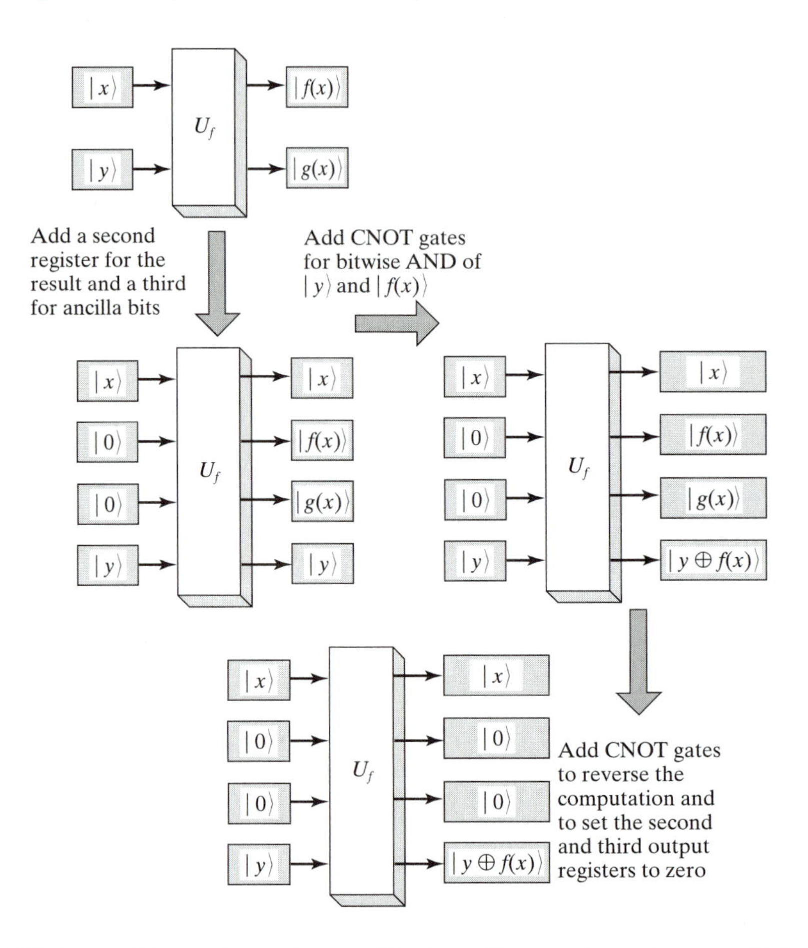

FIGURE 4.27: A reversible quantum gate array. We start with a quantum gate array with two input registers for $| x\rangle$ and $| y\rangle$. As a first step towards reversibility we add two new input registers, one for the result and one for the ancilla bits. Then we add CNOT gates to produce the bitwise AND of $| y\rangle$ and $| f(x)\rangle$. Finally, we add CNOT gates to reverse the computation and set the second and third output registers to zero. We end up with the input $| x\rangle$ and with $| y \oplus f(x)\rangle$. When $| y\rangle = | 0\rangle$ then the second output is $| f(x)\rangle$.

starts in some state $| y\rangle$. This four-register circuit is used to compute $| f(x)\rangle$ and carries out the following transformation

$$| x\, 0\, 0\, y\rangle \mapsto | x f(x) g(x) y\rangle.$$

In the next step, we augment the quantum circuit with CNOT gates to add bitwise $| f(x)\rangle$ obtained in the second register to the fourth register $| y\rangle$. As a result we reach the state

$$| x\, 0\, 0\, y\rangle \mapsto | x f(x) g(x) y\rangle \mapsto | x f(x) g(x) y \oplus f(x)\rangle.$$

But all the steps to compute $| f(x)\rangle$ are reversible and we can return the second and third register (the work register containing the garbage $| g(x)\rangle$) to zero and finally reach the state

$$| x\,0\,0\,y\rangle \mapsto | x f(x) g(x) y\rangle \mapsto | x f(x) g(x) y \oplus f(x)\rangle \mapsto | x\,0\,0\,(y \oplus f(x))\rangle.$$

In a compact form the previous result becomes

$$| x\,y\rangle \mapsto | x(y \oplus f(x))\rangle.$$

Thus, in theory we can construct a reversible quantum circuit computing an arbitrary function $| f(x)\rangle$ of its input state $| x\rangle$.

4.18 ERRORS, UNIFORMITY CONDITIONS, AND TIME COMPLEXITY

Let us briefly review what we already know about the errors in quantum computations. *Closed* quantum systems are systems isolated from the environment. *All state transformations of closed quantum systems are unitary and the superposition states are maintained during the evolution of a closed system.* All the qubits of a quantum system evolve simultaneously as described by the Hamiltonian of the system, as discussed in Chapter 2.

Unfortunately, it is unfeasible to maintain a system in total isolation from the environment and the physical reality forces us to consider *open* systems where a certain degree of interaction of the system with the environment occurs. The evolution of an open system is not unitary. *In an open system, the superposition states decay very quickly and we witness the decoherence phenomena* discussed in Chapter 2. This explains why in our everyday life we rarely observe the superposition states and also hints at potential problems in quantum computing.

Errors due to decoherence are inevitable in a quantum computer. If no measures are taken, and an error of order $1/e$ is introduced when a gate transforms a quantum state, then after $\mathcal{O}(e)$ such state transformations, the quantum state becomes so noisy that a wrong answer is almost certain [19]. Thus, *error correction becomes a necessity for quantum computers*. Fault-tolerant quantum computing is even more intricate, it requires measurements during the evolution of a computation, while the quantum circuit model defers all measurements to the end of a computation.

A *uniform family of circuits* is a set of circuits with one circuit for each number of bits. If a Turing Machine can compute a function F in $T(n)$ number of steps for an input of size n, then there is a family of uniform circuits such that the circuit with n input bits has at most $k f(n)^2$ elements, with k depending upon the complexity of the Turing Machine.

In our discussion of quantum circuits we assumed a constant, as opposed to a variable number of bits for the input, a condition we refer to as *circuit uniformity*. This is a very reasonable assumption inherited from classical circuits and traditional computer architecture. A 32-bit microprocessor accepts as input 32-bit integer and floating point representations, its internal registers, arithmetic and logic unit (ALU), as well as data buses are 32-bit wide. It is conceptually feasible to allow arbitrary length inputs but in that case one could hide a non-computable function in the design of the quantum circuits for each input length.

For *non-uniform quantum circuits* we allow at most a polynomial amount of extra information hidden into the circuit design.

We follow Shor's approach [125] regarding the class of functions that can be computed in polynomial time on a quantum computer and his definition of the *Bounded-Error Quantum Polynomial (BQP)* time complexity class:

> *The Bounded-Error Quantum Polynomial (BQP) functions are all functions, having their domain the set of binary strings, computable by uniform quantum circuits whose number of gates is polynomial in the number of input qubits, and which give the correct answer at least 2/3 of the time.*

The corresponding family of functions when we allow non-uniform quantum circuits and a polynomial amount of extra information hidden in the circuit design is called *BQP/Poly*.

4.19 SUMMARY AND FURTHER READINGS

A quantum computer consists of quantum circuits. Quantum circuits are built by interconnecting quantum gates. A quantum gate maps q input qubits into precisely q output qubits, a necessary, but not a sufficient condition for reversibility.

The input state $| V\rangle$ of a quantum computer is the tensor product of n two-dimensional vectors, each one of them representing the state of one qubit. The input to a quantum computer is classical information represented as a binary string of say $k \leqslant n$ bits. This string is mapped to an input vector with the last $n - k$ bits set to zero.

There are several one-qubit gates: the identity gate `I` leaves a qubit unchanged; the `X` or `NOT` gate with the transfer matrix σ_x transposes the components of an input qubit; the `Z` gate with the transfer matrix σ_z flips the sign, changes the phase, of a qubit; the `Hadamard` gate `H` when applied to a pure state, $| 0\rangle$ or $| 1\rangle$, creates a superposition state, $| 0\rangle \mapsto [1/\sqrt{2}](| 0\rangle + | 1\rangle)$ and $| 1\rangle \mapsto [1/\sqrt{2}](| 0\rangle - | 1\rangle)$.

The two-qubit `CNOT` gate has as input a control qubit and a target qubit. The control input is transferred directly to the control output of the gate. The target output is equal to the target input if the control input is $| 0\rangle$ and it is flipped if the control input is not $| 0\rangle$.

The three-qubit `Fredkin` gate has two regular inputs, a, b, and a control input, c, and three outputs, a', b', and c'. The control input c is transferred directly to the output, $c' = c$. The `Fredkin` gate is reversible; it is a *conservative logic gate* (i.e., it conserves the number of 1s at its input) and it is universal (i.e., it can simulate an `AND` gate and a `NOT` gate).

The `Toffoli` gate, another three-qubit gate, has two control inputs, a and b, and one target input, c. The outputs are: $a' = a$, $b' = b$ and c', see Figure 4.7(a). The `Toffoli` gate is a universal gate and it is reversible.

Several limitations are imposed in the realization of quantum circuits. The circuits are acyclic, feedback from one part of the circuit to another is not allowed, there are no loops. We cannot copy qubits.

The quantum algorithms discussed in the next section are specified in terms of quantum circuits.

The model of a quantum computer has several attributes [98]:

1. An n-qubit quantum computer operates on a 2^n-dimensional Hilbert space $\mathcal{H}_{2^n}$ with computational basis states $| x_1, x_2, \ldots, x_n \rangle$ with $x_i = 0$ or $x_i = 1$.
2. Any computational basis state $| x_1, x_2, \ldots, x_n \rangle$ can be prepared in at most n steps.
3. We are able to perform measurements in the computational basis of one or more of the qubits.
4. A universal family of gates can be implemented. For example, the `Hadamard`, `phase`, `CNOT`, and $\pi/8$ gates are universal, any unitary operation can be approximated by a circuit using only these gates. The gates belonging to such a set of universal gates may be applied to any number of qubits to produce the desired unitary transformation.
5. A quantum computer is an ensemble consisting of classical and quantum components. All classical computations may be carried out by a quantum circuit, but it is more convenient to perform them on the classical component of a quantum computer.

Reference [5] provides the background material for one-qubit and multiple-qubit controlled operations. Chapter 4 in the book of M. A. Nielsen and I. L. Chuang [98] covers the subject in depth and it is accessible to a large audience.

4.20 EXERCISES AND PROBLEMS

4.1. Prove by Boolean algebra manipulation rather than truth tables that the two expressions in Section 4.1 giving the *CarryOut* are equivalent.

4.2. Prove de Morgan's laws

$$\overline{a + b} = \bar{a}\bar{b}$$

$$\overline{ab} = \bar{a} + \bar{b}$$

using the truth table method. a and b are two Boolean variables.

4.3. Consider the matrix $G = [g_{ij}], 1 \leqslant i, j \leqslant 2$, defined in Section 4.2. Derive the four equations relating the real and imaginary parts of g_{ij} implied by the condition that G is unitary.

4.4. Verify that the transfer matrices of the `X`, `Z`, and `H` gates are unitary. Verify that the outputs of the three gates are the ones given in Section 4.2.

4.5. Prove that

$$P_{\theta_1} P_{\theta_2} = P_{\theta_1 + \theta_2}$$

where P_θ is the phase shift transformation.

4.6. Prove the following relationship between the `CNOT` gate and the `I` and `X` single-qubit gates

$$G_{\text{CNOT}} = | 0 \rangle \langle 0 | \otimes I + | 1 \rangle \langle 1 | \otimes X.$$

4.7. The state of a qubit transformed by a Fredkin gate is

$$| \phi \rangle = \alpha_{000} \mid 000\rangle + \alpha_{001} \mid 001\rangle + \alpha_{010} \mid 010\rangle + \alpha_{011} \mid 011\rangle + \alpha_{100} \mid 100\rangle + \alpha_{101} \mid 101\rangle + \alpha_{110} \mid 110\rangle + \alpha_{111} \mid 111\rangle.$$

Show that the normalization condition holds

$$| \alpha_{000} |^2 + | \alpha_{001} |^2 + | \alpha_{010} |^2 + | \alpha_{011} |^2 + | \alpha_{100} |^2 + | \alpha_{101} |^2 + | \alpha_{110} |^2 + | \alpha_{111} |^2 = 1.$$

4.8. Prove the following relationship between the Fredkin gate, and the swap and I gates:

$$G_{\text{Fredkin}} = I \otimes \mid 0\rangle\langle 0 \mid + G_{\text{swap}} \otimes \mid 1\rangle\langle 1 \mid$$

with

$$G_{\text{swap}} = \mid 00\rangle\langle 00 \mid + \mid 01\rangle\langle 10 \mid + \mid 10\rangle\langle 01 \mid + \mid 11\rangle\langle 11 \mid.$$

4.9. Prove the following relationship between the Toffoli gate and the CNOT and I gates

$$G_{\text{Toffoli}} = \mid 0\rangle\langle 0 \mid \otimes I + \mid 1\rangle\langle 1 \mid \otimes G_{\text{CNOT}}.$$

4.10. Construct the transfer matrix of the quantum circuit in Figure 4.28.

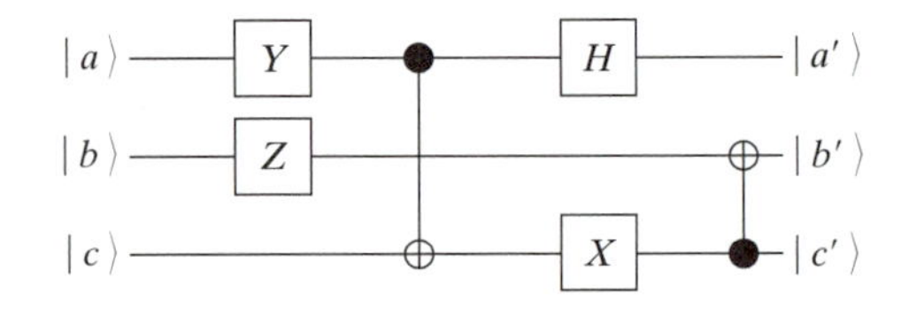

FIGURE 4.28: X, Y, Z are quantum gates implementing the transformations given by the Pauli matrices σ_x, σ_y and σ_z and H is the gate implementing the Walsh-Hadamard Transform.

4.11. A Gray code of length n, G_n, is a sequence of binary numbers such that adjacent members of the list differ in exactly one position. Gray codes can be defined recursively

$$G_1 = \{0, 1\},$$
$$G_n = \{0G_{n-1}, 1G_{n-1}\}.$$

For example,

$$G_2 = \{00, 01, 11, 10\}.$$

(a) Construct $G_3 = \{g_0, g_1, g_2, g_3, g_4, g_5, g_6, g_7\}$.
(b) Construct the operator $\mathbf{U}_{\text{Gray}}$ performing the circular transformations

$$\begin{aligned} \mid g_i\rangle &\mapsto \mid g_{i+1}\rangle \\ \mid g_{i+1}\rangle &\mapsto \mid g_{i+2}\rangle \\ &\vdots \\ \mid g_{i-1}\rangle &\mapsto \mid g_i\rangle. \end{aligned}$$

(c) Compute the transfer matrix of the gate implementing the transformation $\mathbf{U}_{\text{Gray}}$.

4.12. Write a Java simulator for the `CNOT`, the `Fredkin`, and the `Toffoli` gates. The simulator should have a GUI allowing the user to select the gate and then the inputs to that gate. Once the gate is selected, the GUI should display the gate, allow the user to specify the input to that gate, and then display the output of the gate. The simulator should be written as an applet.

4.13. Write a Java simulator for the full one-qubit adder constructed with `Toffoli` and `CNOT` gates in Figure 4.10. Display the values of the individual qubits at every stage of the circuit, as discussed in Section 4.10.

CHAPTER 5
Quantum Algorithms

We discuss quantum algorithms in this chapter. First, we introduce the computational model proposed by David Deutsch, the Quantum Turing Machine. Then, we analyze the relationship between computational complexity and entanglement in an attempt to understand the roots of the immense power of quantum computing.

We present an overview of three classes of quantum algorithms, algorithms to find the periodicity of a function, search algorithms, and algorithms for simulation of quantum systems and then we discuss quantum parallelism and Deutsch's problem.

We introduce the Quantum Fourier Transform (QFT), discuss tensor product factorization, present a circuit for QFT, and dissect the case of a three-qubit QFT. Then we present Shor's algorithm for integer factorization, and Simon's algorithm for phase estimation. Appendices B, C, and D provide some details regarding modular arithmetic necessary for understanding the factoring algorithms, the Walsh-Hadamard Transform, and the Fourier Transform.

The next few sections require some understanding of group theory and we encourage the reader to study Appendix A before reading these sections. First, we discuss the Fourier Transform on an Abelian group and survey the applications of the QFT for determining the period of a periodic function. Then we present the discrete logarithm evaluation problem, and, finally, we cover the hidden subgroup problem. We conclude this chapter with a brief discussion of quantum search algorithms.

5.1 FROM CLASSICAL TO QUANTUM TURING MACHINES

A general-purpose computer is a system capable of storing and transforming information. Information is a primitive concept and as such, it cannot be rigourously defined; we defer an in-depth discussion of this concept for the second volume of this lecture series. For the time being we consider familiar objects such as strings of characters, images, or numbers and refer to them collectively as *information*.

A *Turing Machine (TM)* is an abstraction, a system with a finite number of internal states, a read-write head, and a moving tape consisting of individual cells. Each cell contains a symbol. The Turing Machine starts in a certain state, looks at the symbol currently under the head and, depending upon the internal state of the machine and the current symbol, it may either erase that symbol and replace it with another symbol, or leave it as is, and then move the tape left or right and change its internal state. Let Q_i and S_i denote the state and the symbol read at time t; then Q_j, S_j, and D, the new state, the symbol written, and the direction of movement, respectively, are given by

$$Q_j = F(Q_i, S_i), \quad S_j = G(Q_i, S_i), \quad D = D(Q_i, S_i).$$

The evolution of a computation carried out by a Turing Machine is given by the set of quintuples (Q_i, S_i, Q_j, S_j, D) and is completely specified by the original tape and the set (Q_i, S_i, Q_j, S_j, D).

There is a Universal Turing Machine capable of mimicking any other Turing Machine. Assume that the set of quintuples describing the computation carried out by the original Turing Machine is available. We feed the Universal Turing Machine the description of the original Turing Machine (the set of quintuples), as well as the input tape of the original Turing Machine and the indication where to start and where to end.

Let us turn our attention to the functions that can be computed by a Turing Machine. We expect that the process of finding a solution to some problems cannot be automated. Formulas to calculate the integrals of many functions exist [65], but no general rules to obtain the analytic expression of the integral of an arbitrary function (which has an analytic expression for its integral) are known. Proving theorems and solving Euclidean geometry problems is an art, no general rules to prove theorems and to construct the solution of an Euclidean geometry problem exist.

On the other hand, whenever a function has a derivative, the analytic expression of its derivative can be obtained by applying a finite set of rules to the original expression. Analytical geometry reduces Euclidean geometry problems to problems in a branch of algebra.

There are "effective procedures" to solve some problems while "effective procedures" may not exist for other classes of problems. Having an "effective procedure" to carry out some computation amounts to finding a Turing Machine able to carry out the same computation. Let $f(x)$ be a function of x. The function $f(x)$ is *Turing computable* if there is a Turing Machine T_f with a tape containing a description of x and which will eventually halt with a description of $f(x)$ on tape.

Understanding how a computer works is a non-trivial task, while a description of a Turing Machine needs only a few paragraphs and, yet, it allows us to reason about fundamental computational issues such as the "halting problem."[1]

The distinction between computable and non-computable function is not sufficient for a quantitative analysis of computability. The *complexity theory* addresses the problem of efficiency to compute a specific function $f(x)$. The computational complexity theory is concerned with time and space complexity of algorithms. Yet, as we shall see in the next chapter, when addressing the problem of the resources necessary for computations, besides time and space, we have to examine also the energy consumption by the physical devices performing a computation.

A *Probabilistic Turing Machine (PTM)* is a Turing Machine in which some transitions are random choices among finitely many alternatives. A PTM is equivalent to a Turing Machine except that it has an instruction that allows it to randomly choose an execution path. An example of such an instruction would be a "write" instruction where the value of the write is random and equally distributed among the characters in the Turing Machine's alphabet. Unlike a Turing Machine, a Probabilistic Turing Machine can have stochastic results; a computation executed repeatedly on a given input and instruction state machine may have different run times, or it may not halt at all. A *Non-deterministic Turing Machine* is like a probabilistic one; it has more than one next state per computational step, but the probability distribution of all allowed states is known. It guesses the correct answer (if that applies) every time.

A classical computation carried out by a PTM can be represented as a tree where each node corresponds to a state of the PTM. The root of the tree reflects the starting configuration and each level corresponds to a step of the computation. An edge represents a transition from one state to another; a state can be reached from a parent state (situated one level up in the tree) with a certain probability. The same state may appear multiple times at any level of the tree, or at different tree levels. The probability of reaching a state appearing multiple times at level i of the tree, is the sum of the probabilities of reaching individual instances of that state. The probability of reaching an instance of a state of the PTM is equal to the product of probabilities assigned to all edges leading from the root to the node corresponding to that state. The sum of the probabilities of all states at any given level of the tree should be one.

A computation is said to be *well-defined* if and only if the probabilities on the edges from a parent node and the children states do not depend upon the position of the node in the tree. Well-defined configurations can also be represented as

[1]For some input x a Turing Machine may not halt. It is not possible to construct a computable function which predicts whether or not the Turing Machine T_f with x as input will ever halt.

Markov chains. A Markov chain has a stochastic matrix associated with it whose entries are the transition probabilities between pairs of states.

The quantum model of computation proposed by David Deutsch, the *Quantum Turing Machine (QTM)*, is similar to the PTM model, but considerably more powerful. The traditional laws of probabilities obeyed by classical systems are replaced by different laws specific to quantum systems [127].

A computation on a QTM is represented by a computational tree where each edge has an associated *amplitude*, a complex number with a magnitude at most 1. The amplitude associated with a node (state) is the *product* of the amplitudes of all edges on the path from the root to that node. The amplitude corresponding to a state appearing multiple times at any level i of the tree is the sum of the amplitudes of all nodes corresponding to that state at that particular level of the tree. The probability of an instance of a state is the square of the amplitude of the corresponding node. The sum of the probabilities of all states at any given level of the tree should be one. For example, "the probability of a particular final state is the square of the sum (not the sum of the squares) of all leaf nodes corresponding to that state" [127].

The condition that the squares of the amplitudes associated with all edges emerging from a given node sum to unity is no longer a sufficient condition for a QTM. We require that a QTM always execute unitary steps. The evolution of a QTM is unitary and reversible. At each step, the amplitudes of possible states are determined by the amplitude of the current state according to a fixed, local, unitary transformation similar to a stochastic matrix of a Markov process.

5.2 COMPUTATIONAL COMPLEXITY AND ENTANGLEMENT

The *Church-Turing principle*, stating that "all computing devices can be simulated by a Turing Machine," has profound implications for the computability theory. This remarkable thesis tells us that in order to study computability it is sufficient to restrict ourselves to a single abstract model, the Turing Machine, instead of investigating a potentially infinite set of physical computing devices. To establish if a function $f(x)$ is computable or not, we have to find a TM able to carry out the computation prescribed by the function $f(x)$. If such a machine exists we can be assured that the function is computable.

The distinction between polynomial and exponential use of computer resources represents the cornerstone of the computational complexity. For example, to compute the Discrete Fourier Transform (DFT) for a set of n integers or complex numbers requires $T(n)$ time steps with $T(n)$ bounded by a quadratic polynomial and we say that this algorithm is $\mathcal{O}(n^2)$.[2]

A more general formulation of the problem addressed by the computational complexity is expressed in terms of a language $\mathcal{L}$ consisting of a subset of all possible binary strings (strings consisting only of 0s and 1s).[3] Consider the task

[2] A considerably faster algorithm, due to Cooley and Tukey [36], the so-called Fast Fourier Transform (FFT), requires only $\mathcal{O}(n \log_2 n)$ time steps.

[3] In our example the language consists of the set of n strings representing the moduli of the complex numbers of the DFT of the signal.

of recognizing if a string σ belongs to $\mathcal{L}$ or not. We say that the language $\mathcal{L}$ is in the complexity class $\mathcal{P}$ if there is a recognition algorithm which requires a number of steps $T(n)$ bounded by a polynomial. Otherwise, we say that the language requires an exponential time.

There are also algorithms based upon a probabilistic approach. The *Bounded-Error Probabilistic Polynomial Time* (BPP) group consists of polynomial time algorithms which correctly classify the input string σ with a probability at least $2/3$ (or other value strictly between $1/2$ and 1). We may get the wrong answer when we run such an algorithm once, but, by repeatedly running the algorithm and then taking a majority vote, the probability of obtaining the correct answer can be made arbitrarily close to 1, while maintaining the polynomial running time.

A *quantitative version of the Church-Turing principle* allows us to relate the behavior of the abstract model of computation provided by the TM concept with the physical computing devices used to carry out a computation. This thesis can be formulated as "any physical computing device can be simulated by a Turing Machine in a number of steps polynomial in the resources used by the computing device" [101]. While no one has been able to find counter-examples for this thesis, the search has been limited to systems that are based upon the laws of classical mechanics.

Yet, our universe is essentially quantum mechanical, therefore, there is a possibility that the computing power of quantum mechanical computing devices might be greater than the computing power of classical computing devices [121]. If this is true, then problems such as factoring integers, or finding discrete logarithms for which no polynomial time algorithms are known, could be solved in polynomial time by a quantum device. Therefore, the investigation of quantum computing devices and of quantum algorithms is well motivated.

Quantum computers (i.e., computers based upon quantum circuits that behave according to the laws of quantum mechanics) allow us to transcend the well-defined boundary between polynomial and exponential computations. We suspect that the source of this incredible power of quantum computers is the phenomenon of entanglement, characteristic of quantum systems [75]. Quantum entanglement does not have a classical counterpart.

In Section 5.4 we show that if we are given a quantum circuit able to compute in polynomial time a function $f(j)$, with j a binary string of length n, then we are able to compute a superposition of all 2^n possible values of the function f in polynomial time with a single copy of the quantum circuit. An alternative formulation of this statement is that only a polynomial effort is required to compute f. In contrast, if we wish to calculate the 2^n values of the function f using a classical circuit which computes f in polynomial time, then we need an exponential number of copies of the circuit.

This "magic" effect associated with a quantum computer is called *quantum parallelism*. To achieve this remarkable result we construct a quantum circuit able to compute f in polynomial time and prepare its input in a superposition state, as shown in Figure 5.1. The input to the quantum gate array is the result of transforming the state obtained as the tensor product of the n qubits in state $| 0\rangle$ by a quantum circuit whose transfer matrix is the tensor product of the n

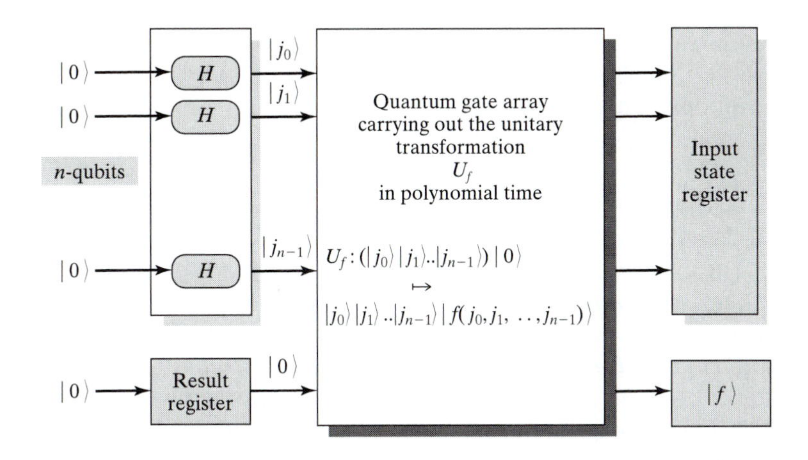

FIGURE 5.1: A quantum computer with two input registers, one, register j, with n qubits $|\ j_0\rangle, |\ j_1\rangle, \ldots, |\ j_{n-1}\rangle$, and the other a result register initially set to zero, $|\ 0\rangle$. There are also two output registers. The quantum gate array implements a function $f : \{0, 1\}^n \mapsto \{0, 1\}$ mapping an n-bit input string to a binary value, 0 or 1 in polynomial time. The n qubits are initially in state $|\ 0\rangle$ and are transformed by n Hadamard gates. The input to the quantum gate array consists of the tensor product of n input qubits in superposition states created by n Hadamard gates. There are 2^n possible values of the input string and the final result is a superposition of the 2^n possible evaluations of the function $f(j)$.

transfer matrices of the Hadamard gates

$$H \otimes H \ldots \otimes H(|\ 0\rangle\ |\ 0\rangle \ldots |\ 0\rangle) = \frac{1}{2^{n/2}}(|\ 0\rangle + |\ 1\rangle)^n = \frac{1}{2^{n/2}} \sum_{k=0}^{2^n-1} |\ k\rangle$$

with

$$H = \frac{1}{\sqrt{2}} \begin{pmatrix} 1 & 1 \\ 1 & -1 \end{pmatrix}.$$

In the preceding expression the notation $|\ k\rangle$ represents the k-th basis vector[4]

$$|\ k\rangle = |\ 00 \ldots 1 \ldots 00\rangle.$$

The quantum gate array implements a function $f : \{0, 1\}^n \mapsto \{0, 1\}$ mapping an n-bit input string $j = (j_0, j_1, \ldots, j_{n-1})$ to a binary value, 0 or 1, in polynomial time. The transformation carried out by the quantum gate array is

$$U_f : |\ j_0\rangle\ |\ j_1\rangle \ldots |\ j_{n-1}\rangle\ |\ 0\rangle \quad \mapsto \quad |\ j_0\rangle\ |\ j_1\rangle \ldots |\ j_{n-1}\rangle\ |\ f(j_0, j_1, \ldots, j_{n-1})\rangle$$

If the input is a superposition state created by the n Hadamard gates, then the result is a superposition of all possible values of function f

$$|\ f\rangle = \sum_{j=0}^{2^n-1} |\ j\rangle\ |\ f(j)\rangle.$$

This relatively simple quantum circuit setup makes us wonder if the superposition is the source of the enormous power of computers based upon quantum effects. The state of a system consisting of n bits is a vector in a 2^n-dimensional vector

[4]The k-th basis vector in $\mathcal{H}_{2^n}$ has a single 1 in the $(k + 1)$-th position, $0 \leq k \leq 2^n - 1$.

space. If we have n qubits, then the state of the system is also a vector, but in the 2^n-dimensional Hilbert space. This state space is the tensor product of n, two-dimensional Hilbert spaces[5]

$$\mathcal{H}_{2^n} = (\mathcal{H}_2)^n.$$

There is an isomorphism between the state space of a system of n qubits $(\mathcal{H}_2)^n$ and the state space of a single particle with 2^n states, $\mathcal{H}_{2^n}$.

The crucial observation is that *in addition to the classical states corresponding to all possible binary vectors of length n, the state space $\mathcal{H}_{2^n}$ includes also entangled states. The entangled states reflect the critical distinction between the Cartesian product in an Euclidean space and the tensor product in a Hilbert space.*

We now ask ourselves if a single classical system with 2^n states could emulate the behavior of a quantum system of n qubits. We can construct a system where each qubit is represented by a classical wave system and select two modes of vibration to represent the basis states $| 0\rangle$ and $| 1\rangle$. We can then construct the superposition corresponding to $| 0\rangle + | 1\rangle$. Finally, we use n copies of such systems to obtain a product state.

For example, we may use an elastic string with fixed end points; then the two modes used to represent the two basis states could be the lowest energy modes. A set of n such elastic strings will represent a product state. Jozsa argues that in such an experiment the joint state of the n strings is always a product state of n separate vibrations [75]. It may be possible for the 2^n modes of a classical vibrating system to emulate the behavior of n qubits and exhibit entangled states, but we would need to expend an exponential amount of energy to emulate the behavior of entangled quantum particles.

Another alternative may be feasible. A physical system with infinitely many discrete energy levels could be used to represent the superposition of exponentially many modes, using a constant energy. In this case, the levels will be exponentially crowded together and we would need an exponential amount of energy to distinguish among them.

We conclude this discussion with the observation that the state of n qubits requires an energy that grows linearly with n, while the energy required by a classical system which mimics the behavior of the n-qubit system grows exponentially with n. It should also be very clear to us that a rigorous assessment of computational complexity must consider all computational resources, including energy, as opposed to the traditional complexity theory focused only on time and space complexity.

5.3 CLASSES OF QUANTUM ALGORITHMS

In a recent paper [126], Peter Shor classifies the quantum algorithms known to offer a significant speed-up over their classical counterparts into three broad categories:

---

[5]Sometimes in addition to the notations used in this book for the tensor product of n two-dimensional Hilbert spaces, $(\mathcal{H}_2)^n$, we find in the literature two different ones, $\mathcal{H}_2^{\otimes n}$ and $\bigotimes_{k=1}^{n} H_2^{(k)}$.

1. Algorithms that find the periodicity of a function using Fourier Transforms, Simon's algorithm [127], Shor's algorithms for factoring and for computing discrete logarithms [123], and Hallgren's algorithm to solve Bell's Equation are all members of this class.
2. Search algorithms which can perform an exhaustive search of N items in $\sqrt{N}$ time. Grover's algorithms [67, 68, 69] belong to this class.
3. Algorithms for simulating quantum systems, as suggested by Feynman. This is a potentially large class of algorithms, but not many algorithms in this class have been developed so far. Once quantum computers become a reality we should expect the development of a considerable number of programs to simulate quantum systems of interest.

However, quantum algorithms have not witnessed the same effervescent developments we have seen in quantum information theory and, to a lesser extent, in quantum complexity. Clearly, the development of quantum algorithms requires very different thinking than the one we are accustomed to in the case of classical computers. To offer a considerable speed-up, a quantum algorithm must rely on superposition states and this is a foreign concept for those unaccustomed to quantum mechanical concepts.

In addition to this obvious reason, Peter Shor speculates [126] that quantum algorithms may offer a substantial speedup over classical algorithms, but may be very limited. To see the spectacular speedups we have to concentrate on problems not in the classical computational class $\mathcal{P}$. Many believe that quantum algorithms solving $\mathcal{NP}$-complete problems in polynomial time do not exist, even though no proof of this assertion is available at this time. Shor argues that if we assume that no polynomial time quantum algorithms exist for solving $\mathcal{NP}$-hard problems, then the class of problems we have to search for is neither $\mathcal{NP}$-hard, nor $\mathcal{P}$, and the population of this class is relatively small.

The algorithms used to simulate quantum mechanical systems follow the evolution in time of a fairly large number of quantum particles. The result of such a simulation reflects what the outcome of a measurement of the physical quantities of the quantum system would reveal.

There is no simple algorithmic formulation for the simulation of quantum mechanical systems. Moreover, some of the physical quantities of interest are not known to be accessible even on a quantum computer [82]. For example, the question of what is the ground energy state of a quantum system whose Hamiltonian is known does not have an easy answer. Recall that the Hamiltonian is the operator corresponding to the energy observable of a quantum system. Finding a simple solution to the problem of determining the ground energy state of a quantum system would lead to efficient algorithms for "hard" computational problems such as optimal scheduling, or the well-known travelling salesman problem.

Classical Monte Carlo algorithms are widely used today to simulate quantum physical systems. In Monte Carlo simulations, state amplitudes are represented by the expected values of random variables calculated during the simulation.

Such simulations produce large statistical errors that can only be reduced by repeating the calculations many times. The advantage of using quantum algorithms for simulations of quantum physical systems is that they allow us to determine the value of relevant physical properties with polynomial bounded statistical errors.

The current challenge in the development of quantum simulation algorithms is to identify physical simulations requiring a small number of qubits, a number considerably less than 100 qubits. Such a simulation would require a small number of quantum gates and would be feasible in the immediate future.

5.4 QUANTUM PARALLELISM

Consider a classical circuit able to compute the values of the function $f(x)$ for a given input x in a single time step. Let x be a binary string of length n. Then, we have two extreme choices for computing the value of the function $f(x)$ for the 2^n possible values of the argument x:

1. Use a single copy of the circuit repeatedly and obtain all the values of the function after 2^n time steps, or
2. Use 2^n copies of the circuit, each with a different input x, and obtain all the values in a single time step.

Quantum circuits have the unique ability to perform the function described above, namely to compute the values of the function f for all possible values of the input x, in *a single time step and with a single copy of the circuit.* This remarkable property of quantum circuits is called *quantum parallelism* and can be traced back to the superposition principle and to entanglement. Indeed, the state of a quantum system is a superposition of the basis states of the corresponding Hilbert space.

Let us start with $| x\rangle$, a single qubit with two possible values $| 0\rangle$ and $| 1\rangle$. The possible values of the function are $| f(x)\rangle = | 0\rangle$ or $| f(x)\rangle = | 1\rangle$. We wish to construct a circuit whose output is a superposition of $| f(0)\rangle$ and $| f(1)\rangle$.

In Section 4.17 we have seen that given a function $f(x)$ we can construct a quantum circuit consisting of Fredkin gates only, capable of transforming two qubits, $| x\rangle$ and $| y\rangle$, into $| x(y \oplus f(x))\rangle$, as shown in Figure 5.2(a). We stress the fact that the function $f(x)$ is hardwired into the circuit.

If the second qubit is set to $| 0\rangle$, as seen in Figure 5.2(b), then the transformation carried out by the circuit is

$$| x\, 0\rangle \quad \mapsto \quad | x\, f(x)\rangle.$$

Recall that this abbreviated notation shows that the input is the tensor product of the two vectors representing the states $| x\rangle$ and $| 0\rangle$ and this state is transformed into a state represented by the tensor product of the vectors $| x\rangle$ and $| f(x)\rangle$. Note also that an equivalent representation using `ket` vectors is

$$| x\rangle\ | 0\rangle \quad \mapsto \quad | x\rangle\ | f(x)\rangle.$$

Now, instead of $| x\rangle$ consider that the qubit in state $| 0\rangle$ is applied first to a Hadamard gate, and the output of the Hadamard gate, $(| 0\rangle + | 1\rangle)/\sqrt{2}$, is then

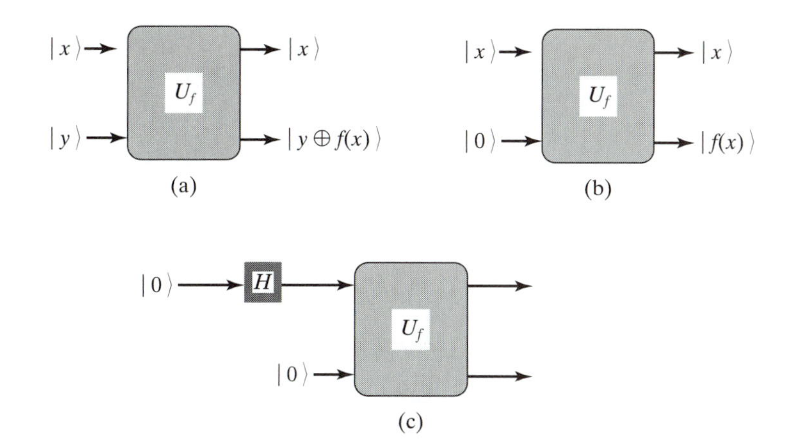

FIGURE 5.2: (a) A reversible quantum circuit transforming $|\, xy\rangle \mapsto |\, x(y \oplus f(x))\rangle$. (b) The circuit in (a) with $|\, y\rangle = |\, 0\rangle$. (c) The first qubit is in a superposition state, $(|\, 0\rangle + |\, 1\rangle)/\sqrt{2}$ and the second qubit is $|\, 0\rangle$. The resulting state of the circuit is $(|\, 0f(0)\rangle + |\, 0f(1)\rangle + |\, 1f(0)\rangle + |\, 1f(1)\rangle)/2$.

applied to the circuit in Figure 5.2(c). The two output qubits of the circuit are

$$(|\, 0\rangle + |\, 1\rangle)/\sqrt{2}$$

and

$$f\Big((|\, 0\rangle + |\, 1\rangle)/\sqrt{2}\Big) = (|\, f(0)\rangle + |\, f(1)\rangle)/\sqrt{2}.$$

The output state of the quantum system is obtained as the tensor product of these two vectors:

$$(|\, 0f(0)\rangle + |\, 0f(1)\rangle + |\, 1f(0)\rangle + |\, 1f(1)\rangle)/2.$$

The output state contains information about $|\, f(0)\rangle$ and about $|\, f(1)\rangle$. This remarkable property of quantum circuits is called quantum parallelism.

This result can be extended to an input consisting of say m qubits. This time we need a quantum circuit to transform an m-dimensional vector $|\, x\rangle$ into $|\, f(x)\rangle$. Moreover, by analogy with the previous case, the m qubits have to be applied in a superposition state using m Hadamard gates.

The circuit transforming the m qubits is a quantum gate array. Figure 5.3 illustrates a quantum gate array characterized by a linear transformation given by U_f. The inputs to this gate array are $|\, x\rangle \in \mathcal{H}_{2^m}$, a 2^m-dimensional vector acting as a control input, and $|\, y\rangle \in \mathcal{H}_{2^k}$, a 2^k-dimensional vector acting as work or ancilla qubit. The outputs are $|\, x\rangle$ and $|\, y \oplus f(x)\rangle \in \mathcal{H}_{2^n}$, with $n = m + k$. When $|\, y\rangle = |\, 0\rangle$, the second output becomes $|\, y \oplus f(x)\rangle = |\, f(x)\rangle$.

Now, assume that the input vector $|\, x\rangle$ is in a superposition state and can be expressed as a linear combination of 2^m vectors forming an orthonormal basis in $\mathcal{H}_{2^m}$. The gate array performs a linear transformation. Henceforth, the transformation is applied to all basis vectors used to express the input superposition simultaneously, and it generates a superposition of results. In

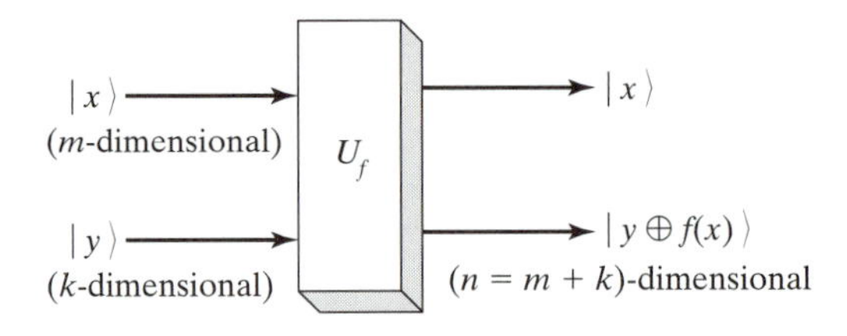

FIGURE 5.3: The transformation $U_f : | xy\rangle \rightarrow | x(y \oplus f(x))\rangle$ performed by the quantum gate array.

other words, the values of the function $f(x)$ for the 2^m possible values of its argument x are computed simultaneously. The quantum parallelism justifies our statement in Section 1.3 that quantum computers can provide an exponential increase of computational space for a linear increase in physical space.

Quantum parallelism allows us to construct the entire truth table of a quantum gate array having 2^n entries in a single time step. In a classical system we can compute the truth table in one time step with 2^n gate arrays running in parallel, or we need 2^n time steps with a single gate array.

Typically, we start with n qubits, each in state $| 0\rangle$ and we apply a Walsh-Hadamard Transform. Each qubit is transformed by a Hadamard gate; recall that the Hadamard gate transforms a $| 0\rangle$ as follows

$$H : | 0\rangle \rightarrow (| 0\rangle + | 1\rangle) / \sqrt{2}.$$

Thus,

$$(H \otimes H \ldots \otimes H) | 00 \ldots 0\rangle = \left[\frac{(| 0\rangle + | 1\rangle) \otimes (| 0\rangle + | 1\rangle) \ldots \otimes (| 0\rangle + | 1\rangle)}{(\sqrt{2})^n} \right]$$

$$= \frac{1}{\sqrt{2^n}} \sum_{j=0}^{2^n-1} | j\rangle.$$

The output of the gate array when we add a k-bit register to the superposition state of integers in the range 0 to $2^n - 1$ with $n = m + k$ is

$$U_f \left(\frac{1}{\sqrt{2^n}} \sum_{x=0}^{2^n-1} | x\, 0\rangle \right) = \frac{1}{\sqrt{2^n}} \sum_{x=0}^{2^n-1} U_f(| x\, 0\rangle) = \frac{1}{\sqrt{2^n}} \sum_{x=0}^{2^n-1} | x\, f(x)\rangle.$$

When we measure the output of the quantum gate array we can only observe one value. Henceforth, we need some level of algorithmic sophistication to exploit the quantum parallelism.

5.5 DEUTSCH'S PROBLEM

Quantum parallelism is best illustrated by the solution to the so-called "Deutsch's problem" formulated below. Consider a black box characterized by a transfer function that maps a single input bit x into an output, $f(x)$. The transformation

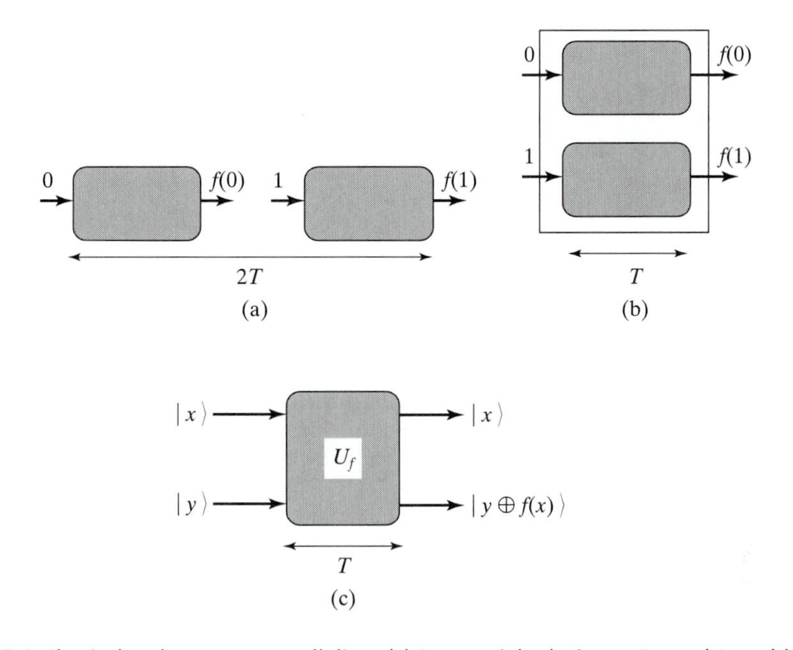

FIGURE 5.4: Classical and quantum parallelism. (a) Sequential solution to Deutsch's problem using a single copy of a classical circuit to compute $f(x)$. Compute $f(0)$, then compute $f(1)$, and finally compare the results. (b) A parallel solution to Deutsch's problem using two copies of the classical circuit to compute $f(x)$. (c) The quantum black box with a transfer function U_f. It evaluates $f(0)$ and $f(1)$ simultaneously.

performed by the black box, $f(x)$, is a general function and might not be invertible. We assume that it takes the same amount of time, T, to carry out each of the four possible mappings performed by the transfer function $f(x)$ of the black box:

$$f(0) = 0, \quad f(0) = 1, \quad f(1) = 0, \quad f(1) = 1$$

and it takes no time to compare the results. The problem posed by David Deutsch is to distinguish if $f(0) = f(1)$ or $f(0) \neq f(1)$.

When using a classical computer, one alternative is to compute sequentially $f(0)$ and $f(1)$ and then compare the results, for a total time $2T$ (see Figure 5.4(a)). A classical parallel solution is illustrated in Figure 5.4(b) where we have two replicas of the circuit and we feed 0 as input to one of the replicas of the black box and 1 to the other, and then compare the partial results. In this case, we obtain the answer after time T but we need two copies of the system, rather than one.

Consider now a quantum computer with a transfer function U_f that takes as input two qubits $| x\rangle$ (control) and $| y\rangle$ (target) and two outputs, $| x\rangle$ and $| y \oplus f(x)\rangle$. We have the choice of selecting the states of the two qubits $| x\rangle$ and $| y\rangle$. We choose the first qubit in the state $| x\rangle = (| 0\rangle + | 1\rangle)/\sqrt{2}$ and the second qubit in the state $| y\rangle = (| 0\rangle - | 1\rangle)/\sqrt{2}$.

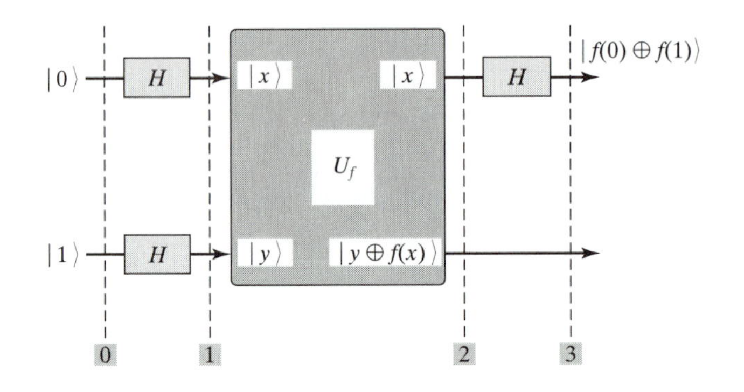

FIGURE 5.5: A quantum circuit for solving Deutsch's problem. The first output qubit of the circuit is $|\, f(0) \oplus f(1)\rangle$. Thus if $f(0) = f(1)$ then this qubit is $|\,0\rangle$ and if $f(0) \neq f(1)$ then this qubit is $|\,1\rangle$. The input state is $|\,\xi_0\rangle$, the state after the Hadamard gate is $|\,\xi_1\rangle$, the output state of the circuit implementing U_f is $|\,\xi_2\rangle$, and the output state of the circuit is $|\,\xi_3\rangle$.

We show that the first output qubit of the circuit is $|\, f(0) \oplus f(1)\rangle$. Thus when $f(0) = f(1)$ this qubit is $|\,0\rangle$ (indeed, $0 \oplus 0 = 0$ and $1 \oplus 1 = 0$) and when $f(0) \neq f(1)$ this qubit is $|\,1\rangle$ (indeed, $0 \oplus 1 = 1$ and $1 \oplus 0 = 1$).

Figure 5.5 illustrates the quantum circuit for solving Deutsch's problem. To compute the output state of this circuit we consider the four stages of this circuit and we examine the four vectors, $|\,\xi_0\rangle, |\,\xi_1\rangle, |\,\xi_2\rangle$ and $|\,\xi_3\rangle$ describing the state of the system at each stage. The input vector is

$$|\,\xi_0\rangle = |\,0\,1\rangle = \begin{pmatrix} 1 \\ 0 \end{pmatrix} \otimes \begin{pmatrix} 0 \\ 1 \end{pmatrix} = \begin{pmatrix} 0 \\ 1 \\ 0 \\ 0 \end{pmatrix}.$$

The transfer matrix of the first stage is

$$G_1 = H \otimes H = \frac{1}{\sqrt{2}} \begin{pmatrix} 1 & 1 \\ 1 & -1 \end{pmatrix} \otimes \frac{1}{\sqrt{2}} \begin{pmatrix} 1 & 1 \\ 1 & -1 \end{pmatrix} = \frac{1}{2} \begin{pmatrix} 1 & 1 & 1 & 1 \\ 1 & -1 & 1 & -1 \\ 1 & 1 & -1 & -1 \\ 1 & -1 & -1 & 1 \end{pmatrix}.$$

Now,

$$|\,\xi_1\rangle = G_1\,|\,\xi_0\rangle = \frac{1}{2} \begin{pmatrix} 1 & 1 & 1 & 1 \\ 1 & -1 & 1 & -1 \\ 1 & 1 & -1 & -1 \\ 1 & -1 & -1 & 1 \end{pmatrix} \begin{pmatrix} 0 \\ 1 \\ 0 \\ 0 \end{pmatrix} = \frac{1}{2} \begin{pmatrix} 1 \\ -1 \\ 1 \\ -1 \end{pmatrix},$$

or

$$|\,\xi_1\rangle = \frac{1}{2}(|\,00\rangle - |\,01\rangle + |\,10\rangle - |\,11\rangle) = \left[(|\,0\rangle + |\,1\rangle)/\sqrt{2}\right]\left[(|\,0\rangle - |\,1\rangle)/\sqrt{2}\right].$$

The two qubits applied to the input of the black box are

$$|\,x\rangle = (|\,0\rangle + |\,1\rangle)/\sqrt{2}, \quad \text{and} \quad |\,y\rangle = (|\,0\rangle - |\,1\rangle)/\sqrt{2}.$$

We know that $|0 \oplus f(x)\rangle = |f(x)\rangle$ thus:

$$|y \oplus f(x)\rangle = (|f(x)\rangle - |1 \oplus f(x)\rangle)/\sqrt{2}.$$

But $|1 \oplus f(x)\rangle$ is equal to $|0\rangle$ when $f(x) = 1$ and it is equal to $|1\rangle$ when $f(x) = 0$ thus

$$|y \oplus f(x)\rangle = \begin{cases} (|0\rangle - |1\rangle)/\sqrt{2} & \text{if } f(x) = 0 \\ -(|0\rangle - |1\rangle)/\sqrt{2} & \text{if } f(x) = 1. \end{cases}$$

Then,

$$|y \oplus f(x)\rangle = (-1)^{f(x)}(|0\rangle - |1\rangle)/\sqrt{2}.$$

It is easy to see that when $f(0) = f(1)$, the state of the two output qubits of the black box, $|x\rangle \otimes |y \oplus f(x)\rangle$, is

$$|\xi_2\rangle = \begin{cases} \left[\frac{|0\rangle + |1\rangle}{\sqrt{2}}\right]\left[\frac{|0\rangle - |1\rangle}{\sqrt{2}}\right] \mapsto \frac{1}{2}\begin{pmatrix} 1 \\ -1 \\ 1 \\ -1 \end{pmatrix} & \text{if } f(0) = f(1) = 0 \\ -\left[\frac{|0\rangle + |1\rangle}{\sqrt{2}}\right]\left[\frac{|0\rangle - |1\rangle}{\sqrt{2}}\right] \mapsto -\frac{1}{2}\begin{pmatrix} 1 \\ -1 \\ 1 \\ -1 \end{pmatrix} & \text{if } f(0) = f(1) = 1. \end{cases}$$

We leave as an exercise the proof that if $f(0) \neq f(1)$, then the output of the black box is

$$|\xi_2\rangle = \begin{cases} \left[\frac{|0\rangle - |1\rangle}{\sqrt{2}}\right]\left[\frac{|0\rangle - |1\rangle}{\sqrt{2}}\right] \mapsto \frac{1}{2}\begin{pmatrix} 1 \\ -1 \\ -1 \\ 1 \end{pmatrix} & \text{if } f(0) = 0 \text{ and } f(1) = 1 \\ -\left[\frac{|0\rangle - |1\rangle}{\sqrt{2}}\right]\left[\frac{|0\rangle - |1\rangle}{\sqrt{2}}\right] \mapsto -\frac{1}{2}\begin{pmatrix} 1 \\ -1 \\ -1 \\ 1 \end{pmatrix} & \text{if } f(0) = 1 \text{ and } f(1) = 0. \end{cases}$$

Combining these two results we have

$$|\xi_2\rangle = \begin{cases} \pm\left[\frac{|0\rangle + |1\rangle}{\sqrt{2}}\right]\left[\frac{|0\rangle - |1\rangle}{\sqrt{2}}\right] \mapsto \pm\frac{1}{2}\begin{pmatrix} 1 \\ -1 \\ 1 \\ -1 \end{pmatrix} & \text{if } f(0) = f(1) \\ \pm\left[\frac{|0\rangle - |1\rangle}{\sqrt{2}}\right]\left[\frac{|0\rangle - |1\rangle}{\sqrt{2}}\right] \mapsto \pm\frac{1}{2}\begin{pmatrix} 1 \\ -1 \\ -1 \\ 1 \end{pmatrix} & \text{if } f(0) \neq f(1). \end{cases}$$

The transfer matrix of the third stage of the quantum circuit in Figure 5.5 is

$$G_3 = H \otimes I = \frac{1}{\sqrt{2}}\begin{pmatrix} 1 & 1 \\ 1 & -1 \end{pmatrix} \otimes \begin{pmatrix} 1 & 0 \\ 0 & 1 \end{pmatrix} = \frac{1}{\sqrt{2}}\begin{pmatrix} 1 & 0 & 1 & 0 \\ 0 & 1 & 0 & 1 \\ 1 & 0 & -1 & 0 \\ 0 & 1 & 0 & -1 \end{pmatrix}.$$

If $f(0) = f(1)$ then

$$| \xi_3 \rangle = \pm \frac{1}{\sqrt{2}}\begin{pmatrix} 1 & 0 & 1 & 0 \\ 0 & 1 & 0 & 1 \\ 1 & 0 & -1 & 0 \\ 0 & 1 & 0 & -1 \end{pmatrix} \frac{1}{2}\begin{pmatrix} 1 \\ -1 \\ 1 \\ -1 \end{pmatrix} = \pm \frac{1}{\sqrt{2}}\begin{pmatrix} 1 \\ -1 \\ 0 \\ 0 \end{pmatrix} = \pm | 0 \rangle \frac{| 0 \rangle - | 1 \rangle}{\sqrt{2}}.$$

If $f(0) \neq f(1)$ then

$$| \xi_3 \rangle = \pm \frac{1}{\sqrt{2}}\begin{pmatrix} 1 & 0 & 1 & 0 \\ 0 & 1 & 0 & 1 \\ 1 & 0 & -1 & 0 \\ 0 & 1 & 0 & -1 \end{pmatrix} \frac{1}{2}\begin{pmatrix} 1 \\ -1 \\ -1 \\ 1 \end{pmatrix} = \pm \frac{1}{\sqrt{2}}\begin{pmatrix} 0 \\ 0 \\ 1 \\ -1 \end{pmatrix} = \pm | 1 \rangle \frac{| 0 \rangle - | 1 \rangle}{\sqrt{2}}.$$

Thus, by examining the first output qubit of the circuit, we decide that $f(0) = f(1)$ when the qubit is $| 0 \rangle$ and $f(0) \neq f(1)$ when the qubit is $| 1 \rangle$. We observe that

$$f(0) \oplus f(1) = \begin{cases} 0 & \text{if} \quad f(0) = f(1) \\ 1 & \text{if} \quad f(0) \neq f(1). \end{cases}$$

Finally, we rewrite $| \xi_3 \rangle$

$$| \xi_3 \rangle = \pm | f(0) \oplus f(1) \rangle \left[\frac{| 0 \rangle - | 1 \rangle}{\sqrt{2}} \right].$$

This expression tells us that by measuring the first output qubit of the circuit in Figure 5.5 we are able to determine $f(0) \oplus f(1)$ after performing a single evaluation of the function $f(x)$.

5.6 QUANTUM FOURIER TRANSFORM

Let us turn our attention to the Quantum Fourier Transform. The *Quantum Fourier Transform (QFT)* is a linear operator that transforms an orthonormal basis

$$\{| 0 \rangle, | 1 \rangle \ldots, | j \rangle, \ldots, | k \rangle, \ldots, | N - 1 \rangle\}$$

as follows

$$| j \rangle \quad \mapsto \quad \frac{1}{\sqrt{N}} \sum_{k=0}^{N-1} e^{i2\pi jk/N} \, | k \rangle.$$

QFT transforms a state $| v \rangle$ of a quantum system into another state $| w \rangle$

$$| v \rangle \quad \mapsto \quad | w \rangle,$$

with

$$| v\rangle = \sum_{j=0}^{N-1} v_j \mid j\rangle,$$

and

$$| w\rangle = \sum_{k=0}^{N-1} w_k \mid k\rangle.$$

Here, the amplitudes w_k are the *Discrete Fourier Transforms* (DFT) of the amplitudes v_j

$$w_k = \mathrm{DFT}(v_j) = \frac{1}{\sqrt{N}} \sum_{j=0}^{N-1} v_j \, e^{i2\pi jk/N}$$

When $N = 2^n$, we consider the binary representation of integers j and k and obtain another expression for the QFT

$$j = j_0 2^{n-1} + j_1 2^{n-2} + \ldots + j_{n-2}2^1 + j_{n-1}2^0.$$

$$k = k_0 2^{n-1} + k_1 2^{n-2} + \ldots + k_{n-2}2^1 + k_{n-1}2^0.$$

Then, the definition of the QFT can be rewritten as:

$$| j_0 j_1 \cdots j_{n-1}\rangle \mapsto \frac{1}{2^{n/2}} \sum_{k_0=(0,1)} \sum_{k_1=(0,1)} \cdots \sum_{k_{n-1}=(0,1)} e^{i2\pi j \Sigma_{m=0}^{n-1} k_m 2^{-m}} \mid k_0 k_1 \ldots k_m \ldots k_{n-1}\rangle,$$

$$| j_0 j_1 \cdots j_{n-1}\rangle \mapsto \frac{1}{2^{n/2}} \sum_{k_0=(0,1)} \sum_{k_1=(0,1)} \cdots \sum_{k_{n-1}=(0,1)} \bigotimes_{m=0}^{n-1} e^{i2\pi jk_m 2^{-m}} \mid k_m\rangle,$$

$$| j_0 j_1 \cdots j_{n-1}\rangle \mapsto \frac{1}{2^{n/2}} \bigotimes_{m=0}^{n-1} \left\{ \sum_{k_m=(0,1)} e^{i2\pi jk_m 2^{-m}} \mid k_m\rangle \right\}.$$

The bit k_m may only take two values, 0 and 1. We note that $e^{i2\pi jk_m} = 1$ when $k_m = 0$, thus

$$| j_0 j_1 \cdots j_{n-1}\rangle \mapsto \frac{1}{2^{n/2}} \bigotimes_{m=0}^{n-1} \{| 0\rangle + e^{i2\pi j 2^{-m}} \mid 1\rangle\}.$$

But

$$| 0\rangle + e^{i2\pi j2^{-m}} \mid 1\rangle = \begin{pmatrix} 1 \\ 0 \end{pmatrix} + \begin{pmatrix} 0 \\ e^{i2\pi j2^{-m}} \end{pmatrix} = \begin{pmatrix} 1 \\ e^{i2\pi(j/2^m)} \end{pmatrix}$$

The transformation of the input $| j\rangle$ can be rewritten as

$$| j\rangle \mapsto \frac{1}{2^{n/2}} \begin{pmatrix} 1 \\ e^{i2\pi(j/2^0)} \end{pmatrix} \otimes \begin{pmatrix} 1 \\ e^{i2\pi(j/2^1)} \end{pmatrix} \otimes \begin{pmatrix} 1 \\ e^{i2\pi(j/2^2)} \end{pmatrix} \cdots \otimes \begin{pmatrix} 1 \\ e^{i2\pi(j/2^{n-1})} \end{pmatrix}$$

As an example, we consider the case $n = 3$. For $j = 0$, the basis vector $|\,000\rangle$ is transformed as follows:

$$|\,000\rangle \mapsto \frac{1}{2^{3/2}}\begin{pmatrix}1\\1\end{pmatrix}\otimes\begin{pmatrix}1\\1\end{pmatrix}\otimes\begin{pmatrix}1\\1\end{pmatrix}=\frac{1}{(\sqrt{2})^3}\begin{pmatrix}1\\1\\1\\1\\1\\1\\1\\1\end{pmatrix}$$

or

$$|\,000\rangle \mapsto \frac{1}{(\sqrt{2})^3}\big[|\,000\rangle + |\,001\rangle + |\,010\rangle + |\,011\rangle + |\,100\rangle + |\,101\rangle + |\,110\rangle + |\,111\rangle\big].$$

For $j = 1$, the basis vector $|\,001\rangle$ is transformed as:

$$|\,001\rangle \mapsto \frac{1}{2^{3/2}}\begin{pmatrix}1\\1\end{pmatrix}\otimes\begin{pmatrix}1\\e^{i2\pi(1/2)}\end{pmatrix}\otimes\begin{pmatrix}1\\e^{i2\pi(1/4)}\end{pmatrix} = \frac{1}{2^{3/2}}\begin{pmatrix}1\\1\end{pmatrix}\otimes\begin{pmatrix}1\\-1\end{pmatrix}\otimes\begin{pmatrix}1\\i\end{pmatrix}$$

or

$$|\,001\rangle \mapsto \frac{1}{(\sqrt{2})^3}\begin{pmatrix}1\\i\\-1\\-i\\1\\i\\-1\\-i\end{pmatrix}.$$

Thus,

$$|\,001\rangle \mapsto \frac{1}{(\sqrt{2})^3}\Big[|\,000\rangle - |\,010\rangle + |\,100\rangle - |\,110\rangle + i\,(|\,001\rangle - |\,011\rangle + |\,101\rangle - |\,111\rangle)\Big].$$

The transformations for the other basis vectors can be computed using the same procedure.

5.7 TENSOR PRODUCT FACTORIZATION

In Section 5.2 we mentioned the existence of a faster algorithm to compute the Discrete Fourier Transform. The *Fast Fourier Transform (FFT)* is an algorithm proposed by Cooley and Tukey [36] in 1965 which reduces the number of operations to compute the Discrete Fourier Transform of N integer or complex numbers from $2N^2$ to $2N \log_2 N$. The algorithm, based on a factorization known to Gauss in 1805, decomposes recursively the transformation for an integer $N = 2^n$ into two transforms of length $N/2$ using the identity:

$$\sum_{j=0}^{N-1} a_j \, e^{-i2\pi jk/N}$$

$$= \sum_{j=0}^{N/2-1} a_{2j} \, e^{-i2\pi(2j)k/N} + \sum_{j=0}^{N/2-1} a_{2j+1} e^{-i2\pi(2j+1)k/N}$$

$$= \sum_{j=0}^{N/2-1} a_j^{even} e^{-i2\pi jk/(N/2)} + e^{-i2\pi k/N} \sum_{j=0}^{N/2-1} a_j^{odd} e^{-i2\pi jk/(N/2)}.$$

A similar idea will be used for the Quantum Fourier Transform. Let G be a $2^n \times 2^n$ unitary matrix and $| V\rangle$ and $| W\rangle$ denote two column vectors with 2^n elements, $| V\rangle, | W\rangle \in \mathcal{H}_{2^n}$. We consider a quantum system in state $| V\rangle$ and a quantum circuit with the transfer matrix G. When we compute the state of a system after the transformation $| W\rangle$ given by

$$| W\rangle = G \, | V\rangle$$

by direct matrix multiplication we must perform $\mathcal{O}((2^n)^2)$ operations.

Let us now assume that the space is a tensor product of n two-dimensional spaces denoted as $\mathcal{H}_2^{(j)}, 1 \leqslant j \leqslant n$

$$\mathcal{H}_{2^n} = \mathcal{H}_2^{(1)} \otimes \mathcal{H}_2^{(2)} \ldots \otimes \mathcal{H}_2^{(j)} \ldots \otimes \mathcal{H}_2^{(n)}$$

and that the matrix G can be decomposed into a tensor product of n matrices $G^{(j)}, 1 \leqslant j \leqslant n$, each of size 2×2

$$G = G^{(1)} \otimes G^{(2)} \ldots \otimes G^{(j)} \ldots \otimes G^{(n)}.$$

with $G^{(j)}$ acting on the respective components of $\mathcal{H}_2^{(j)}$.

Let us now examine the j-th component of $W = (W_1, W_2, \ldots, W_j, \ldots, W_{2^n})$ with the index j expressed in binary as $j = (j_1, j_2, \ldots, j_n)$

$$W_j = \sum_{k_1, k_2, \ldots, k_n} \left[G^{(1)}_{j_1 k_1} G^{(2)}_{j_2 k_2} \ldots G^{(n)}_{j_n k_n} \right] V_{k_1 k_2 \ldots k_n}.$$

Each application of $G^{(j)}$ requires a fixed number of operations and to compute each W_j we need $\mathcal{O}(n)$ operations because in the previous expression we have a product of n such matrices. There are 2^n components W_j, thus, the total number of operations is $\mathcal{O}(n2^n)$. We conclude that the *tensor product factorization reduces the number of operations and leads to an exponential speed up* compared with the direct matrix multiplication which requires $\mathcal{O}((2^n)^2)$ operations. We shall revisit this subject in Section 5.13 after we discuss a circuit for the Quantum Fourier Transform.

5.8 A CIRCUIT FOR QUANTUM FOURIER TRANSFORM

Now, we write the expression giving the QFT in a new format and derive the quantum circuit able to carry out the transformation based upon this new expression. Recall from Section 5.6 that the Quantum Fourier Transform of $| j\rangle = | j_0 j_1 \dots j_{n-1}\rangle$ is

$$| j_0 j_1 \dots j_{n-1}\rangle \mapsto \frac{1}{2^{n/2}} \bigotimes_{m=0}^{n-1} \{| 0\rangle + e^{i2\pi j 2^{-m}} | 1\rangle\}.$$

We use regular vector multiplication of `ket` vectors instead of tensor products:

$$\bigotimes_{m=0}^{n-1} \{| 0\rangle + e^{i2\pi j 2^{-m}} | 1\rangle\} = \prod_{m=0}^{n-1} \left(| 0\rangle + e^{i2\pi j 2^{-m}} | 1\rangle\right).$$

Thus,

$$| j_0 j_1 \dots j_{n-1}\rangle \mapsto$$
$$\frac{1}{2^{n/2}}(| 0\rangle + e^{i2\pi(j/2^0)} | 1\rangle)(| 0\rangle + e^{i2\pi(j/2^1)} | 1\rangle)$$
$$\dots(| 0\rangle + e^{i2\pi(j/2^{n-1})} | 1\rangle).$$

Recall that $j = j_0 2^{n-1} + j_1 2^{n-2} + \dots j_{n-2}2 + j_{n-1}$ and if we use the following notation [98]

$$0.j_m j_{m+1} \cdots j_{n-1} = j_m 2^{-1} + j_{m+1} 2^{-2} + \dots + j_{n-1} 2^{-(n-m)},$$

the transformation becomes

$$| j_0 j_1 \cdots j_{n-1}\rangle \mapsto$$
$$\frac{1}{2^{n/2}}(| 0\rangle + e^{i2\pi 0.j_{n-1}} | 1\rangle)(| 0\rangle + e^{i2\pi 0.j_{n-2} j_{n-1}} | 1\rangle)$$
$$\dots(| 0\rangle + e^{i2\pi 0.j_0 j_1 \dots j_{n-1}} | 1\rangle).$$

Figure 5.6 shows a circuit for the Quantum Fourier Transform based upon this new expression for the transformation. Each basis vector is first transformed into a superposition state by a Hadamard gate and then it goes through one or more R_k gates. Recall from Section 4.12 that the R_k gate transforms a qubit by multiplying its projection on $| 1\rangle$ by $e^{i2\pi/2^k}$

$$R_k = \begin{pmatrix} 1 & 0 \\ 0 & e^{i2\pi/2^k} \end{pmatrix}.$$

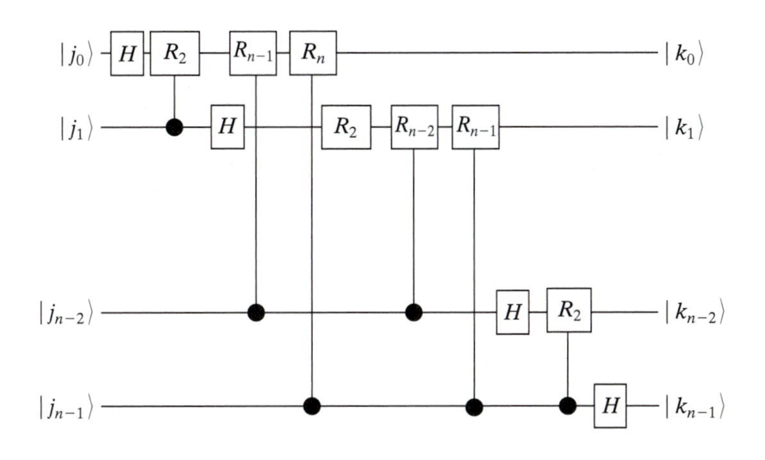

FIGURE 5.6: A circuit for Quantum Fourier Transform.

An equivalent expression for the QFT can be obtained if we denote by $H_p, 0 \leqslant p \leqslant n - 1$, the Walsh-Hadamard Transform applied to qubit p and by $S_{p,q}$ with $0 \leqslant p, q \leqslant n - 1, p \neq q$, the joint transformation of qubits p and q as

$$S_{p,q} = \begin{pmatrix} 1 & 0 & 0 & 0 \\ 0 & 1 & 0 & 0 \\ 0 & 0 & 1 & 0 \\ 0 & 0 & 0 & e^{i\pi/2^{p-q}} \end{pmatrix}.$$

The matrices $S_{p,q}$ do not change the values of any qubits, they only change their phases. If both qubits p and q are equal to 1 then this matrix adds the value π to the phase of qubit p. The QFTs of the n qubits are given by the following products of matrices, respectively

$$H_0(S_{0,1}S_{0,2}S_{0,3}, S_{0,4} \ldots, S_{0,n-1})$$
$$H_1(S_{1,2}S_{1,3}S_{1,4}S_{1,5} \ldots, S_{1,n-1})$$
$$H_2(S_{2,3}S_{2,4}S_{2,5}S_{2,6} \ldots, S_{2,n-1})$$
$$\vdots$$
$$H_{n-3}(S_{n-3,n-2}S_{n-3,n-1})$$
$$H_{n-2}(S_{n-2,n-1})$$
$$H_{n-1}.$$

The matrices must be multiplied left to right and while the H_p transforms only qubit p, all $S_{p,q}$ transform qubit p as a target and use q as a control qubit.

This transformation is followed by a bit reversal. A qubit $| \, b\rangle$ is transformed into a qubit $| \, \overline{b}\rangle$ with $\overline{b}$ the bit reversal of b. If

$$b = b_0 2^{n-1} + b_1 2^{n-2} + \ldots b_{n-3} 2^2 + b_{n-2} 2 + b_{n-1}$$

then

$$\overline{b} = b_{n-1} 2^{n-1} + b_{n-2} 2^{n-2} + b_{n-3} 2^{n-3} + \ldots b_2 2^2 + b_1 2 + b_0.$$

The circuit shown in Figure 5.6 does not carry out the bit reversal. In Shor's algorithm and in other applications the QFT is followed by a measurement and, in that case, the bit reversal can be done using classical algorithms.

Let us now follow the state transformations produced by the circuit in Figure 5.6. The `Hadamard` gate applied to the first qubit produces the following change of state

$$| j_0 j_1 j_2 \cdots j_{n-1}\rangle \mapsto \frac{1}{2^{1/2}}(| 0\rangle + e^{i2\pi 0.j_0} | 1\rangle) | j_1 j_2 \cdots j_{n-1}\rangle$$

with

$$e^{i2\pi 0.j_0} = e^{i2\pi j_0/2} = \begin{cases} +1 & \text{if } j_0 = 0 \\ -1 & \text{if } j_0 = 1. \end{cases}$$

Each `controlled-`R gate applied to the first qubit adds an extra bit to the phase of the projection on $| 1\rangle$. As a result, we observe the successive changes of state

$$\frac{1}{2^{1/2}}(| 0\rangle + e^{i2\pi 0.j_0} | 1\rangle) \ | j_1 j_2 \dots j_{n-1}\rangle \mapsto \text{first } R_k \text{ gate} \mapsto$$

$$\frac{1}{2^{1/2}}(| 0\rangle + e^{i2\pi 0.j_0 j_1} | 1\rangle) \ | j_1 j_2 \dots j_{n-1}\rangle \mapsto \text{second } R_k \text{ gate} \mapsto$$

$$\frac{1}{2^{1/2}}(| 0\rangle + e^{i2\pi 0.j_0 j_1 j_2} | 1\rangle) \ | j_1 j_2 \dots j_{n-1}\rangle \dots \mapsto \text{third } R_k \text{ gate} \mapsto$$

$$\dots \frac{1}{2^{1/2}}(| 0\rangle + e^{i2\pi 0.j_0 j_1 j_2 \dots j_{n-1}} | 1\rangle) \ | j_1 j_2 \cdots j_{n-1}\rangle$$

after the n-th R_k gate.

The transformations of the second qubit are

$$\frac{1}{2^{2(1/2)}}(| 0\rangle + e^{i2\pi 0.j_0 j_1 \dots j_{n-1}} | 1\rangle)(| 0\rangle + e^{i2\pi 0.j_1} | 1\rangle) | j_2 \cdots j_{n-1}\rangle$$

$$\mapsto \text{first } R_k \text{ gate} \mapsto$$

$$\frac{1}{2^{2(1/2)}}(| 0\rangle + e^{i2\pi 0.j_0 j_1 \dots j_{n-1}} | 1\rangle)(| 0\rangle + e^{i2\pi 0.j_1 j_2} | 1\rangle) | j_2 \cdots j_{n-1}\rangle$$

$$\mapsto \text{second } R_k \text{ gate} \mapsto$$

$$\cdots \frac{1}{2^{2(1/2)}}(| 0\rangle + e^{i2\pi 0.j_0 j_1 \dots j_{n-1}} | 1\rangle) \ (| 0\rangle + e^{i2\pi 0.j_1 j_2 \dots j_{n-1}} | 1\rangle) \ | j_2 \cdots j_{n-1}\rangle.$$

If $N \neq 2^n$ then the QFT gives only approximate results. From the previous expressions and from Figure 5.6 it is easy to see that the total number of gates required by a QFT with $N = 2^n$ is

$$\text{Total Number of Gates}_{(\text{QFT with } N=2^n)} = \frac{n(n + 1)}{2}.$$

5.9 A CASE STUDY—A THREE-QUBIT QFT

Let us now give a simple, yet non-trivial example of a QFT calculation. We consider the case $N = 2^3$ and sketch the path to construct the transfer matrix of the circuit. Recall that

$$e^{i\theta} = \cos\theta + i\sin\theta,$$

$$\cos(2k+1)\frac{\pi}{2} = 0 \quad \sin(2k+1)\frac{\pi}{2} = 1,$$

and

$$e^{i\pi/2} = i,$$

$$e^{i\pi/4} = \sqrt{i}.$$

If we denote $\omega = \sqrt{i}$ we see that $\omega^2 = i$ and $\omega^4 = -1$ and we can write

$$R_2 = \begin{pmatrix} 1 & 0 \\ 0 & e^{i2\pi/2^2} \end{pmatrix} = \begin{pmatrix} 1 & 0 \\ 0 & \omega^2 \end{pmatrix} \qquad R_3 = \begin{pmatrix} 1 & 0 \\ 0 & e^{i2\pi/2^3} \end{pmatrix} = \begin{pmatrix} 1 & 0 \\ 0 & \omega \end{pmatrix}.$$

The transfer matrix of a `controlled`-R_2 gate is based upon the derivation in Section 4.12

$$G_{\text{controlled}-R_2} = \begin{pmatrix} 1 & 0 & 0 & 0 \\ 0 & 1 & 0 & 0 \\ 0 & 0 & 1 & 0 \\ 0 & 0 & 0 & e^{i2\pi/2^2} \end{pmatrix} = \begin{pmatrix} 1 & 0 & 0 & 0 \\ 0 & 1 & 0 & 0 \\ 0 & 0 & 1 & 0 \\ 0 & 0 & 0 & \omega^2 \end{pmatrix}.$$

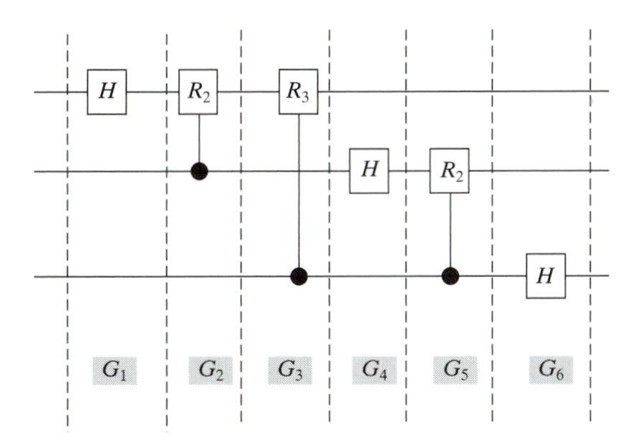

FIGURE 5.7: A circuit for Quantum Fourier Transform for $N = 2^3$ consists of six stages. The transfer matrix of individual stages are G_1, G_2, G_3, G_4, G_5, and G_6.

Figure 5.7 shows the circuit obtained from the one in Figure 5.6 when $n = 3$. We see that the circuit consists of six stages and we compute the transfer matrix of each stage. To maintain consistency with the previous chapters we denote a 2×2 identity matrix as I. The 4×4 identity matrix is denoted as I_4. For the first stage, we have

$$G_1 = H \otimes I \otimes I = \frac{1}{\sqrt{2}}\begin{pmatrix} 1 & 1 \\ 1 & -1 \end{pmatrix} \otimes \begin{pmatrix} 1 & 0 \\ 0 & 1 \end{pmatrix} \otimes \begin{pmatrix} 1 & 0 \\ 0 & 1 \end{pmatrix}$$

or

$$G_1 = \frac{1}{\sqrt{2}} \begin{pmatrix} 1 & 0 & 1 & 0 \\ 0 & 1 & 0 & 1 \\ 1 & 0 & -1 & 0 \\ 0 & 1 & 0 & -1 \end{pmatrix} \otimes \begin{pmatrix} 1 & 0 \\ 0 & 1 \end{pmatrix},$$

$$G_1 = \frac{1}{\sqrt{2}} \begin{pmatrix} I_4 & I_4 \\ I_4 & -I_4 \end{pmatrix}, \quad I_4 = \begin{pmatrix} 1 & 0 & 0 & 0 \\ 0 & 1 & 0 & 0 \\ 0 & 0 & 1 & 0 \\ 0 & 0 & 0 & 1 \end{pmatrix}.$$

For the second stage, we have

$$G_2 = G_{\text{controlled}-R_2} \otimes I = \begin{pmatrix} 1 & 0 & 0 & 0 \\ 0 & 1 & 0 & 0 \\ 0 & 0 & 1 & 0 \\ 0 & 0 & 0 & \omega^2 \end{pmatrix} \otimes \begin{pmatrix} 1 & 0 \\ 0 & 1 \end{pmatrix} = \begin{pmatrix} I_4 & 0 \\ 0 & M_{(2,4)} \end{pmatrix},$$

with

$$M_{(2,4)} = \begin{pmatrix} 1 & 0 & 0 & 0 \\ 0 & 1 & 0 & 0 \\ 0 & 0 & \omega^2 & 0 \\ 0 & 0 & 0 & \omega^2 \end{pmatrix}.$$

For the third stage we observe that the circuit performs the following mapping

$$\begin{aligned} &|\,000\rangle \mapsto |\,000\rangle, \ |\,001\rangle \mapsto |\,001\rangle, \quad |\,010\rangle \mapsto |\,010\rangle, \ |\,011\rangle \mapsto |\,011\rangle \\ &|\,100\rangle \mapsto |\,100\rangle, \ |\,101\rangle \mapsto \omega\,|\,101\rangle, \ |\,110\rangle \mapsto |\,110\rangle, \ |\,111\rangle \mapsto \omega\,|\,111\rangle. \end{aligned}$$

Thus, the transfer matrix of the third stage is

$$G_3 = \begin{pmatrix} I_4 & 0 \\ 0 & M_{(3,4)} \end{pmatrix},$$

with

$$M_{(3,4)} = \begin{pmatrix} 1 & 0 & 0 & 0 \\ 0 & \omega & 0 & 0 \\ 0 & 0 & 1 & 0 \\ 0 & 0 & 0 & \omega \end{pmatrix}.$$

The transfer matrix of the fourth stage is

$$G_4 = I \otimes H \otimes I = \frac{1}{\sqrt{2}} \begin{pmatrix} 1 & 0 \\ 0 & 1 \end{pmatrix} \otimes \begin{pmatrix} 1 & 1 \\ 1 & -1 \end{pmatrix} \otimes \begin{pmatrix} 1 & 0 \\ 0 & 1 \end{pmatrix}$$

or

$$G_4 = \frac{1}{\sqrt{2}} \begin{pmatrix} 1 & 1 & 0 & 0 \\ 1 & -1 & 0 & 0 \\ 0 & 0 & 1 & 1 \\ 0 & 0 & 1 & -1 \end{pmatrix} \otimes \begin{pmatrix} 1 & 0 \\ 0 & 1 \end{pmatrix}.$$

Thus, the transfer matrix of the fourth stage is

$$G_4 = \begin{pmatrix} M_{(4,4)} & 0 \\ 0 & M_{(4,4)} \end{pmatrix},$$

with

$$M_{(4,4)} = \begin{pmatrix} 1 & 0 & 1 & 0 \\ 0 & 1 & 0 & 1 \\ 1 & 0 & -1 & 0 \\ 0 & 1 & 0 & -1 \end{pmatrix}.$$

The transfer matrix of the fifth stage is

$$G_5 = I \otimes G_{\text{controlled}-R_2} = \begin{pmatrix} 1 & 0 \\ 0 & 1 \end{pmatrix} \otimes \begin{pmatrix} 1 & 0 & 0 & 0 \\ 0 & 1 & 0 & 0 \\ 0 & 0 & 1 & 0 \\ 0 & 0 & 0 & \omega^2 \end{pmatrix} = \begin{pmatrix} M_{(5,4)} & 0 \\ 0 & M_{(5,4)} \end{pmatrix},$$

with $M_{(5,4)} = G_{\text{controlled}-R_2}$.

Finally, for the last stage we have

$$G_6 = I \otimes I \otimes H = I \otimes H = \frac{1}{\sqrt{2}} \begin{pmatrix} 1 & 0 & 0 & 0 \\ 0 & 1 & 0 & 0 \\ 0 & 0 & 1 & 0 \\ 0 & 0 & 0 & 1 \end{pmatrix} \otimes \begin{pmatrix} 1 & 1 \\ 1 & -1 \end{pmatrix}$$

$$= \frac{1}{\sqrt{2}} \begin{pmatrix} M_{(6,4)} & 0 \\ 0 & M_{(6,4)} \end{pmatrix},$$

with

$$M_{(6,4)} = \begin{pmatrix} 1 & 1 & 0 & 0 \\ 1 & -1 & 0 & 0 \\ 0 & 0 & 1 & 1 \\ 0 & 0 & 1 & -1 \end{pmatrix}.$$

Now we show that the transfer matrix of the circuit in Figure 5.7 is

$$G_{\text{QFT}_3} = \frac{1}{(\sqrt{2})^3} \begin{pmatrix} 1 & 1 & 1 & 1 & 1 & 1 & 1 & 1 \\ 1 & \omega^1 & \omega^2 & \omega^3 & \omega^4 & \omega^5 & \omega^6 & \omega^7 \\ 1 & \omega^2 & \omega^4 & \omega^6 & 1 & \omega^2 & \omega^4 & \omega^6 \\ 1 & \omega^3 & \omega^6 & \omega^1 & \omega^4 & \omega^7 & \omega^2 & \omega^5 \\ 1 & \omega^4 & 1 & \omega^4 & 1 & \omega^4 & 1 & \omega^4 \\ 1 & \omega^5 & \omega^2 & \omega^7 & \omega^4 & \omega^1 & \omega^6 & \omega^3 \\ 1 & \omega^6 & \omega^4 & \omega^2 & 1 & \omega^6 & \omega^4 & \omega^2 \\ 1 & \omega^7 & \omega^6 & \omega^5 & \omega^4 & \omega^3 & \omega^2 & \omega^1 \end{pmatrix}.$$

Matrices $G_1, G_2, ..., G_6$ are

$$G_1 = \frac{1}{\sqrt{2}} \begin{pmatrix} 1&0&0&0&1&0&0&0 \\ 0&1&0&0&0&1&0&0 \\ 0&0&1&0&0&0&1&0 \\ 0&0&0&1&0&0&0&1 \\ 1&0&0&0&-1&0&0&0 \\ 0&1&0&0&0&-1&0&0 \\ 0&0&1&0&0&0&-1&0 \\ 0&0&0&1&0&0&0&-1 \end{pmatrix}, \quad G_2 = \begin{pmatrix} 1&0&0&0&0&0&0&0 \\ 0&1&0&0&0&0&0&0 \\ 0&0&1&0&0&0&0&0 \\ 0&0&0&1&0&0&0&0 \\ 0&0&0&0&1&0&0&0 \\ 0&0&0&0&0&1&0&0 \\ 0&0&0&0&0&0&\omega^2&0 \\ 0&0&0&0&0&0&0&\omega^2 \end{pmatrix},$$

$$G_3 = \begin{pmatrix} 1&0&0&0&0&0&0&0 \\ 0&1&0&0&0&0&0&0 \\ 0&0&1&0&0&0&0&0 \\ 0&0&0&1&0&0&0&0 \\ 0&0&0&0&1&0&0&0 \\ 0&0&0&0&0&\omega&0&0 \\ 0&0&0&0&0&0&1&0 \\ 0&0&0&0&0&0&0&\omega \end{pmatrix}, \quad G_4 = \frac{1}{\sqrt{2}} \begin{pmatrix} 1&0&1&0&0&0&0&0 \\ 0&1&0&1&0&0&0&0 \\ 1&0&-1&0&0&0&0&0 \\ 0&1&0&-1&0&0&0&0 \\ 0&0&0&0&1&0&1&0 \\ 0&0&0&0&0&1&0&1 \\ 0&0&0&0&1&0&-1&0 \\ 0&0&0&0&0&1&0&-1 \end{pmatrix},$$

$$G_5 = \begin{pmatrix} 1&0&0&0&0&0&0&0 \\ 0&1&0&0&0&0&0&0 \\ 0&0&1&0&0&0&0&0 \\ 0&0&0&\omega^2&0&0&0&0 \\ 0&0&0&0&1&0&0&0 \\ 0&0&0&0&0&1&0&0 \\ 0&0&0&0&0&0&1&0 \\ 0&0&0&0&0&0&0&\omega^2 \end{pmatrix}, \quad G_6 = \frac{1}{\sqrt{2}} \begin{pmatrix} 1&1&0&0&0&0&0&0 \\ 1&-1&0&0&0&0&0&0 \\ 0&0&1&1&0&0&0&0 \\ 0&0&1&-1&0&0&0&0 \\ 0&0&0&0&1&1&0&0 \\ 0&0&0&0&1&-1&0&0 \\ 0&0&0&0&0&0&1&1 \\ 0&0&0&0&0&0&1&-1 \end{pmatrix}.$$

An input $| \, v\rangle$, to the circuit illustrated in Figure 5.7 is transformed to $G_1 \, | \, v\rangle$ and then to $G_2 G_1 \, | \, v\rangle$, etc. After passing through the six stages, it becomes

$$| \, w\rangle = G_6 G_5 G_4 G_3 G_2 G_1 \, | \, v\rangle = G_t \, | \, v\rangle.$$

We denote

$$G_t = G_6 G_5 G_4 G_3 G_2 G_1.$$

Thus,

$$G_t = \frac{1}{(\sqrt{2})^3} \begin{pmatrix} 1&1&1&1&1&1&1&1 \\ 1&-1&1&-1&1&-1&1&-1 \\ 1&\omega^2&-1&-\omega^2&1&\omega^2&-1&-\omega^2 \\ 1&-\omega^2&-1&\omega^2&1&-\omega^2&-1&\omega^2 \\ 1&\omega^1&\omega^2&\omega^3&-1&-\omega^1&-\omega^2&-\omega^3 \\ 1&-\omega^1&\omega^2&-\omega^3&-1&\omega^1&-\omega^2&\omega^3 \\ 1&\omega^3&-\omega^2&-\omega^5&-1&-\omega^3&\omega^2&\omega^5 \\ 1&-\omega^3&-\omega^2&\omega^5&-1&\omega^3&\omega^2&-\omega^5 \end{pmatrix}.$$

We notice that $-1 = e^{i2\pi/2} = \omega^4$ and G_t becomes

$$G_t = \frac{1}{(\sqrt{2})^3}\begin{pmatrix} 1 & 1 & 1 & 1 & 1 & 1 & 1 & 1 \\ 1 & \omega^4 & 1 & \omega^4 & 1 & \omega^4 & 1 & \omega^4 \\ 1 & \omega^2 & \omega^4 & \omega^6 & 1 & \omega^2 & \omega^4 & \omega^6 \\ 1 & \omega^6 & \omega^4 & \omega^2 & 1 & \omega^6 & \omega^4 & \omega^2 \\ 1 & \omega^1 & \omega^2 & \omega^3 & \omega^4 & \omega^5 & \omega^6 & \omega^7 \\ 1 & \omega^5 & \omega^2 & \omega^7 & \omega^4 & \omega^1 & \omega^6 & \omega^3 \\ 1 & \omega^3 & \omega^6 & \omega^1 & \omega^4 & \omega^7 & \omega^2 & \omega^5 \\ 1 & \omega^7 & \omega^6 & \omega^5 & \omega^4 & \omega^3 & \omega^2 & \omega^1 \end{pmatrix}.$$

We observe that the permutation matrix

$$P = \begin{pmatrix} 1 & 0 & 0 & 0 & 0 & 0 & 0 & 0 \\ 0 & 0 & 0 & 0 & 1 & 0 & 0 & 0 \\ 0 & 0 & 1 & 0 & 0 & 0 & 0 & 0 \\ 0 & 0 & 0 & 0 & 0 & 0 & 1 & 0 \\ 0 & 1 & 0 & 0 & 0 & 0 & 0 & 0 \\ 0 & 0 & 0 & 0 & 0 & 1 & 0 & 0 \\ 0 & 0 & 0 & 1 & 0 & 0 & 0 & 0 \\ 0 & 0 & 0 & 0 & 0 & 0 & 0 & 1 \end{pmatrix},$$

transforms G_t into the matrix G_{QFT_3}

$$G_{\mathrm{QFT}_3} = PG_t = \frac{1}{(\sqrt{2})^3}\begin{pmatrix} 1 & 1 & 1 & 1 & 1 & 1 & 1 & 1 \\ 1 & \omega^1 & \omega^2 & \omega^3 & \omega^4 & \omega^5 & \omega^6 & \omega^7 \\ 1 & \omega^2 & \omega^4 & \omega^6 & 1 & \omega^2 & \omega^4 & \omega^6 \\ 1 & \omega^3 & \omega^6 & \omega^1 & \omega^4 & \omega^7 & \omega^2 & \omega^5 \\ 1 & \omega^4 & 1 & \omega^4 & 1 & \omega^4 & 1 & \omega^4 \\ 1 & \omega^5 & \omega^2 & \omega^7 & \omega^4 & \omega^1 & \omega^6 & \omega^3 \\ 1 & \omega^6 & \omega^4 & \omega^2 & 1 & \omega^6 & \omega^4 & \omega^2 \\ 1 & \omega^7 & \omega^6 & \omega^5 & \omega^4 & \omega^3 & \omega^2 & \omega^1 \end{pmatrix}.$$

One can verify the expression for G_{QFT_3} directly from the definition of the QFT:

$$\begin{aligned} |\, j_0 j_1 j_2\rangle \mapsto \left(\frac{1}{\sqrt{2}}\right)^3 & \left(|\,0\rangle + e^{i2\pi 0\cdot j_2}\,|\,1\rangle\right) \\ & \left(|\,0\rangle + e^{i2\pi 0\cdot j_1 j_2}\,|\,1\rangle\right) \\ & \left(|\,0\rangle + e^{i2\pi 0\cdot j_0 j_1 j_2}\,|\,1\rangle\right). \end{aligned}$$

In our case, $n = 3$ and the three input qubits are:

$|\,\alpha\rangle = \alpha_0\,|\,0\rangle + \alpha_1\,|\,1\rangle$, $|\,\beta\rangle = \beta_0\,|\,0\rangle + \beta_1\,|\,1\rangle$, and $|\,\gamma\rangle = \gamma_0\,|\,0\rangle + \gamma_1\,|\,1\rangle$.

Thus, the initial state of the system is

$$| v \rangle = | \alpha\rangle \otimes | \beta\rangle \otimes | \gamma\rangle = \begin{pmatrix} \alpha_0\beta_0\gamma_0 \\ \alpha_0\beta_0\gamma_1 \\ \alpha_0\beta_1\gamma_0 \\ \alpha_0\beta_1\gamma_1 \\ \alpha_1\beta_0\gamma_0 \\ \alpha_1\beta_0\gamma_1 \\ \alpha_1\beta_1\gamma_0 \\ \alpha_1\beta_1\gamma_1 \end{pmatrix}.$$

Then $\omega = e^{2\pi i/8}$ and

$$0.j_2 = \frac{j_2}{2} = \frac{4j_2}{8}, \quad 0.j_1j_2 = \frac{j_1}{2} + \frac{j_2}{4} = \frac{4j_1 + 2j_2}{8}, \quad 0.j_0j_1j_2 = \frac{4j_0 + 2j_1 + j_2}{8}.$$

When the input state is $| v\rangle$ the right side of the general equation for QFT becomes (ignoring the $(1/\sqrt{2})^3$ factor)

$$\begin{pmatrix} \alpha_0 \\ \omega^{4j_2}\alpha_1 \end{pmatrix} \otimes \begin{pmatrix} \beta_0 \\ \omega^{4j_1+2j_2}\beta_1 \end{pmatrix} \otimes \begin{pmatrix} \gamma_0 \\ \omega^{4j_0+2j_1+j_2}\gamma_1 \end{pmatrix}$$

which is equal to

$$\begin{pmatrix} \alpha_0\beta_0 \\ \omega^{4j_1+2j_2}\alpha_0\beta_1 \\ \omega^{4j_2}\alpha_1\beta_0 \\ \omega^{4j_2}\omega^{4j_1+2j_2}\alpha_1\beta_1 \end{pmatrix} \otimes \begin{pmatrix} \gamma_0 \\ \omega^{4j_0+2j_1+j_2}\gamma_1 \end{pmatrix} = \begin{pmatrix} \alpha_0\beta_0\gamma_0 \\ \omega^{4j_0+2j_1+j_2}\alpha_0\beta_0\gamma_1 \\ \omega^{4j_1+2j_2}\alpha_0\beta_1\gamma_0 \\ \omega^{4j_1+2j_2}\omega^{4j_0+2j_1+j_2}\alpha_0\beta_1\gamma_1 \\ \omega^{4j_2}\alpha_1\beta_0\gamma_0 \\ \omega^{4j_2}\omega^{4j_0+2j_1+j_2}\alpha_1\beta_0\gamma_1 \\ \omega^{4j_1+6j_2}\alpha_1\beta_1\gamma_0 \\ \omega^{4j_1+6j_2}\omega^{4j_0+2j_1+j_2}\alpha_1\beta_1\gamma_1 \end{pmatrix}.$$

or, in a compact form

$$\begin{pmatrix} \alpha_0\beta_0\gamma_0 \\ \omega^{4j_0+2j_1+j_2}\alpha_0\beta_0\gamma_1 \\ \omega^{4j_1+2j_2}\alpha_0\beta_1\gamma_0 \\ \omega^{4j_0+6j_1+3j_2}\alpha_0\beta_1\gamma_1 \\ \omega^{4j_2}\alpha_1\beta_0\gamma_0 \\ \omega^{4j_0+2j_1+5j_2}\alpha_1\beta_0\gamma_1 \\ \omega^{4j_1+6j_2}\alpha_1\beta_1\gamma_0 \\ \omega^{4j_0+6j_1+7j_2}\alpha_1\beta_1\gamma_1 \end{pmatrix}.$$

The following table shows the powers of ω in each row of the previous matrix for different j_0, j_1, and j_2:

Row	0+0+0	0+0+ j_2	0+ j_1 +0	0+ j_1 + j_2	j_0 +0+0	j_0 +0+ j_2	j_0 + j_1 +0	j_0 + j_1 + j_2
1	ω^0	ω^0	ω^0	ω^0	ω^0	ω^0	ω^0	ω^0
2	ω^0	ω^1	ω^2	ω^3	ω^4	ω^5	ω^6	ω^7
3	ω^0	ω^2	ω^4	ω^6	ω^0	ω^2	ω^4	ω^6
4	ω^0	ω^3	ω^6	ω^1	ω^4	ω^7	ω^2	ω^5
5	ω^0	ω^4	ω^0	ω^4	ω^0	ω^4	ω^0	ω^4
6	ω^0	ω^5	ω^2	ω^7	ω^4	ω^1	ω^6	ω^3
7	ω^0	ω^6	ω^4	ω^2	ω^0	ω^6	ω^4	ω^2
8	ω^0	ω^7	ω^6	ω^4	ω^3	ω^2	ω^1	ω^0

Thus, the transformation performed by the QFT is:

$$|v\rangle = \begin{pmatrix} \alpha_0\beta_0\gamma_0 \\ \alpha_0\beta_0\gamma_1 \\ \alpha_0\beta_1\gamma_0 \\ \alpha_0\beta_1\gamma_1 \\ \alpha_1\beta_0\gamma_0 \\ \alpha_1\beta_0\gamma_1 \\ \alpha_1\beta_1\gamma_0 \\ \alpha_1\beta_1\gamma_1 \end{pmatrix} \mapsto$$

$$|w\rangle = \frac{1}{(\sqrt{2})^3} \begin{pmatrix} 1 & 1 & 1 & 1 & 1 & 1 & 1 & 1 \\ 1 & \omega^1 & \omega^2 & \omega^3 & \omega^4 & \omega^5 & \omega^6 & \omega^7 \\ 1 & \omega^2 & \omega^4 & \omega^6 & 1 & \omega^2 & \omega^4 & \omega^6 \\ 1 & \omega^3 & \omega^6 & \omega^1 & \omega^4 & \omega^7 & \omega^2 & \omega^5 \\ 1 & \omega^4 & 1 & \omega^4 & 1 & \omega^4 & 1 & \omega^4 \\ 1 & \omega^5 & \omega^2 & \omega^7 & \omega^4 & \omega^1 & \omega^6 & \omega^3 \\ 1 & \omega^6 & \omega^4 & \omega^2 & 1 & \omega^6 & \omega^4 & \omega^2 \\ 1 & \omega^7 & \omega^6 & \omega^5 & \omega^4 & \omega^3 & \omega^2 & \omega^1 \end{pmatrix} \begin{pmatrix} \alpha_0\beta_0\gamma_0 \\ \alpha_0\beta_0\gamma_1 \\ \alpha_0\beta_1\gamma_0 \\ \alpha_0\beta_1\gamma_1 \\ \alpha_1\beta_0\gamma_0 \\ \alpha_1\beta_0\gamma_1 \\ \alpha_1\beta_1\gamma_0 \\ \alpha_1\beta_1\gamma_1 \end{pmatrix}.$$

Taking into account the fact that $\omega^4 = -1$, the transfer matrix for a three-qubit QFT can be rewritten as:

$$G_{\text{QFT}_3} = \frac{1}{(\sqrt{2})^3} \begin{pmatrix} 1 & 1 & 1 & 1 & 1 & 1 & 1 & 1 \\ 1 & \omega^1 & \omega^2 & \omega^3 & -1 & -\omega^1 & -\omega^2 & -\omega^3 \\ 1 & \omega^2 & -1 & -\omega^2 & 1 & \omega^2 & -1 & -\omega^2 \\ 1 & \omega^3 & -\omega^2 & \omega^1 & -1 & -\omega^3 & \omega^2 & -\omega^1 \\ 1 & -1 & 1 & -1 & 1 & -1 & 1 & -1 \\ 1 & -\omega^1 & \omega^2 & -\omega^3 & -1 & \omega^1 & -\omega^2 & \omega^3 \\ 1 & -\omega^2 & -1 & \omega^2 & 1 & -\omega^2 & -1 & \omega^2 \\ 1 & -\omega^3 & -\omega^2 & -\omega^1 & -1 & \omega^3 & \omega^2 & \omega^1 \end{pmatrix}.$$

If the input state is

$$\begin{aligned} | v\rangle \;=\; & v_0 \mid 000\rangle + v_1 \mid 001\rangle + v_2 \mid 010\rangle + v_3 \mid 011\rangle \\ & + v_4 \mid 100\rangle + v_5 \mid 101\rangle + v_6 \mid 110\rangle + v_7 \mid 111\rangle \end{aligned}$$

then its transform is

$$\begin{aligned} | w\rangle \;=\; & \frac{1}{(\sqrt{2})^3} (\omega_0 \mid 000\rangle + \omega_1 \mid 001\rangle + \omega_2 \mid 010\rangle + \omega_3 \mid 011\rangle \\ & + \omega_4 \mid 100\rangle + \omega_5 \mid 101\rangle + \omega_6 \mid 110\rangle + \omega_7 \mid 111\rangle) \end{aligned}$$

with

$$\begin{aligned} w_0 &= v_0 + v_1 + v_2 + v_3 + v_4 + v_5 + v_6 + v_7 \\ \omega_1 &= (v_0 - v_4) + (v_1 - v_5)\omega^1 + (v_2 - v_6)\omega^2 + (v_3 - v_7)\omega^3 \\ w_2 &= (v_0 - v_2 + v_4 - v_6) + (v_1 - v_3 + v_5 - v_7)\omega^2 \\ w_3 &= (v_0 - v_4) + (v_3 - v_7)\omega^1 - (v_2 - v_6)\omega^2 + (v_1 - v_5)\omega^3 \\ w_4 &= v_0 - v_1 + v_2 - v_3 + v_4 - v_5 + v_6 - v_7 \\ w_5 &= (v_0 - v_4) - (v_1 - v_5)\omega^1 + (v_2 - v_6)\omega^2 - (v_3 - v_7)\omega^3 \\ w_6 &= (v_0 - v_2 + v_4 - v_6) - (v_1 - v_3 + v_5 - v_7)\omega^2 \\ w_7 &= (v_0 - v_4) - (v_3 - v_7)\omega^1 - (v_2 - v_6)\omega^2 - (v_1 - v_5)\omega^3. \end{aligned}$$

5.10 SHOR'S FACTORING ALGORITHM AND ORDER FINDING

We now address possibly the best known application of quantum computing and we discuss algorithms for *integer factorization*. In 1994, Peter Shor found a polynomial time algorithm for the factorization of n-bit numbers on quantum computers [119]. His discovery generated a wave of enthusiasm for quantum computing, for two major reasons: the intrinsic intellectual beauty of the algorithm and the fact that efficient integer factorization is a very important practical problem. The security of widely used cryptographic protocols is based upon the conjectured difficulty of the factorization of large integers.[6]

Like most factorization algorithms, Shor's algorithm reduces the factorization problem to the problem of finding the period of a function, but uses quantum parallelism to find a superposition of all values of the function in one step. Then the algorithm calculates the Quantum Fourier Transform of the function, which sets the amplitudes into multiples of the fundamental frequency, the reciprocal of the period. To factor an integer, Shor's algorithm measures the period of the function.[7]

The relationship between integer factorization algorithms and the Quantum Fourier Transform is quite convoluted. We show that integer factorization

---

[6] A paper presenting a polynomial time algorithm which determines if a number is prime or composite was posted on a Web site on August 6, 2002 [3].

[7] A powerful version of the technique used by Shor is the *phase-estimation algorithm* of Kitaev [81].

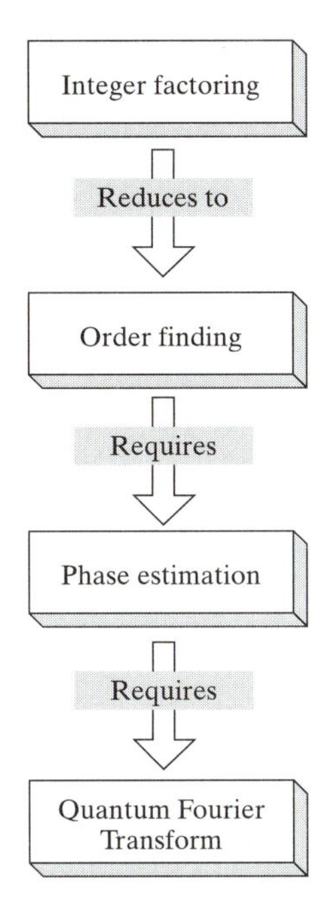

FIGURE 5.8: The relationship between integer factorization, order finding, phase estimation, and Quantum Fourier Transform.

reduces to another problem, namely *order finding*. We shall construct an algorithm that allows us to determine the prime factors of N based upon the procedure to determine the *order of integers* $x \bmod N$ for $x < N$.

But, for our factorization algorithm to be efficient, we have to choose wisely the integers x, we cannot simply try all $x < N$. Now the *phase estimation* comes into picture. *Phase estimation* is the name given to the algorithm that allows us to estimate the eigenvalue associated with an eigenvector of a unitary operator. Phase estimation is the cornerstone of several algorithms in quantum computing. In turn, the Quantum Fourier Transform is used by the phase estimation algorithm. The relationships between integer factoring, order finding, phase estimation and Quantum Fourier Transform algorithms are summarized in Figure 5.8.

Before presenting the factorization algorithm we discuss a few concepts from number theory presented in more detail in Appendix B. To factor an integer N means to write N as a product of prime numbers

$$N = p \times q_1 \times q_2 \ldots \times q_n.$$

The integer p is said to be a "proper factor" of N if three conditions are satisfied:

1. another integer q exists such that $N = pq$,
2. $p \neq 1$, and
3. $p \neq N$.

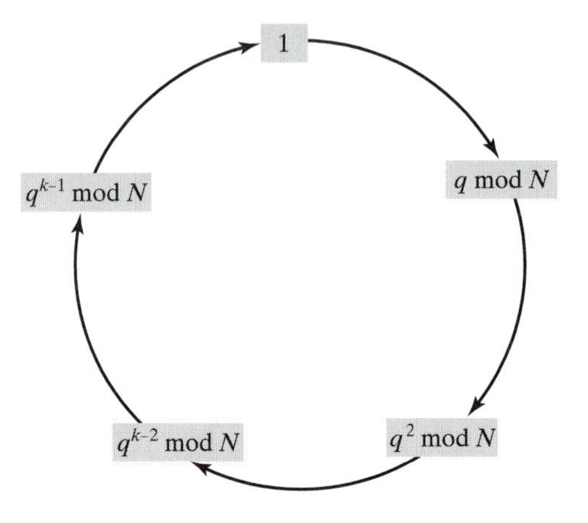

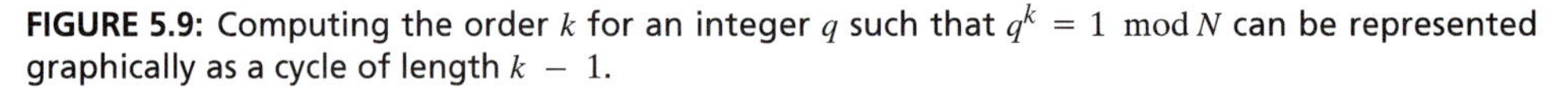

FIGURE 5.9: Computing the order k for an integer q such that $q^k = 1 \bmod N$ can be represented graphically as a cycle of length $k - 1$.

It is trivial to factor a small integer. The proper factors of 14 are $p = 2$ and $q = 7$, while the proper factors of 39 are $p = 3$ and $q = 13$. The problem gets harder and harder as the integer to be factored becomes larger and larger. For example, it is practically unfeasible to factor the integer 2841877 without the use of a calculator, or even better, a computer program. By trial and error we could find that $p = 19$ is a proper factor, thus $2841877 = 19 \times 149573$ but then we have to factor the integer 149573.

Given two integers r and N, *the order of r modulo N* is the smallest integer k such that $r^k = 1 \bmod N$ with two additional conditions

1. $r^1 \neq 1 \bmod N$, and
2. $(r^{k-1} + r^{k-2} + \ldots + r^2 + r + 1) \neq 1 \bmod N$.

Computing the order k modulo N of an integer q ($q^k = 1 \bmod N$) can be represented graphically as a cycle of length $k - 1$, as shown in Figure 5.9.

Several examples of finding the order of an integer modulo N by direct search are given below. First, we compute the order of 11 modulo 21 (i.e., we compute the smallest integer k such that $11^k = 1 \bmod 21$). We start with $k = 2$ and continue with $k = 3, 4, 5, 6$.

$$
\begin{array}{llll}
k = 2 - 11^2 = 121 & = 5 \times 21 + 16 \mapsto & & 11^2 = 16 \bmod 21 \\
k = 3 - 11^3 = 11 \times 11^2 & = 11 \times 16 \bmod 21 = 176 \bmod 21 & & \\
 & = 8 \bmod 21 \mapsto & & 11^3 = 8 \bmod 21 \\
k = 4 - 11^4 = 11 \times 11^3 & = 11 \times 8 \bmod 21 = 88 \bmod 21 & & \\
 & = 4 \bmod 21 \mapsto & & 11^4 = 4 \bmod 21 \\
k = 5 - 11^5 = 11 \times 11^4 & = 11 \times 4 \bmod 21 = 44 \bmod 21 & & \\
 & = 2 \bmod 21 \mapsto & & 11^5 = 2 \bmod 21 \\
k = 6 - 11^6 = 11 \times 11^5 & = 11 \times 2 \bmod 21 = 22 \bmod 21 & & \\
 & = 1 \bmod 21 \mapsto & & 11^6 = 1 \bmod 21
\end{array}
$$

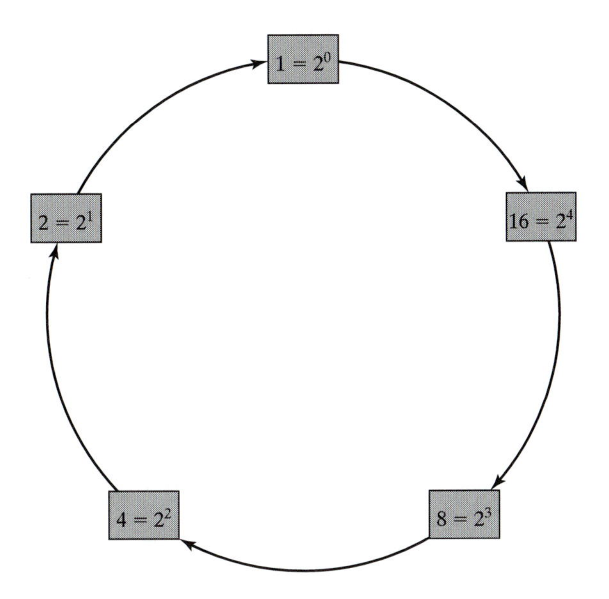

FIGURE 5.10: The multiplicative cycle for $11 \mod 21$. Each number on the cycle is obtained by multiplying the previous one by $11 \mod 21$.

Thus, the order of $11 \mod 21$ is $k = 6$. Indeed, the length of the cycle is $k - 1 = 5$, as shown in Figure 5.10. Let us now show that the order of 5 modulo 21 is also $k = 6$.

$$
\begin{array}{ll}
5^1 = 5 \mod 21 & \mapsto 5^1 = 5 \mod 21 \\
5^2 = 25 = 1 \times 21 + 4 & \mapsto 5^2 = 4 \mod 21 \\
5^3 = 125 = (5 \times 21 + 20) \mod 21 & \mapsto 5^3 = 20 \mod 21 \\
5^4 = 5 \times 5^3 = 5 \times 20 \mod 21 = 100 \mod 21 = 4 \times 21 + 16 & \mapsto 5^4 = 16 \mod 21 \\
5^5 = 5 \times 5^4 = 5 \times 16 \mod 21 = 80 \mod 21 = 3 \times 21 + 17 & \mapsto 5^5 = 17 \mod 21 \\
5^6 = 5 \times 5^5 = 5 \times 17 \mod 21 = 85 \mod 21 = 4 \times 21 + 1 & \mapsto 5^6 = 1 \mod 21
\end{array}
$$

Computing r^k for a wide range of k values is computationally expensive. But it is easy to see that we only need several powers of r to compute r^k for any value of k. Indeed, we can express any integer k as

$$k = k_{m-1}2^{m-1} + k_{m-2}2^{m-2} + \dots k_i 2^i + k_1 2^1 + k_0$$

with $k_i = \{0, 1\}, 0 \leqslant i \leqslant m - 1$. Then,

$$r^k = r^{k_{m-1}2^{m-1}} \times r^{k_{m-2}2^{m-2}} \times \dots r^{k_2 2^2} \times r^{k_1 2^1} \times r^{k_0}.$$

To compute r^k we need at most $m - 1$ exponentiations

$$r^{2^1}, r^{2^2}, r^{2^3}, \dots, r^{2^{m-1}}.$$

For example, when we wish to compute 17^{29} we can write:

$$29 = 16 + 8 + 4 + 1 = 2^4 + 2^3 + 2^2 + 2^0.$$

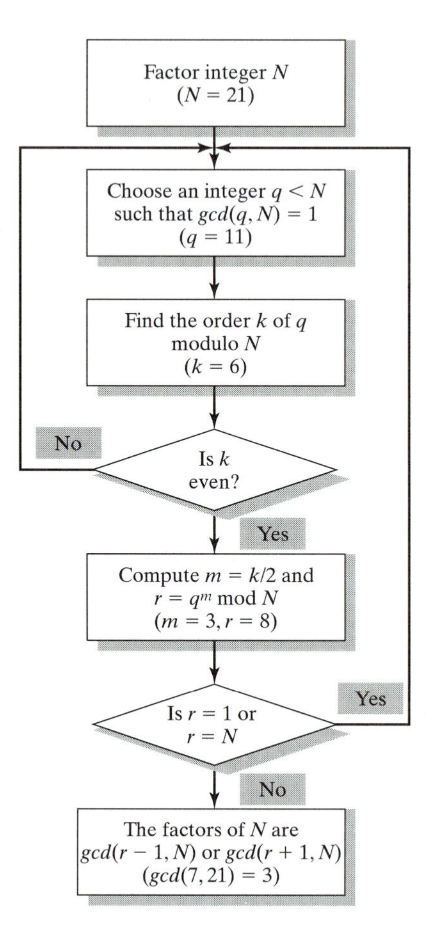

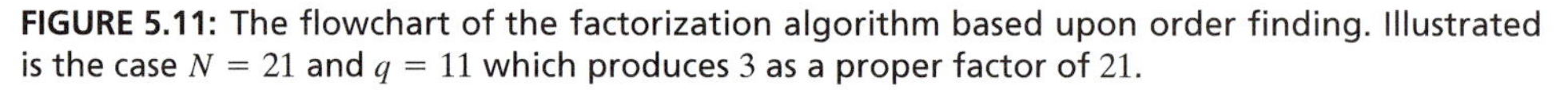

FIGURE 5.11: The flowchart of the factorization algorithm based upon order finding. Illustrated is the case $N = 21$ and $q = 11$ which produces 3 as a proper factor of 21.

Thus,

$$17^{29} = 17^{16} \times 17^{8} \times 17^{4} \times 17^{1}.$$

The pseudo code to carry out this computation is

```
power:= 1
for i=0 to m-1
        if(r^{k_i} == 1)then
            power := power × r^{2^i} mod N
    endif
endfor
```

The condition $r^k = 1 \bmod N$ implies that $r^k - 1$ is divisible by N. Equivalently, this means that an integer m exists such that

$$r^k - 1 = (r - 1)(r^{k-1} + r^{k-2} + \ldots + r^2 + r + 1) = mN.$$

This equation shows that either $(r - 1)$ or $(r^{k-1} + r^{k-2} + \ldots + r^2 + r + 1)$ share a common factor with N. Assume that $r - 1$ is such a factor, or shares a common factor with N. We show that the greatest common divisor (gcd) of N and $r - 1$ is indeed a proper factor of N, namely, $\gcd(N, r - 1) \neq 1$ and $\gcd(N, r - 1) \neq N$. Consider two cases:

1. if $r - 1 < N$, then the common factor of $r - 1$ and N cannot be N.

2. if $r - 1 \geqslant N$, then $\gcd(N, r - 1)$ could be N if $r - 1$ is a multiple of N. But $r - 1$ is not a multiple of N due to the condition $r^1 \neq 1 \bmod N$ that can also be written as $(r - 1) \neq 0 \bmod N$.

Thus the $\gcd(N, r - 1)$ is a proper factor of N. Moreover, efficient algorithms to compute the greatest common divisor (gcd) of two integers exist (e.g., Euclid's algorithm presented in Appendix B).

We conclude that *the problem of finding the factors of N is reduced to the problem of order finding*. To clarify this idea, let us give several examples. First, we consider the easier case when N is neither even, nor the power of a prime. In this case we are looking for an integer r with the property that $r^2 - 1$ is a multiple of N, but neither $r - 1$ nor $r + 1$ are multiples of N (i.e., $r \neq 1 \bmod N$ and $r \neq -1 \bmod N$).

For example, when $N = 21$ a possible choice for r is $r = 8$. Indeed, $r^2 - 1 = 64 - 1 = 63 = 3 \times 21$. As expected, $r - 1 = 8 - 1 = 7$ is a proper factor of $N = 21$. When $q = 5$, we find again that $k = 6$. Now $r = 5^6 \bmod 21 = 20$, $r - 1 = 19$ and $r + 1 = 21$. We see that 21 is a factor, but not a proper factor of 21.

The pseudocode for finding a proper factor p for an integer N, based upon the order finding algorithm is:

```
Step 1. If N is even then return  p=2.
Step 2. If N is a power k of a prime integer p then return p.
Step 3. Randomly pick q, 1 < q < N-1.
        If p = gcd(q,N) > 1 then return p. Else go to Step 4.
Step 4. Determine the order k of q modulo N.
        If k is not even then go to Step 3.
Step 5. Let k = 2 m and determine r, the m-th power of
        q modulo N with 1 < r < N.
        If 1 < p = gcd(r-1, N) < N   then return p.
        If 1 < p = gcd(r+1, N) < N   then return p.
        Else (if we fail to find a proper factor of q)
        go to Step 3.
```

It is easy to see that when $k = 2m$ the condition $q^k = 1 \bmod N$ implies that the factors of N are either $\gcd(r - 1, N)$ or $\gcd(r + 1, N)$ with $r = q^m \bmod N$. Indeed,

$$q^{2m} = 1 \bmod N \Longrightarrow q^{2m} - 1 = (q^m - 1)(q^m + 1) \text{ must be divisible by N.}$$

The flowchart of the algorithm is shown in Figure 5.11. In our example the order of 11 mod 21 is $k = 6$. But $6 = 2 \times 3$. Thus, $m = 3$ and $r = q^m \bmod N = 11^3 \bmod 21 = 8$. Then, $r - 1 = 8 - 1 = 7$ and $p = \gcd(r - 1, N) = \gcd(7, 21) = 7 < 21$ is a factor of $N = 21$.

As we pointed out previously, the key to the efficiency of the algorithm is the choice of q.

5.11 A QUANTUM CIRCUIT FOR COMPUTING $f(x)$ MODULO 2^m

The factorization algorithm discussed in the previous section relies on our ability to perform modular arithmetic. It is thus necessary to present quantum circuits for modular arithmetic and we begin our discussion with a circuit for computing $f(x)$ modulo 2^m.

Figure 5.12 illustrates a circuit with two `Hadamard` gates and a `controlled-`R_k gate. Recall that

$$G_{\text{controlled}-R_k} = \begin{pmatrix} 1 & 0 & 0 & 0 \\ 0 & 1 & 0 & 0 \\ 0 & 0 & 1 & 0 \\ 0 & 0 & 0 & e^{i2\pi/2^k} \end{pmatrix}$$

The successive transformations of the input state carried out by the circuit in Figure 5.12(a) can be expressed as products of `ket` vectors:

$$| 0\rangle \, | u\rangle \mapsto \frac{1}{\sqrt{2}}(| 0\rangle + | 1\rangle) \, | u\rangle \mapsto \frac{1}{\sqrt{2}}(| 0\rangle + e^{i2\pi/2^k} \, | 1\rangle) \, | u\rangle$$

But, the function performed by the `controlled-`R_k may be applied to multiple qubits and map $\mathcal{H}_{2^n} \mapsto \mathcal{H}_{2^m}$, see Figure 5.12(b). The circuit implements the transformation

$$f : \{0, 1\}^n \mapsto \{0, 1\}^m.$$

The transformation performed by this special gate array can again be expressed as a product of `ket` vectors

$$| x\rangle \, | y\rangle \quad \mapsto \quad | x\rangle \, | y + f(x) \bmod 2^m\rangle.$$

This implies that the content of the second register depends upon the state of the first register. The second register is transformed as

$$| y\rangle \quad \mapsto \quad | y + f(x) \bmod 2^m\rangle.$$

Let us now consider the case when the content of the first register is the result of applying the QFT to the input state $| 11 \ldots 1\rangle$, as shown in Figure 5.12(c). Then

$$| u\rangle = \frac{1}{2^m} \sum_{y=0}^{2^m-1} \exp\left(-i\frac{2\pi}{2^m} y\right) | y\rangle.$$

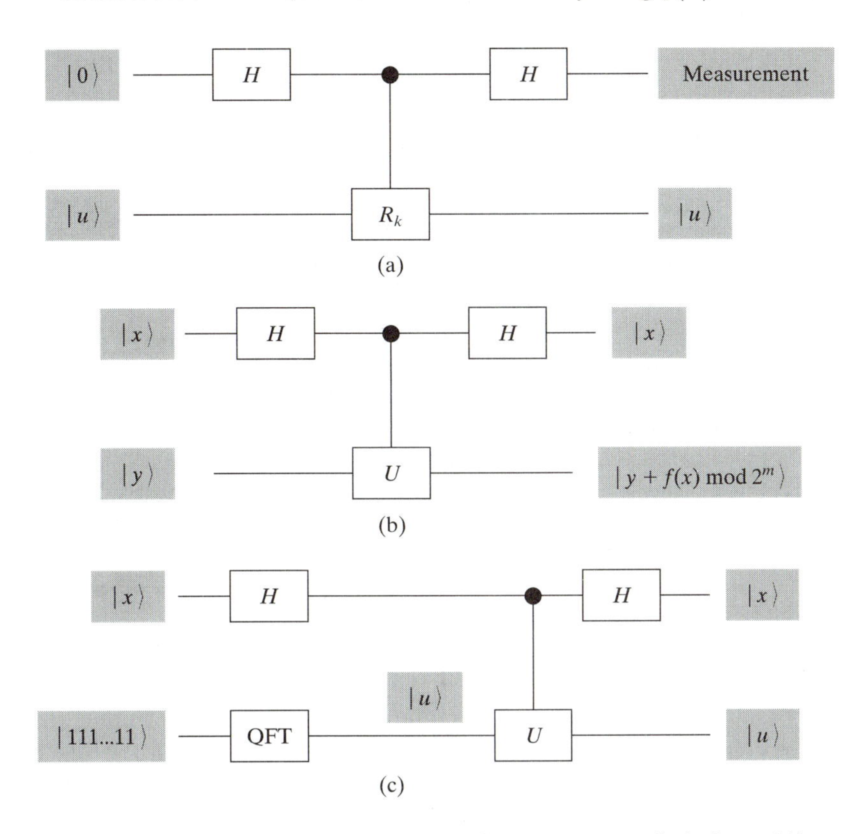

FIGURE 5.12: (a) A two-qubit quantum circuit performing a controlled phase shift operation $|0\rangle \; |u\rangle \mapsto (1/\sqrt{2})(|0\rangle + e^{i2\pi/2^k} \; |1\rangle) \; |u\rangle$. (b) A circuit for mapping multiple-qubit inputs $|x\rangle \; |y\rangle \mapsto |x\rangle \; |y + f(x) \mod 2^m\rangle$. (c) A circuit which introduces the phase factor $\varphi(x) = (2\pi/2^m)f(x)$ to the first register x. The transformation carried out by this circuit is $|x\rangle \; |u\rangle \mapsto e^{(-i(2\pi/2^m)f(x))} \; |x\rangle \; |y\rangle$.

Now

$$
\begin{aligned}
|x\rangle \, |u\rangle &= \frac{1}{2^{m/2}} \, |x\rangle \sum_{y=0}^{2^m-1} \exp\left(-i\frac{2\pi}{2^m}\right) |y\rangle \\
&\mapsto \frac{1}{2^{m/2}} \, |x\rangle \sum_{y=0}^{2^m-1} \exp\left(-i\frac{2\pi}{2^m}\right) |(f(x) + y)\rangle \\
&\mapsto \frac{e^{i(2\pi/2^m)f(x)}}{2^{m/2}} \, |x\rangle \sum_{y=0}^{2^m-1} \exp\left(-i\frac{2\pi}{2^m}(f(x) + y)\right) |(f(x) + y)\rangle \\
&\mapsto \frac{e^{i(2\pi/2^m)f(x)}}{2^{m/2}} \, |x\rangle \sum_{y=0}^{2^m-1} \exp\left(-i\frac{2\pi}{2^m}y\right) |y\rangle \\
&\mapsto e^{i(2\pi/2^m)f(x)} \, |x\rangle \, |u\rangle.
\end{aligned}
$$

Following [52], we have re-labeled the summation index in

$$\sum_{y=0}^{2^m-1} \exp\left(-i\frac{2\pi}{2m}(f(x) + y)\right) | (f(x) + y)\rangle = \sum_{y=0}^{2^m-1} \exp\left(-i\frac{2\pi}{2m}y\right) | y\rangle.$$

5.12 SIMON'S ALGORITHM FOR PHASE ESTIMATION

The factoring method presented earlier requires an efficient algorithm to determine the period of a periodic function. This problem, also known as "phase estimation," is the topic of this section.

Consider a periodic function f given as a "black box." The function f maps binary n-tuples to binary m-tuples. We assume that the domain of this function consists of an integral number of periods. We are not concerned with how efficient this mapping performed by this function is.

We are able to observe the results of the computation carried out by the "black box," but we cannot examine the code of the function. We wish to determine the period, s, of $f(x)$:

$$f(x) = f(y) \Longleftrightarrow x \equiv y \bmod s.$$

This is an example of a so-called "oracle" problem. We provide a pair x, s, the oracle returns the result of the evaluation of the functions $f(x)$ and $f(x + s)$, we compare them and if the values are not equal, then we provide a different value of s and the process continues until we can determine the value of s such that $f(x) = f(x + s)$.[8]

The best known algorithm for determining the period of a function using a classical computer takes an exponential time. Indeed, x and s may take any of the 2^n possible values. At least half of the time one must try more than half of the possible values of s, thus we need $\mathcal{O}(2^n)$ evaluations of the function f.

Finding an efficient phase estimation algorithm is extremely important for several quantum algorithms. To solve problems such as integer factorization we need to perform repeatedly the phase estimation procedure.

In 1994, Dan Simon created an algorithm for phase estimation that requires quadratic execution time on a quantum computer. Simon formulates the phase estimation problem as follows [127]: "We are given a function $f : \{0,1\}^n \mapsto \{0,1\}^m$ with $m > n$ and we know that either

(a) f is a one-to-one function, or
(b) there exists a non-trivial s such that

$$\forall x \neq x' \quad f(x) = f(x') \Longleftrightarrow x' = x \oplus s.$$

We want to determine which one of the two statements, (a) or (b) is true. If (b) is true we wish to find the value s."

The following algorithm provides a solution with time complexity $\mathcal{O}(nT_f(n) + G(n))$ with $T_f(n)$—the time to compute the function f for an

---

[8] In the general case we need to test the value of s for a number of arguments x, x', x'', etc.

input of size n and $G(n)$—the time to solve a linear system of $n \times n$ equations over $\mathbb{C}^2$.

Given a function $f(x)$ we wish to establish whether $f(x)$ is a periodic function and, if so, to determine its period. The following steps are taken:

Step 1. Apply the Quantum Fourier Transform (QFT) to a register of n qubits in state $\mid 0\rangle$. The result is

$$2^{-n/2} \sum_x \mid x\rangle.$$

Step 2. Compute $f(x)$ and concatenate the result with $\mid x\rangle$ to produce

$$2^{-n/2} \sum_x \mid x f(x)\rangle.$$

Step 3. Apply the QFT on $\mid x\rangle$ to produce

$$2^{-n/2} \sum_y \sum_x (-1)^{x \cdot y} \mid y f(x)\rangle.$$

Let us assume first that (a) is true, namely f is one-to-one. Then, every time we carry out Steps 1–3 we obtain a superposition state $\mid y f(x)\rangle$ with amplitude 2^{-n} resulting from 2^n distinct states each occurring with probability 2^{-2n}. Assume that we repeat Steps 1–3 k-times. Then, the k values obtained in Step 3 have the same amplitude.

Assume now that (b) is true, $f(x)$ is periodic, $f(x) = f(y)$ with $x \equiv y \bmod s$. For each pair y, x the states $\mid y f(x)\rangle$ and $\mid y f(x \oplus s)\rangle$ are identical and the amplitude of this state, $\alpha(x, y)$ is

$$\alpha(x, y) = 2^{-n}((-1)^{x \cdot y} + (-1)^{(x \oplus s) \cdot y}).$$

If $y \cdot s \equiv 0 \bmod 2$ then

$$\begin{aligned}(x \oplus s) \cdot y \bmod 2 &= (x \cdot y) \oplus (s \cdot y) \bmod 2 \\ &= (x \cdot y) \oplus 0 \bmod 2 = (x \cdot y) \bmod 2.\end{aligned}$$

It follows that when

$$(x \oplus s) \cdot y \bmod 2 = (x \cdot y) \bmod 2$$

then the amplitude $\alpha(x, y)$ of the identical states $\mid y f(x)\rangle$ and $\mid y f(x \oplus s)\rangle$ is non-zero

$$\alpha(x, y) = 2^{-n}((-1)^{x \cdot y} + (-1)^{x \cdot y}) = 2^{-n+1}(-1)^{x \cdot y}.$$

In this expression $x = (x_0\ x_1 \ldots x_{n-1})$ and $y = (y_0\ y_1 \ldots y_{n-1})$ are binary strings of length n and $x \cdot y = x_0 y_0 + x_1 y_1 + \ldots + x_{n-1} y_{n-1}$. When $(x \oplus s) \cdot y \bmod 2 \neq (x \cdot y) \bmod 2$, then the two terms in the sum below have opposite signs and their sum is 0

$$\alpha(x, y) = 2^{-n}((-1)^{x \cdot y} + (-1)^{(x \oplus s) \cdot y}) = 0.$$

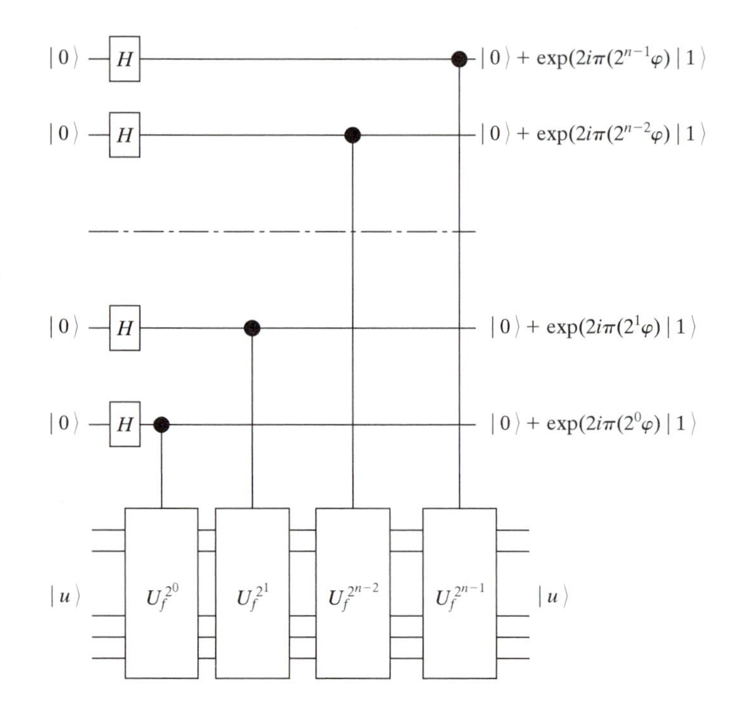

FIGURE 5.13: A quantum circuit for phase estimation. The second register holding $|\,u\rangle$ acts as control. The normalization factors $(1\sqrt{2})$ for the n target qubits are omitted.

Assume that we repeat the algorithm $\mathcal{O}(n)$ times. Then we obtain many independent values of y whose dot products with an unknown string, say s^*, are even. We can determine s^* by solving the linear system of equations.

Therefore, the time complexity of the algorithm reflects the sum of the time to repeat n times the Steps 1–3, $\mathcal{O}(nT_f(n))$ with the time to solve a linear system of $n \times n$ equations over $\mathbb{C}^2$, $\mathcal{O}(G(n))$.

An equivalent formulation of the phase estimation problem is: "Given a unitary operator U_f with an eigenvector $|\,u\rangle$ and an eigenvalue $e^{i2\pi\varphi}$, with φ unknown, estimate φ" [98].

The operator $\mathbf{U}_f$ implementing the transformation described by the function f is related to its eigenvector $|\,u\rangle$ and eigenvalue $e^{i2\pi\varphi}$ through the relation

$$\mathbf{U}_f \,|\, u\rangle = e^{i2\pi\varphi} \,|\, u\rangle.$$

Figure 5.13 describes the quantum circuit used for phase estimation. The circuit has two registers, the first one holds n qubits, initially in state $|\,0\rangle$. The choice of n depends upon the desired accuracy for the estimation of the phase φ. The second register holds $|\,u\rangle$.

Recall from Section 4.12 that the role of the control and target qubits can be reversed by a change of basis. Thus, the second register of the circuit in Figure 5.13 acts as control and the first as target. Each of the boxes representing the unitary transformations $U_f^{2^q}$ contributes with a phase shift of the second component of

$$e^{i2\pi(2^q \varphi)}.$$

Thus, the final state of the first register is

$$\frac{1}{2^{n/2}}(|0\rangle + e^{i2\pi(2^{n-1}\varphi)}|1\rangle)(|0\rangle + e^{i2\pi(2^{n-2}\varphi)}|1\rangle)$$

$$\ldots(|0\rangle + e^{i2\pi(2^1\varphi)}|1\rangle)(|0\rangle + e^{i2\pi(2^0\varphi)}|1\rangle).$$

An additional stage applies an inverse QFT to the results in the first register. Finally, we read out the state of the first register by doing a measurement in the new computational basis.

5.13 THE FOURIER TRANSFORM ON AN ABELIAN GROUP

We have seen earlier that we can reduce the number of operations required to calculate a new state of a quantum system if we know the current state of a quantum system and the transfer matrix of a circuit operating upon the system using the method discussed in Section 5.7. This method is based upon the decomposition of a matrix into a tensor product. We now ask ourselves how to apply this technique to QFT.

The next few sections require some basic concepts in group theory discussed in Appendix A. Here we only provide basic definitions and give an example of a group. A *group* G is a set equipped with one binary operation "$\cdot$", multiplication (the operation may also be addition, "+"), which satisfies several properties:

1. Associative law: $\forall(a, b, c) \in G, \quad a \cdot (b \cdot c) = (a \cdot b) \cdot c.$
2. Identity Element: There is an identity element $e \in G$ and $\forall a \in G \; a \cdot e = e \cdot a = a.$
3. Inverse element property: $\forall a \in G, \exists a^{-1}$ such that $a \cdot a^{-1} = a^{-1} \cdot a = e.$

A group G whose operation satisfies the commutative law (i.e., $a \cdot b = b \cdot a$) is called *commutative* or *Abelian*. The *cardinality of a group* G, denoted by $|G|$, represents the number of elements of the group G. A *finite group* G has a finite number of elements.

It is trivial to show that the integers modulo 4, $\{0, 1, 2, 3\}$, with the usual addition operation, form a finite Abelian group with 4 elements. In this group 0 is the identity element, and each element has an inverse, 3 is the inverse of 1 and 2 is the inverse of 2 (indeed, $3 + 1 = 0 \bmod 4$ and $2 + 2 = 0 \bmod 4$). In Appendix A, we discuss a group of transformations in a plane and show that this group is isomorphic with the integers modulo 4 under addition.

Let G be a finite Abelian group and $\mathbb{C}^*$ be the set of non-zero complex numbers. Let χ be an isomorphism from the additive group G to the multiplicative group $\mathbb{C}^*$

$$\chi : G \mapsto \mathbb{C}^*.$$

It can be shown [60] that the transformation χ has the following properties:

1. There are exactly $N = |G|$ functions χ_j satisfying the condition
$$\chi_j(g_a + g_b) = \chi_j(g_a)\chi_j(g_b) \quad \forall g_a, g_b \in G.$$
$|G|$ is the cardinality of the group G.
Thus, the functions χ_k may be labeled by the elements in G as $\chi_g, g \in G$.
2. Two such functions, χ_j and χ_k, $\forall j, k \in G$ are orthogonal and
$$\frac{1}{N}\sum_{g \in G} \chi_j(g)\overline{\chi_k(g)} = \delta_{jk}.$$
3. Any value $\chi(g)$ is an N-th root of unity. If we denote by $\mathcal{S}^1$ the set of complex numbers $c = a + ib$ with modulus equal to one, $\sqrt{(a^2 + b^2)} = 1$ then $\chi(g)$ is a group homomorphism
$$\chi(g) : G \mapsto \mathcal{S}^1.$$

The Fourier Transform on the group G is described by an $N \times N$ matrix $\mathcal{F} = [F_{jk}]$, $j, k \in G, 1 \leqslant j, k \leqslant N$ whose elements are
$$F_{jk} = \frac{1}{\sqrt{N}}\chi_j(k) \quad j, k \in G.$$

Let us now consider a 2^n-dimensional Hilbert space, $\mathcal{H}_{2^n}$, with basis vectors $|b\rangle$ labeled by the elements of G, $b \in G$. Let $U(k)$ be a shift operator, a transformation which maps the basis vector $|b\rangle$ into $|b + k\rangle$
$$U(k) : |b\rangle \mapsto |b + k\rangle \quad b, k \in G.$$

The states
$$\chi_k = \frac{1}{\sqrt{N}}\sum_{b \in G} \overline{\chi_k(b)}\,|b\rangle$$
form an orthonormal basis, called the *Fourier basis*. This follows from property (2) of transformation χ. This basis consists of the common eigenvectors of the shift operators $U(k)$, $\forall k \in G$. Thus,
$$U(b)\,|\chi_k\rangle = e^{\chi_k(b)}\,|\chi_k\rangle.$$

We can now describe the Fourier Transform as a unitary transformation given by
$$\mathcal{F}\,|\chi_k\rangle = |b\rangle.$$

Let $N = 2^n$ and let $u = (u_0u_1 \ldots u_{n-1})$ and $v = (v_0v_1 \ldots v_{n-1})$ be binary strings of length n, where $u, v \in \mathbb{Z}_{2^n}$ with $\mathbb{Z}_{2^n}$ the group of all n-bit binary strings. Let
$$u \cdot v = u_0v_0 + u_1v_1 + \ldots + u_{n-1}v_{n-1} \bmod 2.$$

Ekert and Jozsa [51] show that in this case the Fourier Transform coincides with the Walsh-Hadamard Transform and the elements of the Fourier Transform matrix are

$$F_{uv} = \frac{1}{\sqrt{2^n}} e^{i2\pi(u \cdot v)/2} = \frac{1}{\sqrt{2^n}} (-1)^{u \cdot v}.$$

When $G = \mathbb{Z}_{2^n}$ we obtain the Discrete Fourier Transform modulo 2^n and the elements of the Fourier Transform matrix are

$$F_{kb} = \frac{1}{\sqrt{2^n}} e^{i2\pi(kb/2^n)} \quad k, b = 0, 1, \ldots, 2^n - 1.$$

The Fourier Transform is then a function $f : G \mapsto \mathbb{C}^*$ which transforms a vector $f(b_1), f(b_2), \ldots, f(b_{2^n})$ as follows

$$\tilde{f} = \sum_{b \in G} F_{kb} f(b) = \frac{1}{\sqrt{2^n}} \sum_{b \in G} \chi_k(b) f(b) \quad k \in G.$$

Let H be a subgroup of G, $H \subset G$ with index

$$I_{G/H} = \frac{|G|}{|H|}$$

with $|G|$ the cardinality of G and $|H|$ the cardinality of H. Let

$$c_1(H), c_2(H), \ldots, c_{I_{G/H}}(H)$$

be a complete list of cosets of H. Recall that given a subgroup H of a group G and an element $x \in G$, we define xH to be the set $\{xh : h \in H\}$ and Hx to be the set $\{hx : h \in H\}$. A subset of G of the form xH for some $x \in G$ is said to be a *left coset of H* and a subset of the form Hx is said to be a *right coset* of H. Then the set G consists of $I_{G/H}$ disjoint equivalence classes

$$G = \bigcup_{j=1}^{I_{G/H}} c_j(H).$$

All elements $b \in G$ may be written in terms of cosets. Then, the Fourier Transform can be re-written as

$$\tilde{f} = \frac{1}{\sqrt{2^n}} \sum_{j=1}^{I_{G/H}} \sum_{h \in H} f(k_j + h) \chi_l(k_j + h) = \frac{1}{\sqrt{2^n}} \sum_{j=1}^{I_{G/H}} \chi_l(k_j) \sum_{h \in H} f_j(h) \chi_l(h).$$

If we call $N_H = |H|$ the cardinality of the subgroup H, then the number of operations required to evaluate the last expression of the Fourier Transform is

$$\mathbb{O}[N(N_H + I_{G/H})]$$

compared with $\mathbb{O}(N^2)$ operations for the matrix multiplication required by a direct evaluation of the Fourier Transform. Indeed, the Fourier Transform on the set G is reduced to the evaluations of $I_{G/H}$ Fourier Transforms on the subset H. Then the $I_{G/H}$ results are summed together with coefficients $\chi_l(k_j)$, for each $l \in G$.

We can construct what Ecker and Jozsa [51] call *a tower of subgroups* of increasingly smaller cardinality

$$\mathbb{Z}_{2^n} \supset \mathbb{Z}_{2^{n-1}} \supset \mathbb{Z}_{2^{n-2}} \supset \ldots \mathbb{Z}_{2^2} \supset \ldots \mathbb{Z}_2$$

with $\mathbb{Z}_{2^{n-1}} = \{0, 2, 4, 6, 8 \ldots\}$ the set of all even integers, $\mathbb{Z}_{2^{n-2}} = \{0, 4, 8, 12 \ldots\}$ the set of all integers multiple of 4, and so on. Then, the irreducible representations of the group $\mathbb{Z}_{2^n}$ are

$$\chi_j(k) = (\omega^j)^k \quad j, k = 0, 1, 2 \ldots 2^n - 1$$

with $\omega = e^{i2\pi/2^n}$. The expression giving the Fourier Transform becomes

$$\tilde{f}(j) = \frac{1}{\sqrt{2^n}} \sum_{k=0}^{2^n-1} f(k)\chi_j(k)$$

or

$$\tilde{f}(j) = \frac{1}{\sqrt{2}} \left(\sum_{k=0}^{2^{n-1}-1} f(2k) \frac{\omega^{2jk}}{\sqrt{2^{n-1}}} + \omega^j \sum_{k=0}^{2^{n-1}-1} f(2k+1) \frac{\omega^{2jk}}{\sqrt{2^{n-1}}} \right).$$

The two sums above are the Fourier Transforms of size 2^{n-1} for the even and, respectively, odd values of the arguments of the function $f(j)$.

The entangled state of n qubits is

$$| f\rangle = \sum_{j_0=0,1} \sum_{j_1=0,1} \cdots \sum_{j_{n-2}=0,1} \sum_{j_{n-1}=0,1} f(j_{n-1} j_{n-2} \cdots j_1 j_0)$$
$$| j_{n-1}\rangle \, | j_{n-2}\rangle \ldots | j_1\rangle \, | j_0\rangle.$$

We could use a single circuit for a Fourier Transform of size 2^{n-1} operating on $(n-1)$ qubits $\{1, 2, \ldots (n-1)\}$ to transform the even and odd components and then combine the results. To combine the results we have to:

1. Apply to each qubit $j = 0, 1 \ldots, 2^{n-1}$ the unitary transformation

$$\frac{1}{\sqrt{2}} \begin{pmatrix} 1 & \omega^j \\ 1 & -\omega^j \end{pmatrix}.$$

But, we can express this transformation as

$$\frac{1}{\sqrt{2}} \begin{pmatrix} 1 & \omega^j \\ 1 & -\omega^j \end{pmatrix} = \frac{1}{\sqrt{2}} \begin{pmatrix} 1 & 1 \\ 1 & -1 \end{pmatrix} \begin{pmatrix} 1 & 0 \\ 0 & \omega^j \end{pmatrix} = HR_j.$$

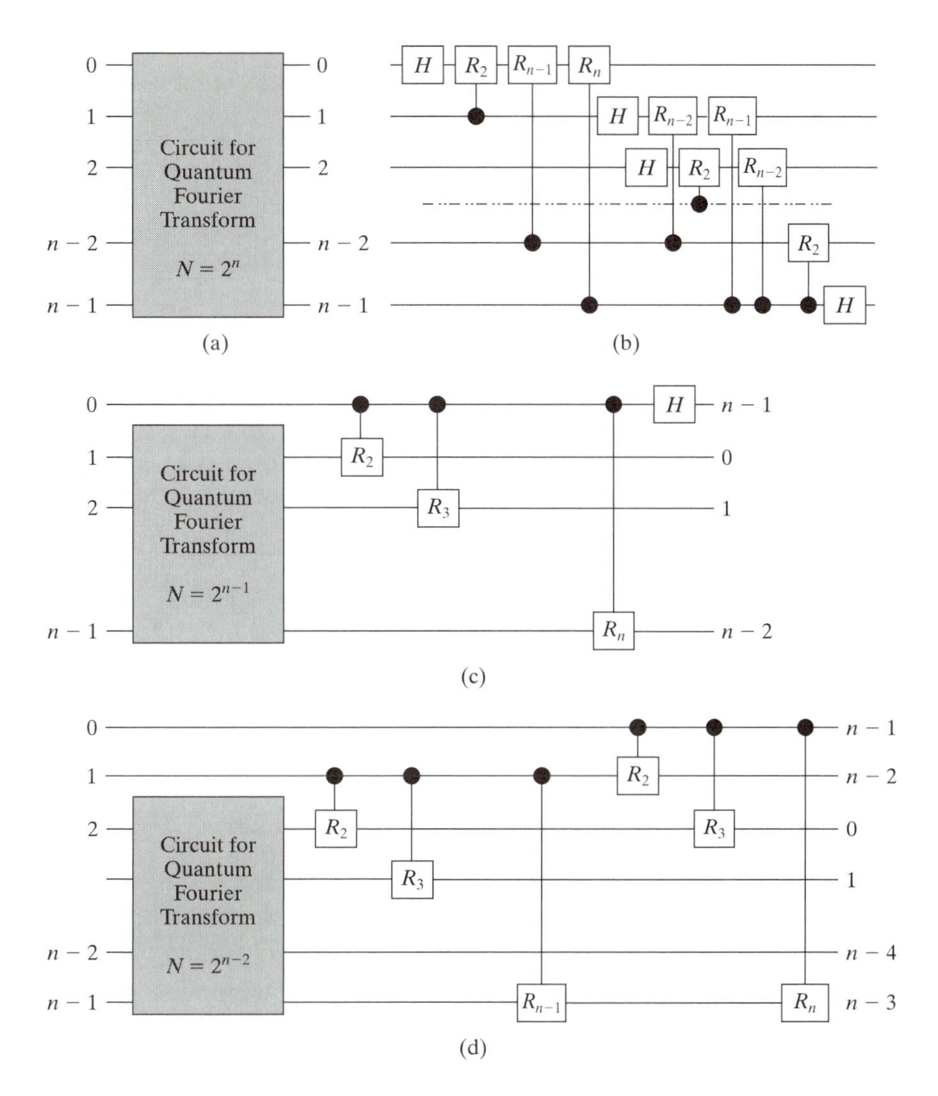

FIGURE 5.14: (a) The block diagram for the Quantum Fourier Transform of n qubits. (b) The quantum circuit for the QFT of n qubits. (c) The decomposition of the QFT for n qubits into a QFT for $n - 1$ qubits and $\mathcal{O}(n)$ extra operations. (d) The decomposition of the QFT for $n - 1$ qubits into a QFT for $n - 2$ qubits and $\mathcal{O}(n)$ extra operations. The labels of the n qubits are shown at the input and the output of the quantum circuits.

Here R_j is applied in dimensions $(2j, 2j + 1)$. This transformation leaves the probability amplitude in the even dimension unchanged and applies a phase shift of ω^j to the probability amplitude of the odd dimension. Instead of performing exponentially many operations for each of the 2^{n-1} qubits we can use the entanglement effect for all values of j simultaneously by applying a two-qubit gate to qubits 1 and q for $q = 1, 2, \dots, (n - 1)$. The matrix of the conditional shift is then

$$R_q = \begin{pmatrix} 1 & 0 & 0 & 0 \\ 0 & 1 & 0 & 0 \\ 0 & 0 & 1 & 0 \\ 0 & 0 & 0 & \omega^{2^{q-1}} \end{pmatrix}.$$

2. Reorder the results according to the permutation

$$(2j, 2j + 1) \mapsto (j, j + 2^{n-1}).$$

These transformations are illustrated in Figures 5.14(c) and (d) where we show the decomposition of a QFT for n qubits into a QFT for $n - 1$ qubits and $\mathcal{O}(n)$ extra operations followed by the decomposition of the QFT for $n - 1$ qubits into a QFT for $n - 2$ qubits and $\mathcal{O}(n)$ extra operations. The process continues recursively until we have only two qubits and then we construct the QFT for $\mathcal{H}_2$.

5.14 PERIODICITY AND THE QUANTUM FOURIER TRANSFORM

In this section, we investigate how to use the Quantum Fourier Transform to determine the period of a function whose domain is the additive group of integers modulo N, where N is a positive integer. Recall that $\mathbb{Z}_N$ denotes the additive group of integers modulo N and $\mathbb{Z}$ is the additive group of integers. We are given a periodic function $f : \mathbb{Z}_N \mapsto \mathbb{Z}$ with period T

$$f(t) = f(t + T) \quad \forall t \in \mathbb{Z}_N, T \in \mathbb{Z}_N.$$

Our goal is to determine the period T. We assume that N is a multiple of the period, $N = KT$.

A classical algorithm evaluates pairs of values

$$[f(t), f(t + T)] \quad \forall\, (t, T) \in \{0, 1, \ldots, N - 1\}$$

in an attempt to obtain two equal results. To determine the period T of $f(t)$ using a classical algorithm we must carry out $\mathcal{O}(N)$ random trials; each random trial requires two evaluations of the function f. We show now that a quantum algorithm using the QFT needs only $\mathcal{O}((\log N)^2)$ evaluations of the function f, thus, it archives an exponential speedup over the classical algorithm.

We assume that a quantum circuit like the one in Figure 5.15(a) able to carry out the state transformation

$$\mid t\rangle \mid 0\rangle \mapsto \mid t\rangle \mid f(t)\rangle$$

exists. We can apply an equal superposition state of the original state $\mid t\rangle$ as the input of the quantum circuit and transform it into another superposition state

$$\frac{1}{\sqrt{N}} \sum_{t=0}^{N-1} \mid t\rangle \mid 0\rangle \mapsto \frac{1}{\sqrt{N}} \sum_{t=0}^{N-1} \mid t\rangle \mid f(t)\rangle.$$

Let $f(t_0)$ be the value of the function for a randomly selected argument $0 \leqslant t_0 \leqslant T - 1$. Then, we see in Figure 5.15(b) that

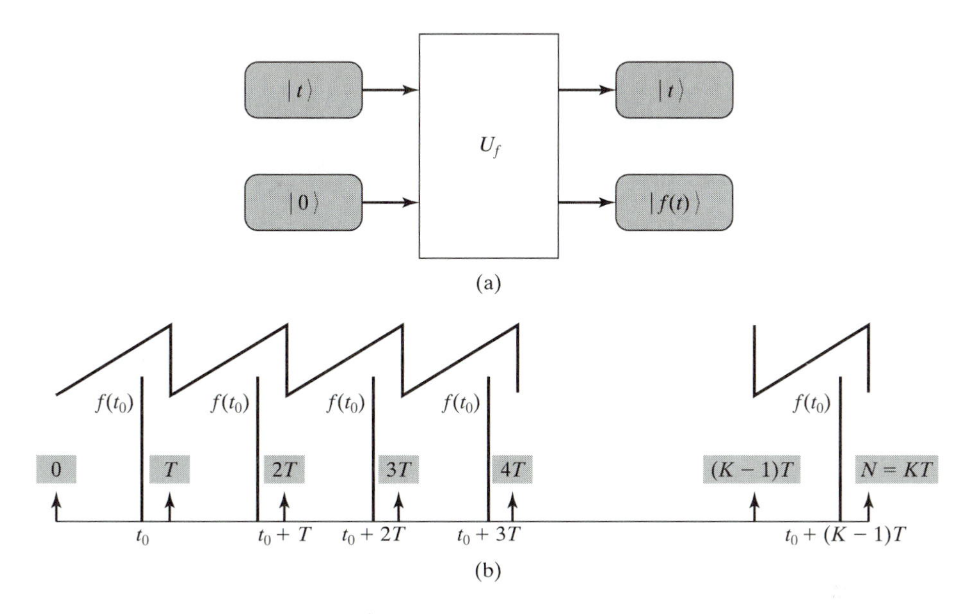

FIGURE 5.15: (a) The circuit transforming the state $| t\rangle \ | 0\rangle$ into $| t\rangle \ | f(t)\rangle$. (b) The function $f(t)$ is periodic with period T and $N = KT$. Then $f(t_0) = f(t_0 + kT), k = 0, 1, 2, \ldots, K - 1$.

$f(t_0) = f(t_0 + kT)$, $k = 0, 1, 2, \ldots, K - 1$. Call $| \psi\rangle$ the superposition of all states $| t\rangle$ such that $f(t) = f(t_0)$. Then

$$| \psi\rangle = \frac{1}{\sqrt{K}} \sum_{k=0}^{K-1} | t_0 + kT\rangle.$$

This superposition gives no information about the period T. When we measure the state $| \psi\rangle$ we obtain a random integer uniformly distributed in the interval $(0, N - 1)$, if t_0 is a random integer in the interval $0 \leqslant t_0 \leqslant T - 1$.

In Appendix D, we discuss the fact that the Fourier Transform allows us to identify periodic patterns in a data set. We use the same idea and apply the Fourier Transform of $| \psi\rangle$ as the input to the circuit, instead of the original state $| \psi\rangle$ [78].

Recall that in Section 5.6 we introduced the Discrete Fourier Transform (DFT) of the additive Abelian group $\mathbb{Z}_N$ and the QFT of the state $| \psi\rangle$ is

$$\mathcal{F} | \psi\rangle = \frac{1}{\sqrt{N}} \sum_{j=0}^{N-1} \left(e^{i2\pi \frac{t_0}{T}} \right)^j | j\frac{N}{T}\rangle.$$

It is easy to see that

$$\sum_{j=0}^{N-1} \left(e^{i2\pi \frac{t_0}{T}} \right)^j = \begin{cases} N & \text{if } t_0 = kT \\ 0 & \text{if } t_0 \neq kT. \end{cases}$$

Indeed, when $t_0 = kT$, each one of the N terms of the sum is

$$\left(e^{i2\pi k} \right)^j = (\cos 2\pi k + i \sin 2\pi k)^j = 1.$$

When $t_0 \neq kT$,

$$\sum_{j=0}^{N-1} \left(e^{i2\pi \frac{t_0}{T}}\right)^j = \frac{1 - (e^{i2\pi \frac{t_0}{T}})^N}{1 - e^{i2\pi \frac{t_0}{T}}} = \frac{1 - (e^{i2\pi \frac{t_0}{T}})^{KT}}{1 - e^{i2\pi \frac{t_0}{T}}} = \frac{1 - e^{i2\pi K t_0}}{1 - e^{i2\pi \frac{t_0}{T}}} = 0.$$

Thus,

$$\mathcal{F} \mid \psi\rangle = \frac{1}{\sqrt{N}} \sum_{j=0}^{N-1} \left(e^{i2\pi \frac{t_0}{T}}\right)^j \mid j\frac{N}{T}\rangle = \frac{1}{\sqrt{T}} \sum_{j=0}^{T-1} \left(e^{i2\pi \frac{t_0}{T}}\right)^j \mid j\frac{N}{T}\rangle.$$

Now we read the label of the state $\mathcal{F} \mid \psi\rangle$ obtained as a result of a QFT applied to the state $\mid \psi\rangle$ and call it c. This value is a multiple of N/T as we can see from the expression of $\mathcal{F} \mid \psi\rangle$

$$c = j\frac{N}{T}$$

with $0 \leqslant j \leqslant (T - 1)$. Therefore,

$$\frac{c}{N} = \frac{j}{T}.$$

If j and T do not have any common factors, we can determine T by transforming c/N to an irreducible fraction. A theorem from number theory gives the number of coprimes less than or equal to a large integer T [78]

$$n_T \approx e^{\gamma} \frac{T}{\log \log T}$$

with $\gamma = 1.781072$ Euler's constant. Thus, the probability that a randomly selected j is coprime to T is $\mathcal{O}(1/\log(\log N))$. If we repeat the process of randomly selecting j, $\mathcal{O}(\log(\log N))$ times, we succeed to determine T with a probability arbitrarily close to 1.

We conclude that the quantum algorithm to find the period of a function f needs only $\mathcal{O}((\log N)^2)$ evaluations of the function. Thus, the quantum algorithm archives an exponential speedup over the classical algorithm that requires $\mathcal{O}(N)$ evaluations of the function f.

5.15 THE DISCRETE LOGARITHMS EVALUATION PROBLEM

Let $\mathbb{Z}_p^*$ be the multiplicative group consisting of the set of integers $\{1, 2, \ldots, p - 1\}$ with the multiplication modulo p as the group operation. We assume that p is a prime number.

In Appendix A, we show that the set of integers modulo m, $\mathbb{Z}_m^* = \{1, 2, \ldots, m - 1\}$ is not a group unless m is a prime number. When m is not prime, some elements of the set do not have a multiplicative inverse. For example, in $\mathbb{Z}_9$ the elements 3 and 6 do not have a multiplicative inverse.[9]

[9]See Section A.7 for the multiplication table of $\mathbb{Z}_g$. Indeed, we see that $[2][5] = [1]$ but there is no element $[\chi]$ such that $[3][\chi] = 1$.

The multiplicative inverse in $\mathbb{Z}_p^*$ can be computed efficiently using Euclid's algorithm. Given $x \in \mathbb{Z}_p^*$, the $\gcd(x, p) = 1$; thus, there are integers a and b such that $ax + bp = 1$, or $ax \equiv 1 \bmod p$. This tells us that a is the inverse of x in $\mathbb{Z}_p^*$.

An element q of $\mathbb{Z}_p^*$ is called a *generator of the group* if the powers of q generate all the elements of the group

$$\mathbb{Z}_p^* = \{q^0 = 1, q^1, q^2, \ldots, q^{p-2}\}.$$

Thus, every $x \in \mathbb{Z}_p^*$ may be expressed as

$$x = q^y \quad y \in \mathbb{Z}_{p-1}.$$

We call y *the discrete logarithm* of x *in base* p if

$$y = \log_p x.$$

If $x_1 = q^{y_1}$ and $x_2 = q^{y_2}$, then $x_1 x_2 = q^{y_1 + y_2}$. Thus,

$$\log_p x_1 = y_1 \text{ and } \log_p x_2 = y_2 \implies \log_p x_1 x_2 = y_1 + y_2.$$

Given p and a generator q of $\mathbb{Z}_p^*$ we wish to compute $y = \log_p x, \ \forall\, x \in \mathbb{Z}_p^*$.

Recall that $\mathbb{Z}_{p-1}$ is the additive group of integers modulo $(p - 1)$ and Z_p^* is the group of integers modulo p. A generator $q \in \mathbb{Z}_p^*$ allows us to represent $\mathbb{Z}_p^*$ as $\mathbb{Z}_{p-1}$.

Now we reformulate our problem, namely to compute $y = \log_p(x), \forall x \in \mathbb{Z}_p^*$ as the problem to find the period of a function f. Consider a function f which for given p, q, x maps pairs of integers

$$f : (\mathbb{Z}_{p-1} \times \mathbb{Z}_{p-1}) \mapsto \mathbb{Z}_p^*$$

such that,

$$f(a, b) = q^a x^{-b} \bmod p \quad \forall (a, b, x) \in \mathbb{Z}_{p-1}.$$

If $y = \log_q(x)$, then $x = q^y$ and

$$f(a, b) = q^a x^{-b} \bmod p = q^a (q^y)^{-b} \bmod p = q^{a-yb} \bmod p.$$

It follows that

$$f(a_1, b_1) = q^{a_1 - yb_1} \bmod p,$$
$$f(a_2, b_2) = q^{a_2 - yb_2} \bmod p,$$

and

$$f(y, 1) = q^{y-y} = 1 \bmod p.$$

Then,

$$f(a_2, b_2) = f(a_1, b_1) f(y, 1) \bmod p.$$

It follows that

$$f(a_1, b_1) = f(a_2, b_2) \text{ if and only if } (a_2, b_2) = (a_1, b_1) + \lambda(y, 1), \; \lambda \in \mathbb{Z}_{p-1}$$

and that *the pair* $(y, 1)$ *is the period of function* f *on the product domain.*

To solve this problem with a quantum computer we apply the algorithm to determine the period of a function. We use a circuit similar to the one in Figure 5.15. The circuit is able to compute the function f in polynomial time.

Following Jozsa [78] we label the orthonormal basis vectors as pairs of integers (e.g., (a, b), rather than a single integer, (a)). We work in a Hilbert space $\mathcal{H}$ with an orthonormal basis labeled by pairs of integers modulo $(p - 1)$,

$$| a\rangle \; | b\rangle \quad a, b \in \mathbb{Z}_{p-1}.$$

$\mathbb{Z}_{p-1}$ is a finite group with $p - 1$ elements and we create a superposition of all $p - 1$ possible values of $f(a, b)$

$$| f\rangle = \frac{1}{p - 1} \sum_{(a,b)} | a\rangle \; | b\rangle \; | f(a, b)\rangle.$$

If we measure the output result register and observe the value $k_0 = f(a_0, b_0)$ then the output's first register contains the periodic state

$$| \psi\rangle = \frac{1}{\sqrt{p - 1}} \sum_{k=0}^{p-2} | a_0 + ky\rangle \; | b_0 + k\rangle.$$

Consider the functions

$$\chi_{k_1,k_2}(a, b) = \exp i2\pi \left(\frac{ak_1 + bk_2}{p - 1} \right).$$

For $a = y$ and $b = 1$ we have

$$\chi_{k_1,k_2}(y, 1) = \exp i2\pi \left(\frac{yk_1 + k_2}{p - 1} \right).$$

When $yk_1 + k_2 = 0 \bmod (p - 1)$ which translates into $k_2 = -yk_1 \bmod (p - 1)$

$$\chi_{k_1,-yk_1}(y, 1) = \exp i2\pi \left(\frac{yk_1 - yk_1}{p - 1} \right) = 1.$$

The Fourier Transform $\mathcal{F} \otimes \mathcal{F} \, | \psi\rangle$ results in a superposition of the states with labels (k_1, k_2) and with $k_2 \equiv -yk_1 \bmod (p - 1)$

$$\mathcal{F} \otimes \mathcal{F} \, | \psi\rangle = \frac{1}{\sqrt{p - 1}} \sum_{k_1=0}^{p-2} \exp i2\pi \left(\frac{a_0k_1 - b_0yk_1}{p - 1} \right) | k_1\rangle \; | -yk_1\rangle.$$

Keep in mind that we want to compute $y = \log_p x$ knowing the pair (k_1, k_2) which labels the states. If the value selected at random, $k_1 \in \mathbb{Z}_{p-1}$, is coprime with $p - 1$, then we can find the multiplicative inverse of k_1, such that $k_1 k_1^{-1} = 1 \bmod (p - 1)$ and then compute

$$y = -k_1^{-1} k_2.$$

If k_1 is not coprime with $p - 1$, then we cannot uniquely determine y from the pair (k_1, k_2). The probability that a randomly chosen k_1 is coprime with $p - 1$ is of order $1/\log(\log p - 1)$ thus, to find y with high probability, we need to repeat the selection of y for $\mathcal{O}(\log(\log p - 1))$ times.

If n is the number of digits of p, then the fastest classical algorithm to compute $y = \log_p(x)$, $\forall x \in \mathbb{Z}_p^*$ runs in time $\mathcal{O}(\exp(n^{1/3}(\log n)^{2/3}))$ while the quantum algorithm requires less than $\mathcal{O}(n^3)$ time steps.

5.16 THE HIDDEN SUBGROUP PROBLEM

Some of the problems discussed in this chapter are particular instances of a more general problem called the *hidden subgroup* problem. The hidden subgroup problem can be formulated in terms of group theory as follows:

Consider a finite additive group G and let f be a function from G to the finite set X. Let K be a subgroup of G, $K \subset G$, such that

$$K = \{k \in G : f(k + g) = f(g) \quad \forall g \in G\}.$$

We are given a black box that computes f and we wish to determine the hidden subgroup K in time $\mathcal{O}(poly(\log(\mid G \mid)))$. To determine the hidden subgroup K means to either compute a set of generators of the subgroup K, or to output a randomly chosen element of K. The subgroup K is called the *stabilizer* or the *symmetry group* of f.

This condition stated above means that f is constant on the cosets of the subgroup K. All the elements $g_i + K$ of one coset map to the same value, $f(g_i + K) = constant$. Moreover, different cosets map to different values, $f(g_i + K) \neq f(g_j + K)$ when $g_i \neq g_j$. The function f partitions G into disjoint subsets of equal size $\mid K \mid$.

The function f can be viewed as the composition of the homomorphism from G to a subgroup H [97], see Figure 5.16,

$$u : G \mapsto H$$

and a one-to-one mapping from the subgroup H to the set X

$$v : H \mapsto X.$$

Thus,

$$f = u \circ v.$$

The quantum black box performs the unitary transformation

$$U \mid g\rangle \mid x\rangle = \mid g\rangle \mid x \oplus f(g)\rangle \quad g \in G, x \in X.$$

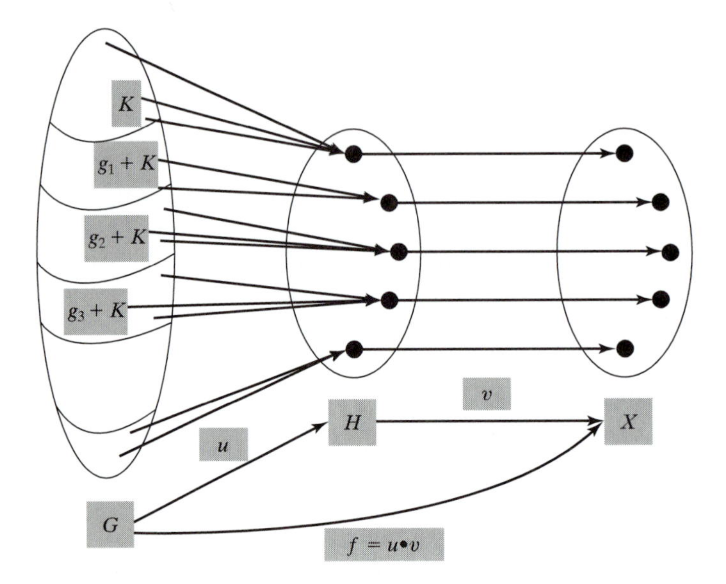

FIGURE 5.16: Graphical representation of the hidden subgroup problem. f is a function from the finite group G to the finite set X. K is the hidden subgroup we want to find. f is constant on the cosets of the subgroup K and different cosets map to different values. The function f is a composition of the homomorphism from G to a subgroup H, $u : G \mapsto H$ and a one-to-one mapping from the subgroup H to the set X, $v : H \mapsto X$.

We examine the relationship of some of the problems discussed earlier with the hidden subgroup problem (see Table 5.1). First, consider Deutsch's problem. In this case, $G = \mathbb{Z}_2 = \{0, 1\}$ with addition modulo two ($\oplus$). Recall that $\mathbb{Z}_r$ is the group of integers modulo r. The function f is a mapping

$$f : \{0, 1\} \mapsto \{0, 1\}.$$

with $f(x) = f(y)$ if and only if $x - y \in K$. Now K can be either $K = \{0, 1\}$ and then f is constant, or $K = \{0\}$ and, in that case, f is one-to-one, it is balanced.

Let us now examine Simon's problem. In this case, $G = \mathbb{Z}_2^n = \{0, 1\}^n$ with addition modulo two. The function f is a mapping from the set of binary strings of length n to any finite set X. The subgroup K consists of binary strings of length n and the function f is characterized by the property that $f(x \oplus s) = f(x)$.

Consider now the discrete logarithms with $G = \mathbb{Z}_{p-1} \times \mathbb{Z}_{p-1}$ where p is a prime number. Let X be the subgroup generated by some element of a group $H, q \in H$ such that $q^{p-1} = 1$. If $(q, y) \in G$ and $y = q^x$ we define f as follows

$$f : (a, b) \mapsto q^a x^{-b}.$$

The hidden subgroup of G gives us the logarithm of y in base $p - 1$. This subgroup is

$$K = \{(k_1, -k_1 x) \mid k_1 = 0, 1, 2, \ldots, p - 1\} = \langle(1, -x)\rangle.$$

Thus, $K = \langle(1, -k_1)\rangle$ is the subgroup generated by $(1, -k_1)$.

TABLE 5.1: Several Formulations of the Hidden Subgroup Problem

Problem	Group G	Finite set X	Hidden subgroup K	Function specification $f(x)$
Deutsch	$\{0, 1\}, \oplus$	$\{0, 1\}$	$\{0, 1\}$	$\begin{cases} f(x) = 0 \\ f(x) = 1 \end{cases}$
			$\{0\}$	$\begin{cases} f(x) = x \\ f(x) = 1 - x \end{cases}$
Simon	$\{0, 1\}^n, \oplus$	any	$\{0, s\}$ $s \in \{0, 1\}^n$	$f(s \oplus x) = f(s)$
Period finding	$\mathbb{Z}, +$	any	$\{0, r, 2r, \ldots\}$ $r \in G$	$f(x + r) = f(x)$
Order finding	$\mathbb{Z}, +$	$\{a^j\}, j \in \mathbb{Z}_r$ $a^r = 1$	$\{0, r, 2r, \ldots\}$ $r \in G$	$f(x) = a^x$ $f(x + r) = f(x)$
Discrete log	$\mathbb{Z}_{p-1} \times \mathbb{Z}_{p-1}, +$	$\{q^j\}, j \in \mathbb{Z}_{p-1}$ $q^p = 1$	$\{(k_1, -ak_1)\}$ $(a, k_1) \in \mathbb{Z}_{p-1}$	$f(a, b) = q^{k_1 a + b}$ $f(a + k_1, b - km)$ $= f(a, b)$

5.17 QUANTUM SEARCH ALGORITHMS

We begin the discussion of quantum search algorithms with a review of the relationship between classical probabilistic algorithms and quantum algorithms, followed by an abstract formulation of the search problem. Then, we outline the quantum search algorithm of Grover, present a circuit for this algorithm, and discuss a graphical interpretation of the state transformations required by the algorithm.

Probabilistic algorithms assume a certain probability distribution over the state space of a system, $\mathcal{S}$. In other words, instead of knowing with certainty that at time t the system is in state $j \in \mathcal{S}$ we know the probability distribution $p(t)$ and we can compute $p_j(t)$, the probability of the system being in state j at time t. The system is characterized now by the *state probability vector* $P(t) = (p_1(t), p_2(t), \ldots, p_j(t), \ldots)$ and by the *state transition matrix*, $\mathcal{T} = [\tau_{jk}]$ with τ_{jk}, $j, k \in \mathcal{S}$ being the probability of a transition from state j to state k. Clearly, $\sum_{j \in \mathcal{S}} p_j(t) = 1$ and $\sum_{k \in \mathcal{S}} \tau_{jk}(t) = 1$.

The evolution of the system in time is obtained by multiplying the state probability vector with the state transition matrix. If we know $P(t)$ at time $t = 0$, then we can calculate $P(t)$ for any t.

A quantum system is characterized by a state probability vector, but this characterization is incomplete, we also need to know the amplitude of each state. The amplitude is a complex number characterized by its magnitude and phase. The probability of a state is equal to the square of the absolute value of the amplitude of that state. The phase does not have any analogy for classical probabilistic algorithms.

To illustrate these ideas we consider two possible states of a qubit

$$| \phi_0 \rangle = (| 0 \rangle + | 1 \rangle)/\sqrt{2} \quad \text{and} \quad | \phi_1 \rangle = (| 0 \rangle - | 1 \rangle)/\sqrt{2}.$$

The probabilities of observing the two basis states $| \, 0\rangle$ and $| \, 1\rangle$ are the same for $| \, \phi_0\rangle$ and for $| \, \phi_1\rangle$

$$p_0 = p_1 = | \pm \frac{1}{\sqrt{2}} |^2 = \frac{1}{2}$$

even though the amplitude of $| \, \phi_1\rangle$ in state $| \, 1\rangle$ is $-1/\sqrt{2}$ which means that its phase is inverted.

The two states, $| \, \phi_0\rangle$ and $| \, \phi_1\rangle$, are obtained by a *fair coin flipping* of the two basis states, $| \, 0\rangle$ and $| \, 1\rangle$. To flip a fair coin means to apply the Walsh-Hadamard Transform H to the two basis states

$$| \, \phi_0\rangle = H \, | \, 0\rangle \qquad | \, \phi_1\rangle = H \, | \, 1\rangle$$

with

$$H = \frac{1}{\sqrt{2}} \begin{pmatrix} 1 & 1 \\ 1 & -1 \end{pmatrix}.$$

Both states $| \, \phi_0\rangle$ and $| \, \phi_1\rangle$ have the same probability distribution $(1/2, 1/2)$, but the amplitude vectors for the two states are different, $(1/\sqrt{2}, 1/\sqrt{2})$ and $(1/\sqrt{2}, -1/\sqrt{2})$, respectively. If we apply the Walsh-Hadamard Transform H to the state $| \, \phi_0\rangle$ the system ends up in the state $| \, 0\rangle$, while H applied to the state $| \, \phi_1\rangle$ leads to the state $| \, 1\rangle$.[10]

Now we consider a state vector in $\mathcal{H}_{2^n}$. The state transition matrix is of dimension $2^n \times 2^n$. The n qubits can be transformed by applying the Walsh-Hadamard Transform individually to each of them. When the system starts in state $| \, 00\ldots0\rangle$ the probability distribution for the 2^n states after the application of the Walsh-Hadamard Transform is

$$P_{|00\ldots0\rangle}(t = 0) = \left(\frac{1}{2^n}, \frac{1}{2^n}, \ldots, \frac{1}{2^n} \right)$$

all states are equally likely. The amplitude vector is

$$\left(\frac{1}{2^{n/2}}, \frac{1}{2^{n/2}}, \ldots, \frac{1}{2^{n/2}} \right).$$

If we start in any other state described by an arbitrary binary vector

$$| \, j\rangle = | \, b_0^j b_1^j \ldots b_l^j \ldots b_{n-1}^j\rangle \quad b_l^j \in \{0, 1\}$$

and if we apply the Walsh-Hadamard Transform to each qubit individually, then the system reaches the state

$$| \, k\rangle = | \, b_0^k b_1^k \ldots b_l^k \ldots b_{n-1}^k\rangle \quad b_l^k \in \{0, 1\}$$

[10]Indeed $HH = I$, thus $H \, | \, \phi_0\rangle = H(H \, | \, 0\rangle) = | \, 0\rangle$ and $H \, | \, \phi_1\rangle = H(H \, | \, 1\rangle) = | \, 1\rangle$.

and the amplitude vector is

$$\left(\pm\frac{1}{2^{n/2}}, \pm\frac{1}{2^{n/2}}, \ldots, \pm\frac{1}{2^{n/2}}\right).$$

If $(\mathbf{j} \cdot \mathbf{k})$ denotes the dot product of the binary vectors **j** and **k**, then the sign of the amplitude of state k is given by the parity of the inner product of **j** and **k**. Therefore, the amplitude of the state $| k\rangle$ when the initial state is $| j\rangle$ and we apply a Walsh-Hadamard Transform to each qubit individually is:

$$(-1)^{(\mathbf{j}\cdot\mathbf{k})}\frac{1}{2^{n/2}}.$$

Another transformation which preserves the state probability vector is the selective rotation of the phase of the amplitude in states $1, 2, 3, \ldots, 2^n$ with the angles $\varphi_1, \varphi_2, \varphi_3, \ldots, \varphi_{2^n}$, respectively. This transformation is described by the $2^n \times 2^n$ matrix

$$R = \begin{pmatrix} e^{i\varphi_1} & 0 & 0 & \ldots & 0 \\ 0 & e^{i\varphi_2} & 0 & \ldots & 0 \\ 0 & & e^{i\varphi_3} & \ldots & 0 \\ \vdots & & & & \\ 0 & 0 & 0 & \ldots & e^{i\varphi_{2^n}} \end{pmatrix}$$

with $i = \sqrt{-1}$ and $\varphi_1, \varphi_2, \varphi_3, \ldots, \varphi_{2^n}$ real numbers.

Now let us turn our attention to the parallel search algorithm. Consider a search space $\mathcal{S}_{search} = \{E_j\}$ consisting of $N = 2^n$ elements. Each element $E_j, 1 \leqslant j \leqslant 2^n$, is uniquely identified by a binary n-tuple, j, called *the index* of the element. We assume that $M \leqslant N$ elements satisfy the requirements of a query and we wish to identify *one of them*.

To process a query we repeatedly select an element E_j, decide if the element is a solution to the query, and if so, terminate the search. When the solution to the query is unique, $(M = 1)$, a classical search algorithm requires $\mathcal{O}(2^n)$ iterations.

To abstract this process we consider a function $f(x)$ with $0 \leqslant x \leqslant 2^n - 1$ such that

$$f(x) = \begin{cases} 0 & \text{if } x \text{ is not a solution} \\ 1 & \text{if } x \text{ is a solution.} \end{cases}$$

Consider an oracle, a black box accepting as input n qubits representing the index j of an element $E_j, 1 \leqslant j \leqslant 2^n$. The black box has also an *oracle qubit*, $| q\rangle$, initially set to $| 0\rangle$ and reset to $| 1\rangle$ when the oracle recognizes a solution to the search problem we pose. The black box performs the following transformation

$$| j\rangle \; | q\rangle \mapsto | j\rangle \; | q \oplus f(j)\rangle.$$

The oracle qubit can be initially in the state

$$| q\rangle = (| 0\rangle - | 1\rangle)/\sqrt{2}.$$

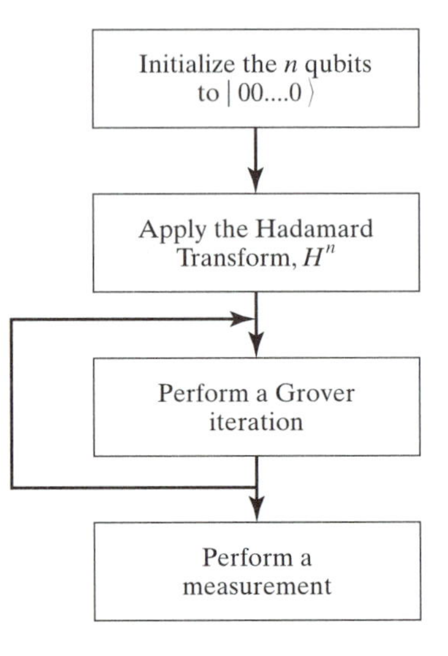

FIGURE 5.17: The flowchart of the quantum search algorithm. H^n denotes the n-dimensional Walsh-Hadamard Transform discussed in Appendix C.

If an input index of the element $E_j \in \mathcal{S}_{search}$, $| j \rangle$, is not a solution to our search problem, then the oracle qubit is unchanged. On the other hand, a solution to the search problem transforms $| 0 \rangle$ into $| 1 \rangle$. Thus, the transformation performed by the black box is

$$| j \rangle (| 0 \rangle - | 1 \rangle)/\sqrt{2} \mapsto \begin{cases} | j \rangle (| 0 \rangle - | 1 \rangle)/\sqrt{2} \text{ if } f(x) = 0 \\ \qquad \Longrightarrow E_j \text{ is not a solution} \\ -| j \rangle (| 0 \rangle - | 1 \rangle)/\sqrt{2} \text{ if } f(x) = 1 \\ \qquad \Longrightarrow E_j \text{ is a solution.} \end{cases}$$

This transformation can be rewritten as

$$| j \rangle (| 0 \rangle - | 1 \rangle)/\sqrt{2} \mapsto (-1)^{f(j)} | j \rangle (| 0 \rangle - | 1 \rangle)/\sqrt{2}.$$

The state of the oracle qubit does not change and can be omitted from the description of the quantum search algorithm. Thus, the transformation performed by the oracle on the input state $| j \rangle$ is:

$$| j \rangle \quad \mapsto \quad (-1)^{f(j)} | j \rangle.$$

Now we outline the algorithm used for the quantum search procedure in Figure 5.17. We start with the n qubits in state $| 00 \ldots 0 \rangle$ and apply an n-dimensional Walsh-Hadamard Transform to create an equal superposition state

$$| \psi \rangle = \frac{1}{\sqrt{2^n}} \sum_{j=0}^{2^n - 1} | j \rangle.$$

Then we carry out repeatedly the *Grover iteration*. The Grover iteration consists of the following four steps:

Step 1. Apply the oracle.

Step 2. Apply the n qubits resulting from the transformation to a circuit H^n performing an n-dimensional Walsh-Hadamard Transform.

Step 3. Perform a conditional phase shift

$$| i\rangle \quad \mapsto \quad (-1)^{f(j)} \mid j\rangle.$$

Step 4. Apply an n-dimensional Walsh-Hadamard Transform, H^n.

Let us now take a closer look at the Grover iteration.

Call G the transformation carried out as a result of one Grover iteration. Then G can be expressed as

$$G = (2 \mid \psi\rangle\langle\psi \mid - I)O = f_{|\psi\rangle} O.$$

In this expression O is the transformation carried out by the oracle. We now show that when applied to a state

$$| \varphi\rangle = \alpha_0 \mid 0\rangle + \alpha_1 \mid 1\rangle + \ldots + \alpha_{N-1} \mid N - 1\rangle = \sum_{k=0}^{N-1} \alpha_k \mid k\rangle$$

the transformation $f_{|\psi\rangle} = (2 \mid \psi\rangle\langle\psi \mid -I)$ produces the state

$$\begin{aligned} | \xi\rangle &= (2 \mid \psi\rangle\langle\psi \mid -I) \mid \varphi\rangle, \\ | \xi\rangle &= \beta_0 \mid 0\rangle + \beta_1 \mid 1\rangle + \ldots + \beta_{N-1} \mid N - 1\rangle \\ &= \sum_{k=0}^{N-1} \beta_k \mid k\rangle \end{aligned}$$

with $\beta_k = -\alpha_k + 2\overline{\alpha}$ where

$$\overline{\alpha} = \frac{1}{N} \sum_{k=0}^{N-1} \alpha_k.$$

Recall that the equal superposition obtained after applying the Walsh-Hadamard Transform to the initial state $| 00 \ldots 0\rangle$ is:

$$| \psi\rangle = \gamma_0 \mid 0\rangle + \gamma_1 \mid 1\rangle + \ldots + \gamma_{N-1} \mid N - 1\rangle = \sum_{k=0}^{N-1} \gamma_k \mid k\rangle$$

with $\gamma_k = 1/\sqrt{N}$ and $N = 2^n$.

The matrix representation of the transformation $f_{|\psi\rangle}$ is

$$2(|\psi\rangle\langle\psi| - I) = \begin{pmatrix} \frac{2}{N} - 1 & \frac{2}{N} & \dots & \frac{2}{N} & \dots & \frac{2}{N} \\ \frac{2}{N} & \frac{2}{N} - 1 & \dots & \frac{2}{N} & \dots & \frac{2}{N} \\ \vdots \\ \frac{2}{N} & \frac{2}{N} & \dots & \frac{2}{N} - 1 & \dots & \frac{2}{N} \\ \vdots \\ \frac{2}{N} & \frac{2}{N} & \dots & \frac{2}{N} & \dots & \frac{2}{N} - 1 \end{pmatrix}.$$

Now

$$|\xi\rangle = 2(|\psi\rangle\langle\psi| - I)|\varphi\rangle$$

$$= \begin{pmatrix} \alpha_0\left(\frac{2}{N} - 1\right) + \alpha_1\frac{2}{N} + \alpha_2\frac{2}{N} + \dots + \alpha_k\frac{2}{N} + \dots + \alpha_{N-1}\frac{2}{N} \\ \alpha_0\frac{2}{N} + \alpha_1\left(\frac{2}{N} - 1\right) + \alpha_2\frac{2}{N} + \dots + \alpha_k\frac{2}{N} + \dots + \alpha_{N-1}\frac{2}{N} \\ \vdots \\ \alpha_0\frac{2}{N} + \alpha_1\frac{2}{N} + \alpha_2\frac{2}{N} + \dots + \alpha_k(\frac{2}{N} - 1) + \dots + \alpha_{N-1}\frac{2}{N} \\ \vdots \\ \alpha_0\frac{2}{N} + \alpha_1\frac{2}{N} + \alpha_2\frac{2}{N} + \dots + \alpha_k\frac{2}{N} + \dots + \alpha_{N-1}(\frac{2}{N} - 1) \end{pmatrix}.$$

It follows that

$$|\xi\rangle = \sum_{k=0}^{N-1} \beta_k |k\rangle$$

with

$$\beta_k = \alpha_0\frac{2}{N} + \alpha_1\frac{2}{N} + \alpha_2\frac{2}{N} + \dots + \alpha_k(\frac{2}{N} - 1) + \dots + \alpha_{N-1}\frac{2}{N} = -\alpha_k + 2\overline{\alpha}.$$

The circuit implementing one iteration of Grover's algorithm is presented in Figure 5.18. An n-qubit register is used to specify the index of an item, $|j\rangle$, and a one-qubit register is needed for the oracle qubit $|q\rangle$. The n qubits produced by the oracle are applied to a circuit consisting of n `Hadamard` gates. The equal superposition state resulting from the H^n circuit is processed by a phase shift circuit, followed once more by a circuit consisting of n `Hadamard` gates. The phase shift leaves $|j\rangle$ unchanged when $f(j) = 0$ and transforms $|j\rangle \mapsto -|j\rangle$ when $f(j) > 0$. Figure 5.19 presents the circuit for $\mathcal{H}_4$ when the cardinality of the search space is $|\mathcal{S}| = 4$.

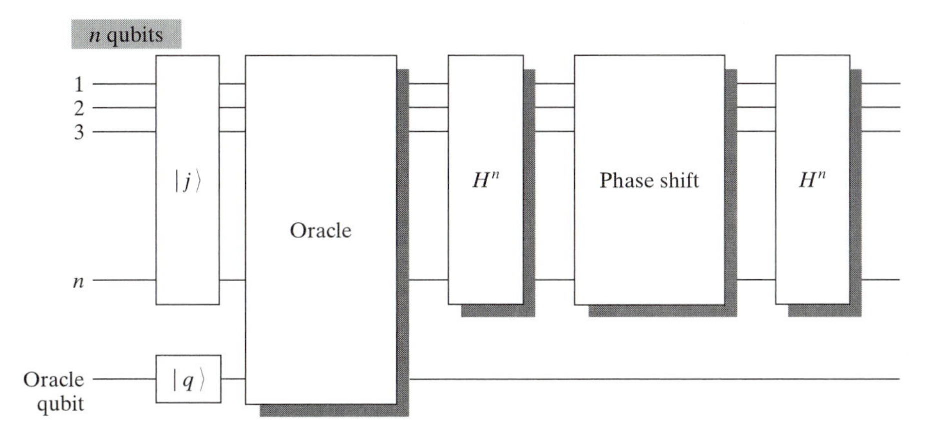

FIGURE 5.18: A quantum circuit for one iteration of Grover's algorithm. An n-qubit register is used to specify $| j\rangle$ and a one-qubit register is used for the oracle qubit $| q\rangle$. The oracle performs the transformation $| j\rangle \mapsto (-1)^{f(j)} | j\rangle$. The n qubits are applied to a circuit consisting of n Hadamard gates to create an equal superposition state, then to a phase shift circuit, and then once more to a circuit consisting of n Hadamard gates. The phase shift leaves $| j\rangle$ unchanged when $f(j) = 0$ and transforms $| j\rangle \mapsto - | j\rangle$ when $f(j) > 0$. The work qubits used by the oracle for intermediate results are omitted.

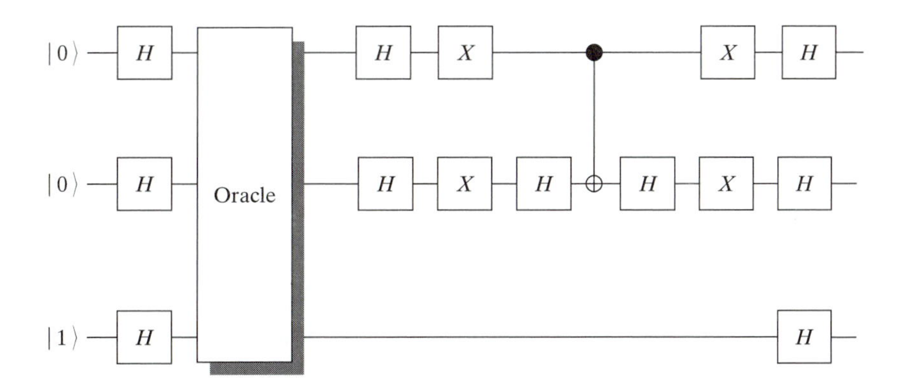

FIGURE 5.19: A quantum circuit performing the initial Walsh-Hadamard Transforms and one iteration of Grover's algorithm when the cardinality of the search space is $| \mathcal{S} |= 4$. The indexes can only be $00, 01, 10$, and 11. The top two qubits are initially in state $| 0\rangle$ and the third is in state $| 1\rangle$.

Let us now discuss a geometric interpretation of the Grover iteration. Let $| b\rangle$ represent the state corresponding to a uniform superposition of all solutions to the search problem and $| \psi\rangle$ the initial state. Then a Grover iteration is a rotation in the two-dimensional space spanned by $| b\rangle$ and $| \psi\rangle$, as shown in Figure 5.20(d).

Figures 5.20(a), (b), and (c) illustrate a reflection in two dimensions. Given two mirror lines $L1$ and $L2$ at an angle θ, we construct the reflection of $L2$ in $L1$ and call it $L2'$, see Figure 5.20(b). Then we construct the reflection of $L2'$ on $L2$

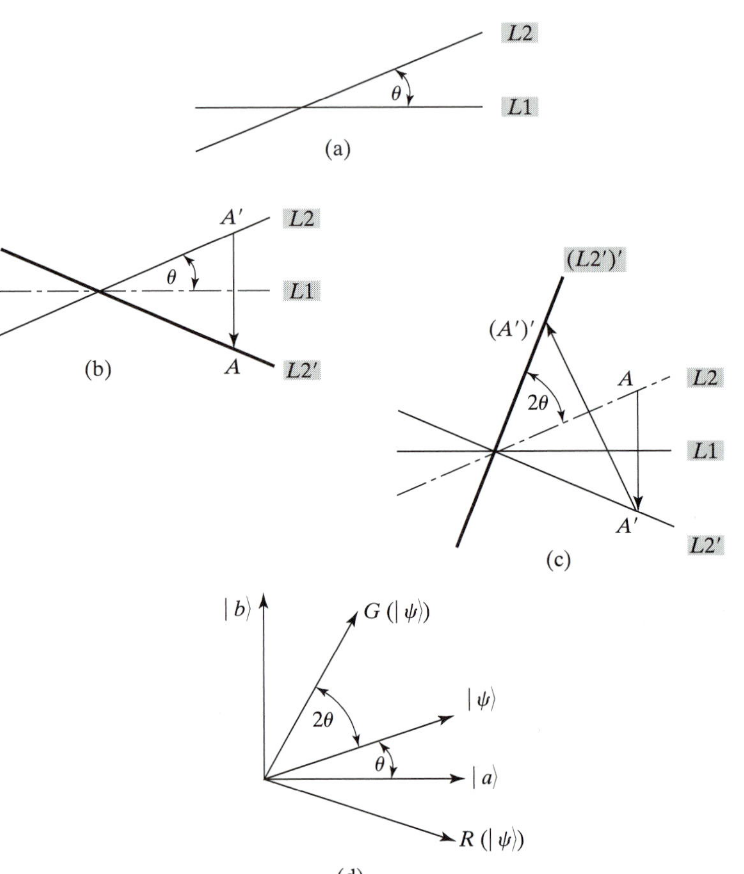

FIGURE 5.20: (a) Two lines $L1$ and $L2$ at an angle θ in a two-dimensional Euclidean plane. (b) $L2'$ is the reflection of line $L2$ about $L1$. (c) $(L2')'$ is the reflection of line $(L2)'$ about $L2$. The two successive reflections correspond to a rotation with an angle 2θ. (d) A graphical presentation of one iteration of Grover's search algorithm. A reflection about $| a\rangle$ transforms the original state $| \psi\rangle$ into $R(| \psi\rangle) = f_{|\psi\rangle}$. Then, this state is transformed into $G(| \psi\rangle)$ by a reflection about $| \psi\rangle$. These two transformations are equivalent to a rotation of the original state $| \psi\rangle$ with an angle 2θ. The state vectors, $| a\rangle$, $| b\rangle$, $| \psi\rangle$, $R(| \psi\rangle)$, and $G(| \psi\rangle)$ are of unit length.

and call it $(L2')'$, see Figure 5.20(c). Simple geometric arguments show that the angle of $(L2')'$ with $L1$ is 3θ. The product of two reflections in a two-dimensional plane is equivalent to a rotation in the plane spanned by the two vectors.

Now let $\mathbf{v}$ be a unit vector perpendicular to the line $L1$ in Figure 5.20(a) and $\mathbf{v}^{\perp}$ be a unit vector perpendicular to $\mathbf{v}$ and collinear with $L1$. Any vector ψ can be expressed as

$$\psi = \alpha_0 \mathbf{v}^{\perp} + \alpha_1 \mathbf{v}.$$

The reflection of ψ in $L1$ results in a vector

$$\psi_{L1} = \alpha_0 \mathbf{v}^{\perp} - \alpha_1 \mathbf{v}.$$

Consider now a complex space of higher dimension. The operation of the oracle to recognize a particular index j_0 can be described as a function which inverts the amplitude of the $| j_0\rangle$ component of an input index $| j\rangle$ to the quantum circuit in Figure 5.18:

$$O = f_{|j_0\rangle}(| j\rangle) = \begin{cases} | j\rangle & \text{if } | j\rangle \neq | j_0\rangle \\ -| j_0\rangle & \text{if } | j\rangle = | j_0\rangle \end{cases}.$$

We interpret f as a reflection in the hyperplane orthogonal to $| j_0\rangle$. In the case of a vector $| \psi\rangle$

$$f_{|\psi\rangle} = 2 | \psi\rangle\langle\psi | - I$$

with I the identity matrix. Jozsa [77] observes that any state $| \xi\rangle$ can be expressed as a sum of components parallel and orthogonal to $f_{|\psi\rangle}$ and proves a number of properties of $f_{|\psi\rangle}$:

1. Given any state $| \xi\rangle$, $f_{|\psi\rangle}(\xi)$ preserves the two-dimensional space spanned by $| \xi\rangle$ and $| \psi\rangle$.
2. For any unitary operator $\mathbf{U}$

$$\mathbf{U} f_{|\psi\rangle} \mathbf{U}^{-1} = f(\mathbf{U} | \psi\rangle).$$

To be more specific, instead of $\mathbf{v}^{\perp}$ and $\mathbf{v}$ we consider two orthogonal vectors $| a\rangle$ and $| b\rangle$ corresponding to two normalized states [98] such that $| b\rangle$ corresponds to a uniform superposition of all solutions to the search problem:

$$| a\rangle = \frac{1}{\sqrt{N - M}} \sum_{k \in \overline{Q}} | k\rangle$$

$$| b\rangle = \frac{1}{\sqrt{M}} \sum_{k \in Q} | k\rangle$$

with Q the set of indices that are solutions to the search problem and $\overline{Q}$ the set of indices that are not solutions to the search problem. Recall that $N = 2^n$ is the number of elements in the search space and M is the number of elements which satisfy the query.

Figure 5.20(d) depicts the transformations of the vector $| \psi\rangle$. The initial state $| \psi\rangle$ is in the space spanned by $| a\rangle$ and $| b\rangle$ and can be expressed as

$$| \psi\rangle = \alpha_0 | a\rangle + \alpha_1 | b\rangle = \sqrt{\frac{N - M}{N}} | a\rangle + \sqrt{\frac{M}{N}} | b\rangle.$$

Thus, we can denote

$$\alpha_0 = \cos\frac{\theta}{2} = \sqrt{\frac{N - M}{N}} = \sqrt{1 - \frac{M}{N}} \quad \text{and}$$

$$\alpha_1 = \sin\frac{\theta}{2} = \sqrt{\frac{M}{N}}.$$

The transformation of the state $| \psi\rangle$ has the following property

$$f_{(|\psi\rangle)} = f(\alpha_0 \mid a\rangle + \alpha_1 \mid b\rangle) = \alpha_0 \mid a\rangle - \alpha_1 \mid b\rangle.$$

Recall that the transformation $2 \mid \psi\rangle\langle\psi \mid -I$ represents a reflection in the plane defined by $| a\rangle$ and $| b\rangle$ about the vector $| \psi\rangle$ and that the product of two reflections is a rotation. This means that for any number m of Grover iterations *the resulting state* $G^m \mid \psi\rangle$ *remains in the plane spanned by* $| a\rangle$ *and* $| b\rangle$. We can also compute the rotation angle between the original vector $| \psi\rangle$ and $G(| \psi\rangle)$, the vector obtained after the two composite reflections. From Figure 5.20(d) we see that

$$G \mid \psi\rangle = \cos\frac{3\theta}{2} \mid a\rangle + \sin\frac{3\theta}{2} \mid b\rangle.$$

After m Grover iterations

$$G^m \mid \psi\rangle = \cos\left(\frac{2m+1}{2}\theta\right) \mid a\rangle + \sin\left(\frac{2m+1}{2}\theta\right) \mid b\rangle.$$

This graphical representation provides the intuition behind Grover iterations. *Repeated iterations bring the resulting vector* $G^m \mid \psi\rangle$ *closer and closer to* $| b\rangle$*, therefore produces with high probability one of the outcomes superposed by* $| b\rangle$.

The angle θ the state vector $| \psi\rangle$ is rotated with at one iteration towards the vector $| b\rangle$ representing a superposition of all solutions to the search problem, behaves like $\sqrt{\frac{M}{N}}$. Thus only, $\mathcal{O}\left(\sqrt{\frac{M}{N}}\right)$ iterations are required to rotate $| \psi\rangle$ close to $| b\rangle$, and this explains the efficiency of the algorithm.

5.18 HISTORICAL NOTES

In the early 1980s, Peter Benioff of IBM Research established the connection between quantum mechanics and computations [8, 9, 10]. He showed that quantum mechanics leads to a computational model as powerful as Turing Machines. A few years later, Feynman [54, 55] suggested that quantum mechanics might be even more powerful computationally than Turing Machines. He showed that simulating quantum mechanical systems on a classical computer is unfeasible. In 1980, Yurii Manin published a paper (in Russian) arguing along the same line (see [89]).

In 1985, David Deutsch recognized that a quantum computer has capabilities well beyond a classical computer and suggested that such capabilities can be exploited by cleverly crafted algorithms. Deutsch realized that a quantum computer can evaluate a function $f(x)$ for many values of x simultaneously and called this striking new feature *quantum parallelism* [42].

In 1992, Deutsch and Jozsa [43] and Berthiaume and Brassard [21, 22] showed that some problems that can only be solved with high probability using a random number generator, can be solved exactly with quantum computers. The next step was to show that there are problems that can be solved on a quantum computer in polynomial time, while only classical exponential time algorithms for solving these problems on a classical computer are known. This milestone was achieved by Peter Shor.

In 1994, Peter Shor [119] found a polynomial time algorithm for factorization of n-bit numbers on quantum computers.

5.19 SUMMARY AND FURTHER READINGS

Recent research in quantum computing and quantum information theory enforces our conviction that the power of quantum computing devices is greater than the computing power of classical computing devices. Therefore, problems such as factoring large integers or finding discrete logarithms for which no polynomial time algorithms are known can be solved in polynomial time by a quantum device.

This very dense chapter provides a concise introduction to quantum algorithms. The development of quantum algorithms requires a very different thinking than the one we are accustomed to for classical computers. A quantum algorithm must rely on superposition states and this is a foreign concept for those unfamiliar with quantum mechanics concepts.

There are three broad classes of quantum algorithms. In this chapter, we focus on algorithms that find the periodicity of a function using Fourier Transform methods and search algorithms; we only mention briefly algorithms for simulating quantum systems.

Quantum parallelism allows us to construct the entire truth table of a quantum gate array having 2^n entries, in a single time step. In a classical system we can compute the truth table in one time step with 2^n circuits running in parallel, or we need 2^n time steps with a single gate array. Quantum parallelism is best illustrated by the solution to Deutsch's problem.

To factor an integer N means to write N as a product of prime numbers. Efficient integer factorization is a very important problem because the security of widely used cryptographic protocols is based upon the conjectured difficulty of large integers factorization.

The expression $r^k = 1 \bmod N$ implies that $r^k - 1$ is divisible by N. This means that an integer m exists such that

$$r^k - 1 = (r - 1)(r^{k-1} + r^{k-2} + \dots + r^2 + r + 1) = mN.$$

Either $(r - 1)$ or $(r^{k-1} + r^{k-2} + \dots + r^2 + r + 1)$ share a common factor with N. Assume that $r - 1$ is such a factor, or shares a common factor with N. Then the greatest common divisor of N and $r - 1$ is indeed a proper factor of N. Euclid's algorithm discussed in Appendix B can be used to compute the greatest common divisor of two integers.

Like most factorization algorithms, Shor's algorithm reduces the factorization problem to the problem of finding the period of a function, but uses quantum parallelism to find a superposition of all values of the function in one step. Then, the algorithm calculates the QFT of the function, which sets the amplitudes into multiples of the fundamental frequency, the reciprocal of the period. To factor an integer, the Shor's algorithm calculates the period of the function.

The relationship between integer factorization and QFT is quite convoluted. Quantum integer factorization reduces to order finding, which in turn requires an algorithm for phase estimation, and this problem can be solved efficiently using the QFT.

The phase estimation is used by many quantum algorithms. In the general case, we need to perform the phase estimation procedure many times to solve problems such as integer factorization.

The best known algorithm for determining the period of a function takes exponential time on a classical computer. In 1994, Dan Simon created an algorithm for phase estimation that requires quadratic execution time on a quantum computer.

Another important quantum algorithm is the one to compute the logarithm of p. If n is the number of digits of p, then the fastest known classical algorithm to compute $y = \log_p(x)\ \forall x \in \mathbb{Z}_p^*$ runs in time $\mathcal{O}(\exp(n^{1/3}(\log n)^{2/3}))$ while the quantum algorithm requires less than $\mathcal{O}(n^3)$ time steps.

The essence of Grover's search algorithm is concentrated in Grover's iteration. The intuition behind Grover's iteration is captured by Figure 5.20. The operation of the oracle to recognize a particular index j_0 can be described as a function which inverts the amplitude of the $| j_0\rangle$ component

$$f_{|j_0\rangle}(| j\rangle) = \begin{cases} | j\rangle & \text{if } | j\rangle \neq | j_0\rangle \\ -| j_0\rangle & \text{if } | j\rangle = | j_0\rangle. \end{cases}$$

We consider two orthogonal vectors $| a\rangle$ and $| b\rangle$ corresponding to two normalized states

$$| a\rangle = \frac{1}{\sqrt{N - M}} \sum_{k \in \overline{Q}} | k\rangle$$

$$| b\rangle = \frac{1}{\sqrt{M}} \sum_{k \in Q} | k\rangle$$

with Q the set of indices that are solutions to the search problem and $\overline{Q}$ the set of indices that are not solutions to the search problem.

The transformation of the state $| \psi\rangle$ caused by a Grover iteration has the following property

$$f(\alpha_0 | a\rangle + \alpha_1 | b\rangle) = \alpha_0 | a\rangle - \alpha_1 | b\rangle.$$

The transformation $2 | \psi\rangle\langle\psi | - I$ represents a reflection in the plane defined by $| a\rangle$ and $| b\rangle$ about the vector $| \psi\rangle$ and that the product of two reflections is a rotation. This means that for any number m of Grover iterations the resulting state $G^m | \psi\rangle$ remains in the plane spanned by $| a\rangle$ and $| b\rangle$. After m Grover iterations

$$G^m | \psi\rangle = \cos\left(\frac{2m + 1}{2}\theta\right) | a\rangle + \sin\left(\frac{2m + 1}{2}\theta\right) | b\rangle.$$

Repeated iterations bring the resulting vector $G^m \mid \psi\rangle$ closer and closer to $\mid b\rangle$, therefore produces with high probability one of the outcomes superposed by $\mid b\rangle$. If the search space consists of M items and if there are $M < N$ solutions, we need $\mathcal{O}\left(\sqrt{\frac{N}{M}}\right)$ iterations to obtain a solution.

Several papers by Shor [125, 126], Jozsa [75, 79] and Ekert et. al. [51, 52] provide a very good introduction to the subject of quantum algorithms. A discussion of quantum Turing Machines can be found in the paper by Deutsch [42]. Shor's papers on integer factorization [119, 121, 123], Grover's papers on quantum search algorithms [67, 68, 69], and Simon's paper [127] on phase estimation are the most authoritative references for the respective subjects.

Several papers by Jozsa [76, 77, 78] are very helpful in understanding the Quantum Fourier Transform, the searching algorithms, the discrete logarithms, and the hidden subgroup problem.

5.20 EXERCISES AND PROBLEMS

5.1. Show that if $f(0) \neq f(1)$, then the output of the black box in Deutsch's problem is

$$\mid \xi_2\rangle = \begin{cases} \left[\frac{\mid 0\rangle - \mid 1\rangle}{\sqrt{2}}\right]\left[\frac{\mid 0\rangle - \mid 1\rangle}{\sqrt{2}}\right] \mapsto \frac{1}{2}\begin{pmatrix} 1 \\ -1 \\ -1 \\ 1 \end{pmatrix} & \text{if } f(0)=0 \quad \text{and} \quad f(1)=1 \\ -\left[\frac{\mid 0\rangle - \mid 1\rangle}{\sqrt{2}}\right]\left[\frac{\mid 0\rangle - \mid 1\rangle}{\sqrt{2}}\right] \mapsto -\frac{1}{2}\begin{pmatrix} 1 \\ -1 \\ -1 \\ 1 \end{pmatrix} & \text{if } f(0)=1 \quad \text{and} \quad f(1)=0 \end{cases}$$

5.2. In 1995, Griffiths and Niu [66] proposed a network of only one-qubit gates for performing Quantum Fourier Transforms. Analyze and discuss the merits of the solution proposed by the two researchers.

5.3. Construct a quantum circuit to perform the Quantum Fourier Transform on four qubits. Calculate its transfer matrix.

5.4. Construct the permutation matrix for the QFT circuit from the previous problem.

5.5. Read the paper *A Pseudo-Simulation of Shor's Quantum Factoring Algorithm* by J.F. Schneiderman, M. E. Stanley, and P.K. Aravind and construct a Java program able to simulate Shor's algorithm.

5.6. Given three integers a, b, c prove the following identities

$$a \cdot b = \gcd(a, b) \cdot \text{lcm}(a, b) \quad \text{if } a, b \geqslant 0$$

$$\gcd[(a, b) \cdot c] = \gcd(a \cdot c, b \cdot c) \quad \text{if } c \geqslant 0$$

$$\text{lcm}[(a, b) \cdot c] = \text{lcm}(a \cdot c, b \cdot c) \quad \text{if } c \geqslant 0$$

$$\gcd[a, \gcd(b, c)] = \gcd[b, \gcd(a, c)] = \gcd[\gcd(a, b), c]$$

$$\text{lcm}[a, \gcd(b, c)] = \text{lcm}[b, \gcd(a, c)] = \text{lcm}[\gcd(a, b), c]$$

$$\gcd[\text{lcm}(a, b), \text{lcm}(a, c)] = \text{lcm}[a, \gcd(b, c)]$$

$$\text{lcm}[\gcd(a, b), \gcd(a, c)] = \gcd[a, \text{lcm}(b, c)].$$

In these expressions, $\gcd(a, b)$ denotes the greatest common divisor of integers a and b while $\text{lcm}(a, b)$ the least common multiple of the two integers (see Appendix B).

5.7. Compute the order of 13 mod 25.

5.8. Given a set of "moduli" $\{m_1, m_2, \ldots, m_r\}$ that have no common factors, we can represent the integer a as $a = (a_1, a_2, \ldots, a_{r-1}, a_r)$ using its residues

$$a_1 = a \bmod m_1, a_2 = a \bmod m_2, \ldots, a_r = a \bmod m_r.$$

Show that if we have two integers a and b with the internal representation

$$a = (a_1, a_2, \ldots, a_{r-1}, a_r)$$

and

$$b = (b_1, b_2, \ldots, b_{r-1}, b_r)$$

then the following three equations are true

$$\begin{aligned} a + b &= (a_1, a_2, \ldots, a_{r-1}, a_r) + (b_1, b_2, \ldots, b_{r-1}, b_r) \\ &= ((a_1 + b_1) \bmod m_1, (a_2 + b_2) \bmod m_2, \\ &\quad \ldots, (a_{r-1} + b_{r-1}) \bmod m_{r-1}, (a_r + b_r) \bmod m_r). \\ a - b &= (a_1, a_2, \ldots, a_{r-1}, a_r) - (b_1, b_2, \ldots, b_{r-1}, b_r) \\ &= ((a_1 - b_1) \bmod m_1, (a_2 - b_2) \bmod m_2, \\ &\quad \ldots, (a_{r-1} - b_{r-1}) \bmod m_{r-1}, (a_r - b_r) \bmod m_r. \\ a \cdot b &= (a_1, a_2, \ldots, a_{r-1}, a_r) \cdot (b_1, b_2, \ldots, b_{r-1}, b_r) \\ &= ((a_1 \cdot b_1) \bmod m_1, (a_2 \cdot b_2) \bmod m_2, \\ &\quad \ldots, (a_{r-1} \cdot b_{r-1}) \bmod m_{r-1}, (a_r \cdot b_r) \bmod m_r. \end{aligned}$$

(`Hint:` $a \bmod m_i = b \bmod m_i$ if and only if $a = b$.)

5.9. Let $\mathbf{d}$ be a binary q-tuple and let $\mathbf{c}$ be a binary 2^q-tuple. Let $\mathbf{R}(\mathbf{c}) = (-1)^{\mathbf{c}(\mathbf{b})}$ be a 2^q-tuple with entries either $+1$, or -1 as defined earlier. Then the *Walsh-Hadamard Transform* of $\mathbf{R}(\mathbf{c})$ is

$$\hat{\mathbf{R}}(\mathbf{d}) = \sum_{\mathbf{b} \in B_q} (-1)^{\mathbf{d} \cdot \mathbf{b}} \mathbf{R}(\mathbf{c})$$

or

$$\hat{\mathbf{R}}(\mathbf{d}) = \sum_{\mathbf{b} \in B_q} (-1)^{\mathbf{d} \cdot \mathbf{b} + \mathbf{c}(\mathbf{b})}.$$

Show that a direct computation of the Walsh-Hadamard Transform requires $2^q(2^{q+1} - 1)$ operations. (`Hint:` Multiplication of $\mathbf{R}$ with a column of H requires 2^q multiplications and $2^q - 1$ additions.)

5.10. Prove that $\hat{\mathbf{R}}$ is the number of 0s minus the number of 1s in $\mathbf{t} = \mathbf{c} + \sum_{i=1}^{q} d_i v_i$. (`Hint`: Consult Appendix C for the discussion of the Walsh-Hadamard Transform.)

5.11. Show that a computation of the Fast Hadamard Transform requires $3 \times 2^q \times q$ operations. (`Hint`: to compute the product $\mathbf{R}M_{2^q}^{(1)}$ we need 2 multiplications and one addition for each of the 2^q columns. Consult Appendix C for the discussion of the Walsh-Hadamard Transform.)

5.12. Factor the integer 39 following the algorithm described Figure 5.11.

5.13. Figure 5.19 shows a quantum circuit performing the initial Walsh-Hadamard Transform and one iteration of Grover's algorithm when the cardinality of the search space is $\mid \mathcal{S} \mid = 4$. Illustrate how the search algorithm works in this particular case and construct the quantum circuit for the oracle.

5.14. How many iterations of the algorithm are necessary to find the solution for the previous problem?

CHAPTER 6

The "Entanglement" of Computing and Communication with Quantum Mechanics—Reversible Computations

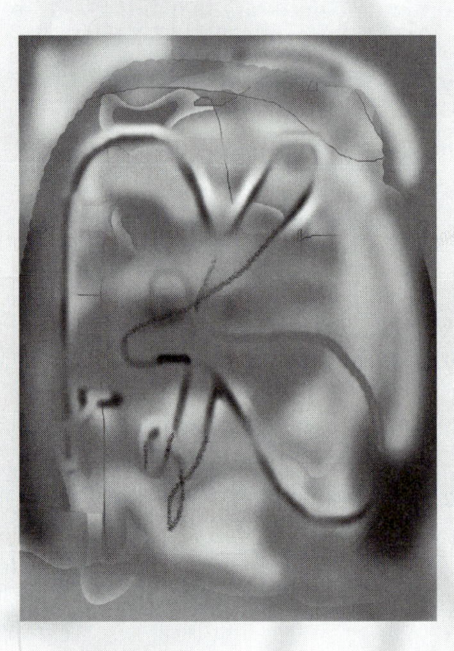

"Entangled" in abstractions, we have moved further and further from the physical realities. Let us now turn back from abstractions to properties of matter. We are concerned with the relationships between the wonderful abstractions used by the theoretical computer science and the physical systems surrounding us. We want to understand the physical support of information and determine the energy required to store and to transform information. We ask ourselves what is the relationship between information and energy and matter.

In the first part of this chapter, we switch our focus from quantum computing to quantum communication. We discuss applications of communication involving quantum particles such as teleportation, dense coding, and quantum key distribution. Then, in the second part of this chapter we address the problem of reversible computations.

In a science fiction context teleportation means: making an object or person disintegrate in one place and have it reembodied as the same object or person somewhere else. In the context of quantum information theory teleportation means:[1] "a way to scan out part of the information from an object A, which one wishes to teleport, while causing the remaining, unscanned part of the information to pass, via the Einstein-Podolsky-Rosen effect, into another object C which has never been in contact with A. Later, by applying to C a treatment depending on the scanned-out information, it is possible to maneuver C into exactly the same state as A was in before it was scanned." In this process, the original state is destroyed.

Dense coding allows us to transmit two bits of classical information by transmitting a single qubit. This form of data compression requires a classical and a quantum communication. Quantum key distribution enables us to detect with a very high probability the presence of an eavesdropper trying to intercept encryption keys sent over a quantum communication channel.

The remarkable properties of communication involving quantum states exploit the properties of ensembles of entangled particles. To fully understand the physical phenomena involved in communication with quantum particles, we revisit the subject of entanglement. The entanglement of quantum particles seems to violate the principle of locality and to allow instant interactions among distant agents, in blatant violation of the limits imposed by the finite speed of light.

We also address the problem of the measurements of quantum states and, in this context, we discuss briefly Bell's inequality. Bell's inequality is derived based upon common sense assumptions, the principles of locality and realism. We show that quantum systems do not obey Bell's inequality and conclude that locality is probably not a sound assumption.

Then we turn our attention to the problem of energy consumption during a computation. In the previous chapter, we discussed models of computations and computing machines that abstract properties of the thought process necessary to solve a problem automatically. Turing Machines and effective procedures are familiar examples of such models. These models view a computation as a unidirectional process taking a system from an initial state to a final state, transforming some input data into results. As such, a computation is an irreversible process inexorably condemned to consume energy, in strong contrast with physical processes which are reversible.

According to Landauer's Principle when a computer erases a bit of information the amount of energy dissipated into the environment is at least $k_B T \ln(2)$. Here k_B is the Boltzmann's constant and T is the temperature of the environment.

---

[1] Definition from http://www.research.ibm.com/quantuminfo/teleportation.

6.1 COMMUNICATION, ENTROPY, AND QUANTUM INFORMATION

In the late 1940s, Claude Shannon of Bell Labs developed an abstract model of communication. In this model three entities a source of information, a communication channel, and a receiver interact with each other. The source of information and the receiver share a common alphabet, $\mathcal{A}$, a finite set of symbols that can be transmitted over the channel (e.g., the English alphabet). The symbols of the alphabet $\mathcal{A} = \{\alpha_1, \alpha_2, \ldots, \alpha_n\}$ are selected according to probabilities $p_1, p_n, \ldots, p_n$.

Shannon wanted to define a measure of the quantity of information a source could generate. Earlier, in 1927, another fellow scientist of Bell Labs, Ralph Hartley, had proposed to take the logarithm of the total number of possible messages as a measure of the amount of information in a message, arguing that the logarithm tells us how many digits or characters are required to convey the message.

Shannon took a hint from classical thermodynamics (see Section 6.12) and decided to relate the information content with the probabilities of a message. His motivation is intuitive: assume that Bob wants to transmit a message to Alice with the information that "no meteorite struck Tibet in the last 24 hours." A meteorite striking the Earth is a very rare event, thus Bob's message conveys little information. It only enforces our expectation that a rare event did not happen. On the other hand, the message that "an earthquake struck Tibet in the last 24 hours," contains a larger amount of information because it reveals that another rare event, an earthquake, did occur.

The entropy is a measure of the uncertainty of a single random variable X before it is observed. It can also be considered the average uncertainty removed by observing the outcome of a random process. This quantity is called entropy due to its similarity to the thermodynamic entropy.

The *entropy* of a random variable X with a probability density function $p_X(x)$ is

$$H(X) = -\sum_x p_X(x) \cdot \log_2 p_X(x).$$

The entropy of a random variable is a positive real number. Indeed, the probability is a positive real number between 0 and 1. Therefore, $\log_2 p_X(x) \leqslant 0$ and $H(X) \geqslant 0$.

A binary random variable X can only take two values, $x = 0$ and $x = 1$. The entropy of a binary random variable is measured in bits.

Consider a binary random variable X and let $p = p_X(x = 1)$ be the probability that the random variable X assumes the value 1. Then the entropy of X is

$$H(X) = -p \cdot \log_2(p) - (1 - p) \cdot \log_2(1 - p).$$

Table 6.1 shows some values for $H(X)$ for $0.0001 \leqslant p \leqslant 0.5$. The probability that the random variable X assumes the value 1 is $p = p_X(x = 1)$. Figure 6.1 shows $H(X)$ function of p. The entropy has a maximum of 1 bit when $p = 1/2$ and goes to zero when $p = 0$ or $p = 1$. Intuitively, we expect the entropy to be

TABLE 6.1: $H(X)$, the entropy of a binary random variable X for $0.0001 \leqslant p \leqslant 0.5$

p	$H(X)$
0.0001	0.001
0.001	0.011
0.01	0.081
0.1	0.469
0.2	0.722
0.3	0.881
0.4	0.971
0.5	1.000

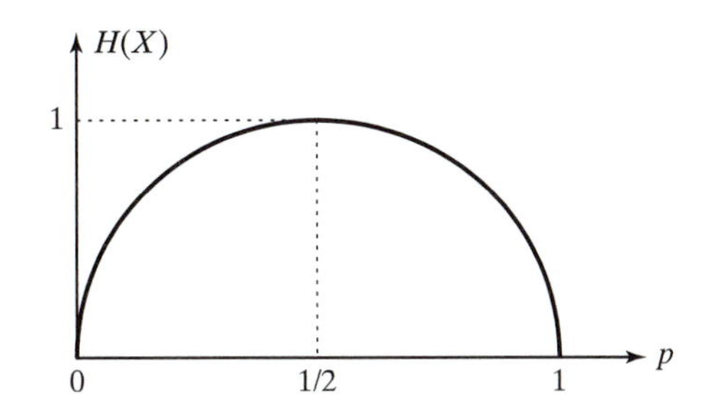

FIGURE 6.1: $H(X)$, the entropy of a binary random variable X function of p, the probability of an outcome. The entropy has the following properties: (i) $H(X) > 0$ for $0 < p < 1$; (ii) $H(X)$ is symmetric about $p = 0.5$; (iii) $\lim_{p \to 0} H(X) = \lim_{p \to 1} H(X) = 0$; (iv) $H(X)$ is increasing for $0 < p < 0.5$, decreasing for $0.5 < p < 1$ and has a maximum for $p = 0.5$.

zero when the outcome is certain and reach its maximum when both outcomes are equally likely.

The following example provides additional insight into the relationship between the entropy and the amount of data needed to convey information. A group of eight cars takes part in several Formula I races. The probabilities of winning for each of the eight cars are respectively

$$\frac{1}{2}, \frac{1}{4}, \frac{1}{8}, \frac{1}{16}, \frac{1}{64}, \frac{1}{64}, \frac{1}{64}, \frac{1}{64}.$$

The entropy of the random variable X indicating the winner after one of the races is

$$H(X) = -\frac{1}{2} \cdot \log_2 \frac{1}{2} - \frac{1}{4} \cdot \log_2 \frac{1}{4} - \frac{1}{8} \cdot \log_2 \frac{1}{8} - \frac{1}{16} \cdot \log_2 \frac{1}{16} - \frac{4}{64} \cdot \log_2 \frac{1}{64} = 2 \text{ bits}.$$

If we want to send a binary message to reveal the winner of a particular race, we could encode the identity of the winning car in several ways. For example, we can use three bits and encode the identity of each car as the binary representation of integers 0 to 7 as, 000, 001, 010, 011, 100, 101, 110, and 111.

An *optimal encoding* means that the average number of bits transmitted is minimal. Given the individual probability of winning a race the optimal encoding of the identities of individual cars is

$$0, 10, 110, 1110, 111100, 111101, 111110, 111111$$

and the corresponding lengths of the strings encoding the identity of each car are

$$l_1 = 1, l_2 = 2, l_3 = 3, l_4 = 4, l_5 = l_6 = l_7 = l_8 = 6.$$

To prove that this encoding is optimal we have to show that the expected length of the string designating the winner over a large number of car races is less than the "obvious" encoding presented previously. The probabilities of winning the race are: $p_1 = \frac{1}{2}$ for the car encoded as 0; $p_2 = \frac{1}{4}$ for 10, and so on. The expected value of the length of the string we have to send to communicate the winner is

$$\mathbf{E}(l) = \sum_{i=1}^{8} l_i \cdot p_i = 1 \cdot \frac{1}{2} + 2 \cdot \frac{1}{4} + 3 \cdot \frac{1}{8} + 4 \cdot \frac{1}{16} + 4 \cdot \left(6 \cdot \frac{1}{64}\right) = 2 \text{ bits}.$$

The expected length of the string identifying the winner of a race for this particular encoding scheme is equal to the entropy. This example shows that indeed the entropy provides the average information obtained by observing an outcome, or the average uncertainty removed by observing X.

The applications of quantum effects to communication revolve around yet another primitive concept which will be discussed in depth in the second volume and cannot be rigourously defined, *quantum information*. We have already accepted the fact that a quantum state $| \psi \rangle$ is a carrier of information and transformations of the quantum state performed by quantum circuits imply transformation of information.

The quantum state $| \psi \rangle$ may be viewed as a carrier of classical information, namely the *identity of the quantum state*. The classical problem posed by a noisy communication channel with the input alphabet $\alpha_1, \alpha_2, \ldots, \alpha_n$ is that the receiver should correctly identify the actual symbol sent when the probability of incorrectly decoding the input symbol α_i as α_j is non-zero, $p_{ji} \neq 0$.

In this case, the sender of information encodes the two binary values $i = 0$ and $i = 1$ as two non-orthogonal states $| \psi_0 \rangle$ and $| \psi_1 \rangle$. The receiver is expected to determine the value of i from $| \psi_i \rangle$. Call $p_{j|i}$ the probability that the receiver decodes the state $| \psi_i \rangle$ as $| \psi_j \rangle$ and p_i the probability that the state $| \psi_i \rangle$ was sent. The task of the receiver is to minimize the probability of an error, namely

$$p_0 p_{1|0} + p_1 p_{0|1}.$$

In this expression, p_0 is the probability that the sender sends the symbol 0 and $p_{1|0}$ is the probability that the receiver incorrectly decodes a 0 as a 1.

If $\langle \psi_i \mid \psi_j \rangle \neq 0$ the minimum error probability cannot be zero and no physical process allows a perfect, errorless identification of the symbol sent.

We can view the quantum state $| \psi\rangle$ as a carrier of quantum information. *Quantum information is a concept considerably more profound than the ability to identify, or label a state.*

6.2 INFORMATION ENCODING

Communication between a sender and a receiver involves a mapping called *encoding* done at the sender's site and an inverse mapping called *decoding* done at the receiver's site. The encoding transforms an original message into an encoded one, and the decoding, done at the receiving end of the communication channel, restores the original message.

The problem of transforming, repackaging, or encoding information is a major concern in modern communication. Encoding is used to:

1. make transmission resilient to errors,
2. reduce the amount of data transmitted through a communication channel, and
3. ensure information confidentiality.

A first reason for information encoding is *error control.* Information transmitted over a noisy communication channel may be altered, or even lost. The *error control* mechanisms transform a noisy communication channel into a noiseless one; they are built into communication protocols to eliminate the effect of transmission errors. An error occurs when an input symbol is distorted during transmission and interpreted at the destination as another symbol from the alphabet. *Coding theory* is concerned with this aspect of information theory.

Another reason for encoding is the desire to reduce the amount of data transferred through communication channels and to eliminate redundant or less important information. The discipline covering this type of encoding is called *data compression.* In this case, encoding and decoding are called *data compression* and *data decompression*, respectively.

Last, but not least, we want to ensure information confidentiality, to restrict access to information to only those individuals who have the proper authorization. The discipline covering this facet of encoding is called *cryptography* and the processes of encoding and decoding are called *encryption and decryption*, respectively.

6.3 QUANTUM TELEPORTATION WITH MAXIMALLY ENTANGLED PARTICLES

Can we transport quantum states? This intriguing question was posed relatively early by scientists involved in quantum computing and communication. Many decades earlier, the science-fiction literature had coined the term teleportation meaning the instantaneous transport of matter from one location to another.

In 1992, a group of scientists were discussing the impact of entanglement upon information transmission with application to the distribution of encryption keys. An embellished version of the problem they were attempting to solve follows. Assume that Alice and Bob are given as a wedding present a pair of entangled particles called "particle 1" and "particle 2". After several years, Bob alone takes part in an expedition on K2 and takes "particle 2" with him, while Alice remains in London to take care of their newborn infant and keeps "particle 1" with her. A third party, Carol, asks Alice to deliver a secret message to Bob. The message is encoded as the state of "particle 3."

Alice cannot send the quantum state of "particle 3" directly over a quantum communication channels being afraid that in the process the state might be altered. She is looking for alternative means to transfer the quantum state.

Someone confined to classical thinking might suggest to Alice to perform a measurement of "particle 3" and deliver the information to Bob via a classical communication channel. Then Bob could reconstruct the quantum state by manipulating a particle similar to "particle 3". This is not going to work because Alice can only get partial information about the state of "particle 3" as a result of a measurement. For example, let us assume that the information is in the polarization of a photon. Alice may be given a photon polarized at 45° and, not knowing the orientation, she might measure for horizontal polarization. In this case, Alice would not only get the wrong answer, but she will alter the quantum state of the photon.

The solution for Alice is to perform a *joint measurement* on her own half of the EPR pair, "particle 1" and on the particle given by Carol, "particle 3." Then she sends to Bob over a classical communication channel the result of her measurement. At his end, upon receiving Alice's results, Bob performs upon his own particle, "particle 2," one of the four types of transformations, in particular, the one communicated by Alice. Bob transforms his qubit using an I, X, Y, or a Z gate. The last three transformations are in fact 180° rotations around one of the x, y, z axes. As a result of these transformations "particle 2" will be a perfect replica of "particle 3."

At first sight, this seems to violate the principle of no-cloning discussed in Chapter 5. A closer analysis shows that when Alice measures the joint state of "particle 1" and "particle 3," the state of "particle 3" is altered; thus, the original state of the particle given to Alice by Carol is destroyed.

In summary, Alice is able to transfer the quantum state, *not the actual particle* and to send only classical information to Bob. The transfer of quantum information appears to happen instantly, though Bob needs to first receive classical information regarding the result of Alice's measurement before validating his own result.

Let us now dissect the process described above to fully understand the transformations of quantum states involved in teleportation. As we have seen earlier there are several entangled states of two particles. Let us assume first that the particles in the possession of Alice and Bob ("particle 1" and "particle 2") are in the maximally entangled state

$$| \beta_{00} \rangle = \frac{| 00 \rangle + | 11 \rangle}{\sqrt{2}}$$

and that "particle 3" is in state

$$| c \rangle = \alpha_0 | 0 \rangle + \alpha_1 | 1 \rangle \text{ with } | \alpha_0 |^2 + | \alpha_1 |^2 = 1.$$

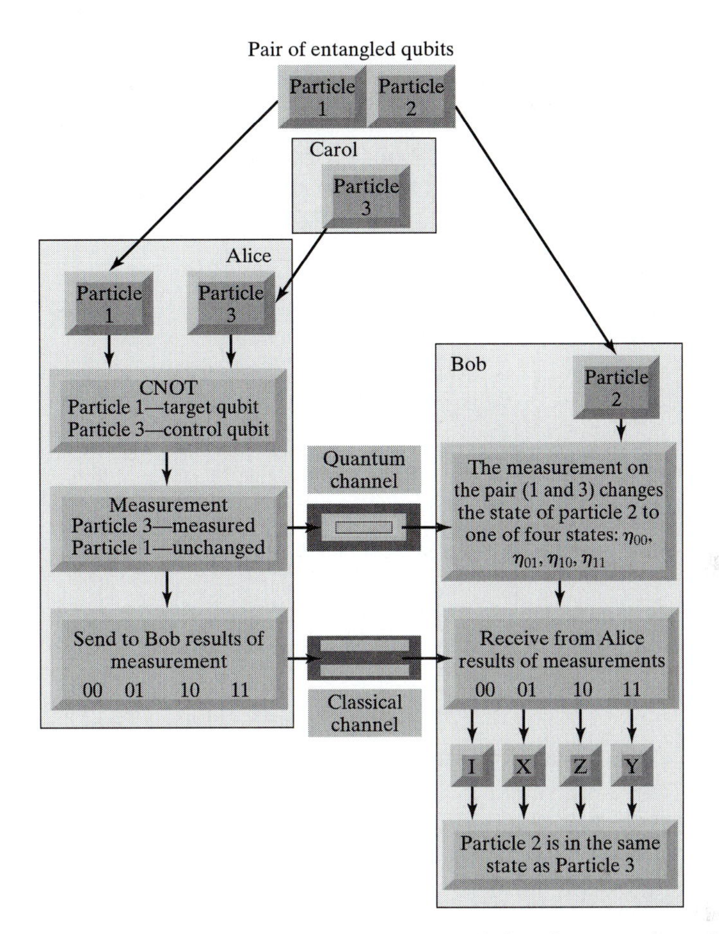

FIGURE 6.2: Carol wants to send Bob a secret message encoded as the state of "particle 3." Alice and Bob share a pair of entangled particles, "particle 1" and "particle 2." Carol gives to Alice "particle 3." Alice performs a CNOT on the pair consisting of "particle 1" and "particle 3," using "particle 3" as the control qubit and "particle 1" as the target qubit. Alice then measures "particle 1" and sends over the classical communication channel the result of the measurement, 00,01,10, or 11. Then Bob transforms the state of "particle 2" by applying one of the four transformations presented in the discussion of the algorithm.

The joint state of "particle 3" and "particle 1", $| \xi\rangle$, is a vector in $\mathcal{H}_8$ because "particle 1" is entangled with "particle 2"

$$| \xi\rangle = | c\rangle \otimes | \beta_{00}\rangle = \begin{pmatrix} \alpha_0 \\ \alpha_1 \end{pmatrix} \otimes \frac{1}{\sqrt{2}} \begin{pmatrix} 1 \\ 0 \\ 0 \\ 1 \end{pmatrix} = \frac{1}{\sqrt{2}} \begin{pmatrix} \alpha_0 \\ 0 \\ 0 \\ \alpha_0 \\ \alpha_1 \\ 0 \\ 0 \\ \alpha_1 \end{pmatrix},$$

or

$$| \xi\rangle = 1/\sqrt{2}(\alpha_0 \,| 000\rangle + \alpha_0 \,| 011\rangle + \alpha_1 \,| 100\rangle + \alpha_1 \,| 111\rangle).$$

Alice carries out two operations on the two qubits in her possession, the particle from Carol and her own half of the entangled pair:

1. Alice applies a `CNOT` to the pair; she uses Carol's qubit as a control and her own as a target. She applies the $G_{CNOT} \otimes I$ to the state $\mid \xi\rangle$ and transforms the qubits to a new state $\mid \kappa\rangle$.

$$\mid \kappa\rangle = (G_{CNOT} \otimes I)(\mid \xi\rangle) = \begin{pmatrix} 1&0&0&0 \\ 0&1&0&0 \\ 0&0&0&1 \\ 0&0&1&0 \end{pmatrix} \otimes \begin{pmatrix} 1&0 \\ 0&1 \end{pmatrix} \mid \xi\rangle,$$

$$\mid \kappa\rangle = \begin{pmatrix} 1&0&0&0&0&0&0&0 \\ 0&1&0&0&0&0&0&0 \\ 0&0&1&0&0&0&0&0 \\ 0&0&0&1&0&0&0&0 \\ 0&0&0&0&0&0&1&0 \\ 0&0&0&0&0&0&0&1 \\ 0&0&0&0&1&0&0&0 \\ 0&0&0&0&0&1&0&0 \end{pmatrix} \frac{1}{\sqrt{2}} \begin{pmatrix} \alpha_0 \\ 0 \\ 0 \\ \alpha_0 \\ \alpha_1 \\ 0 \\ 0 \\ \alpha_1 \end{pmatrix} = \frac{1}{\sqrt{2}} \begin{pmatrix} \alpha_0 \\ 0 \\ 0 \\ \alpha_0 \\ 0 \\ \alpha_1 \\ \alpha_1 \\ 0 \end{pmatrix}.$$

Thus,

$$\mid \kappa\rangle = 1/\sqrt{2}(\alpha_0 \mid 000\rangle + \alpha_0 \mid 011\rangle + \alpha_1 \mid 101\rangle + \alpha_1 \mid 110\rangle).$$

2. Alice measures the first qubit and leaves the second (the one entangled with Bob's) untouched. This means that she applies to $\mid \kappa\rangle$ the transformation $H \otimes I \otimes I$ and transforms the state of the qubit to $\mid \xi\rangle$:

$$\mid \zeta\rangle = (H \otimes I \otimes I) \mid \kappa\rangle = \frac{1}{\sqrt{2}} \begin{pmatrix} 1&1 \\ 1&-1 \end{pmatrix} \otimes \begin{pmatrix} 1&0 \\ 0&1 \end{pmatrix} \otimes \begin{pmatrix} 1&0 \\ 0&1 \end{pmatrix} \mid \kappa\rangle,$$

$$\mid \zeta\rangle = \frac{1}{\sqrt{2}} \begin{pmatrix} 1&0&0&0&1&0&0&0 \\ 0&1&0&0&0&1&0&0 \\ 0&0&1&0&0&0&1&0 \\ 0&0&0&1&0&0&0&1 \\ 1&0&0&0&-1&0&0&0 \\ 0&1&0&0&0&-1&0&0 \\ 0&0&1&0&0&0&-1&0 \\ 0&0&0&1&0&0&0&-1 \end{pmatrix} \frac{1}{\sqrt{2}} \begin{pmatrix} \alpha_0 \\ 0 \\ 0 \\ \alpha_0 \\ 0 \\ \alpha_1 \\ \alpha_1 \\ 0 \end{pmatrix} = \frac{1}{2} \begin{pmatrix} \alpha_0 \\ \alpha_1 \\ \alpha_1 \\ \alpha_0 \\ \alpha_0 \\ -\alpha_1 \\ -\alpha_1 \\ \alpha_0 \end{pmatrix}.$$

Thus,

$$\begin{aligned} \mid \zeta\rangle = & \frac{1}{2}[\alpha_0(\mid 000\rangle + \mid 011\rangle + \mid 100\rangle + \mid 111\rangle) \\ & +\alpha_1(\mid 001\rangle + \mid 010\rangle - \mid 101\rangle - \mid 110\rangle)]. \end{aligned}$$

We can rewrite the expression for $\mid \zeta\rangle$ and isolate the first two qubits in the expression for the new state

$$\begin{aligned} \mid \zeta\rangle = & 1/2[\mid 00\rangle(\alpha_0 \mid 0\rangle + \alpha_1 \mid 1\rangle) + \mid 01\rangle(\alpha_0 \mid 1\rangle) + \alpha_1 \mid 0\rangle) \\ & + \mid 10\rangle(\alpha_0 \mid 0\rangle - \alpha_1 \mid 1\rangle) + \mid 11\rangle(\alpha_0 \mid 1\rangle - \alpha_1 \mid 0\rangle)]. \end{aligned}$$

From this expression it follows that when Alice performs a joint measurement of her two qubits, she gets one of the four equally probable results $| \, 00\rangle$, or $| \, 01\rangle$, or $| \, 10\rangle$, or $| \, 11\rangle$. As we already know, the measurement forces the pair of qubits to one of the four basis states, and transforms quantum information into classical one. Then, she sends Bob the result of her measurement, 00, 01, 10, or 11, over a classical communication channel.

At the same time, the measurement performed by Alice forces the qubit in Bob's possession to change to one of four states:

1. $| \, \eta_{00}\rangle = \alpha_0 \, | \, 0\rangle \, + \, \alpha_1 \, | \, 1\rangle$ when the result is 00.
2. $| \, \eta_{01}\rangle = \alpha_0 \, | \, 1\rangle \, + \, \alpha_1 \, | \, 0\rangle$ when the result is 01.
3. $| \, \eta_{10}\rangle = \alpha_0 \, | \, 0\rangle \, - \, \alpha_1 \, | \, 1\rangle$ when the result is 10.
4. $| \, \eta_{11}\rangle = \alpha_0 \, | \, 1\rangle \, - \, \alpha_1 \, | \, 0\rangle$ when the result is 11.

Bob applies to his qubit the transformations performed by one-qubit gates depending upon the string received from Alice, according to the following table:

String received by Bob	*Gate applied by Bob*	*The qubit in Bob's possession becomes*
00	I	$I \, | \, \eta_{00}\rangle = | \, c\rangle$
01	X	$X \, | \, \eta_{01}\rangle = | \, c\rangle$
10	Z	$Z \, | \, \eta_{10}\rangle = | \, c\rangle$
11	Y	$Y \, | \, \eta_{11}\rangle = | \, c\rangle$

Now the qubit in Bob's possession is in the same state as the original qubit of Carol's. As pointed out earlier, the state of Carol's qubit has been altered, henceforth the no cloning theorem is not violated.

Note that the new state obtained as a result of the transformation of the original state of "particle 1" and "particle 3" can also be written as:

$$| \, \zeta\rangle = (H \otimes I \otimes I)(G_{CNOT} \otimes I)(| \, \xi\rangle).$$

The discussion of teleportation with maximally entangled particles gives us the opportunity to observe some of the subtleties of handling entangled particles. We already know that the state of a pair of entangled particles is a vector in $\mathcal{H}_4$. Alice possesses only two particles, the one from Carol and her own half of the entangled pair. Note that the state of two particles in the possession of Alice is a vector in $\mathcal{H}_8$; this explains why she applies $H \otimes I \otimes I$ to measure the first qubit.

Figure 6.3 shows a circuit involving several `CNOT` and `Hadamard` gates able to perform teleportation. The circuit has three inputs, $| \, a\rangle$, $| \, b\rangle$, and $| \, c\rangle$, and three outputs, $| \, a'\rangle$, $| \, b'\rangle$, and $| \, c'\rangle$. The unknown input state $| \, c\rangle$ appears at output $| \, b'\rangle = | \, c\rangle$.

Recall that $| \, c\rangle = \alpha_0 \, | \, 0\rangle \, + \, \alpha_1 \, | \, 1\rangle$. We identify 10 stages for the circuit in Figure 6.3. The corresponding states vectors at the input and respectively the output of each stage are $\xi_1, \xi_2, \ldots, \xi_{10}$. We compute the state vectors stage by

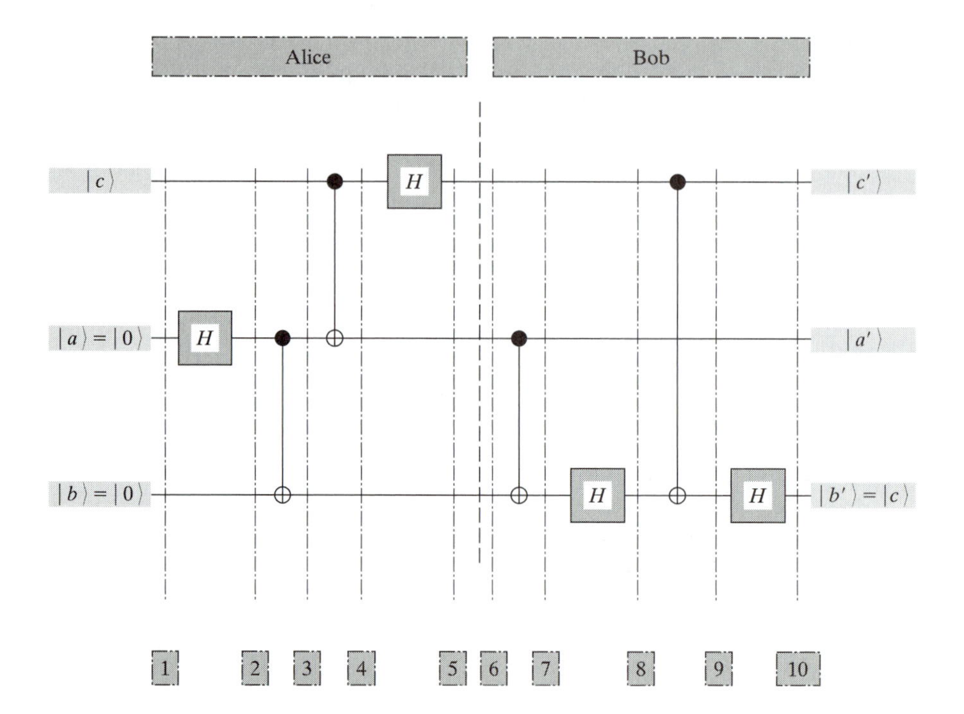

FIGURE 6.3: A quantum teleportation circuit. The left side shows Alice's transformations, the right side Bob's. Carol's particle, "particle 3" (shown at the top) initially is in the state $|c\rangle$. Bob's particle, "particle 2" (shown at the bottom) initially is in the state $|b\rangle$. Alice's particle, "particle 1", (shown in the middle) initially is in state $|a\rangle$. The final state of Bob's particle is identical with the initial state of Carol's particle, $|b'\rangle = |c\rangle$.

stage as follows:

$$|\xi_1\rangle = \begin{pmatrix} \alpha_0 \\ \alpha_1 \end{pmatrix} \otimes \begin{pmatrix} 1 \\ 0 \end{pmatrix} \otimes \begin{pmatrix} 1 \\ 0 \end{pmatrix} = \begin{pmatrix} \alpha_0 \\ 0 \\ 0 \\ 0 \\ \alpha_1 \\ 0 \\ 0 \\ 0 \end{pmatrix}.$$

For the second stage,

$$|\xi_2\rangle = (I \otimes H \otimes I)\,|\xi_1\rangle = \frac{1}{\sqrt{2}} \begin{pmatrix} 1 & 0 & 1 & 0 & 0 & 0 & 0 & 0 \\ 0 & 1 & 0 & 1 & 0 & 0 & 0 & 0 \\ 1 & 0 & -1 & 0 & 0 & 0 & 0 & 0 \\ 0 & 1 & 0 & -1 & 0 & 0 & 0 & 0 \\ 0 & 0 & 0 & 0 & 1 & 0 & 1 & 0 \\ 0 & 0 & 0 & 0 & 0 & 1 & 0 & 1 \\ 0 & 0 & 0 & 0 & 1 & 0 & -1 & 0 \\ 0 & 0 & 0 & 0 & 0 & 1 & 0 & -1 \end{pmatrix} \begin{pmatrix} \alpha_0 \\ 0 \\ 0 \\ 0 \\ \alpha_1 \\ 0 \\ 0 \\ 0 \end{pmatrix}.$$

Thus,

$$|\xi_2\rangle = \frac{1}{\sqrt{2}} \begin{pmatrix} \alpha_0 \\ 0 \\ \alpha_0 \\ 0 \\ \alpha_1 \\ 0 \\ \alpha_1 \\ 0 \end{pmatrix}.$$

For the third stage,

$$|\xi_3\rangle = (I \otimes G_{CNOT})\,|\xi_2\rangle = \begin{pmatrix} 1 & 0 & 0 & 0 & 0 & 0 & 0 & 0 \\ 0 & 1 & 0 & 0 & 0 & 0 & 0 & 0 \\ 0 & 0 & 0 & 1 & 0 & 0 & 0 & 0 \\ 0 & 0 & 1 & 0 & 0 & 0 & 0 & 0 \\ 0 & 0 & 0 & 0 & 1 & 0 & 0 & 0 \\ 0 & 0 & 0 & 0 & 0 & 1 & 0 & 0 \\ 0 & 0 & 0 & 0 & 0 & 0 & 0 & 1 \\ 0 & 0 & 0 & 0 & 0 & 0 & 1 & 0 \end{pmatrix} \frac{1}{\sqrt{2}} \begin{pmatrix} \alpha_0 \\ 0 \\ \alpha_0 \\ 0 \\ \alpha_1 \\ 0 \\ \alpha_1 \\ 0 \end{pmatrix}.$$

Thus,

$$|\xi_3\rangle = \frac{1}{\sqrt{2}} \begin{pmatrix} \alpha_0 \\ 0 \\ 0 \\ \alpha_0 \\ \alpha_1 \\ 0 \\ 0 \\ \alpha_1 \end{pmatrix}.$$

For the fourth stage,

$$|\xi_4\rangle = (G_{CNOT} \otimes I)\,|\xi_3\rangle = \begin{pmatrix} 1 & 0 & 0 & 0 & 0 & 0 & 0 & 0 \\ 0 & 1 & 0 & 0 & 0 & 0 & 0 & 0 \\ 0 & 0 & 1 & 0 & 0 & 0 & 0 & 0 \\ 0 & 0 & 0 & 1 & 0 & 0 & 0 & 0 \\ 0 & 0 & 0 & 0 & 0 & 0 & 1 & 0 \\ 0 & 0 & 0 & 0 & 0 & 0 & 0 & 1 \\ 0 & 0 & 0 & 0 & 1 & 0 & 0 & 0 \\ 0 & 0 & 0 & 0 & 0 & 1 & 0 & 0 \end{pmatrix} \frac{1}{\sqrt{2}} \begin{pmatrix} \alpha_0 \\ 0 \\ 0 \\ \alpha_0 \\ \alpha_1 \\ 0 \\ 0 \\ \alpha_1 \end{pmatrix}.$$

Thus,

$$|\xi_4\rangle = \frac{1}{\sqrt{2}} \begin{pmatrix} \alpha_0 \\ 0 \\ 0 \\ \alpha_0 \\ 0 \\ \alpha_1 \\ \alpha_1 \\ 0 \end{pmatrix}.$$

Then for the fifth stage

$$|\,\xi_5\rangle = (H \otimes I \otimes I)\,|\,\xi_4\rangle = \frac{1}{\sqrt{2}}\begin{pmatrix} 1&0&0&0&1&0&0&0\\ 0&1&0&0&0&1&0&0\\ 0&0&1&0&0&0&1&0\\ 0&0&0&1&0&0&0&1\\ 1&0&0&0&-1&0&0&0\\ 0&1&0&0&0&-1&0&0\\ 0&0&1&0&0&0&-1&0\\ 0&0&0&1&0&0&0&-1 \end{pmatrix}\frac{1}{\sqrt{2}}\begin{pmatrix}\alpha_0\\0\\0\\\alpha_0\\0\\\alpha_1\\\alpha_1\\0\end{pmatrix}.$$

Thus,

$$|\,\xi_5\rangle = \frac{1}{2}\begin{pmatrix}\alpha_0\\\alpha_1\\\alpha_1\\\alpha_0\\\alpha_0\\-\alpha_1\\-\alpha_1\\\alpha_0\end{pmatrix},$$

or

$$\begin{aligned}|\,\xi_5\rangle \;=\; &\frac{1}{2}(\alpha_0\,|\,000\rangle + \alpha_1\,|\,001\rangle + \alpha_1\,|\,010\rangle + \alpha_0\,|\,011\rangle\\ &+\alpha_0\,|\,100\rangle - \alpha_1\,|\,101\rangle - \alpha_1\,|\,110\rangle + \alpha_0\,|\,111\rangle).\end{aligned}$$

Note that ξ_5 can be written as

$$\begin{aligned}|\,\xi_5\rangle \;=\; &\tfrac{1}{2}(|\,00\rangle(\alpha_0\,|\,0\rangle + \alpha_1\,|\,1\rangle) + |\,01\rangle(\alpha_0\,|\,1\rangle + \alpha_1\,|\,0\rangle)\\ &+|\,10\rangle(\alpha_0\,|\,0\rangle - \alpha_1\,|\,1\rangle) + |\,11\rangle(\alpha_0\,|\,1\rangle - \alpha_1\,|\,0\rangle)).\end{aligned}$$

Now,

$|\,\xi_6\rangle = |\,\xi_5\rangle.$

$$|\,\xi_7\rangle = (I \otimes G_{\text{CNOT}})\,|\,\xi_6\rangle = \begin{pmatrix} 1&0&0&0&0&0&0&0\\ 0&1&0&0&0&0&0&0\\ 0&0&0&1&0&0&0&0\\ 0&0&1&0&0&0&0&0\\ 0&0&0&0&1&0&0&0\\ 0&0&0&0&0&1&0&0\\ 0&0&0&0&0&0&0&1\\ 0&0&0&0&0&0&1&0 \end{pmatrix}\frac{1}{2}\begin{pmatrix}\alpha_0\\\alpha_1\\\alpha_1\\\alpha_0\\\alpha_0\\-\alpha_1\\-\alpha_1\\\alpha_0\end{pmatrix} = \frac{1}{2}\begin{pmatrix}\alpha_0\\\alpha_1\\\alpha_0\\\alpha_1\\\alpha_0\\-\alpha_1\\\alpha_0\\-\alpha_1\end{pmatrix},$$

$$|\xi_8\rangle = (I \otimes I \otimes H)\,|\xi_7\rangle = \frac{1}{\sqrt{2}}\begin{pmatrix} 1 & 1 & 0 & 0 & 0 & 0 & 0 & 0 \\ 1 & -1 & 0 & 0 & 0 & 0 & 0 & 0 \\ 0 & 0 & 1 & 1 & 0 & 0 & 0 & 0 \\ 0 & 0 & 1 & -1 & 0 & 0 & 0 & 0 \\ 0 & 0 & 0 & 0 & 1 & 1 & 0 & 0 \\ 0 & 0 & 0 & 0 & 1 & -1 & 0 & 0 \\ 0 & 0 & 0 & 0 & 0 & 0 & 1 & 1 \\ 0 & 0 & 0 & 0 & 1 & 0 & 0 & 1 & -1 \end{pmatrix} \frac{1}{2}\begin{pmatrix} \alpha_0 \\ \alpha_1 \\ \alpha_0 \\ \alpha_1 \\ \alpha_0 \\ -\alpha_1 \\ \alpha_0 \\ -\alpha_1 \end{pmatrix}.$$

or,

$$|\xi_8\rangle = \frac{1}{2\sqrt{2}}\begin{pmatrix} \alpha_0 + \alpha_1 \\ \alpha_0 - \alpha_1 \\ \alpha_0 + \alpha_1 \\ \alpha_0 - \alpha_1 \\ \alpha_0 - \alpha_1 \\ \alpha_0 + \alpha_1 \\ \alpha_0 - \alpha_1 \\ \alpha_0 + \alpha_1 \end{pmatrix}.$$

We use a result from Section 4.10 giving the transfer matrix of the gate involved in this stage, G_Q

$$|\xi_9\rangle = G_Q\,|\xi_8\rangle = \begin{pmatrix} 1 & 0 & 0 & 0 & 0 & 0 & 0 & 0 \\ 0 & 1 & 0 & 0 & 0 & 0 & 0 & 0 \\ 0 & 0 & 1 & 0 & 0 & 0 & 0 & 0 \\ 0 & 0 & 0 & 1 & 0 & 0 & 0 & 0 \\ 0 & 0 & 0 & 0 & 0 & 1 & 0 & 0 \\ 0 & 0 & 0 & 0 & 1 & 0 & 0 & 0 \\ 0 & 0 & 0 & 0 & 0 & 0 & 0 & 1 \\ 0 & 0 & 0 & 0 & 0 & 0 & 1 & 0 \end{pmatrix} \frac{1}{2\sqrt{2}}\begin{pmatrix} \alpha_0 + \alpha_1 \\ \alpha_0 - \alpha_1 \\ \alpha_0 + \alpha_1 \\ \alpha_0 - \alpha_1 \\ \alpha_0 - \alpha_1 \\ \alpha_0 + \alpha_1 \\ \alpha_0 - \alpha_1 \\ \alpha_0 + \alpha_1 \end{pmatrix}$$

or,

$$|\xi_9\rangle = \frac{1}{2\sqrt{2}}\begin{pmatrix} \alpha_0 + \alpha_1 \\ \alpha_0 - \alpha_1 \\ \alpha_0 + \alpha_1 \\ \alpha_0 - \alpha_1 \\ \alpha_0 + \alpha_1 \\ \alpha_0 - \alpha_1 \\ \alpha_0 + \alpha_1 \\ \alpha_0 - \alpha_1 \end{pmatrix}.$$

For the tenth stage,

$$|\xi_{10}\rangle = (I \otimes I \otimes H)\,|\xi_9\rangle$$

$$= \frac{1}{\sqrt{2}}\begin{pmatrix} 1 & 1 & 0 & 0 & 0 & 0 & 0 & 0 \\ 1 & -1 & 0 & 0 & 0 & 0 & 0 & 0 \\ 0 & 0 & 1 & 1 & 0 & 0 & 0 & 0 \\ 0 & 0 & 1 & -1 & 0 & 0 & 0 & 0 \\ 0 & 0 & 0 & 0 & 1 & 1 & 0 & 0 \\ 0 & 0 & 0 & 0 & 1 & -1 & 0 & 0 \\ 0 & 0 & 0 & 0 & 0 & 0 & 1 & 1 \\ 0 & 0 & 0 & 0 & 1 & 0 & 0 & 1 & -1 \end{pmatrix} \frac{1}{2\sqrt{2}}\begin{pmatrix} \alpha_0 + \alpha_1 \\ \alpha_0 - \alpha_1 \\ \alpha_0 + \alpha_1 \\ \alpha_0 - \alpha_1 \\ \alpha_0 + \alpha_1 \\ \alpha_0 - \alpha_1 \\ \alpha_0 + \alpha_1 \\ \alpha_0 - \alpha_1 \end{pmatrix}.$$

Thus,

$$| \xi_{10} \rangle = \frac{1}{2} \begin{pmatrix} \alpha_0 \\ \alpha_1 \\ \alpha_0 \\ \alpha_1 \\ \alpha_0 \\ \alpha_1 \\ \alpha_0 \\ \alpha_1 \end{pmatrix}$$

or

$$| \xi_{10} \rangle = \frac{1}{2}[| 00\rangle(\alpha_0 | 0\rangle + \alpha_1 | 1\rangle) + | 01\rangle(\alpha_0 | 0\rangle + \alpha_1 | 1\rangle) + | 10\rangle(\alpha_0 | 0\rangle + \alpha_1 | 1\rangle) + | 11\rangle(\alpha_0 | 0\rangle + \alpha_1 | 1\rangle)].$$

Finally,

$$| \xi_{10} \rangle = \frac{1}{2}(\alpha_0 | 0\rangle + \alpha_1 | 1\rangle)(| 00\rangle + | 01\rangle + | 10\rangle + | 11\rangle).$$

The previous equation gives the state at the output of the quantum teleportation circuit in Figure 6.3. We see that we have a single qubit in state $(\alpha_0 | 0\rangle + \alpha_1 | 1\rangle)$ and a pair of entangled qubits, in state $1/2(| 00\rangle + | 01\rangle + | 10\rangle + | 11\rangle)$. The final state of Bob's particle (first qubit) denoted as $| b'\rangle$ is identical with the initial state of Carol's particle, $| c\rangle = \alpha_0 | 0\rangle + \alpha_1 | 1\rangle$.

We have successfully replicated the state of one particle using another particle located possibly light-years away. We did not violate the no-cloning principle, the state of the original particle has been altered.

So far, successful teleportation experiments have been reported only for short distances (a few meters), rather than light years.

6.4 ANTI-CORRELATION AND TELEPORTATION

Two entangled qubits could be in any EPR state. Now we discuss briefly the anti-correlated state. Consider the following two-qubit state called a *spin singlet*:

$$| \beta_{11} \rangle = \frac{| 01\rangle - | 10\rangle}{\sqrt{2}}.$$

This is an entangled state and we shall see later that a measurement reveals that the two particles involved are in opposite states. For example, if we measure the spin of the first particle and find it to be "up," then the spin of the second is "down" and vice versa. Miraculously, the second qubit changes its state, as if knowing the result of the measurement on the first qubit. This state is distinguished from the three other EPR states; it changes sign when "particle 1" and "particle 2" of an entangled pair are interchanged.

Let us analyze the anti-correlation case described above. For the sake of clarity we consider the three particles to be photons and use the symbol $\rightarrow$ for horizontal polarization and $\uparrow$ for vertical polarization. The corresponding basis vectors are $|\rightarrow\rangle$ and $|\uparrow\rangle$. The subscript identifies the particle in the pair. The entangled pair $(1, 2)$ forms a single quantum system with a shared state:

$$|\beta_{11}\rangle_{12} = \frac{1}{\sqrt{2}}(|\rightarrow\rangle_1 |\uparrow\rangle_2 - |\uparrow\rangle_1 |\rightarrow\rangle_2).$$

This expression shows that particles 1 and 2 are in a $|\beta_{11}\rangle$ state.

This entangled state indicates only that the two particles are in opposite states, yet provides no information about the state of each particle of the pair. Once we make a measurement of one of the particles by projecting it onto one of the basis vectors, say $|\rightarrow\rangle$, then the state of the other particle becomes instantaneously $|\uparrow\rangle$. If "particle 1" is measured and found to have vertical polarization, then "particle 2" will have horizontal polarization.

Now we perform a *specific joint measurement* on "particle 1" and "particle 3" which projects them onto the entangled state:

$$|\beta_{11}\rangle_{13} = \frac{1}{\sqrt{2}}(|\rightarrow\rangle_1 |\uparrow\rangle_3 - |\uparrow\rangle_1 |\rightarrow\rangle_3).$$

Now "particle 1" and "particle 3" are anti-correlated. If "particle 3" has horizontal polarization it forces "particle 1" to have vertical polarization. In turn, "particle 1" forces "particle 2" to have opposite polarization, thus "particle 2" ends up with the same polarization as "particle 3."

A demonstration of quantum teleportation was carried out in 1997 at the University of Rome by Francesco de Martini based upon an idea of Sandu Popescu and at about the same time at Innsbruck by Anton Zeillinger. In both experiments, the quantum state was teleported a few meters.

The experiment of de Martini is illustrated in Figure 6.4 [29]. In this experiment, the information is double encoded into a single photon instead of two. The source generates two parametric downconverted[2] photons with opposite polarization, "photon 1," with horizontal polarization, h, for Alice and "photon 2," with vertical polarization, v, for Bob. The polarization entanglement of the two photons sent to Alice and Bob is converted into an entanglement of the paths followed by the two photons. A calcite crystal performs this conversion.

If "photon 1" travels to Alice via path A then "photon 2" travels to Bob via path C; if "photon 1" travels via path B, then "photon 2" travels via path D. Carol encodes her message in the polarization of the photon sent to Alice, "photon 1." Alice measures the polarization of the photon she receives from the source and sends the classical result to Bob. Finally, Bob performs the measurement suggested by Alice's result and he gets a photon with the polarization imposed by Carol.

In this experiment, the polarizer forces a certain polarization on "photon 1" and because of the anti-correlation of "photon 1" and "photon 2" the latter is forced to an opposite polarization.

---

[2]A parametric downconversion source uses a UV laser beam which upon an interaction with a non-linear medium, a crystal of ammonium dihydrogen phosphate (ADP), generates two photons for one input photon.

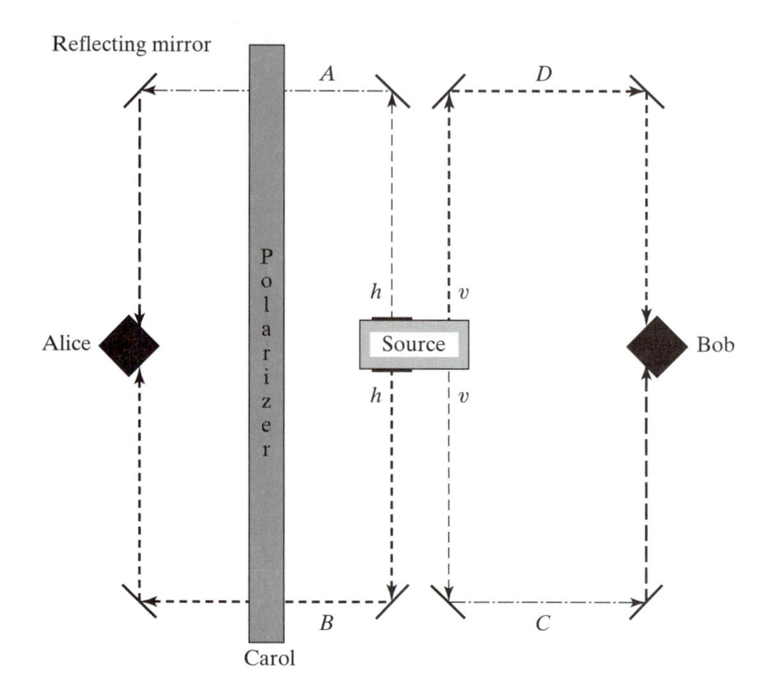

FIGURE 6.4: The teleportation experiment at University of Rome. The source generates a photon with horizontal polarization h for Alice and one with vertical polarization v for Bob. The entanglement is in the path selection. If Alice gets her photon via path A, then Bob gets his via path C; if Alice gets the photon via path B then Bob gets his via path D. Carol encodes her quantum information using a polarizer. Alice measures the polarization of the photon she receives and sends this classical information to Bob.

6.5 DENSE CODING

Coding is the process of transforming information during a communication process. The sender of a message encodes the message, then transmits the encoded information over a classical communication channel. The recipient of the message decodes the encoded information. The question we address now is whether there is an advantage in exchanging quantum information (qubits over a quantum communication channel), instead of sending classical bits over a classical communication channel.

The main characters of the following example are also Alice and Bob. Alice and Bob have been married for some time and Alice, inspired by the wonderful pictures taken by Bob during his last trip to K2 decides to join a new expedition to that remote part of the world. They want to exchange daily messages and compress them as much as possible to reduce communication costs. Prior to Alice's departure they agree to exchange daily information about the temperature and the cloud cover on K2. They decide that Alice will construct two bit messages. The first bit will describe the temperature (0 if the temperature is below 0° C, 1 if it is above 0° C); the second bit will describe the cloud cover (0 if there are no clouds and 1 if there is a degree of cloud cover. The sentence "At

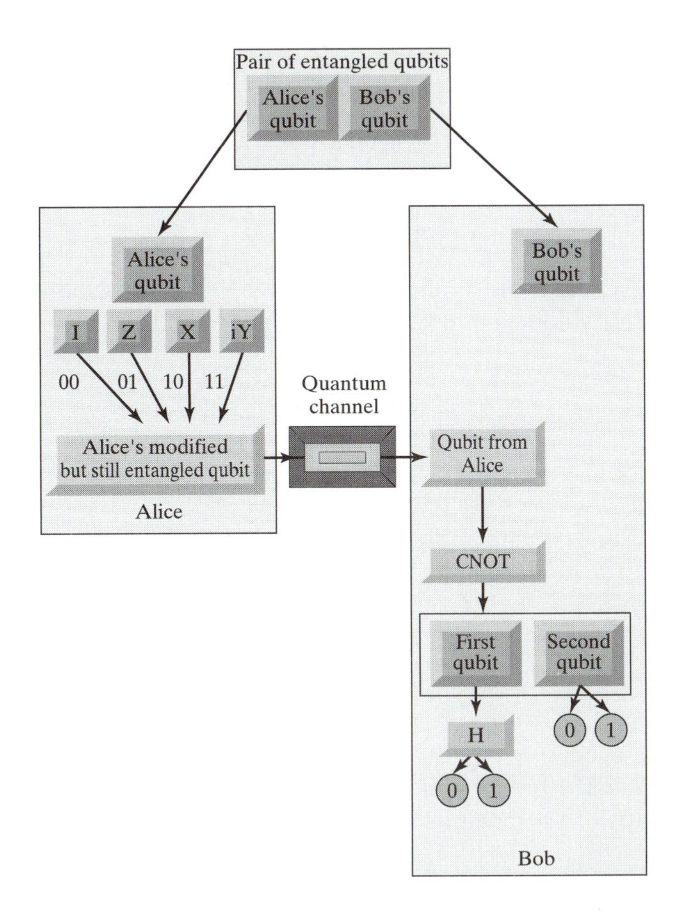

FIGURE 6.5: Dense coding. Alice sends to Bob one qubit instead of two bits.

noon today the temperature on the summit of K2 is below zero and there are no clouds. Love, Alice." will be encoded as the binary string 00. The other possible messages are 01 ("...below zero...clouds..."), 10 ("...above zero...no clouds..."), and 11 ("...above zero...clouds...").

To send one of the four messages Alice must transmit two bits of classical information. But Alice and Bob are already acquainted with quantum information. They decide to exchange a single qubit of information over a quantum channel to encode and decode the four possible messages. Here is the intricate story of how they succeed.

Assume that there is a source of entangled particles and Alice gets one qubit (one of the two entangled particles) and Bob gets the other particle of the entangled pair, see Figure 6.5. The two entangled particles are in the state

$$| \beta_{00} \rangle = \frac{| 00 \rangle + | 11 \rangle}{\sqrt{2}} \quad \mapsto \frac{1}{\sqrt{2}} \begin{pmatrix} 1 \\ 0 \\ 0 \\ 1 \end{pmatrix}.$$

Alice transforms the first qubit of the pair, the one in her possession, using the Pauli matrices, I, X, Y and Z. Recall from Section 4.2 that these matrices are

$$I = \begin{pmatrix} 1 & 0 \\ 0 & 1 \end{pmatrix}, \quad X = \begin{pmatrix} 0 & 1 \\ 1 & 0 \end{pmatrix}, \quad Y = \begin{pmatrix} 0 & -i \\ i & 0 \end{pmatrix}, \quad Z = \begin{pmatrix} 1 & 0 \\ 0 & -1 \end{pmatrix}.$$

Alice prepares her qubit as follows:

1. To send 00, she applies to her qubit the transformation produced by I, the identity matrix and transmits a qubit in state

$$| \varphi_{00} \rangle = | \beta_{00} \rangle = \frac{| 00 \rangle + | 11 \rangle}{\sqrt{2}} \quad \mapsto \quad \frac{1}{\sqrt{2}} \begin{pmatrix} 1 \\ 0 \\ 0 \\ 1 \end{pmatrix}.$$

2. To send 01, she applies to her qubit the transformation produced by the Z matrix and transmits one qubit in state

$$| \varphi_{01} \rangle = | \beta_{10} \rangle = \frac{| 00 \rangle - | 11 \rangle}{\sqrt{2}} \quad \mapsto \quad \frac{1}{\sqrt{2}} \begin{pmatrix} 1 \\ 0 \\ 0 \\ -1 \end{pmatrix}.$$

3. To send 10, she applies to her qubit the transformation produced by the X matrix and transmits one qubit in state

$$| \varphi_{10} \rangle = | \beta_{01} \rangle = \frac{| 01 \rangle + | 10 \rangle}{\sqrt{2}} \quad \mapsto \quad \frac{1}{\sqrt{2}} \begin{pmatrix} 0 \\ 1 \\ 1 \\ 0 \end{pmatrix}.$$

4. To send 11, she applies to her qubit the transformation produced by the iY matrix and transmits one qubit in state

$$| \varphi_{11} \rangle = | \beta_{11} \rangle = \frac{| 01 \rangle - | 10 \rangle}{\sqrt{2}} \quad \mapsto \quad \frac{1}{\sqrt{2}} \begin{pmatrix} 0 \\ 1 \\ -1 \\ 0 \end{pmatrix}.$$

Let us now take a closer look at the transformations performed by Alice to send the four strings 00, 01, 10, and 11.

We already know that `X, Y, Z`, and `I` are one-qubit gates used to transform vectors in $\mathcal{H}_2$, but this time we have an entangled qubit, a vector in $\mathcal{H}_4$. Each one of the four matrices used to transform the first qubit of the entangled pair, the qubit in possession of Alice, is obtained as the tensor product of the corresponding single qubit gate and the identity matrix. We transform the first qubit and leave the second qubit of the entangled pair untouched.

The transfer matrices and the outputs for the four transformations Alice is expected to carry on her qubit are G_{00}, G_{01}, G_{10}, and G_{11}.

To send the string 00 Alice uses the transformation given by G_{00}.

$$00 \mapsto G_{00} = I \otimes I = \begin{pmatrix} 1 & 0 \\ 0 & 1 \end{pmatrix} \otimes \begin{pmatrix} 1 & 0 \\ 0 & 1 \end{pmatrix} = \begin{pmatrix} 1 & 0 & 0 & 0 \\ 0 & 1 & 0 & 0 \\ 0 & 0 & 1 & 0 \\ 0 & 0 & 0 & 1 \end{pmatrix}.$$

Thus,

$$|\,\varphi_{00}\rangle = \begin{pmatrix} 1 & 0 & 0 & 0 \\ 0 & 1 & 0 & 0 \\ 0 & 0 & 1 & 0 \\ 0 & 0 & 0 & 1 \end{pmatrix} \frac{1}{\sqrt{2}} \begin{pmatrix} 1 \\ 0 \\ 0 \\ 1 \end{pmatrix} = \frac{1}{\sqrt{2}} \begin{pmatrix} 1 \\ 0 \\ 0 \\ 1 \end{pmatrix}.$$

To send the string 01 Alice uses the transformation given by G_{01}

$$01 \mapsto G_{01} = Z \otimes I = \begin{pmatrix} 1 & 0 \\ 0 & -1 \end{pmatrix} \otimes \begin{pmatrix} 1 & 0 \\ 0 & 1 \end{pmatrix} = \begin{pmatrix} 1 & 0 & 0 & 0 \\ 0 & 1 & 0 & 0 \\ 0 & 0 & -1 & 0 \\ 0 & 0 & 0 & -1 \end{pmatrix}.$$

Thus,

$$|\,\varphi_{01}\rangle = \begin{pmatrix} 1 & 0 & 0 & 0 \\ 0 & 1 & 0 & 0 \\ 0 & 0 & -1 & 0 \\ 0 & 0 & 0 & -1 \end{pmatrix} \frac{1}{\sqrt{2}} \begin{pmatrix} 1 \\ 0 \\ 0 \\ 1 \end{pmatrix} = \frac{1}{\sqrt{2}} \begin{pmatrix} 1 \\ 0 \\ 0 \\ -1 \end{pmatrix}.$$

To send 10 Alice uses the transformation given by G_{10}

$$10 \mapsto G_{10} = X \otimes I = \begin{pmatrix} 0 & 1 \\ 1 & 0 \end{pmatrix} \otimes \begin{pmatrix} 1 & 0 \\ 0 & 1 \end{pmatrix} = \begin{pmatrix} 0 & 0 & 1 & 0 \\ 0 & 0 & 0 & 1 \\ 1 & 0 & 0 & 0 \\ 0 & 1 & 0 & 0 \end{pmatrix}.$$

Thus,

$$|\,\varphi_{10}\rangle = \begin{pmatrix} 0 & 0 & 1 & 0 \\ 0 & 0 & 0 & 1 \\ 1 & 0 & 0 & 0 \\ 0 & 1 & 0 & 0 \end{pmatrix} \frac{1}{\sqrt{2}} \begin{pmatrix} 1 \\ 0 \\ 0 \\ 1 \end{pmatrix} = \frac{1}{\sqrt{2}} \begin{pmatrix} 0 \\ 1 \\ 1 \\ 0 \end{pmatrix}.$$

To send 11 Alice uses the transformation given by G_{11}

$$11 \mapsto G_{11} = iY \otimes I = \begin{pmatrix} 0 & -i \\ i & 0 \end{pmatrix} \otimes \begin{pmatrix} 1 & 0 \\ 0 & 1 \end{pmatrix} = \begin{pmatrix} 0 & 0 & -i & 0 \\ 0 & 0 & 0 & -i \\ i & 0 & 0 & 0 \\ 0 & i & 0 & 0 \end{pmatrix}.$$

Thus,

$$| \varphi_{11} \rangle = i \begin{pmatrix} 0 & 0 & -i & 0 \\ 0 & 0 & 0 & -i \\ i & 0 & 0 & 0 \\ 0 & i & 0 & 0 \end{pmatrix} \frac{1}{\sqrt{2}} \begin{pmatrix} 1 \\ 0 \\ 0 \\ 1 \end{pmatrix},$$

or

$$| \varphi_{11} \rangle = \frac{i}{\sqrt{2}} \begin{pmatrix} 0 \\ -i \\ i \\ 0 \end{pmatrix} = \frac{1}{\sqrt{2}} \begin{pmatrix} 0 \\ 1 \\ -1 \\ 0 \end{pmatrix}.$$

After getting the qubit from Alice, Bob is in possession of both qubits of the entangled pair. Bob decodes the message as follows. First, he applies a CNOT gate to the pair and gets the following results for each of the four cases, when the pair is in the states $| \varphi_{00} \rangle, | \varphi_{01} \rangle, | \varphi_{10} \rangle$, or $| \varphi_{11} \rangle$:

1. $00 \mapsto | \xi_{00} \rangle = G_{\text{CNOT}} \, | \varphi_{00} \rangle = \begin{pmatrix} 1 & 0 & 0 & 0 \\ 0 & 1 & 0 & 0 \\ 0 & 0 & 0 & 1 \\ 0 & 0 & 1 & 0 \end{pmatrix} \frac{1}{\sqrt{2}} \begin{pmatrix} 1 \\ 0 \\ 0 \\ 1 \end{pmatrix} = \frac{1}{\sqrt{2}} \begin{pmatrix} 1 \\ 0 \\ 1 \\ 0 \end{pmatrix}$

or

$$| \xi_{00} \rangle = \frac{| 00 \rangle + | 10 \rangle}{\sqrt{2}} = \frac{| 0 \rangle + | 1 \rangle}{\sqrt{2}} | 0 \rangle.$$

The two qubits of the pair are

$$| \xi^{I}_{00} \rangle = \frac{| 0 \rangle + | 1 \rangle}{\sqrt{2}} \mapsto \frac{1}{\sqrt{2}} \begin{pmatrix} 1 \\ 1 \end{pmatrix} \qquad | \xi^{II}_{00} \rangle = | 0 \rangle \mapsto \begin{pmatrix} 1 \\ 0 \end{pmatrix}.$$

2. $01 \mapsto | \xi_{01} \rangle = G_{\text{CNOT}} \, | \varphi_{01} \rangle = \begin{pmatrix} 1 & 0 & 0 & 0 \\ 0 & 1 & 0 & 0 \\ 0 & 0 & 0 & 1 \\ 0 & 0 & 1 & 0 \end{pmatrix} \frac{1}{\sqrt{2}} \begin{pmatrix} 1 \\ 0 \\ 0 \\ -1 \end{pmatrix} = \frac{1}{\sqrt{2}} \begin{pmatrix} 1 \\ 0 \\ -1 \\ 0 \end{pmatrix},$

or

$$| \xi_{01} \rangle = \frac{| 00 \rangle - | 10 \rangle}{\sqrt{2}} = \frac{| 0 \rangle - | 1 \rangle}{\sqrt{2}} | 0 \rangle.$$

The two qubits of the pair are

$$| \xi^{I}_{01} \rangle = \frac{| 0 \rangle - | 1 \rangle}{\sqrt{2}} \mapsto \frac{1}{\sqrt{2}} \begin{pmatrix} 1 \\ -1 \end{pmatrix} \qquad | \xi^{II}_{01} \rangle = | 0 \rangle \mapsto \begin{pmatrix} 1 \\ 0 \end{pmatrix}.$$

3. $10 \mapsto | \xi_{10} \rangle = G_{\text{CNOT}} \, | \varphi_{10} \rangle = \begin{pmatrix} 1 & 0 & 0 & 0 \\ 0 & 1 & 0 & 0 \\ 0 & 0 & 0 & 1 \\ 0 & 0 & 1 & 0 \end{pmatrix} \frac{1}{\sqrt{2}} \begin{pmatrix} 0 \\ 1 \\ 1 \\ 0 \end{pmatrix} = \frac{1}{\sqrt{2}} \begin{pmatrix} 0 \\ 1 \\ 0 \\ 1 \end{pmatrix},$

or

$$| \xi_{10} \rangle = \frac{| 01 \rangle + | 11 \rangle}{\sqrt{2}} = \frac{| 0 \rangle + | 1 \rangle}{\sqrt{2}} | 1 \rangle.$$

The two qubits of the pair are

$$| \xi_{10}^{I} \rangle = \frac{| 0 \rangle + | 1 \rangle}{\sqrt{2}} \mapsto \frac{1}{\sqrt{2}} \begin{pmatrix} 1 \\ 1 \end{pmatrix} \quad | \xi_{10}^{II} \rangle = | 1 \rangle \mapsto \begin{pmatrix} 0 \\ 1 \end{pmatrix}.$$

4. $11 \mapsto | \xi_{11} \rangle = G_{\text{CNOT}} \, | \varphi_{11} \rangle = \begin{pmatrix} 1 & 0 & 0 & 0 \\ 0 & 1 & 0 & 0 \\ 0 & 0 & 0 & 1 \\ 0 & 0 & 1 & 0 \end{pmatrix} \frac{1}{\sqrt{2}} \begin{pmatrix} 0 \\ 1 \\ -1 \\ 0 \end{pmatrix} = \frac{1}{\sqrt{2}} \begin{pmatrix} 0 \\ 1 \\ 0 \\ -1 \end{pmatrix},$

or

$$| \xi_{11} \rangle = \frac{| 01 \rangle - | 11 \rangle}{\sqrt{2}} = \frac{| 0 \rangle - | 1 \rangle}{\sqrt{2}} | 1 \rangle.$$

The two qubits of the pair are

$$| \xi_{11}^{I} \rangle = \frac{| 0 \rangle - | 1 \rangle}{\sqrt{2}} \mapsto \frac{1}{\sqrt{2}} \begin{pmatrix} 1 \\ -1 \end{pmatrix}, \quad | \xi_{11}^{II} \rangle = | 1 \rangle \mapsto \begin{pmatrix} 0 \\ 1 \end{pmatrix}.$$

Amazingly enough, Bob can now measure the second qubit without affecting the state of the entangled pair. If the second qubit is $| 0 \rangle$, it means that Alice sent either the string 00 or 01. If the second qubit is $| 1 \rangle$, it means that Alice sent either the string 10 or 11.

Bob applies the `Hadamard` gate to the first qubit. The results are

$$00 \mapsto | \zeta_{00} \rangle = H \, | \xi_{00}^{I} \rangle = \frac{1}{\sqrt{2}} \begin{pmatrix} 1 & 1 \\ 1 & -1 \end{pmatrix} \frac{1}{\sqrt{2}} \begin{pmatrix} 1 \\ 1 \end{pmatrix} = \frac{1}{2} \begin{pmatrix} 2 \\ 0 \end{pmatrix} = \begin{pmatrix} 1 \\ 0 \end{pmatrix},$$

$$| \zeta_{00} \rangle = | 0 \rangle.$$

$$01 \mapsto | \zeta_{01} \rangle = H \, | \xi_{01}^{I} \rangle = \frac{1}{\sqrt{2}} \begin{pmatrix} 1 & 1 \\ 1 & -1 \end{pmatrix} \frac{1}{\sqrt{2}} \begin{pmatrix} 1 \\ -1 \end{pmatrix} = \frac{1}{2} \begin{pmatrix} 0 \\ 2 \end{pmatrix} = \begin{pmatrix} 0 \\ 1 \end{pmatrix},$$

$$| \zeta_{01} \rangle = | 1 \rangle.$$

$$10 \mapsto | \zeta_{10} \rangle = H \, | \xi_{10}^{I} \rangle = \frac{1}{\sqrt{2}} \begin{pmatrix} 1 & 1 \\ 1 & -1 \end{pmatrix} \frac{1}{\sqrt{2}} \begin{pmatrix} 1 \\ 1 \end{pmatrix} = \frac{1}{2} \begin{pmatrix} 2 \\ 0 \end{pmatrix} = \begin{pmatrix} 1 \\ 0 \end{pmatrix},$$

$$| \zeta_{10} \rangle = | 0 \rangle.$$

$$11 \mapsto | \zeta_{11} \rangle = H \, | \xi_{11}^{I} \rangle = \frac{1}{\sqrt{2}} \begin{pmatrix} 1 & 1 \\ 1 & -1 \end{pmatrix} \frac{1}{\sqrt{2}} \begin{pmatrix} 1 \\ -1 \end{pmatrix} = \frac{1}{2} \begin{pmatrix} 0 \\ 2 \end{pmatrix} = \begin{pmatrix} 0 \\ 1 \end{pmatrix},$$

$$| \zeta_{11} \rangle = | 1 \rangle.$$

Now Bob knows precisely which one of the four messages was sent based upon the value of the second qubit and the result of applying the Hadamard Transform to the first qubit. The following table summarizes the decision process followed by Bob:

Second qubit	First qubit	Message
$\vert 1\rangle$	$\vert 1\rangle$	00
$\vert 1\rangle$	$\vert 0\rangle$	01
$\vert 0\rangle$	$\vert 1\rangle$	10
$\vert 0\rangle$	$\vert 0\rangle$	11

6.6 QUANTUM KEY DISTRIBUTION

System security is a critical concern in the design of networked computer systems. *Confidentiality* is a property of a system that guarantees that only agents with proper credentials have access to information. Confidentiality can be compromised during transmission over insecure communication channels or while being stored on sites that allow multiple agents to modify it. The common method to support confidentiality is based on *encryption*. Data (or *plaintext* in cryptographic terms) is mathematically transformed into *ciphertext* and only agents with the proper key are able to decrypt the ciphertext and transform it back into plaintext.

The algorithms used to transform plaintext into ciphertext and back form a *cipher*. A *symmetric cipher* uses the same key for encryption and decryption. *Asymmetric or public key ciphers* involve a *public key* that can be freely distributed and a secret *private key*. Data is encrypted by the sender using the public key of the intended recipient of the message and it is decrypted using the private key of the recipient, see Figure 6.6.

The problem of distributing the key of a symmetric cipher is known as the *key distribution* problem. Decryption using an asymmetric cipher is time consuming. Typically, the parties involved use the public key system to exchange a *session key*, and then they use a symmetric cipher based upon this key to communicate. To make it harder for a third party to break the cipher the encryption key should be changed as frequently as possible.

When, in addition to the classical communication channel, a quantum communication channel is available, then the key distribution problem has an ingenious solution. We show that the two parties involved in the exchange of the cipher key can detect with a very high probability if a third party eavesdrops. As before, the main characters are Alice who wants to send the encryption key to Bob and Carol who attempts to eavesdrop.

The exchange without a third party is captured in Figure 6.7(a). Alice sends over the quantum communication channel n qubits encoded randomly in one of two bases, $(\vert \leftrightarrow\rangle, \vert \updownarrow\rangle)$ and $(\vert \swarrow\!\!\!\nearrow\rangle, \vert \nwarrow\!\!\!\searrow\rangle)$. Bob randomly chooses one of the two bases when receiving and measuring one qubit. Bob ends up guessing correctly the bases for about $n/2$ of the qubits, but he does not know which qubit is measured correctly. Then Alice uses the classical communication channel to send Bob information about the bases she used for each qubit and Bob sends Alice the bases he guessed for each qubit. Now, both Alice and Bob know

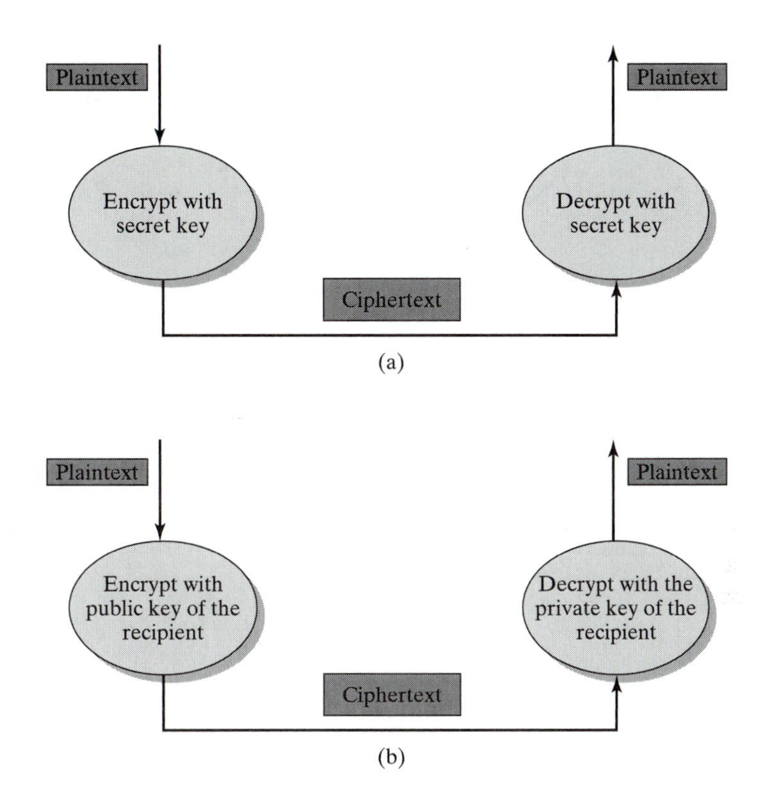

FIGURE 6.6: (a) Secret key cryptography. (b) Public key cryptography.

precisely the qubits they have agreed upon. They use these approximately $n/2$ qubits as the encryption key.

The exchange when Carol eavesdrops on both the quantum and classical communication channels is illustrated in Figure 6.7(b). Carol intercepts each of the n qubits. She randomly picks up one of the two bases to measure each qubit and then she resends the qubit to Bob. Approximately $n/2$ of the qubits received by Bob have their state altered by Carol. When Alice and Bob exchange information over the classical communication channel they realize that they agree in considerably less than $n/2$ of qubits. Thus they detect the presence of Carol.

We give without proof the following proposition: *To distinguish between two non-orthogonal quantum states we can only gain information by introducing additional disturbance to the system.*

Protocols for quantum key distribution are based upon the idea of transmitting non-orthogonal qubit states and then checking for the disturbance in their transmitted states. To prove the correctness of a quantum key distribution (QKD) protocol we have to show that Alice and Bob agree on a key about which Carol can obtain only an exponentially small amount of information.

Eavesdropping must be distinguished from the noise on the communication channel. To do so, "check" bits must be interspaced randomly among the "data" bits used to construct the encryption key.

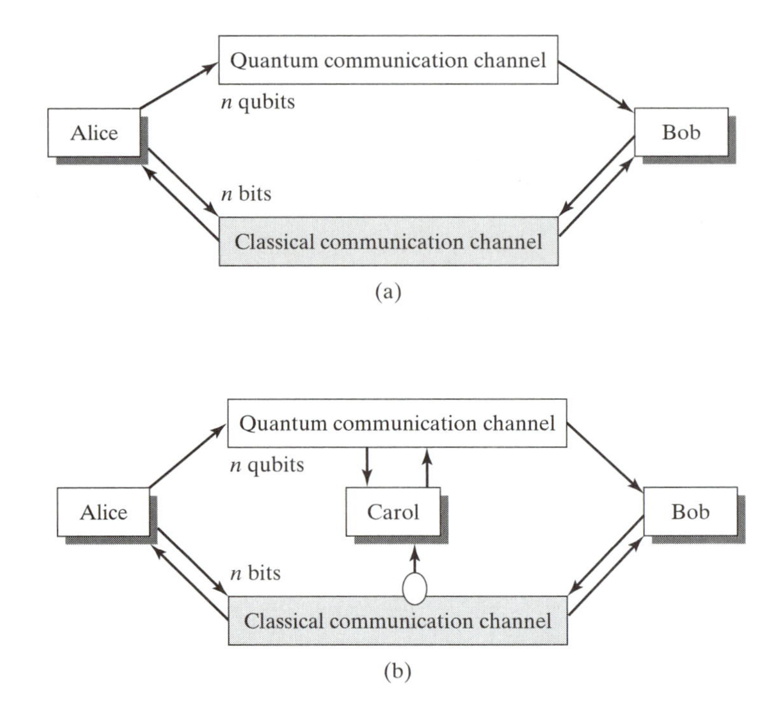

FIGURE 6.7: (a) Alice sends over the quantum communication channel n qubits encoded randomly in one of two bases. Bob randomly chooses one of the two bases when receiving one qubit. Bob ends up guessing correctly the bases for about $n/2$ of the qubits. Then Alice and Bob exchange over the classical communication channel information about the base used by each one of them for each qubit. Now both Alice and Bob know precisely on which qubits they have agreed and use them as the encryption key. (b) Carol intercepts each of the n qubits. She randomly picks up one of the two bases to measure each qubit and then she re-sends the qubit to Bob. Approximately $n/2$ of the qubits received by Bob have their state altered by Carol. Bob guesses the correct bases for only about half of the qubits whose state has not been altered by Carol. When Alice and Bob exchange information over the classical communication channel they realize that they agree on considerably less than $n/2$ qubits. Thus, they detect the presence of Carol.

We now outline the first QKD protocol, *BB84*, proposed by Bennett and Brassard in 1984 [13]. A proof of the security of this protocol can be found in [124]. Here we follow the description of the protocol in [98]:

1. Alice selects n, the approximate length of the desired encryption key. Then she generates two random strings of bits a and b of length $(4n + \delta)$.
2. Alice encodes the bits in string a using the bits in string b to choose the basis (either X or Z) for each qubit in a. She generates $| \psi\rangle$, a block of $(4n + \delta)$ qubits

$$| \psi\rangle = \bigotimes_{k=1}^{4n+\delta} | \psi_{a_k b_k}\rangle$$

where a_k and b_k are the k-th bit of strings a and b respectively. Each qubit is in one of four pure states in two bases, $[| 0\rangle, | 1\rangle]$ and

$[(1/\sqrt{2})(|0\rangle + |1\rangle), (1/\sqrt{2})(|0\rangle - |1\rangle)]$. The four states used are

$$\psi_{00} = |0\rangle \quad \psi_{10} = |1\rangle \quad \psi_{01} = (1/\sqrt{2})(|0\rangle + |1\rangle)$$
$$\psi_{11} = (1/\sqrt{2})(|0\rangle - |1\rangle).$$

3. If $\mathscr{E}$ describes the combined effect of the channel noise and Carol's interference, then the block of qubits received by Bob is $\mathscr{E}(|\psi\rangle\langle\psi|)$.
4. Bob constructs a random string of bits, b', of length $(4n + \delta)$. He then measures every qubit either in basis X or in basis Z depending upon the value of the corresponding bit of b'. As a result of his measurement, he constructs the binary string a'. He tells Alice over the classical channel that he now expects information about b.
5. Alice uses the classical channel to disclose b.
6. Alice and Bob exchange information over the classical channel and keep only the bits in the set $\{a, a'\}$ for which the corresponding bits of the strings b and b' are equal. Let us assume that Alice and Bob keep only $2n$ bits. By choosing δ sufficiently large Alice and Bob can ensure that the number of bits kept is close to $2n$ with a very high probability.
7. Alice and Bob perform several tests to determine the level of noise and eavesdropping on the channel. The set of $2n$ bits is split into two subsets of n bits each. One subset will be the *check* bits used to estimate the level of noise and eavesdropping, and the other consists of the *data* bits used for the quantum key. Alice selects n check bits at random and sends the position of the selected bits over the classical channel to Bob. Then Alice and Bob compare the values of the check bits. If more than, say, t bits disagree then they abort and re-try the protocol.

In summary, the attempt of an intruder to eavesdrop increases the level of disturbance of a signal on a quantum communication channel. The two parties wishing to communicate in a secure manner establish an upper bound for the level of disturbance tolerable. They use a set of check bits to estimate the level of noise, or eavesdropping. Then they reconcile their information and distill a shared secret key.

6.7 EPR PAIRS AND BELL STATES

Some of the applications presented in this chapter are based upon the "miraculous" properties of entangled particles, the so-called *EPR pairs*. An EPR pair can be in one of four states called *Bell states*. These states form a normal basis:

$$|\beta_{00}\rangle = \frac{|00\rangle + |11\rangle}{\sqrt{2}}; \qquad |\beta_{01}\rangle = \frac{|01\rangle + |10\rangle}{\sqrt{2}};$$
$$|\beta_{10}\rangle = \frac{|00\rangle - |11\rangle}{\sqrt{2}}; \qquad |\beta_{11}\rangle = \frac{|01\rangle - |10\rangle}{\sqrt{2}}.$$

The four Bell states are states of maximum entanglement between the two particles; any one of them can be transformed into any other by a purely local

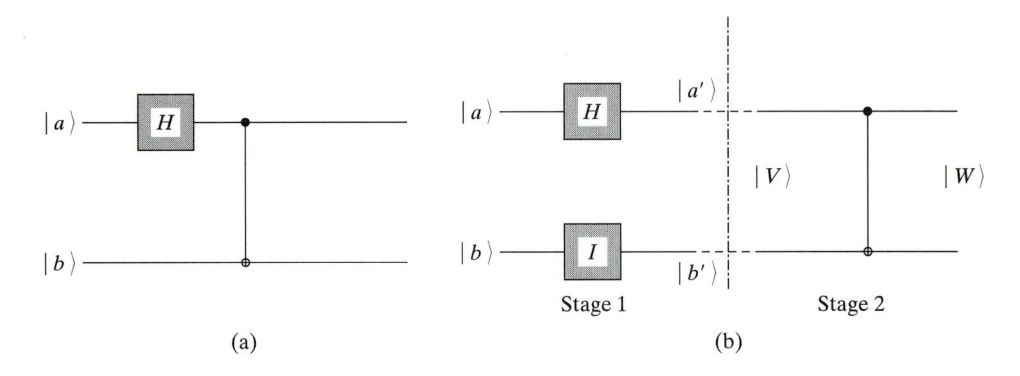

FIGURE 6.8: (a) A quantum circuit to create Bell states. (b) The two stages of the circuit on the left showing the I gate for the second qubit in the first stage.

rotation of one of the particles. The last state in this list, β_{11}, is usually called the *anti-correlated* state.

The circuit in Figure 6.8(a) takes as input two particles in pure states $|\,0\rangle$ and $|\,1\rangle$ and creates a pair of entangled particles. The circuit consists of a `CNOT` gate with a `Hadamard` gate on its control input.

It is easy to show that the truth table of the quantum circuit in Figure 6.8(a) is

In	Out
$\vert\,00\rangle$	$(\vert\,00\rangle + \vert\,11\rangle)/\sqrt{2} = \vert\,\beta_{00}\rangle$
$\vert\,01\rangle$	$(\vert\,01\rangle + \vert\,10\rangle)/\sqrt{2} = \vert\,\beta_{01}\rangle$
$\vert\,10\rangle$	$(\vert\,00\rangle - \vert\,11\rangle)/\sqrt{2} = \vert\,\beta_{10}\rangle$
$\vert\,11\rangle$	$(\vert\,01\rangle - \vert\,10\rangle)/\sqrt{2} = \vert\,\beta_{11}\rangle$

Figure 6.8(b) helps us understand the operation of the circuit used in Figure 6.8(a) to generate entangled states. Here we distinguish two stages:

1. A first stage where the two input qubits are transformed separately. The first qubit, $|\,a\rangle$, is applied to the input of a `Hadamard` gate, H, and produces an output state $|\,a'\rangle$. The second qubit, $|\,b\rangle$, is applied to the input of an identity gate, I, and produces an output $|\,b'\rangle = |\,b\rangle$.
2. A second stage consisting of a `CNOT` gate with input state $|\,V\rangle$ and output state $|\,W\rangle$. Here

$$|\,V\rangle = |\,a'\rangle \otimes |\,b'\rangle,$$

$$|\,W\rangle = G_{\text{CNOT}}\,|\,V\rangle.$$

Thus the output of the circuit is $|\,W\rangle = G_{\text{CNOT}}\,|\,V\rangle = G_{\text{CNOT}}(|\,a'\rangle \otimes |\,b'\rangle)$. Recall that the output of the `Hadamard` gate with $|\,\psi\rangle = \alpha_0\,|\,0\rangle + \alpha_1\,|\,1\rangle$ as input is $(\alpha_0/\sqrt{2})(|\,0\rangle + |\,1\rangle) + (\alpha_1/\sqrt{2})(|\,0\rangle - |\,1\rangle)$.

We now derive the output of the first and the second stage of the circuit in Figure 6.8(b). There are four possible cases depending upon the input qubits $| a\rangle$ and $| b\rangle$:

1. $| a\rangle = | 0\rangle$ and $| b\rangle = | 0\rangle$:

$$| a'\rangle = \frac{| 0\rangle + | 1\rangle}{\sqrt{2}}; \quad | b'\rangle = | 0\rangle.$$

The input state is:

$$| V\rangle = \frac{1}{\sqrt{2}}\begin{pmatrix}1\\1\end{pmatrix} \otimes \begin{pmatrix}1\\0\end{pmatrix} = \frac{1}{\sqrt{2}}\begin{pmatrix}1\\0\\1\\0\end{pmatrix}.$$

The output state is:

$$| W\rangle = G_{\text{CNOT}} | V\rangle = \begin{pmatrix}1&0&0&0\\0&1&0&0\\0&0&0&1\\0&0&1&0\end{pmatrix} \frac{1}{\sqrt{2}}\begin{pmatrix}1\\0\\1\\0\end{pmatrix} = \frac{1}{\sqrt{2}}\begin{pmatrix}1\\0\\0\\1\end{pmatrix} = | \beta_{00}\rangle.$$

2. $| a\rangle = | 0\rangle$ and $| b\rangle = | 1\rangle$:

$$| a'\rangle = \frac{| 0\rangle + | 1\rangle}{\sqrt{2}}; \quad | b'\rangle = | 1\rangle.$$

The input state is:

$$| V\rangle = \frac{1}{\sqrt{2}}\begin{pmatrix}1\\1\end{pmatrix} \otimes \begin{pmatrix}0\\1\end{pmatrix} = \frac{1}{\sqrt{2}}\begin{pmatrix}0\\1\\0\\1\end{pmatrix},$$

and the output state:

$$| W\rangle = G_{\text{CNOT}} V = \begin{pmatrix}1&0&0&0\\0&1&0&0\\0&0&0&1\\0&0&1&0\end{pmatrix} \frac{1}{\sqrt{2}}\begin{pmatrix}0\\1\\0\\1\end{pmatrix} = \frac{1}{\sqrt{2}}\begin{pmatrix}0\\1\\1\\0\end{pmatrix} = | \beta_{01}\rangle.$$

3. $| a\rangle = | 1\rangle$ and $| b\rangle = | 0\rangle$:

$$| a'\rangle = \frac{| 0\rangle - | 1\rangle}{\sqrt{2}}; \quad | b'\rangle = | 0\rangle.$$

The input state is:

$$| V\rangle = \frac{1}{\sqrt{2}}\begin{pmatrix} 1 \\ -1 \end{pmatrix} \otimes \begin{pmatrix} 1 \\ 0 \end{pmatrix} = \frac{1}{\sqrt{2}}\begin{pmatrix} 1 \\ 0 \\ -1 \\ 0 \end{pmatrix}.$$

The output state is:

$$| W\rangle = G_{\text{CNOT}}V = \begin{pmatrix} 1 & 0 & 0 & 0 \\ 0 & 1 & 0 & 0 \\ 0 & 0 & 0 & 1 \\ 0 & 0 & 1 & 0 \end{pmatrix} \frac{1}{\sqrt{2}}\begin{pmatrix} 1 \\ 0 \\ -1 \\ 0 \end{pmatrix} = \frac{1}{\sqrt{2}}\begin{pmatrix} 1 \\ 0 \\ 0 \\ -1 \end{pmatrix} = | \beta_{10}\rangle.$$

4. $| a\rangle = | 1\rangle$ and $| b\rangle = | 1\rangle$:

$$| a'\rangle = \frac{| 0\rangle - | 1\rangle}{\sqrt{2}}; \quad | b'\rangle = | 1\rangle.$$

The input state is:

$$| V\rangle = \frac{1}{\sqrt{2}}\begin{pmatrix} 1 \\ -1 \end{pmatrix} \otimes \begin{pmatrix} 0 \\ 1 \end{pmatrix} = \frac{1}{\sqrt{2}}\begin{pmatrix} 0 \\ 1 \\ 0 \\ -1 \end{pmatrix}.$$

The output state is:

$$| W\rangle = G_{\text{CNOT}}V = \begin{pmatrix} 1 & 0 & 0 & 0 \\ 0 & 1 & 0 & 0 \\ 0 & 0 & 0 & 1 \\ 0 & 0 & 1 & 0 \end{pmatrix} \frac{1}{\sqrt{2}}\begin{pmatrix} 0 \\ 1 \\ 0 \\ -1 \end{pmatrix} = \frac{1}{\sqrt{2}}\begin{pmatrix} 0 \\ 1 \\ -1 \\ 0 \end{pmatrix} = | \beta_{11}\rangle.$$

This completes the derivation of the output of the quantum circuit used to create entangled particles from particles in pure states.

6.8 UNCERTAINTY AND LOCALITY

Earlier, we noted that Heisenberg's Uncertainty Principle reflects the fact that measurements of quantum systems disturb the system being measured. This fundamental law of physics is even more profound. Outcomes of experiments, we call them *observables*, have a minimum level of uncertainty that cannot possibly be removed with the help of a theoretical model.

Let us assume that we are interested in two observables of a quantum particle, call them $\mathscr{A}$ and $\mathscr{B}$, for example, the position and the momentum of an electron. We prepare a large number of quantum systems—in our example electrons—in the identical state $| \psi\rangle$.

We measure first the observable $\mathcal{A}$ and then the observable $\mathcal{B}$ on all particles. All systems are initially in the same state, $|\psi\rangle$, which we choose to be an eigenstate of $\mathcal{A}$. Therefore, in our measurements we obtain the same value for the observable $\mathcal{A}$, while we notice a large standard deviation of the observable $\mathcal{B}$.

Alternatively, we can measure the observable $\mathcal{A}$ on some of the quantum systems and the observable $\mathcal{B}$ on the other quantum systems. Call $\Delta(\mathcal{A})$ and $\Delta(\mathcal{B})$ the standard deviation of the measurements of $\mathcal{A}$ and $\mathcal{B}$, respectively. Heisenberg's Uncertainty Principle states that

$$\Delta(\mathcal{A}) \times \Delta(\mathcal{B}) \geqslant \frac{1}{2} \mid \langle\psi \mid [\mathcal{A}, \mathcal{B}] \mid \psi\rangle \mid$$

with $[\mathcal{A}, \mathcal{B}]$ the commutator of the two operators, $\mathcal{A}$ and $\mathcal{B}$, as discussed in Section 2.17. When $[\mathcal{A}, \mathcal{B}] > 0$—and this is the case of all non-commutative quantum observables—we know with absolute certainty that $\Delta(\mathcal{A}) \times \Delta(\mathcal{B}) > 0$. This simply means that two non-commutative quantum observables have a minimum level of uncertainty that cannot possibly be removed.[3]

Uncertainty is a fundamental assumption regarding quantum systems. The classical interpretation of probabilities as lack of knowledge regarding future events is no longer true for quantum systems. In quantum mechanics, probabilities reflect our limited ability to acquire knowledge about the present.

The following examples illustrate the conceptual differences between classical and quantum interpretation of probabilities. After observing the weather in Florida for many years, we may determine that the conditional probability of rain for a fifth consecutive day given that there were already four consecutive rainy days is, say, one in one hundred. So, after four rainy days, we can assume with high probability—in our example $p = 0.99$—that there will be no rain during the fifth day. But, very seldom, this prediction will be false because we used a historical rather than a phenomenological model. Should we have based our weather prediction on a very accurate model taking into account the temperature, wind strength and direction, humidity, and other atmospheric parameters, collected for every point of a very fine atmospheric grid over a long period of time, we would have been able to predict the weather more accurately. A better theoretical model would reduce the error in our predictions.

On the other hand, for quantum systems, once we have measured the position of a particle there is a certain distribution of its momentum. The momentum cannot be precisely determined now, regardless of our ingenuity.

Many physicists and philosophers believe that we need a theory able to describe accurately nature and that such a theory must obey three fundamental principles:

1. Determinism—like the one supported by Isaac Newton's theory. Variables describing results of observations can be known with great precision.

[3]Keep in mind the fact that Heisenberg's uncertainty does not apply to every pair of quantum observables.

Probabilities are acceptable to describe the outcomes of experiments, but only under special conditions (e.g., when inaccurate or unknown boundary conditions limit our ability to get a complete description of reality).

2. Locality—systems far apart in space can influence each other only by exchanging signals subject to the limitations caused by the finite speed of light.
3. Completeness—the theory must include all pertinent elements of reality, such as position and momentum of a particle.

Leading the camp of those who did not accept the uncertainty at the core of quantum mechanics was no more imposing figure than Albert Einstein. He was one of the pioneers of the quantum theory.

The real test for uncertainty came in 1933, when Albert Einstein imagined the following Gedanken experiment designed to put the matter to rest once and for all. The idea of this experiment is to have two particles related to each other, measure one of them and gather knowledge about the other.

Consider two particles "A" and "B" with known momentum flying towards each other and interacting with each other at a known position, for a very brief period of time. An observer, far away from the place where the two particles interacted with each other, measures the momentum of particle "A" and based upon this measurement is able to deduce the momentum of particle "B." The observer may choose to measure the position of particle "A" instead of its momentum. According to the principles of quantum mechanics this would be a perfectly legitimate proposition, but in flagrant violation of common sense. How could the final state of particle "B" be influenced by a measurement performed on particle "A" long after the physical interaction between the two particles has terminated?

A year later, Einstein, Podolski, and Rosen (thus the abbreviation EPR) wrote a paper published in the Physical Review that stimulated a great interest and seemed at that time to have definitely settled the dispute between Niels Bohr and Albert Einstein in favor of the author of the relativity theory [50]. The brilliance of the EPR experiment is that the position and the momentum of one particle are determined precisely by measurements performed on its "entangled twin." The authors of the EPR paper write "If, without disturbing the system, we can predict with certainty (i.e., with probability one) the value of a physical quantity, then there is an element of physical reality corresponding to this physical quantity." They concluded that the position and the momentum of the "entangled twin" are elements of the physical reality and, since quantum mechanics does not allow both to be part of an unboundedly accurate description of the state of the particle, quantum mechanics is an incomplete theory.

In 1952, David Bohm, a former student of Robert Openheimer[4] at Berkeley, suggested a change of the EPR thought experiment. In Bohm's version of the

[4]J. Robert Openheimer (1904–1967), distinguished American physicist, professor at U.C. Berkeley. He was the Director of the Manhattan project to build the first atomic bomb, the first Chairman of the Atomic Energy Commission, and the Director of the Institute

thought experiment there are two particles with one variable of interest, the spin, with two possible values, "up" and "down," instead of two variables of interest, the position and the momentum. Two entangled particles, "A" and "B" are generated by the same source and move away from each other. The two particles are entangled in their spins and if one has spin "up" the other has spin "down" along the same direction. Once Alice measures the spin of particle "A" along the x-axis and finds it to be "up", the spin measured by Bob on particle "B," along the same direction x, must be "down."

Bell inequality is a constraint on the sum of averages of measured observables, based on assumptions of *local realism* (i.e., assumptions of locality and of existence of hidden variables). Quantum mechanics predicts a value for the sum of averages of observables which is in violation of Bell's Inequality.

Consider two photons with the same polarization, moving apart from one another at the speed of light. The experiment is summarized as follows:[5] "If the coherence of the system can be maintained, a measurement made at one location will assign a single definite state to the system, which should be verified by any subsequent measurements, no matter where they occur. Whether the first photon passes through the polarizer or is absorbed completely determines the polarization of both photons from that point onward. The polarizer can be thought of as asking the system *are you polarized at angle θ?* and accepting only a yes/no answer, which the system is then obliged to repeat consistently thereafter.

In the Copenhagen interpretation,[6] there is no information about the angle of polarization of the photons before the measurement. The angle of polarization has no objective reality until it is measured, and the measurement result is not predetermined by anything; it is intrinsically random. After it is measured, then the polarization state of both photons is completely known and will not change. In a *hidden variable theory*, there is a variable, or variables that determine the real polarization angle of the photons; these variables have definite values from the moment the photons are formed, and they determine what the result of the polarization measurement will be.

Let us make the polarization measurement on the second particle at a different angle from our measurement on the first. Now the two photons have different probabilities of passing through their respective polarizers. We can expect that sometimes photon "A" will pass through its polarizer, but photon "B" will be absorbed by its polarizer. We call such events errors, in the sense that there is a disagreement between the measurements of the two polarizers. Suppose the polarizer at "A" is rotated by an angle θ relative to the polarizer at "B." A certain number of errors $\mathcal{E}$ are produced. Now what happens if we

---

for Advanced Studies at Princeton. He thought and wrote much about the problems of intellectual ethics and morality.

[5]For more details see http://www.telp.com/philosophy/qw3.htm.

[6]In the Copenhagen interpretation, nothing can be said about a property of a quantum system before the property is measured. Through a measurement one cannot learn anything about the past state of the system; the measurement can only provide information about the future state of the system.

rotate the "B" polarizer through an angle θ in the opposite direction, so that the total angle between them is 2θ? We expect more errors, but how much more? Twice as many? Bell demonstrated that the error rate with an angle 2θ must be less than or equal to twice the error rate at angle θ: $\mathscr{E}(2\theta) \leqslant 2\mathscr{E}(\theta)$, provided that the photons have a definite polarization (as postulated by the hidden variable theories) and that the "A" photon cannot affect the "B" photon's state instantaneously (this is the requirement of locality)."

When we obtain two contradictory results using two different theoretical models it means that at least one of the models is wrong. Then, we must turn to experiments to determine if both, or only one of them, are (is) incorrect. In this case, the experimental evidence shows that Bell inequality is violated, therefore, the experimental results increase our confidence that quantum mechanics provides a correct model of the physical world.

6.9 POSSIBLE EXPLANATIONS OF THE EPR EXPERIMENT

The Einstein-Podolsky-Rosen (EPR) Gedanken experiment suggests that if two particles interact they become correlated and, when measuring one particle, we gather information about the wave function of the other.

An EPR Gedanken experiment can be recast in a slightly different setting. We consider a source generating a pair of maximally entangled particles and demonstrate their strange properties with the aid of the main characters of cryptographic texts, Alice and Bob, with Carol in a supporting role. Alice and Bob are at different locations and need to communicate with one another. Carol sends one of the entangled particles to Alice and the other to Bob. The state of these particles is described by the vector $| \psi\rangle = (1/\sqrt{2})(| 00\rangle + | 11\rangle)$ in a system with $| 00\rangle$, $| 01\rangle$, $| 10\rangle$, and $| 11\rangle$ as an orthonormal basis.

Let us assume that when Alice measures her qubit she observes the state $| 0\rangle$. This means that the combined state is $| 00\rangle$ and Bob will observe the same state, $| 0\rangle$ on his qubit. When Alice observes her qubit in state $| 1\rangle$, the combined state is $| 11\rangle$ and Bob observes the state of his qubit to be $| 1\rangle$.

This behavior appears to suggest that non-local effects may occur. Einstein, Podolski, and Rosen proposed a *hidden variable theory*. They argued that the state of the particle is hidden from us; both particles are either in state $| 0\rangle$, or in state $| 1\rangle$, but we do not know which one. Later, John Bell showed that the hidden variable theory predicts that measurements performed on any system must satisfy the so-called Bell inequality [6, 7]. But measurements performed on quantum systems indicate that the Bell inequality is violated. Thus the hidden variable theory must be false.

The results of measurements performed with respect to different bases confirm the fallacy of the hidden variable theory.

Another attempt to explain the EPR experiment is that the measurement on one of the particles affects the other. But this contradicts the relativity theory as we shall see shortly. Imagine two external characters, say Samantha and Hector who are moving relative to each other while observing Alice and Bob. Samantha reports that Alice measures her particle first (Alice may have observed the state

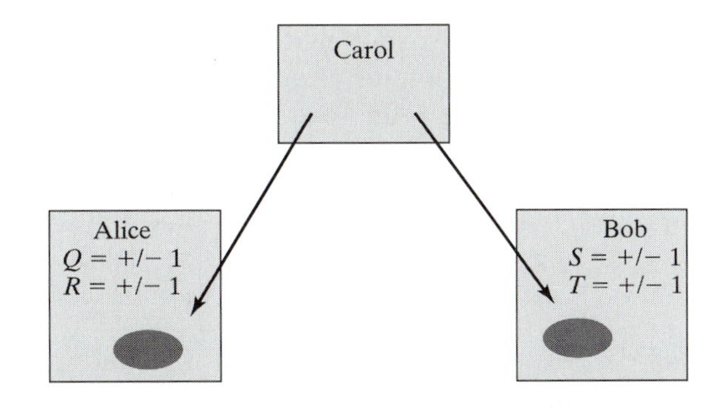

FIGURE 6.9: The experimental set-up for Bell's Inequality. Carol prepares two particles and sends one of them to Alice and the other to Bob. Alice can measure properties $\mathcal{Q}$ and $\mathcal{R}$ of her particle while Bob can measure properties $\mathcal{S}$ and $\mathcal{T}$ of his particle. The results of the measurements can only be $Q, R, S, T = \pm 1$. Bell's inequality requires that $|\, \mathbf{E}(QS) + \mathbf{E}(RS) + \mathbf{E}(RT) - \mathbf{E}(QT)\,| \leq 2$.

$|\,1\rangle$ of her particle forcing the same state on Bob's particle). In turn, Hector reports that Bob measures his particle first (Bob may have measured the state $|\,0\rangle$ of his particle forcing the same state on Alice's particle). Yet, the laws of physics must be independent of the observer's coordinate system. In our case, we must provide equally consistent explanations of the observation reported by both Samantha and Hector. Therefore, causality does not explain the EPR paradox either.

6.10 BELL'S INEQUALITY—LOCAL REALISM

We now describe a Gedanken experiment similar to EPR that sets the stage for deriving *Bell's Inequality*. This new experiment is consistent with common sense and does not involve any reference to quantum mechanics. To derive our results we need two intuitive and common sense assumptions collectively known as *local realism*:

1. *Locality*—measurements of different physical properties of different objects, carried out by different individuals, at distinct locations, cannot influence each other.
2. *Realism*—physical properties are independent of observations.

Assume that Carol prepares a pair of entangled particles and sends one of them to Alice and the other to Bob, see Figure 6.9. There are two physical properties that Alice could measure on her particle, $\mathcal{Q}$ and $\mathcal{R}$. The results of these measurements are the values of the two physical properties, Q and R, respectively. We assume that the results of the measurements can only be $Q, R = \pm 1$.

Similarly, Bob can measure his particle which in turn has two physical properties $\mathcal{S}$ and $\mathcal{T}$. The results of these measurements can be $S, T = \pm 1$. All four properties $(\mathcal{Q}, \mathcal{R}, \mathcal{S}, \mathcal{T})$ are objective and the results of the measurements, (Q, R, S, T), have a well-defined physical interpretation.

Neither Alice, nor Bob knows in advance which property they will measure. They perform the measurements simultaneously. Before a measurement, each one of them tosses a fair coin to decide which property they are going to measure.

Q, R, S, and T are random variables with a joint probability distribution function $p(q, r, s, t) = Prob(Q = q, R = r, S = s, T = t)$. We shall prove that

$$| \mathbf{E}(QS) + \mathbf{E}(RS) + \mathbf{E}(RT) - \mathbf{E}(QT) | \leq 2$$

where $\mathbf{E}(\kappa)$ denotes the expected value of the random variable κ and $| \mathbf{E}(\kappa) |$ means the absolute value of $\mathbf{E}(\kappa)$.

Take κ to be

$$\kappa = QS + RS + RT - QT = S(R + Q) + T(R - Q).$$

It is relatively easy to see that $\kappa = \pm 2$. Indeed, either the first term, $S(R + Q)$, or the second one, $T(R - Q)$, must be equal to zero because $Q, R, S, T = \pm 1$. If one of the terms is zero, then the other one must be equal to ± 2 as shown in the following table:

R	Q	$R + Q$	$R - Q$
+1	+1	+2	0
+1	−1	0	+2
−1	+1	0	−2
−1	−1	−2	0

Now

$$\mathbf{E}(\kappa) = \sum_{(q,r,s,t)} p(q, r, s, t)(qs + rs + rt - qt) \leq 2 \times \sum_{(q,r,s,t)} p(q, r, s, t).$$

But

$$\sum_{(q,r,s,t)} p(q, r, s, t) = 1.$$

The results of the measurements performed by Alice, Q and R, and the ones performed by Bob, S and T, are independent random variables. The pairs of random variables, each pair consisting of one variable from Alice and the other from Bob, QS, RS, RT, and QT, are also independent random variables, thus

$$\mathbf{E}(QS + RS + RT - QT) = \mathbf{E}(QS) + \mathbf{E}(RS) + \mathbf{E}(RT) - \mathbf{E}(QT) \leq 2.$$

This completes our proof of Bell's inequality. It is now time to leave the world of classical systems and turn our attention back to quantum systems. Consider a pair of entangled qubits in the state $| \psi \rangle = (1/\sqrt{2})(| 01 \rangle - | 10 \rangle)$. Alice has one of the particles of the entangled pair and Bob has the other. We have several one-qubit gates that can be used to observe a qubit; among them are the X gate (it transposes the components of a qubit) and the Z gate (it flips the sign of a qubit). Alice and Bob measure the following observables

$$\begin{array}{llll} Alice & \Longrightarrow Q = Z_1 & & R = X_1 \\ Bob & \Longrightarrow S = (1/\sqrt{2})(-Z_2 - X_2) & & T = (1/\sqrt{2})(Z_2 - X_2). \end{array}$$

The observable Q is the output of a Z gate and R the one of an X gate with Alice's particle as input. A similar interpretation holds for Bob's observables.

Let $\langle QS \rangle$ denote the average value of the observable QS. The average values of pairs of observables are

$$\langle QS \rangle = \langle RS \rangle = \langle RT \rangle = \frac{1}{\sqrt{2}} \langle QT \rangle = -\frac{1}{\sqrt{2}}.$$

It follows immediately that

$$\langle QS \rangle + \langle RS \rangle + \langle RT \rangle - \langle QT \rangle = 2\sqrt{2}.$$

This means that quantum mechanics predicts a value for the sum of averages of observables in violation of Bell's Inequality. When we obtain two contradictory results using two different theoretical models, it means that one of the models is wrong. Then we must turn to experiments to determine which one is wrong.

The results of numerous experiments confirm the predictions of quantum mechanics and gives us confidence that quantum mechanics is a correct model of the physical world. No amount of experimentation can prove that a theory is correct, but a single experiment may prove that it is wrong.

The fact that quantum systems violate Bell's Inequality leads us to believe that at least one of the two common sense assumptions needed to derive Bell's Inequality is wrong. Sandu Popescu discusses issues related to non-locality in [106].

6.11 REVERSIBILITY AND ENTROPY

We now turn our attention from communication involving quantum states to the relationship between information, energy, and matter. It seems very sensible to believe that the information must be related to some physical properties of the matter. This is the main theme of the next few sections.

To understand which steps of a computation require energy consumption we discuss the reversibility of physical phenomena and review the basic principles of thermodynamics.

A physical process is said to be *reversible* if it can evolve forwards as well as backwards in time. A reversible system can always be forced back to its original state from a new state reached during an evolutionary process.

Reversibility is a fundamental property of nature. This may not be obvious at a macroscopic level where irreversibility seems to prevail. A crystal glass dropped on the floor breaks, and the pieces cannot be glued back together. Red wine spilled on a silk table cloth cannot be recovered. A missile fired from a launcher cannot be brought back.

Yet, many believe that at the atomic level, all processes are reversible. This means that all equations of quantum mechanics are symmetric in time. If we replace time, t, with $-t$, the equations are not altered.

Macroscopic systems could behave reversibly as well. Of course, if you drop a piano from the 25th floor of an apartment building to the street below, it will break into pieces. But if you have a pulley and lower it slowly, it will reach street level in one piece; its potential energy will be gradually transferred to the counter weight. *The trick is to make slow, subtle changes in the system rather than sudden, dramatic ones. The system must be in equilibrium with its environment at all times.*

Fortunately, some properties of physical systems do not require a detailed knowledge of the structure of matter and they were studied during the nineteenth century, well before the atomic structure of matter was fully understood. The subject of the branch of physics called *thermodynamics* is the statistical behavior of ensembles of molecules. For example, we wish to characterize a gas by familiar macroscopic properties such as, volume, pressure, and temperature without knowing its microscopic properties (e.g., the vectors describing the velocities of individual molecules of the gas). A comprehensive discussion of the subject is well beyond our intentions; the interested reader will certainly enjoy the presentation in physics texts, our favorite being Feynman's *Lectures on Physics* [53]. In this chapter we only present several concepts useful to illustrate the relationship between information and energy dissipation.

6.12 THERMODYNAMICS AND THERMODYNAMIC ENTROPY

Thermodynamics is the study of energy, its ability to carry out work, and the conversion between various forms of energy, such as the internal energy of a system, heat, and work. Thermodynamic laws are derived from statistical mechanics.

There are several equivalent formulations of each of the three laws of thermodynamics. The *First Law of Thermodynamics* states that energy cannot be created, nor destroyed, it can only be transformed from one form to another. An equivalent formulation is that the heat flowing into a system equals the sum of the change of the internal energy and the work done by the system.

The *Second Law of Thermodynamics* states that it is impossible to create a process that has the unique effect of subtracting positive heat from a reservoir and converting it to positive work. An equivalent formulation is that the entropy of a closed system never decreases. A consequence of this law is that no heat engine can have 100% efficiency.

The *Third Law of Thermodynamics* states that the entropy of a system at zero absolute temperature is a well-defined constant. This is due to the fact that many systems at zero temperature are in a ground state and the entropy is determined by the degeneracy of the ground state. For example, crystal lattices have a unique ground state and have zero entropy when the temperature reaches 0° Kelvin.

C. P. Snow[7] summarized the three Laws of Thermodynamics as follows:

1. You cannot win. Matter and energy are conserved, thus you cannot get something for nothing.
2. You cannot break even. You cannot return to the same energy state because there is always an increase in disorder. The entropy always increases.
3. You cannot get out of the game. Absolute zero is unattainable.

Clausius defined *entropy* as a measure of energy unavailable for doing useful work. He discovered the fact that entropy can never decrease in a physical process and can only remain constant in a reversible process. This result became known as the Second Law of Thermodynamics. There is a strong belief among astrophysicists and cosmologists that our Universe started in a state of perfect order and its entropy is steadily increasing, leading some to believe in the possibility of a "heat death." Of course, this sad perspective could be billions or trillions of years away.

Statistical arguments indicate that systems tend to become more disordered. Indeed, our immediate experience and intuition indicate that disordered states vastly outnumber the highly ordered states of a system. After any change of state, a system is more likely to settle into a disordered state, than into an ordered one.

The higher the entropy of a system, the less information we have about the system. *Henceforth, information is a form of negative entropy.* Claude Shannon recognized the relationship between the two and, on von Neumann's advice, he called the negative logarithm of probability of an event, *entropy*.[8]

The following quantities are used in thermodynamics equations: the temperature, T, the pressure, p, the heat, Q, the energy of the system, U, the free energy, F, the work done on the system, W. All of them are statistical averages.

The First Law of Thermodynamics is a conservation law; it states that the total change in the energy of a system is the sum of the heat put into the system and the work done on the system

$$\Delta U = \Delta Q + \Delta W.$$

Throughout this section Δ signifies a finite change while δ is an infinitesimal change of a variable.

The *thermodynamic entropy* of a gas, S, is also defined statistically but it does not reflect a macroscopic property. *The entropy quantifies the notion that a gas is a statistical ensemble and it measures the randomness, or the degree of*

[7]Charles Percy Snow, Baron Snow of Leicester, 1905–1980, English author and physicist. Snow had an active, varied career, including several important positions in the British government. As a novelist, Snow was particularly noted for his series of 11 related novels known collectively as Strangers and Brothers.

[8]It is rumored that von Neumann told Shannon "It is already in use under that name and besides it will give you a great edge in debates because nobody really knows what entropy is anyway" [29].

disorder of the ensemble. The entropy is larger when the vectors describing the individual movements of the molecules of gas are in a higher state of disorder than when all of them are well organized and moving in the same direction with the same speed. Boltzmann postulated that:

$$S = k_B \ln \Omega.$$

where k_B is Boltzmann's constant, $k_B = 1.3807 \times 10^{-23}$ Joules per degree Kelvin and Ω is the number of microstates which are specified by the positions and the momenta of all molecules of gas. A version of this equation is engraved on Ludwig Boltzmann's tombstone.[9]

The Second Law of Thermodynamics tells us that the entropy of an isolated system never decreases. Indeed, differentiating the previous equation we get

$$\delta S = k_B \frac{\delta \Omega}{\Omega} \geqslant 0.$$

It is relatively easy to see that when we compress a volume containing N molecules of gas from V_1 to V_2, maintaining the temperature of the system constant (isothermal compression), the work done on the system is

$$\Delta W = -\int_{V_1}^{V_2} \frac{N k_B T}{V} dV = -N k_B T \ln \frac{V_2}{V_1}.$$

Indeed, the pressure, the variation of volume δV and the work are related by $\delta W = -p \delta V$. But an ideal gas with volume V, at temperature T, and pressure p, obeys the law $pV = N k_B T$.

For an isothermal process $\Delta U = 0$ and $\Delta Q = -\Delta W$ and the entropy defined as $\Delta S = \Delta Q / T$ becomes $\Delta S = -\Delta W / T$. It follows immediately that

$$\Delta S = N k_B \ln \frac{V_2}{V_1}.$$

Let us now consider a gas consisting of a single molecule. When $N = 1$ and we reduce the volume of the gas in half, $V_2 = V_1/2$ the previous expression becomes

$$\Delta S = -k_B \ln 2.$$

The reduction of an ensemble to a single molecule requires a leap of faith; yet it allows us to relate information and entropy.

By reducing the volume where the molecule can be located, we have increased our information about the system and we have decreased the entropy. Now the molecule can hide in a volume twice as small as before.

Was the Second Law of Thermodynamics violated by this experiment? No, because we do not have an isolated system, we have in fact increased the amount of free energy by $k_B T \ln 2$ and decreased the entropy by $k_B \ln 2$.

---

[9] Boltzmann took his life in 1906 not knowing the impact of his findings upon the world of physics.

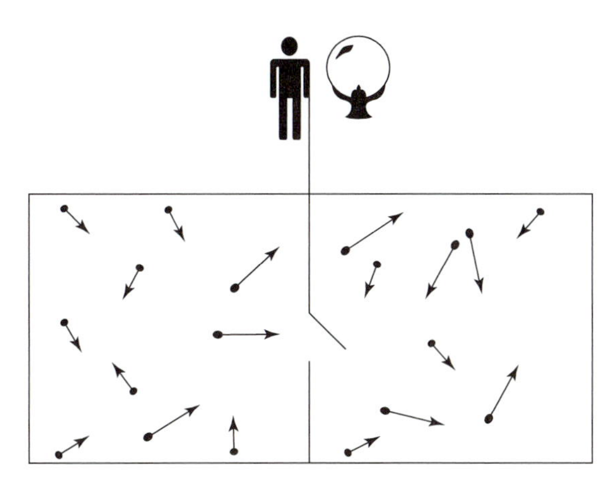

FIGURE 6.10: Maxwell's Demon separates fast-moving molecules of gas from slow-moving ones; the fast moving ones end up on the right and the slow moving ones on the left. The demon separates hot from cold in blatant violation of the Second Law of Thermodynamics, or so it seems.

Now we can use the gas cylinder with a molecule of gas to store information. We reset our "bit" by compressing the volume and then let the volume expand and the molecule will be in one or the other halves of the cylinder depending upon its energy and the bit will be either 0 or 1. But we need to expend energy to compress the gas and this means that erasing information is the moment when we expend energy.

Two terms frequently used in thermodynamics are *adiabatic*, meaning "without transfer of heat," and *isothermal*, meaning "at a constant temperature." For example, if we open the valve of a gas canister, the gas rushes out and expands without having the time to equalize its temperature with the environment. The expansion is adiabatic and the rushing gas feels cool.

6.13 MAXWELL'S DEMON

One of the formulations of the Second Law of Thermodynamics is that the entropy of a system is a non-decreasing function of time. Now we describe an experiment that seems to violate this law.

Imagine the molecules of a gas in a cylinder partitioned in two by a wall with a slit covered with a door controlled by a little demon (see Figure 6.10). The demon examines every molecule of gas and determines its velocity; those of high velocity on the left are allowed to migrate to the right and those with low velocity on the right are allowed to migrate to the left. As a result of these measurements, the demon separates hot from cold in blatant violation of the Second Law of Thermodynamics. The entropy is a measure of the degree of disorder of a system and Maxwell's Demon creates order by separating hot molecules (those with high velocity) from cold molecules (those with low velocity).

For almost a century physicists have been trying to spot the flaw in Maxwell's argument without great success. In 1929, Leo Szilard had the intuition to relate the demon and the binary information. He imagined a simplified version of

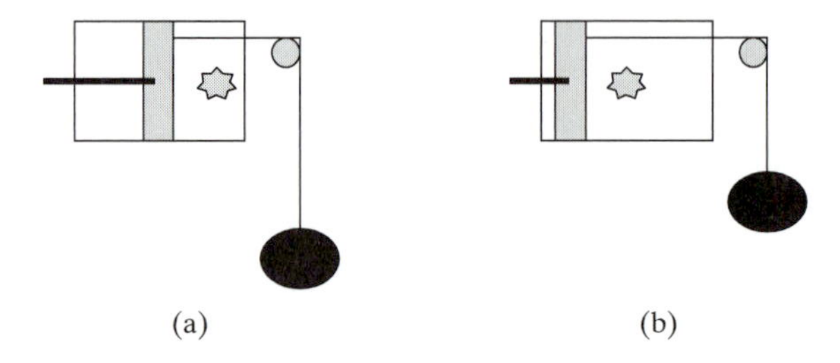

FIGURE 6.11: Erasing information requires energy consumption. Szilard's Gedanken experiment considers a single molecule of gas. The demon performs a measurement of the position of the molecule in the cylinder. (a) The system is in state 0. The molecule is on the right, the piston is pushed halfway without any energy consumption. (b) The system is in state 0 or 1. The molecule moves to the left, the piston is pushed and lifts the weight. The molecule can be anywhere in the cylinder.

Maxwell's thought experiment. Consider a horizontal cylinder with a piston and a single molecule of gas inside the cylinder. If the demon waits until the molecule is on the right side, the piston is pushed halfway without any energy consumption, Figure 6.11(a). As the molecule moves to the left, the piston is pushed and lifts the weight, Figure 6.11(b).

This experiment shows that *the demon has to expand energy to reduce the degree of disorder of the system.* Indeed, the state shown in Figure 6.11(a) corresponds to the binary state 1 and the entropy of the system is low, we know that the particle is confined to the right side of the cylinder. To reach the state in Figure 6.11(b) we erase information. Now that the binary information is either 0 or 1, the position of the molecule is less certain. The molecule could be anywhere in the cylinder. To reach this state the system must consume energy.

The great insight brought by Leo Szilard is that the demon had to expend energy by performing the measurements. Many regard Leo Szilard as the father of information theory. *Szilard identified the measurement, the information, and the memory as critical aspects of the thorny problem posed by Maxwell.*

In 1950, Leon Brillouin advanced the idea that the demon needed a torchlight to see where the molecule is located and that the energy of the photons emitted by the torchlight should exceed the energy of the background photons.

6.14 ENERGY CONSUMPTION—LANDAUER'S PRINCIPLE

Is there an analogy between a heat engine and a computing engine? Is it sensible to think that only irreversible processes in computations require energy consumption? These were the main questions addressed by Rudolf Landauer, a physicist working at IBM Research Laboratories.

The precise physical phenomena leading to energy consumption in a computing device was identified in 1961 by Landauer [87] who discovered that erasing information is the process that requires energy dissipation. At first sight, this seems a strange statement so we need a simple model to justify it. We already know that if a process is irreversible it requires some energy consumption. Thus, we only need to prove that erasing information is an irreversible process. This is

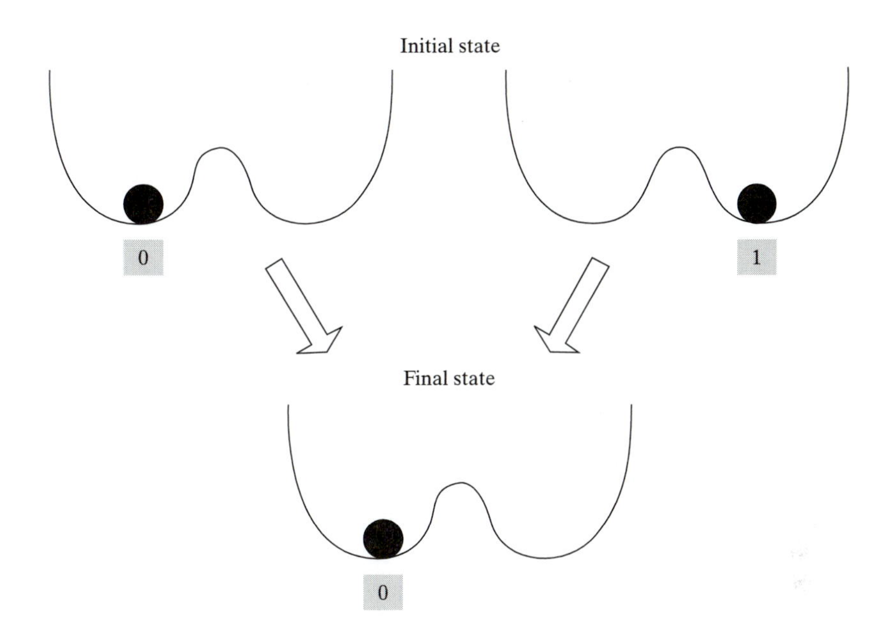

FIGURE 6.12: Landauer's double-well information storage model used to illustrate the fact that erasing information is an irreversible process. The information is provided by the position of the ball. If the ball is in the left well, then we have stored a "0." If the ball is in the right well we have stored a "1." Erasing information means that regardless where the ball happens to be initially, in the final state the ball must be in the left well. This implies that we have a mapping from two different initial states (the ball in the left well, and the ball in the right well) to a single final state (the ball in the left well). Clearly, such a process is irreversible, henceforth erasing information is always associated with energy dissipation.

much more easier to grasp. Once we erased the blackboard, the information on it is lost. We have to back up periodically our files, because once a disk crashes, the information stored on the disk cannot be retrieved.

Let us consider a slightly more formal justification of the statement that erasing information is an irreversible process. Consider a storage system for binary information. The double-well in Figure 6.12 allows us to store a bit of information; if the ball is on the left, then we have a "0" and if it is on the right we have stored a "1." Erasing information means that wherever the ball happens to be in the current state, we should end in a state with the ball on the left side of the well. In other words, we have a mapping from two different initial states, the ball in the left well and the ball in the right well, to a single final state, the ball in the left well. Clearly, such a process is irreversible, henceforth erasing information is always associated with energy dissipation.

There are two equivalent formulations of *Landauer's Principle*, one in terms of energy consumption and the other in terms of entropy:

(a) When a computer erases a bit of information the amount of energy dissipated into the environment is at least $k_B T \ln 2$ with k_B Boltzmann's constant and T the temperature of the environment.

(b) When a computer erases a bit of information the entropy of the environment increases by at least $k_B \ln 2$ with k_B Boltzmann's constant.

Landauer's Principle traces the energy consumption in a computation to the need to erase information. This seems a bit counterintuitive, we would expect that writing information also requires energy dissipation, that some symmetry of these two operations exists.

Landauer provides only a *lower bound* on the energy consumption. It turns out that the logic circuits in microprocessors vintage year 2000 need roughly $500 k_B T \ln 2$ in energy for each logic operation [98]. Since 1970, the computing power of microprocessors has doubled every 18 months following very closely the prediction of Intel's Gordon Moore. To limit the energy consumption of increasingly more powerful microprocessors the energy consumed for every logical operation had to decrease at a rate comparable to, or exceeding the one given by the Moore's Law.

A consequence of Landauer's Principle is that *no strictly positive lower bound on energy consumption exists for a reversible computer. If we could build a reversible computer that does not erase any information, then, in principle, we can compute without any energy loss.* This is, in itself, less shocking than it seems at a first glance; all laws of physics are reversible and if we know the final state of a closed physical system then we can determine its initial state.

After understanding Landauer's Principle, the flaw in the Gedanken experiment proposed by Maxwell's demon is obvious: *the demon has to perform some measurement before allowing molecule of gas approaching the slit to cross over to the other side. The results of measurements must be stored in the demon's memory. But his memory is finite, so the demon must start erasing information after a while.*

According to Landauer's Principle, the entropy of the entire system, including the gas cylinder and the demon, increases as a result of erasing information. This increase should be large enough to compensate for the decrease in entropy caused by the separation of molecules with high velocity from the ones with low velocity and to vindicate the Second Law of Thermodynamics.

In 1929, Leo Szilard noticed the role of the measurement process in this Gedanken experiment. Charles Bennett was the first to point out that the erasure of information and not the measurements are the source of the entropy created in the process of separating high velocity molecules from the low velocity ones in this experiment. His model for the demon's behavior allows the demon to make the measurements with zero energy expenditure: the demon is initially in a state of uncertainty, let us call this state ξ. After measuring the velocity of a molecule, the demon enters a state α for a high velocity molecule approaching, or state β for a low-velocity one and overwrites ξ with either α or β. This can be done without any energy expenditure; the energy is dissipated in the next step when the demon has to erase the α or β and set the value to ξ, in order to prepare for the next measurement.

The fact that erasing information requires consumption of energy is in strict agreement with the Second Law of Thermodynamics. Erasing information reduces our knowledge about the state of a system thus, it leads to an increase of the entropy; we need to consume energy to compensate for the increase of the entropy.

6.15 LOW POWER COMPUTING—ADIABATIC SWITCHING

John von Neumann [141] was the first to reflect on the absolute minimum amount of energy required for an elementary operation of an abstract computing device capable of making binary decisions and of transmitting information. He advanced the idea that this energy is of the order of $k_B T$. At room temperature, $k_B T \approx 4 \times 10^{-21}$ Joules. John von Neumann reasoned that if a capacitor is used to store a bit of information one would need an amount of energy large enough to guarantee that the level of the signal is above the noise level.

As early as 1978, Fredkin and Toffoli discussed a scheme to implement reversible logic circuits using switches, capacitors, and inductors. Under normal circumstances, a capacitor with capacity C charged to voltage V dissipates as heat an amount of energy equal to $(CV^2)/2$ when discharged instantly. In their scheme, a capacitor used to store a bit of information could be discharged without losing the energy. The electric charge was transferred to an inductor and from there to another capacitor. Unfortunately, the scheme is not practical; inductors cannot be accommodated on a silicon substrate.

Adiabatic switching is a term coined for a switching device that does not produce heat. Charles Seitz from Caltech invented a scheme called *hot-clocking* when the energy is saved by varying the power supply voltages. Ralph Merkle from Xerox PARC, William Athas from USC, and Storrs Hall from Rutgers pioneered a reversible adiabatic switching scheme.

6.16 BENNETT'S INFORMATION DRIVEN ENGINE

Bennett imagined an information driven engine. Instead of using electricity, gas, or any other traditional source of energy, Benenett's engine consumes a tape with information stored on it and converts this information into energy (see Figure 6.13). This sounds very exciting, but do not jump to conclusions yet; it may be a while until you'll be able to create an audio or video tape recording your kids in the evening, and then feed the tape next morning to your Honda Civic, instead of filling it up with gas, hydrogen, or electricity.

The engine vaguely resembles a Turing Machine; the input tape contains cells, each cell is similar to the cylinders with one molecule of gas in it. The machine spits out a "randomized tape" where the atom can be anywhere in the cylinder. This system is just the opposite of the one presented earlier where we used energy to force the molecule of gas into one half of the cylinder and reduced the entropy. In this engine the entropy, which is a measure of randomness, increases and this means that some energy is produced.

Let us now carefully describe the setup. The engine itself is submerged into a heat bath. A *heat bath* is a system able to keep constant the temperature T of the engine. When a cell enters the engine, its contents are spilled inside a cylinder with a piston half way into the cylinder. The molecule heats up to the

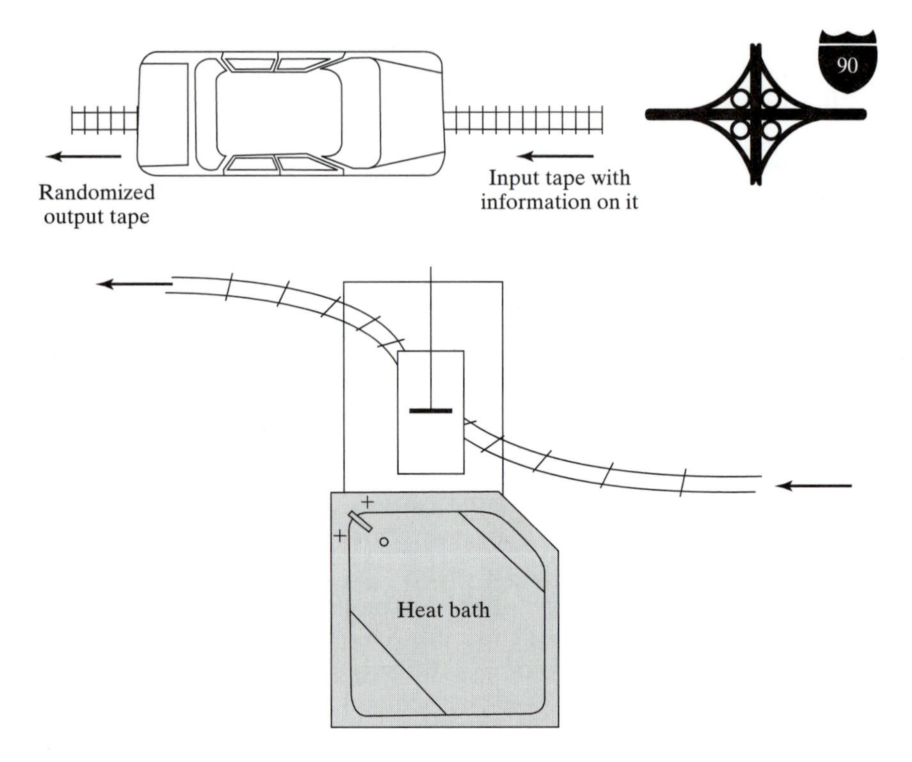

FIGURE 6.13: Bennett's information-driven engine. The input tape contains information; the output tape is randomized. The entropy of the system increases and the energy produced is used to move a piston. The engine itself is submerged into a heat bath. When a tape cell enters the engine, its contents are spilled into a cylinder with a piston half way into the cylinder. The molecule heats up to the temperature of the heat bath. When the system has reached thermal equilibrium, the molecule isothermally pushes the piston and we are able to extract the energy of this movement.

temperature of the heat bath. When the system has reached thermal equilibrium the molecule isothermally pushes the piston and with some imagination we are able to extract the energy of this movement. If we have a tape with n bits of information on it, then the work produced by the engine is $nk_BT \ln 2$.

6.17 LOGICALLY REVERSIBLE TURING MACHINES AND PHYSICAL REVERSIBILITY

An ordinary Turing Machine has a control unit and a read-write head; it performs a sequence of read-write-shift operations on an infinite tape divided into squares. The dynamics of the Turing Machine are described by quintuples (A, T, A', T', σ) of the form

$$AT \mapsto T'\sigma A'.$$

The significance of this notation is that when the control unit is in state A and the symbol currently scanned by the read-write head is T, then the machine will first write T' in place of T and then shift left one square, right one square, or remain on the same square, depending upon the value of σ ($\sigma = -, +, 0$, respectively). The new state of the control unit will be A'. An n-tape Turing Machine is one where T, T', and σ of each quintuple are themselves n-tuples.

A Turing Machine performs a mapping of its entire current state into a successor state given by its transition function. The entire state of a Turing Machine is given by the state of its control unit, the tape contents, and the position of the read-write head. When a Turing Machine traverses a set of states we have a set of mappings associated with this evolution.

A Turing Machine is deterministic if and only if the quintuples defining the mappings have non-overlapping domains. This is guaranteed by requiring that the portion of the tape to the left of the arrow marking the position of the read-write head be different for different quintuples.

A Turing Machine is reversible if and only if the mappings have non-overlapping ranges. An ordinary Turing Machine is not reversible. Indeed, the write and shift operations in a cycle do not commute; the inverse of a read-write-shift cycle is shift-read-write.

The problem of constructing a reversible computing automaton is non-trivial. A tempting solution is to add to the Turing Machine a *history tape*, initially blank, and save on this tape the details of every operation performed. Then we would be able to retrace the steps of the direct computation; starting from the last record on the tape we would determine the previous state and keep going back until we reached the last record on the tape.

The history tape must be left blank at the end, in the same state it was when we started the process. We know by now that erasing information requires energy dissipation. Thus, we require that the reversible computer, if it halts, should erase all intermediate results leaving only the original input and the desired output; hence, the process of building a reversible automaton is more intricate than we have anticipated.

In 1973, Charles Bennett was able to prove that given an ordinary Turing Machine $\mathcal{S}$ one can construct a reversible three tape machine $\mathcal{R}$ which emulates $\mathcal{S}$ on any standard input, and which leaves behind, at the end of its computation only the original input and the desired output. The formal proof of this statement can be found in [11].

Here we present Bennett's informal argument. Imagine that at the end of the original computation which is deterministic and reversible we continue with a stage when the machine uses the inverse of the original transfer function and carries out the entire computation backwards. Like the forward computation, the backwards computation is deterministic and reversible.

We have to be a bit careful and create a copy of the results on a new tape immediately after the completion of the forward computation and before the backward one starts, because the backward computation destroys the results. We have also to stop recording on the history tape during the process of copying the results. The copy operation can be done reversibly if we start with a blank tape.

After this three-stage process the system will consist of a copy of the results obtained during the first phase, the copy being made during the second phase, and the original input tape reconstructed during the third phase. A vast

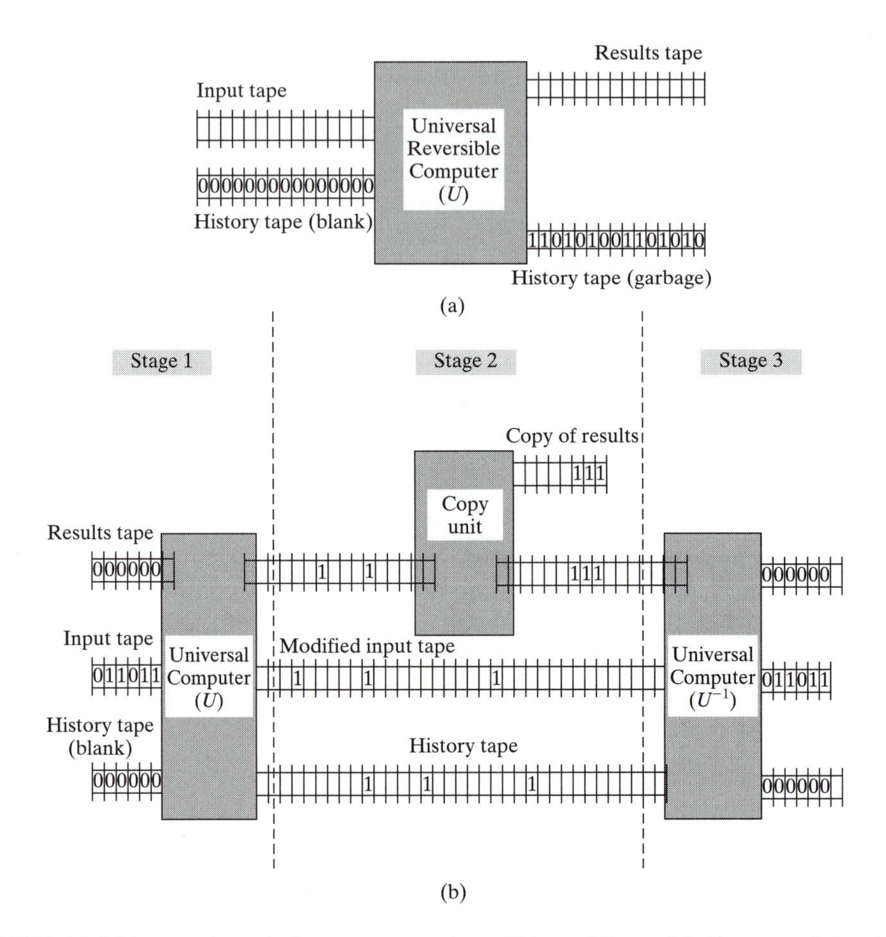

FIGURE 6.14: (a) A tempting solution to constructing a Universal Reversible Computer is to add to a Turing Machine a *history tape*, initially blank, and save on this tape the details of every operation performed. Then we retrace the steps of the direct computation; starting from the last record on the tape we determine the previous state, and keep going back until we reached the last record on the tape. (b) A Zero Entropy Loss Reversible Computer. The history tape must be left blank at the output, in the same state it was when we started the process. Erasing information requires energy dissipation. Thus, we require that the reversible computer, if it halts, should erase all intermediate results leaving only the original input and the desired output.

amount of storage on the history tape was used but returned to its original blank condition after the third stage.

Once convinced that logically reversible automata exist, we can think of thermodynamically reversible physical computers operating very slowly near thermodynamic equilibrium. For example, a chemical reversible computer could consist of DNA encoding logical states and reactants able to change the logical state.

Figures 6.14(a) and (b) depict a Universal Reversible Computer and a Zero Entropy Loss Reversible Computer based upon Bennett's arguments discussed above. They are inspired by Feynman [57].

6.18 HISTORICAL NOTES

In 1905, Albert Einstein, then a little known public servant of the patent office in Bern, published in the *Annalen der Physik* three brief papers that attracted the world's attention: one on the photoelectric effect, one on the special theory of relativity, and one on statistical thermodynamics.[10] In his explanation of the photoelectric effect, Einstein considered the photons as discrete packets carrying an energy $E = h\nu$ with h the Planck constant and ν the frequency of the light. His explanation was in perfect agreement with the experiments and with Max Plank's expression for the energy produced by a light-emitting system, $E = nh\nu$ with $n = 0, 1, 2, ...$

For years Einstein, together with other preeminent scientists of the time, met regularly at Solvay Conferences.[11] Regularly, at these meetings Einstein attempted to illustrate the fallacy of Heisenberg's Uncertainty Principle through the so-called "Gedanken" or "thought" experiments, to the exasperation of his friends and foes alike. Bohr & Co. were forced to find explanations for the apparent contradictions brought forth by Einstein. One of these experiments is outlined below. A box containing radiation is weighed both before and after releasing several photons. The time of the release is measured precisely by a clock controlling the door of the box and the energy of the photons can be deduced precisely using Einstein's formula $E = mc^2$, where $c = 3 \times 10^{10}$ cm/s is the speed of light. "Thus both the time of the release of the photons and their energy can be precisely determined," argued Einstein. It took Niels Bohr a sleepless night in Bruxelles to find the flaws in Einstein's argument. A recent discussion of this scientific duel is found in [136].

The Copenhagen interpretation represents an attempt by Niels Bohr and several other physicists working with him to frame an overall picture of the world based on a theoretical structure compatible with the quantum phenomena. Until his death, in 1955, Albert Einstein argued with Bohr, Heisenberg, Schrödinger, Dirac, and other supporters of the Copenhagen interpretation that non-determinism cannot possibly be the basis for the laws of nature. The famous pronouncement "God does not play dice" reflects Einstein's strong conviction that the quantum theory was missing something; it was an incomplete scientific theory. Einstein believed that some "hidden variables" are probably missing from quantum mechanics and if one could discover the values of these variables the randomness would disappear. Einstein called the Copenhagen interpretation "the Heisenberg-Bohr tranquilizing philosophy- or religion?" and argued that it "provides a gentle pillow for the true believer from which he cannot

[10]The January 8, 2004 issue of the APS (American Physical Society) News reprints an article written by Robert E. Millikan in 1949 for the seventieth birthday of Albert Einstein. Millikan, who got the 1923 Nobel Prize in Physics for his measurement of the charge of the electron and for his work on the photoelectric effect, discusses the impact of the three seminal papers of the 26-year-old Einstein.

[11]Ernest Solvay was an industrialist from Brussels who financed scientific meetings with the hope of having an audience for his own scientific theories.

easily be aroused" and that it "covers up" the true problems of the quantum universe [128].

In June 1824, Sadi Nicolas Leonard Carnot published a book "Réflexions sur la puissance motrice du feu et sur les machines propres a developper cette puissance" [12] showing that a heat engine can behave reversibly. Carnot's engine consists of an idealized gas in a cylinder with a piston. The system can be heated or cooled by placing it in contact with a hot, or a cool reservoir, respectively. In contact with the hot reservoir the gas within the cylinder expands, pushes the piston and does some useful work. In contact with the cool reservoir, the gas contracts, restoring the piston to its most compressed state. *The motion of the piston must be very slow to ensure that the gas in the cylinder is always in equilibrium with its surroundings.*

Carnot came to the conclusion that if an engine is reversible, it makes no difference how the engine is actually designed. The amount of work obtained if the engine absorbs a given amount of heat at temperature T_1 and delivers heat at temperature T_2 *is a property of the world and not of that particular engine* [53].

In 1865, in conjunction with the study of heat engines, the German physicist Rudolph Clausius abandoned the idea that heat was conserved and stated formally the First Law of Thermodynamics. He reconciled the results of Joule with the theories of Sadi Carnot.

James Clerk Maxwell is best known for equations of the electromagnetic field and for the kinetic theory of gases. By treating gases statistically in 1866 he formulated, independently of Ludwig Boltzmann, the Maxwell-Boltzmann kinetic theory of gases. This theory is based on the assumption that temperature and heat involved only molecular movement. In 1871, Maxwell proposed a "thought" experiment with puzzling results, known as "Maxwell's Demon."

Leo Szilard, in 1929, [134] and John von Neumann in a lecture given in 1949 [141] were the first to address the connection between energy dissipation and computations.

6.19 SUMMARY AND FURTHER READINGS

The remarkable properties of communication involving quantum states exploit the properties of ensembles of entangled particles.

In our discussion of teleportation, we show how to transfer the quantum state, not the actual particle, by sending classical information only. The transfer of quantum information appears to happen instantly, though the subject needs to first receive classical information regarding the result of a measurement.

A demonstration of quantum teleportation was carried out in 1997 at the University of Rome. In this experiment, the information is double encoded into a single photon instead of two. The source generates two photons with opposite polarizations "photon 1" with horizontal polarization, h, and "photon 2" with vertical polarization, v. The polarization entanglement of the two photons sent is converted into an entanglement of the paths followed by the two photons (see

[12] Reflections upon the kinetic energy of the fire and upon the machines capable of using this power.

Figure 6.4). If "photon 1" travels via path A, then "photon 2" travels via path C; if "photon 1" travels via path B, then "photon 2" travels via path D. The message is encoded in the polarization of "photon 1."

Then we discuss dense coding and address the question whether there is an advantage in exchanging quantum information (qubits over a quantum communication channel), instead of sending classical bits over a classical communication channel. We show that by exchanging a single qubit of information over a quantum channel we are able to encode and send four possible messages and that would require sending two bits of classical information over a classical communication channel.

When in addition to the classical communication channel, a quantum communication channel is available, then the key distribution problem has an ingenious solution. We show that the two parties involved in the exchange of the cipher key can detect with a very high probability if a third party eavesdrops.

To gain a deeper insight into the behavior of quantum systems we take another look at Heisenberg's Uncertainty Principle. The classical interpretation of probabilities as lack of knowledge regarding future events is no longer true for quantum systems. In quantum mechanics, probabilities reflect our limited ability to acquire knowledge about the present.

Consider two observables of a quantum particle, call them $\mathcal{A}$ and $\mathcal{B}$. We prepare a large number of quantum systems in the identical state $| \psi\rangle$. All systems are initially in the same state, $| \psi\rangle$, which we choose to be an eigenstate of $\mathcal{A}$. We measure first the observable $\mathcal{A}$ and then the observable $\mathcal{B}$ on all systems. In our measurements, we obtain the same value for the observable $\mathcal{A}$, while we notice a large standard deviation of the observable $\mathcal{B}$.

Alternatively, we can measure the observable $\mathcal{A}$ on some of the quantum systems and the observable $\mathcal{B}$ on the other quantum systems. Call $\Delta(\mathcal{A})$ and $\Delta(\mathcal{B})$ the standard deviation of the measurements of $\mathcal{A}$ and $\mathcal{B}$ respectively. Heisenberg's Uncertainty Principle states that

$$\Delta(\mathcal{A}) \times \Delta(\mathcal{B}) \geqslant \frac{1}{2} \mid \langle \psi \mid [\mathcal{A}, \mathcal{B}] \mid \psi \rangle \mid$$

with $[\mathcal{A}, \mathcal{B}]$ the commutator of the two operators.

Many physicists and philosophers believe that a model of the physical world must obey three fundamental principles (i) determinism, (ii) locality, and (iii) completeness. Yet, the entanglement of quantum particles seems to violate the principle of locality and to allow instant interactions among quantum systems in blatant violation of the limits imposed by the finite speed of light.

Quantum mechanics predicts a value for the sum of averages of observables in violation of Bell's Inequality. The results of numerous experiments confirm the predictions of quantum mechanics and give us confidence that quantum mechanics is a correct model of the physical world. The fact that quantum systems violate Bell's Inequality leads us to believe that at least one of the two common sense assumptions needed to derive Bell's Inequality is wrong.

We want to understand the physical support of information and determine the energy required to store and to transform information. We ask ourselves what is the relationship of information with energy and matter. Leo Szilard and John von Neumann were the first to address the connection between energy dissipation and computations. A physical process is said to be reversible if it can evolve forwards as well as backwards in time. A reversible system can always be forced back to its original state from a new state reached during an evolutionary process.

Reversibility is a fundamental property of nature. At the atomic level, all processes are reversible. All equations of classical and quantum mechanics are symmetric in time. Macroscopic systems could behave reversibly as well if one made slowly subtle changes in the system rather than sudden, dramatic ones. The system must be in equilibrium with its environment at all times.

The book by Cover Thomas [40] provides a comprehensive introduction to information theory. Claude Shannon's papers [116, 117, 118] are the most authoritative accounts for the statistical theory of communication. Charles Bennett's papers published between 1984 and 1998 [13, 14, 15, 16] introduced most of the concepts in quantum information, teleportation, and quantum key distribution. The papers of Rolf Landauer [87], Peter Benioff [8, 9], and Charles Bennett [12] were the first to address the thermodynamics of computations and reversibility. Richard Feynman's book [57] contains a very clear and comprehensible discussion of reversibility. John Bell's paper [6] discusses in depth the EPR experiment, while his collected papers on quantum philosophy [7] contain a most insightful discussion of measurements and other topics in the philosophy of physics.

The book of Lawrence Sklar [128] discusses various facets of the philosophy of physics. Two recent popular science books, one authored by Peter Galison [61] and the other by Harald Fritzsch [59] cover Einstein's relativity theory and the position of Einstein versus quantum physics and entanglement.

6.20 EXERCISES AND PROBLEMS

6.1. Read Chapter 2 of [40] and then discuss the relationship between the Second Law of Thermodynamics and the entropy functions defined by the information theory.

6.2. Find out the frequency of occurrence of the individual letters of the alphabet and compute the entropy of the English language.

6.3. Let X be the random variable denoting the result of the NBA finals between two equally matched teams A and B. Assume that the winner of the series is the first team which wins five games. Possible values for X when A wins are $AAAAA$, $\binom{6}{1} = 6$ strings with one occurrence of B, $\binom{7}{2} = 21$ strings with two occurrences of B, $\binom{8}{3} = 56$ strings with three occurrences of B, $\binom{9}{4} = 126$ strings with four occurrences of B, as well as an equal number of strings when B wins. Assume that the games are independent. Let Y denote the number of games played. Compute $H(X)$, $H(Y)$, as well as the conditional entropies $H(Y \mid X)$ and $H(X \mid Y)$.

6.4. Read Chapter 3 of the book *Philosophy of Physics* [128] covering the probabilities in Physics. Discuss the subjective and objective interpretations of probability.

6.5. Discuss the following concepts presented in [128]: Poincaré recurrence, Loschmidt's "reversibility," and the "concentration curve" of a collection of systems.

6.6. Discuss the Ergodic Hypothesis, the Ergodic Theorem, and the Kolmogorov-Arnold-Moser (KAM) Theorem in connection with irreversibility [128].

6.7. In his article *On the electrodynamics of moving bodies* published in *Annalen der Physik* in 1905, Einstein formulates the Relativity Postulate of the General Relativity "there is no way to tell which unaccelerated reference frames is *truly* at rest" [49]. Read the relevant chapters in Galison's book [61] and summarize the relationship between this postulate and the additional assumption required by the relativity theory that light travels at the same speed, 300, 000 kilometers per second in vacuum, regardless of how fast the light source is travelling.

6.8. Read the paper by Tore, et al. [136] and discuss Bohr's arguments showing that the experiment suggested by Einstein and discussed in Section 6.8 does not contradict the uncertainty principle.

6.9. Assume that G_1 is the transfer matrix of "stage 1" consisting of the H and I gate and $G_2 = G_{\text{CNOT}}$ is the transfer matrix of "stage 2" of the quantum circuit in Figure 6.8(b). Construct $G_{\text{Bell}} = G_2 G_1$. Apply G_{Bell} to an input to the circuit in Figure 6.8(a) and calculate the output. Explain the results.

6.10. Can the strategy for encoding and decoding two bits of classical information into one qubit be extended to more than two qubits? If you believe that this is possible, describe an algorithm for dense coding of three bits into one qubit. If not, justify your answer.

6.11. Recall from Chapter 2 that the average value of an observable A of a quantum system in state $\mid \psi\rangle$ is given by

$$\langle A\rangle = \langle \psi \mid A \mid \psi\rangle.$$

Given a two-qubit quantum system in the entangled state

$$\mid \psi\rangle = \frac{\mid 01\rangle - \mid 10\rangle}{\sqrt{2}},$$

the first qubit is sent to Alice and the second qubit is sent to Bob. Alice measures the observables Q and R on her qubit, where

$$Q = Z_1 \quad \text{and} \quad R = X_1$$

are the output of Z and X gates, respectively.

Bob measures the observables S and T on his qubit, where

$$S = -\frac{Z_2 + X_2}{\sqrt{2}} \quad \text{and} \quad T = \frac{Z_2 - X_2}{\sqrt{2}}$$

are the output of two circuits, each using a combination of X and Z gates. Show that

$$\langle QS\rangle = \langle RS\rangle = \langle RT\rangle = \frac{1}{\sqrt{2}}$$

and

$$\langle QT\rangle = -\frac{1}{\sqrt{2}}.$$

APPENDIX A
Elements of Abstract Algebra

The algebraic structures we consider in this book consist of a set of elements, R, equipped with one or more binary laws of composition (operations) denoted as $\top$ and $\perp$ which satisfy a number of properties (axioms). A *binary law* is a rule that associates to an ordered pair of elements in R, say (a, b), an element $c = a \top b \in R$ or $c = a \perp b \in R$.

Usually the term "algebraic structure" refers only to the binary laws and their axioms. The algebraic structures we encounter in this text are: groups, rings (commutative rings), integral domains, fields, and vector spaces. They are the most common but, of course, not all of the algebraic structures in Mathematics. Sometimes, we have to consider an algebraic structure in conjunction with a metric structure on a set R. Informally, a metric structure specifies the "distance" between two elements of the set R. This is the case of Hilbert spaces, of ordered structures such as ordered fields, or both. We are not interested in the general theory of "structures" and we do not discuss structures other than those already mentioned.

A.1 RINGS, COMMUTATIVE RINGS, INTEGRAL DOMAINS, AND FIELDS

Ring. A ring is a set R of elements, $R = \{a, b, c, \ldots\}$, equipped with two binary operations, "+" (addition) and "·" (multiplication) with the following properties:

1. Associative laws: $\forall a, b, c \in R$
$$a + (b + c) = (a + b) + c \quad \text{and} \quad a \cdot (b \cdot c) = (a \cdot b) \cdot c.$$
2. Commutative law: $\forall a, b \in R$
$$a + b = b + a.$$
3. Distributive law: $\forall a, b, c \in R$
$$a \cdot (b + c) = a \cdot b + a \cdot c.$$
4. Zero element: There is a zero additive element $0 \in R$ such that:
$$\forall a \in R \quad a + 0 = a.$$
5. Unity element: There is a multiplicative unit element $1 \in R, 1 \neq 0$ such that:
$$\forall a \in R \quad a \cdot 1 = a.$$

6. Additive inverse: $\forall a \in R$ the equation $a + x = 0$ has a solution $x \in R$.

   The ring is called *commutative* if the multiplication operation "$\cdot$" is commutative.

7. Commutative law: $\forall a, b \in R$

$$a \cdot b = b \cdot a.$$

   For example, the set of $n \times n$ matrices with integer, real, or complex entries equipped with addition and composition/product of matrices as binary operations is a ring. This ring is not commutative since, in general, given two matrices A and B, $A \cdot B$ can be different from $B \cdot A$.

   An integral domain D is a commutative ring in which the following axiom holds:

8. Cancellation law: if $c \neq 0$ and $c \cdot a = c \cdot b$ then $a = b$.

   For example, the set of integers, $\mathbb{Z} = \{\ldots, -3, -2, -1, 0, +1, +2, +3, \ldots\}$ when equipped with the usual (binary) operations, addition and multiplication of integers, is obviously a commutative ring that is an integral domain.

A *subdomain of an integral domain D* is a subset D' of D which is closed to addition and multiplication, contains the additive zero and the multiplicative unit element, and for each a in D' the additive inverse $-a$ lies in D'.

A field F is an integral domain where each element, except zero, has a multiplicative inverse $1/a$, which is unique

$$\forall a \in F,\ a \neq 0,\ \exists a^{-1} \quad \text{and} \quad a^{-1} \cdot a = 1.$$

A subdomain F' of a field F is a *subfield* of F if for any $a \neq 0, a \in F'$, the multiplicative inverse $1/a$ is in F'. In this case, F is also called an *extension of the field F'*.

The sets $\mathbb{Q}$, $\mathbb{R}$, and $\mathbb{C}$ of rational numbers, real numbers, and complex numbers, respectively, equipped with the obvious operations of addition and multiplication are examples of fields. Rational numbers are contained inside the real numbers as a subfield and the real numbers are contained inside the complex numbers as a subfield. The field of complex numbers is an extension of the field of rational, respectively, real numbers.

The first two are *ordered fields*, the last two are *complete fields* (we do not define these concepts, the reader should consult [23]). The third is an *algebraically closed* field. This means that any polynomial equation with coefficients in the field has solutions in the field. $\mathbb{Q}$ and $\mathbb{R}$ are not algebraically closed; the equation $x^2 + 1 = 0$ has no solutions among the real numbers. The property "complete" is necessary for the definition and the study of the concept of continuity of functions $f(x)$ when x varies in the field.

An *equivalence relation* on a set S is a binary relation on S that is reflexive, symmetric and transitive. Equivalence relations are often used to group together objects that are similar in some sense. If $a, b, c \in S$ then the equivalence relation denoted as "$\sim$" has the following properties:

1. **Reflexivity:** $a \sim a$.
2. **Symmetry:** if $a \sim b$ then $b \sim a$.
3. **Transitivity:** if $a \sim b$ and $b \sim c$ then $a \sim c$.

Given a set S and an equivalence relation $\sim$ over S, an *equivalence class* is a subset H of S of the form

$$H \subset S = \{x \in S \mid x \sim a\}$$

with $a \in S$. This equivalence class is usually denoted as $[a]$ and consists of those elements of S which are equivalent to a.

A.2 COMPLEX NUMBERS

The equation $x^2 = -1$ has no root among real numbers. An *imaginary* number i satisfying the equality $i^2 = -1$ was invented. Informally, the complex numbers form the smallest field which contains the field of real numbers as a subfield and the imaginary number i, the "imaginary" solution of the equation $x^2 + 1 = 0$. It can be shown that any polynomial equation of degree n has n solutions (fundamental theorem of algebra).

A more familiar definition of the field $\mathbb{C}$ of complex numbers is the following: The underlying set of $\mathbb{C}$ consists of ordered pairs of real numbers (x, y) with x called the *real component* and y called the *imaginary component* of the complex number z. It is convenient to write a complex number as $z = x + iy$ with $x = \Re(z)$ and $y = \Im(z)$.

Addition and *multiplication* of complex numbers are defined as follows:

$$(x + iy) \pm (x' + iy') = (x \pm x') + i(y \pm y'),$$
$$(x + iy) \cdot (x' + iy') = (x \cdot x' - y \cdot y') + i(x \cdot y' + y \cdot x').$$

It is not hard to check that the complex numbers form a field.

A geometric representation of complex numbers is provided by the points of a Cartesian plane. The complex number $z = x + iy$ is represented by the point with the pair of coordinates (x, y).

The polar coordinates representation suggests that one can represent a complex number, z, as a point P in the "Cartesian plane," by specifying with respect to the Cartesian axes the distance r from the point P to the origin O and the angle θ the segment OP makes with the x axis. These two quantities, the distance r and the angle θ, are usually referred to as *polar coordinates*. With the convention that the angle θ satisfies $0 \leq \theta < 2\pi$ and $r > 0$ any point P different from the origin has a unique pair of polar coordinates. However it is more convenient not to put any restriction on θ and r but identify the pair of polar coordinates (r, θ) and $(r, \theta + 2\pi)$ and (r, θ) and $(-r, \theta + \pi)$.

All pairs $(0, \theta)$ represent the origin or equivalently the complex number zero. When non-negative, r is called the *absolute value* of z and in this case θ is called the *argument* of z

$$r \equiv |z| = (x^2 + y^2)^{1/2},$$

$$\theta \equiv \arg(z) = \arctan \frac{y}{x} = \tan^{-1} \frac{y}{x}.$$

Thus,

$$x = r \cdot \cos\theta \quad y = r \cdot \sin\theta,$$

$$z = r \cdot \cos\theta + i\, r \cdot \sin\theta = r(\cos\theta + i \sin\theta).$$

An amazing observation is the following: If one introduces the "imaginary" powers of the (famous) Euler number e [1] by

$$\cos\theta + i \sin\theta = e^{i\theta}$$

one can write any complex number $z = |z|\, e^{i\theta}$. All familiar rules of manipulations of "powers" and all familiar arithmetic operations the reader knows from calculus with real numbers remain true when applied to complex numbers. This, of course, is something to be proven, but the reader of this book has been most likely exposed to elementary mathematics. In particular,

$$e^{(x+iy)} = e^x \cdot e^{iy}.$$

Here are several expressions useful when manipulating complex numbers:

1. Euler's formulae

$$\sin\theta = \frac{1}{2i}(e^{i\theta} - e^{-i\theta})$$

$$\cos\theta = \frac{1}{2}(e^{i\theta} + e^{-i\theta}).$$

2. Moivre's formulae give the absolute value and the argument of a product of complex numbers z and z'

$$|z \cdot z'| = |z| \cdot |z'| \quad \text{and} \quad \arg(z \cdot z') = \arg(z) + \arg(z').$$

3. The absolute value of a sum of complex numbers is smaller or equal to the sum of their absolute values

$$|z + z'| \leq |z| + |z'|.$$

4. The absolute value of a complex number is a positive real number:

$$|z| > 0 \quad \text{unless } z = 0 \text{ when } |0| = 0.$$

[1] Euler's number $e = 2.71828...$ is the only positive real number with the property that the function $f(x) = e^x$ is the same as its derivative.

5. The "conjugation" has the following properties:

$$(z_1 + z_2)^* = z_1^* + z_2^*,$$
$$(z_1 \cdot z_2)^* = z_1^* \cdot z_2^*,$$
$$(z^*)^* = z,$$
$$|z|^2 = z \cdot z^*,$$
$$z^{-1} = \frac{z^*}{|z|^2}.$$

A.3 ABSTRACT GROUPS AND ISOMORPHISMS

A group G is a set equipped with one binary operation "$\cdot$", called multiplication, which satisfies:

1. Associative law: $\forall (a, b, c) \in G \quad a \cdot (b \cdot c) = (a \cdot b) \cdot c.$
2. Identity element: there is an identity element[2] $e \in G$ and $\forall a \in G \quad a \cdot e = e \cdot a = a.$
3. Inverse element property: $\forall a \in G, \exists a^{-1}$ such that $a \cdot a^{-1} = a^{-1} \cdot a = e.$

It is not hard to check that if $a \in G$ the inverse of a, a^{-1}, is unique. It can be also shown that these properties imply the uniqueness of the identity element e.

A group G whose operation satisfies the commutative law (i.e., $a \cdot b = b \cdot a$) is called *commutative* or *Abelian*. As an example of a non-Abelian group, consider the group of Pauli matrices with multiplicative factors ± 1 and $\pm i$

$$G_1 = \{\pm I, \pm iI, \pm X, \pm iX, \pm Y, \pm iY, \pm Z, \pm iZ\}$$

The general Pauli group on n qubits consists of the n-fold tensor products of Pauli matrices with multiplicative factors ± 1 and $\pm i$.

From this perspective we can restate the definitions of a ring, integral domain, and field. For example, a field is a set F equipped with two binary operations, addition and multiplication, such that:

1. Under addition F is an Abelian group with identity 0 (zero).
2. Under multiplication, the nonzero elements form an Abelian group with identity 1 (one).
3. The distributive law holds, $a(b + c) = ab + ac$.

Given two groups G and G', a *homomorphism* is a rule (transformation) which associates to any element g of G a precise element $\varphi(g)$ in G' such that:

$$\varphi(g_1 \cdot (g_2)^{-1}) = \varphi(g_1) \cdot (\varphi(g_2))^{-1} \qquad \forall g_1, g_2 \in G.$$

[2] The identity element of an algebraic structure is often denoted as e or u. In this context, e is not Euler's number, but the identity element.

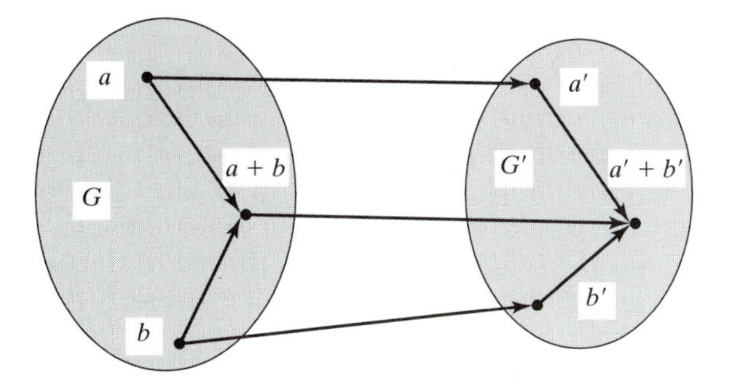

FIGURE A.1: If an isomorphism between two groups G and G' exists, then given two elements $a, b \in G$ and their transforms $a', b' \in G'$, we can: (i) construct the product of $a \cdot b$ and then transform the product $a \cdot b$ to $a' \cdot b' \in G'$, or (ii) first transform the elements into $a', b' \in G'$ and then construct the product $a' \cdot b' \in G'$.

If the homomorphism φ is:

1. *one-to-one* (i.e., $\varphi(g_1) = \varphi(g_2) \quad \Rightarrow \quad g_1 = g_2$) and
2. *onto* (i.e., any g' in G' is of the form $g' = \varphi(g)$ for some g in G)

then the homomorphism φ is called an *isomorphism* (see Figure A.1). The symbol $\Rightarrow$ means "implies."

A non-void subset S of a group G is a *subgroup* iff (if and only if) it is closed to the composition law and to the operation of taking inverse:

1. $a, b \in S \quad \Rightarrow \quad (a \cdot b) \in S,$
2. $a \in S \quad \Rightarrow \quad a^{-1} \in S.$

The *intersection* $S \cap T$ of two subgroups S and T of a group G is a subgroup of G:

$$(S \subset G) \quad \wedge \quad (T \subset G) \quad \Rightarrow \quad (S \cap T) \subset G.$$

The subgroup S is called *normal subgroup* provided that:

$$ghg^{-1} \in H \quad \forall g \in G \quad \text{and} \quad \forall h \in H.$$

A group G is *finite* if the number of elements of G is finite. The *order of a finite group* $\mid G \mid$, also called the *cardinality*, is the number of elements of G.

Lagrange's Theorem. If H is a subgroup of a finite group G, then $\mid H \mid$ divides $\mid G \mid$.

For a subgroup H of a group G and an element $x \in G$, define xH to be the set $\{xh : h \in H\}$ and Hx to be the set $\{hx : h \in H\}$. A subset of G of the form xH for some $x \in G$ is said to be a *left coset of* H and a subset of the form Hx is said to be a *right coset* of H.

For any subgroup H, we can define an equivalence relation "$\equiv$" by $x \equiv y$ if $x = yh$ for some $h \in H$. The equivalence classes of this equivalence relation are exactly the left cosets of H, and an element $x \in G$ is in the *equivalence class* xH. Thus, the left cosets of H form a partition of G.

Any two left cosets of H have the same cardinality and, in particular, every coset of H has the same cardinality as $eH = H$, where e is the identity element. Thus, the cardinality of any left coset of H has cardinality the order of H. The same results are true of the right cosets of G. One can prove that the set of left cosets of H has the same cardinality as the set of right cosets of H.

The *power of an element.* Given $a \in G$ and $m > 0$ a positive integer then

$$a^m = a \cdot a \cdot a \ldots a \text{ (m factors)}, \quad a^0 = e, \quad a^{-m} = (a^{-1})^m.$$

The following holds:

$$a^r \cdot a^s = a^{r+s}.$$

If $a \in G$, the *order* of element a is the least positive power of a such that $a^m = e$. If no positive m exists, then a has order *infinity*. The group G is *cyclic* if it contains one element x whose powers exhaust G. This element is called the *generator* of the group .

The set of elements $\{g_1, g_2, \ldots, g_k\} \in G$ generates the group G if every element of the group can be expressed as a product of the elements of the set. The advantage of using generators to describe a group is that they provide a concise representation of the group. We write $G = \langle g_1, g_2, \ldots, g_k \rangle$.

A.4 MATRIX REPRESENTATION

Let I be the $n \times n$ matrix with all diagonal elements equal to 1 and all off-diagonal elements equal to 0. The set of $n \times n$ matrices with elements from $\mathbb{R}$, or from $\mathbb{C}$, the sets of real and complex numbers, respectively, and with a non-zero determinant forms a multiplicative group with I as the "identity." This group is called the *n-linear group* and it is denoted as $GL(n, \mathbb{R})$ or $GL(n, \mathbb{C})$. Any subgroup $\mathcal{G}$ of $GL(n, \mathbb{R})$ or $GL(n, \mathbb{C})$ is called a *matrix group*.

An n-linear representation of a group G with the identity element denoted by e, or a matrix representation of G is a function

$$\rho : G \mapsto \mathcal{G}$$

which preserves the "identity" element

$$\rho(e) = I$$

and the multiplication

$$\forall g_i, g_j, g_k \in G \quad g_i g_j = g_k \quad \Longrightarrow \quad \rho(g_i)\rho(g_j) = \rho(g_k).$$

Two n-linear representations, ρ_1 and ρ_2, of the group G are *equivalent* if there exists an $n \times n$ matrix A with non-zero determinant such that

$$\forall g \in G \qquad \rho_1(g) = A^{-1}\rho_2(g)A.$$

The *character* χ of a matrix representation is the function

$$\chi_\rho(g) = \mathrm{Tr}[\rho(g)] \qquad \forall g \in G.$$

Here $\mathrm{Tr}(\gamma)$ is the trace of a matrix, the sum of its diagonal elements. The character function has several properties:

$$\chi_\rho(g_1, g_2) = \chi_\rho(g_2, g_1) \quad \forall g_1, g_2 \in G$$

and

$$\chi_\rho(e) = n.$$

If G is finite, the character determines the representation up to an equivalence.

A.5 GROUPS OF TRANSFORMATIONS

A transformation Φ from a set S to a set T, (often called a "map," or, when S and T are sets of numbers, a "function") is a rule which assigns to an element $p \in S$ a unique image element $\Phi(p) \in T$.

$$\Phi : S \mapsto T.$$

The set S is the *domain* and T is the *co-domain* of Φ. The set $\Phi(S)$ of all images under transformation Φ of elements in S is called the *range* of Φ; it may comprise only part of the co-domain T.

A transformation $\Phi : S \mapsto T$ is:

- *onto*—if the co-domain T equals the range of Φ. This means that every $q \in T$ is the image $q = \Phi(p)$ of at least one $p \in S$.
- *one-to-one*—from S into T if Φ carries distinct elements of S into distinct elements of T so that each $q \in T$ is the image of at most one $p \in S$.

Given two transformations Φ and Ψ such that:

$$\Phi : S \mapsto T,$$

$$\Psi : T \mapsto U,$$

the product of two transformations $\Phi\Psi$ is defined as the result of performing them in succession: first Φ then Ψ, provided that the co-domain of Φ is the domain of Ψ. The product $\Phi\Psi$ is the transformation of S into U given by the equation:

$$(\Phi\Psi)(p) = \Phi(p)\Psi.$$

If both Φ and Ψ have the same domain and co-domain, say S, their composition is a transformation with domain and range S so the composition is a binary operation on the set of self transformations of S.

Often in literature the name "transformation" is reserved only for those transformations which are both one to one and onto, for short, *bijective*. The collection of all bijective transformation of a set S with the above binary operation is a group sometimes called *the group of permutations of S* (at least in the case S is a finite set).

A *group of transformations* is any subgroup of the group of bijective transformations. If the set S is a subset of the plane or space, we might consider the transformations of S which preserve the distances between points. Such transformations that are necessarily bijective, are called *symmetries*, and form a group (a subgroup of all bijective transformations of S), the group of symmetries of S. For example, one can suppose that S is a square plane.

A.6 SYMMETRY IN A PLANE

Consider a square in a plane, see Figure A.2. There are eight transformations that preserve distances:

1. Four counter-clockwise rotations with an angle multiple of 90° around its center O:

$$\begin{array}{ll} R = 90^\circ, & (1 \mapsto 2 \;\; 2 \mapsto 3 \;\; 3 \mapsto 4 \;\; 4 \mapsto 1) \\ R' = 180^\circ, & (1 \mapsto 3 \;\; 2 \mapsto 4 \;\; 3 \mapsto 1 \;\; 4 \mapsto 2) \\ R'' = 270^\circ, & (1 \mapsto 4 \;\; 2 \mapsto 1 \;\; 3 \mapsto 2 \;\; 4 \mapsto 3) \\ I = 360^\circ, & (1 \mapsto 1 \;\; 2 \mapsto 2 \;\; 3 \mapsto 3 \;\; 4 \mapsto 4) \end{array}$$

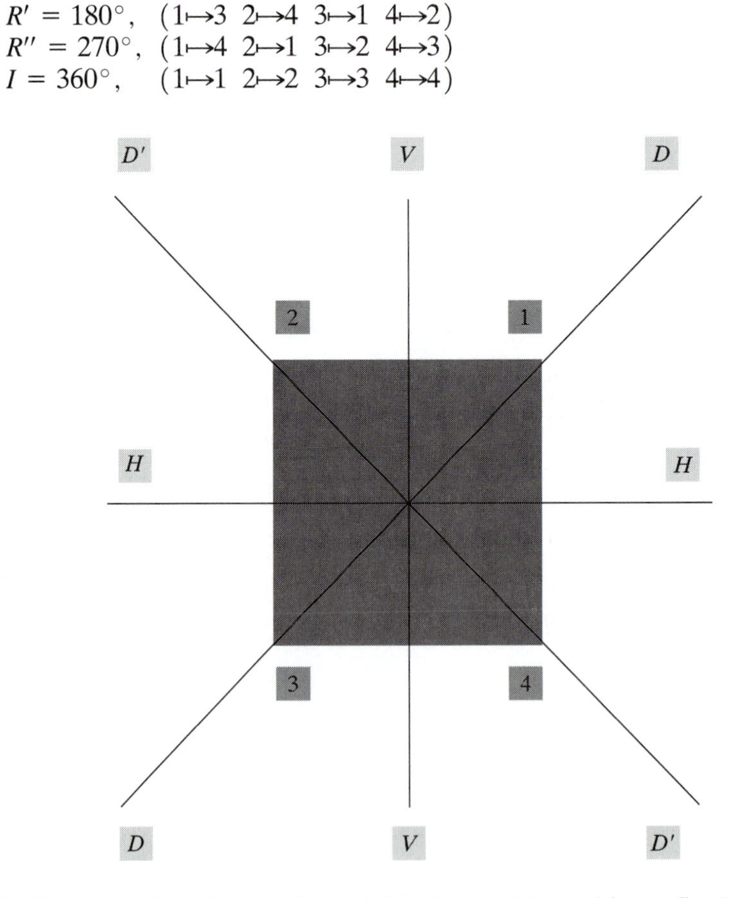

FIGURE A.2: A square in a plane has four rotational symmetries and four reflections. The vertices of the square are labeled 1,2,3, and 4 and the center is O. There are four axes of symmetry V (vertical), H (horizontal), D, and D' (diagonal).

2. Four reflections:

H : reflection on the horizontal axis through O, $(1 \mapsto 4 \;\; 2 \mapsto 3 \;\; 3 \mapsto 2 \;\; 4 \mapsto 1)$
V : reflection on the vertical axis through O, $(1 \mapsto 2 \;\; 2 \mapsto 1 \;\; 3 \mapsto 4 \;\; 4 \mapsto 3)$
D : reflection through the diagonal in quadrants I and III, $(1 \mapsto 1 \;\; 2 \mapsto 4)$ $(3 \mapsto 3 \;\; 4 \mapsto 2)$
D' : reflection through the diagonal in quadrants II and IV, $(1 \mapsto 3 \;\; 2 \mapsto 2)$ $(3 \mapsto 1 \;\; 4 \mapsto 4)$

We can compose symmetry operations; for example, $H \cdot R$ is obtained by first reflecting the square on the horizontal axis then rotating counter-clockwise by $90°$. H sends vertex 1 into 4 and R sends 4 into 1. Thus $H \cdot R$ sends 1 into 1. On the other hand, D sends 1 into 1. After checking for all vertices we can see that

$$H \cdot R = D.$$

Following a similar argument it follows that:

$$R \cdot H = D'.$$

Thus we see that $R \cdot H \neq H \cdot R$, the multiplication is not commutative.

The group of integers modulo 4 with the usual addition operation is isomorphic with the group of rotational symmetries of the square.

A.7 FINITE FIELDS

A field F which contains only a finite number of elements is called a *finite field*.

For example, consider the set of integers modulo n, $\mathbb{Z}_n$, with the standard integer addition and integer multiplication. $\mathbb{Z}_n$ is a commutative ring. It is easy to show that $\mathbb{Z}_n$ is a finite field if and only if n is a prime number.

To prove the "if" side of the proposition, let us assume that n is prime. Then we verify all the properties or axioms of a field. Here we only show that given $0 < a < n$ the multiplicative inverse, a^{-1} exists. The $\gcd(n, a) = 1$ because n is prime; then according to Euclid's algorithm presented in Appendix B, there exist integers s and t such that $s \cdot n + t \cdot a = 1$; this implies that $t \cdot a = 1$ or $t = a^{-1}$. We also know that $\mathbb{Z}_n$ has a finite number of elements, $0, 1, 2, \ldots, (n - 1)$. Thus, if n is prime, then $\mathbb{Z}_n$ is a finite field.

To prove the "only if" side of the proposition, note that if $n = ab$ with $a, b \neq 0 \bmod n$, then b has no multiplicative inverse in $\mathbb{Z}_n$. There is no c such that $bc \equiv 1 \bmod n$, or $b = c^{-1}$. The existence of such c would imply that $n = abc \equiv a \bmod n$. But

$$n = abc = 0 \bmod n \quad \text{or} \quad a \equiv 0 \bmod n$$

contradicting the assumption that $a, b \neq 0 \bmod n$.

A finite field with q elements is also called a *Galois Field* and it is denoted as $GF(q)$. In particular, $\mathbb{Z}_p$ is denoted as $GF(p)$. It is easy to prove that a finite field has p^n elements with p a prime number.

For example, $\mathbb{Z}_5$ consists of the following elements: [0], [1], [2], [3], and [4]. The addition and multiplication tables of $\mathbb{Z}_5$ are

+	[0]	[1]	[2]	[3]	[4]
[0]	[0]	[1]	[2]	[3]	[4]
[1]	[1]	[2]	[3]	[4]	[0]
[2]	[2]	[3]	[4]	[0]	[1]
[3]	[3]	[4]	[0]	[1]	[2]
[4]	[4]	[0]	[1]	[2]	[3]

×	[0]	[1]	[2]	[3]	[4]
[0]	[0]	[0]	[0]	[0]	[0]
[1]	[0]	[1]	[2]	[3]	[4]
[2]	[0]	[2]	[4]	[1]	[3]
[3]	[0]	[3]	[1]	[4]	[2]
[4]	[0]	[4]	[3]	[2]	[1]

It is easy to see from these tables that [3] is the multiplicative inverse of [2] (indeed, $[3] \times [2] = [1]$) and that [1] and [4] are their own multiplicative inverses ($[1] \times [1] = [1]$ and $[4] \times [4] = [1]$).

$\mathbb{Z}_5, \mathbb{Z}_7, \mathbb{Z}_{11}$, and $\mathbb{Z}_{13}$ are finite fields but $\mathbb{Z}_9$ and $\mathbb{Z}_{15}$ are not because 9 and 15 are not prime numbers. Indeed, $\mathbb{Z}_9$ is finite and consists of the following elements: [0], [1], [2], [3], [4], [5], [6], [7], and [8]. The multiplication table of $\mathbb{Z}_9$ is

×	[0]	[1]	[2]	[3]	[4]	[5]	[6]	[7]	[8]
[0]	[0]	[0]	[0]	[0]	[0]	[0]	[0]	[0]	[0]
[1]	[0]	[1]	[2]	[3]	[4]	[5]	[6]	[7]	[8]
[2]	[0]	[2]	[4]	[6]	[8]	[1]	[3]	[5]	[7]
[3]	[0]	[3]	[6]	[0]	[3]	[6]	[0]	[3]	[6]
[4]	[0]	[4]	[8]	[3]	[7]	[2]	[6]	[1]	[5]
[5]	[0]	[5]	[1]	[6]	[2]	[7]	[3]	[8]	[4]
[6]	[0]	[6]	[3]	[0]	[6]	[3]	[0]	[6]	[3]
[7]	[0]	[7]	[5]	[3]	[1]	[8]	[6]	[4]	[2]
[8]	[0]	[8]	[7]	[6]	[5]	[4]	[3]	[2]	[1]

We see immediately that this is not a field because some non-zero elements, e.g., [3] and [6] do not have a multiplicative inverse. The multiplicative inverse of [2] is [5] (indeed, $[2] \times [5] = [1]$), the one of [7] is [4] (indeed, $[7] \times [4] = [1]$), and the one of [1] is [1].

Consider a field F and denote by 1 the multiplicative identity of F. The *characteristic* of F is the smallest positive integer m such that

$$\sum_{i=1}^{m} 1 = 0.$$

Let m be the characteristic of the field F. If $m \neq 0$, then m is a prime.

Assume that $m = ab$ with $a, b > 1$. It follows that there are two integers $t, s \in F$ such that

$$t = \sum_{i=1}^{a} 1, \quad s = \sum_{i=1}^{b} 1, \quad ts = \sum_{i=1}^{ab} 1 = \sum_{i=1}^{m} 1 = 0.$$

Neither t nor s are zero because a and b are both smaller than m and m is the smallest number of times we have to add the multiplicative identity to get zero. Since $t \neq 0$ there is a multiplicative inverse t^{-1}. It follows that $t^{-1}ts = s$. But $ts = 0$ thus $t^{-1}ts = 0$. This is possible only if $s = 0$ which contradicts the minimality of m.

Any finite field of characteristic p contains $\mathbb{Z}_p$ as a subfield and the number of its elements has to be p^n for some n.

If the reader is familiar with the concept of vector space and simple facts about bases of vector spaces the above statement is immediate.

Indeed, F can be considered a vector space over $\mathbb{Z}_p$. Since F is a finite field, the vector space has a finite dimension. If we call this finite dimension n there are n elements of F, $f_1, f_2, \ldots, f_n \in F$, which form a basis for F

$$F = \sum_{i=1}^{n} \lambda_i f_i \quad \lambda_i \in \mathbb{Z}_p.$$

It follows that the finite field F contains p^n elements.

As an example consider $p = 2$, the finite field of binary numbers, $\mathbb{Z}_2$. For $n \geqslant 2$ we construct the set of all 2^n polynomials with binary coefficients

$$\mathbb{Z}_2[x] = \{0, 1, x, 1 + x, 1 + x^2, x^2, x + x^2, 1 + x + x^2, x^3, 1 + x^3,$$
$$1 + x + x^3, 1 + x^2 + x^3, 1 + x + x^2 + x^3, x + x^3, x^2 + x^3 \ldots\}.$$

$\mathbb{Z}_p[x]$, the set of polynomials over the finite field of integers modulo p with regular polynomial addition and multiplication forms a commutative ring with identity. It is easy to prove that $\mathbb{Z}_p[x]$, has all the properties of a field except the existence of a multiplicative inverse.

APPENDIX B

Modular Arithmetic

Factoring large integers into primes is a hard computational problem. This property of large integer factorization is exploited by several cryptosystems. The first known quantum algorithm developed by Peter Shor in 1994, is a polynomial time algorithm for prime factorization [119]. This justifies our interest in modular arithmetic.

Donald Knuth provides a comprehensive discussion of all the topics covered in this section [83, 84]. S.A. Vanstone and P.C. van Oorschot include a brief presentation of the Euclid algorithms and the Chinese Remainder Theorem [140].

B.1 ELEMENTARY NUMBER THEORY CONCEPTS

If x is a real number, then $\lfloor x \rfloor$ denotes the *floor of* x, the greatest integer less then or equal to x. If x and y are real numbers, we define the *remainder* when x is divided by y as the binary relation

$$x \bmod y = \begin{cases} x - y\lfloor x/y \rfloor & \text{if } y \neq 0 \\ x & \text{if } y = 0 \end{cases}$$

such that

$$y > 0 \quad \Longrightarrow \quad 0 \leqslant x \bmod y < y$$

and

$$y < 0 \quad \Longrightarrow \quad 0 \geqslant x \bmod y > y.$$

A *prime number* is an integer greater than 1 which has no integer divisor other than 1 and itself. The "fundamental theorem of arithmetic" states that any positive integer $n > 1$ can be expressed as a product of prime numbers with unique nonnegative exponents $n_2, n_3, \ldots$.

$$n = 2^{n_2} \cdot 3^{n_3} \cdot 5^{n_5} \cdot \ldots = \prod_{p \text{ prime}} p^{n_p}.$$

A proof of this theorem by induction follows. The statement of the theorem is true for $n = 2$. Let us assume that the statement is also true for all integers up to n, namely $2, 3, 4, \ldots, n$ and prove that is also true for $n + 1$. From the definition of a prime number it follows that $n + 1$ is either prime or a product of some positive integers

$$n + 1 = p_1 \cdot p_2 \ldots \cdot p_q.$$

It is obvious that $1 \leqslant p_1, p_2, \ldots, p_q \leqslant n$. According to our assumption $p_1, p_2, \ldots, p_q$ can be expressed as products of prime numbers. It follows that $n + 1$ can be expressed as a product of prime numbers.

Integers n and m are said to be *relatively prime* if they do not have any common factors. Given two integers n and m, their *greatest common divisor*, $\gcd(n, m)$, is the largest integer that divides both n and m. Their *least common multiple*, $\text{lcm}(n, m)$, is the smallest positive integer that is a multiple of both n and m. According to the fundamental theorem of arithmetic,

$$n = \prod_{p \text{ prime}} p^{n_p} \quad \text{and} \quad m = \prod_{p \text{ prime}} p^{m_p}.$$

It follows that

$$\gcd(n, m) = \prod_{p \text{ prime}} p^{\min(n_p, m_p)}$$

$$\text{lcm}(n, m) = \prod_{p \text{ prime}} p^{\max(n_p, m_p)}.$$

Given any integers a, b, c the following identities are true:

$$a \cdot b = \gcd(a, b) \cdot \text{lcm}(a, b) \quad \text{if } a, b \geqslant 0$$

$$\gcd[(a, b) \cdot c] = \gcd(a \cdot c, b \cdot c) \quad \text{if } c \geqslant 0$$

$$\text{lcm}[(a, b) \cdot c] = \text{lcm}(a \cdot c, b \cdot c) \quad \text{if } c \geqslant 0$$

$$\gcd[a, \gcd(b, c)] = \gcd[b, \gcd(a, c)] = \gcd[\gcd(a, b), c]$$

$$\text{lcm}[a, \gcd(b, c)] = \text{lcm}[b, \gcd(a, c)] = \text{lcm}[\gcd(a, b), c]$$

$$\gcd[\text{lcm}(a, b), \text{lcm}(a, c)] = \text{lcm}[a, \gcd(b, c)]$$

$$\text{lcm}[\gcd(a, b), \gcd(a, c)] = \gcd[a, \text{lcm}(b, c)].$$

Consider the set of q non-negative integers $(p_1, p_2, \ldots, p_i, \ldots, p_q)$. The notation $(k \mid m)$ means that k divides m exactly. The *greatest common divisor* (gcd) of $(p_1, p_2, \ldots, p_i, \ldots, p_q)$ is the largest integer that divides all of them

$$\gcd(p_1, p_2, \ldots, p_i, \ldots, p_q) = \max(m : m \mid p_i, 1 \leqslant i \leqslant q).$$

The *least common multiple* (lcm) of $(p_1, p_2, \ldots p_i, \ldots p_q)$ is the smallest integer divisible by all of them

$$\text{lcm}(p_1, p_2, \ldots, p_i, \ldots, p_q) = \min(m : m > 0 \quad \text{and} \quad p_i \mid m, 1 \leqslant i \leqslant q).$$

Euclid's algorithm discussed in the next section allows us to compute the gcd of two integers n and m. We factor n and m into prime numbers, select the common factors at the smallest power, and multiply them. For example, if

$$n = 94829 = 7 \cdot 19 \cdot 23 \cdot 31$$

$$m = 120745 = 5 \cdot 19 \cdot 31 \cdot 41$$

then

$$\gcd(n, m) = \gcd(94829, 120745) = 19 \cdot 31 = 589.$$

Both 19 and 31 divide 94829 and 120745, but $589 = 19 \cdot 31$ is the largest integer dividing both 94829 and 120745.

To compute the lcm of two integers n and m we factor the integers. Then we include every prime number appearing in the factorization of either m or n raised to the largest exponent and compute the product. For example, if

$$94829 = 7 \cdot 19 \cdot 23 \cdot 31$$

$$120745 = 5 \cdot 19 \cdot 31 \cdot 41$$

then

$$\text{lcm}(94829, 120745) = 5 \cdot 7 \cdot 19 \cdot 23 \cdot 31 \cdot 41 = 19439945.$$

Other integers, such as $11450127605 = 94829 \cdot 120745$, and all multiples of it, are divisible by 94829 and by 120745 but 19439945 is the smallest integer divisible both by 94829 and by 120745.

Consider $p_1 = 3$, $p_2 = 5$, $p_3 = 7$. Then

$$\text{lcm}(3, 5, 7) = 3 \cdot 5 \cdot 7 = 105$$

$$\text{lcm}(5, 7) = 5 \cdot 7 = 35 \quad \text{and} \quad \text{lcm}(3, 35) = 3 \cdot 35 = 105$$

$$\text{lcm}(3, 5) = 3 \cdot 5 = 15 \quad \text{and} \quad \text{lcm}(15, 7) = 15 \cdot 7 = 105$$

Given a positive integer q and a pair of positive integers m and n, we say that n and m are *congruent modulo q* if $(n - m)$ is an integer multiple of q. We write

$$n \equiv m \mod q.$$

The *congruence relation* has the following properties:

1. If $n \equiv m$ and $a \equiv b$ then $n \cdot a \equiv m \cdot b$ and $n \pm a \equiv m \pm b$. All congruences are modulo q.
2. If $n \cdot a \equiv m \cdot b \mod q$ and $n \equiv m \mod q$ and if n is relatively prime to q then $a \equiv b \mod q$.
3. $n \equiv m \mod q$ if and only if $m \cdot k \equiv n \cdot k \mod q \cdot k$.
4. If p is relatively prime to q, then $n \equiv m \mod p \cdot q$ if and only if

$$n \equiv m \mod p \quad \text{and} \quad n \equiv m \mod q.$$

Theorem. If p is a prime number, then $m^p \equiv m \mod p$ for any integer m (Fermat's theorem dated 1640).

Proof. We consider only the case when m and p are relatively prime. Since p is prime, this implies that $m \mod p \neq 0$. If m is a multiple of p, then

$$m^p \equiv 0 \mod p \quad \text{and} \quad m \equiv 0 \mod p$$

and the statement of the theorem is trivially satisfied.

First, we observe that the p integers:

$$0 \bmod p \qquad m \bmod p \qquad 2m \bmod p \quad \dots \quad (p-1)m \bmod p$$

are all distinct. Indeed,

$$a \cdot m \equiv b \cdot m \bmod p$$

is only possible if $a \equiv b \bmod p$ and certainly this is not the case. Moreover, all the $(p-1)$ non-zero distinct integers in the sequence above are smaller than p, therefore they must be the set of integers $1, 2, \dots, (p-1)$ in some order. Thus, we have the following congruence relationship

$$m \cdot 2m \cdot 3m \dots (p-1)m \equiv 1 \cdot 2 \cdot 3 \dots (p-1) \quad \bmod p.$$

We multiply both sides of the above relation by m and obtain the result proving Fermat's theorem

$$m^p[1 \cdot 2 \cdot 3 \dots (p-1)] \equiv m[1 \cdot 2 \cdot 3 \dots (p-1)] \bmod p$$

Using similar arguments one could prove the following theorem:

Theorem. Given any positive integer n, let the positive integer a be relatively prime to n. Then $a^{\varphi(n)} \equiv 1 \mod n$ (Euler's theorem).

Here $\varphi(n)$ denotes the number of positive integers m, where $m \leq n$, which are relatively prime to n, i.e. $\gcd(m, n) = 1$.
The function

$$n \mapsto \varphi(n)$$

is called the *Euler function*.

B.2 EUCLID'S ALGORITHM FOR INTEGERS

The algorithm to compute the greatest common divisor of two integers was most likely known a few hundred years before appearing around the year 300 B.C., in Book 7 of Euclid's *Elements*.

Let $a, b \in \mathcal{I}$ be two positive integers. The *quotient*, q, and the *remainder*, r, when a is divided by b are defined by

$$a = q \cdot b + r \quad \text{for} \quad 0 \leqslant r < b.$$

If $r = 0$, then we say that *b divides a*. If $\gcd(a, b) = c$, then there are two integers d and e such that $a = c \cdot d$ and $b = c \cdot e$.

The greatest common divisor of positive integers a and b is the last remainder in the sequence of applications of the division algorithm

$$
\begin{aligned}
a &= q_2 \cdot b + r_2 & \quad 0 < r_2 < b \\
b &= q_3 \cdot r2 + r_3 & \quad 0 < r_3 < r_2 \\
r_2 &= q_4 \cdot r3 + r_4 & \quad 0 < r_4 < r_3 \\
&\dots
\end{aligned}
$$

$$r_{k-3} = q_{k-1} \cdot r_{k-2} + r_{k-1} \qquad 0 < r_{k-1} < r_{k-2}$$
$$r_{k-2} = q_k \cdot r_{k-1} + r_k \qquad 0 < r_k < r_{k-1}$$
$$r_{k-1} = q_{k+1} \cdot r_k + 0.$$

The *extended Euclid algorithm* allows us to determine integers s and t such that

$$s \cdot a + t \cdot b = \gcd(a, b).$$

We start with:

$$s_0 = 1 \quad t_0 = 0 \quad r_0 = a$$
$$s_1 = 0 \quad t_1 = 1 \quad r_1 = b.$$

For $i \geqslant 2$ we apply the division algorithm to r_{i-2} and r_{i-1} to obtain q_i and r_i

$$r_{i-2} = q_i \cdot r_{i-1} + r_i \qquad 0 \leqslant r_i < r_{i-1}$$

Then we compute s_i and t_i

$$s_i = s_{i-2} - q_i \cdot s_{i-1}$$
$$t_i = t_{i-2} - q_i \cdot t_{i-1}.$$

The procedure to compute the greatest common divisor and the least common multiple of two integers a and b using the extended Euclid algorithm consists of the following steps:

1. at the last iteration k when the reminder is non-zero, $r_k \neq 0$, we compute

$$\gcd(a, b) = s_k \cdot a + t_k \cdot b,$$

2. at the next iteration, $k + 1$, when the reminder is zero, $r_{k+1} = 0$, we compute

$$\mathrm{lcm}(a, b) = | s_{k+1} \cdot a | = | t_{k+1} \cdot b | .$$

Let us now give an example. Consider the case $a = 78$ and $b = 63$.

$k = 0 \quad r_0 = 78, \quad s_0 = 1, \quad t_0 = 0.$

$k = 1 \quad r_1 = 63, \quad s_1 = 0, \quad t_1 = 1.$

$k = 2 \quad r_0 = q_2 \cdot r_1 + r2 \Rightarrow 78 = q_2 \cdot 63 + r_2 \quad \Rightarrow \quad q_2 = 1$ and $r_2 = 15$

$s_2 = s_0 - q_2 \cdot s_1 = 1 - 1 \cdot 0 = 1$ and

$t_2 = t_0 - q_2 t_1 = 0 - 1 \cdot 1 = -1.$

$k = 3 \quad r_1 = q_3 \cdot r_2 + r3 \Rightarrow 63 = q_3 \cdot 15 + r_3 \quad \Rightarrow \quad q_3 = 4$ and $r_3 = 3$

$s_3 = s_1 - q_3 \cdot s_2 = 0 - 4 \cdot 1 = -4$ and

$t_3 = t_1 - q_3 t_2 = 1 - 4 \cdot (-1) = 5.$

$$k = 4 \quad r_2 = q_4 \cdot r_3 + r4 \Rightarrow 15 = q_4 \cdot 3 + r_4 \quad \Rightarrow \quad q_4 = 5 \quad \text{and} \quad r_4 = 0$$
$$s_4 = s_2 - q_4 \cdot s_3 = 1 - 5 \cdot (-4) = 21 \quad \text{and}$$
$$t_4 = t_2 - q_4 t_3 = -1 - 5 \cdot (5) = -26.$$

The last iteration with a non-zero reminder $r_3 = 3$, $k = 3$, allows us to compute

$$\gcd(78, 63) = s_3 a + t_3 b = (-4) \cdot 78 + 5 \cdot 63 = -312 + 315 = 3.$$

Indeed, $\gcd(78, 63) = 3$, as expected from the definition of the greatest common divisor of $a = 78 = 2 \cdot 3 \cdot 13$ and $b = 63 = 3^2 \cdot 7$.

The last iteration, $k = 4$, allows us to compute

$$\text{lcm}(78, 63) = | s_4 \cdot a | = 21 \cdot 78 = 1638$$

or

$$\text{lcm}(78, 63) = | t_4 \cdot b | = | (-26) \cdot 63 | = 1638.$$

Indeed, $\text{lcm}(78, 63) = 1638 = 2 \cdot 3^2 \cdot 7 \cdot 13$, as expected from the definition of the least common multiple of $a = 78 = 2 \cdot 3 \cdot 13$ and $b = 63 = 3^2 \cdot 7$.

B.3 THE CHINESE REMAINDER THEOREM AND ITS APPLICATIONS

Consider the following problem posed sometime during the 4th century A.D., by the Chinese philosopher Sun Tsu Suan-Ching [143]: "There are certain things whose number is unknown. Repeatedly divided by 3, the remainder is 2, by 5 the remainder is 3, and by 7 the remainder is 2. What is the number?" Similar problems appear in the Hindu text Brahma-Sphurta-Siddhanta in the 6th century A.D.

The more general formulation of these problems is: *find the integer n that satisfies simultaneously the equations:*

$$n = r_1 \bmod p_1$$
$$n = r_2 \bmod p_2$$
$$\vdots$$
$$n = r_i \bmod p_i$$
$$\vdots$$
$$n = r_q \bmod p_q.$$

We assume that the moduli p_k, $1 \leqslant k \leqslant q$ are prime numbers. First, we show that a solution to this problem exists and it is unique modulo the least common multiple of the integers p_i, $1 \leqslant i \leqslant q$. Then, we present an algorithm to solve the problem.

Theorem. (The Chinese Remainder Theorem) Let $p_1, p_2, \ldots, p_i, \ldots, p_q$ be positive integers that are pairwise coprime (i.e., $\gcd(p_i, p_j) = 1$, if $i \neq j$). Let $P = p_1 \cdot p_2 \ldots \cdot p_i \ldots \cdot p_q$ and let $r_1, r_2, \ldots, r_k, \ldots, r_q$ be integers. Then there is exactly one integer n such that

$$0 \leqslant n \leqslant P \quad \text{and} \quad n = r_i \bmod p_i, \quad 1 \leqslant i \leqslant q.$$

Proof. Let Q_i, $1 \leqslant i \leqslant q$ be integers such that

$$Q_i \equiv 1 \bmod p_i \quad \text{and} \quad Q_i \equiv 0 \bmod p_j, \quad i \neq j.$$

Such integers exist because P and P/p_i are relatively prime, therefore according to Euler's theorem

$$Q_i = (P/p_i)^{\varphi(p_i)}.$$

The integer

$$n = (r_1 Q_1 + r_2 Q_2 + \ldots r_i Q_i + \ldots r_m Q_m) \bmod P$$

is divisible by all p_i $1 \leqslant i \leqslant m$. We leave to the reader the proof of the uniqueness of the solution.

The algorithm for solving the problem posed at the beginning of this section is

Step 1. Compute the product of the moduli

$$P = p_1 \cdot p_2 \ldots \cdot p_k \ldots \cdot p_m.$$

Step 2. Compute the integers Q'_k, $1 \leqslant k \leqslant m$, the inverse modulo p_k of each of the quantities P/p_k, using the following expression

$$Q'_k = (P/p_k)^{p_k - 2} \bmod p_k.$$

Step 3. The solution is

$$n = [(P/p_1) \cdot r_1 \cdot Q'_1 + (P/p_2) \cdot r_2 \cdot Q'_2 + \ldots (P/p_k) \cdot r_k \cdot Q'_k + \ldots + (P/p_m) \cdot r_m \cdot Q'_m] \bmod P.$$

The calculation of quantities Q'_k, in Step 2 is justified by the following arguments. According to Fermat's theorem, $a^{p_k - 1} = 1 \bmod p_k$ for any integer a. This can be written as

$$a \cdot (a^{p_k - 2}) = 1 \bmod p_k \implies a^{-1} \bmod p_k = (a^{p_k - 2}) \bmod p_k.$$

with $a^{-1} \cdot a \equiv 1 \bmod p_k$ (i.e., a^{-1} is the inverse of a modulo p_k).

By definition,

$$Q'_k = (P/p_k)^{-1} \bmod p_k.$$

Thus,

$$Q'_k = (P/p_k)^{p_k-2} \bmod p_k.$$

Let us now apply this algorithm to solve the problem posed at the beginning of this section:

$$n = 2 \bmod 3, \quad n = 3 \bmod 5, \quad n = 2 \bmod 7.$$

In this case, $P = 3 \cdot 5 \cdot 7 = 105$ and we have to compute $Q'_i = [P/p_i]^{p_i-2} \bmod p_i$. The three integers are:

$$Q'_1 = (105/3)^{3-2} \bmod 3 = (35)^1 \bmod 3 = (11 \cdot 3 + 2) \bmod 3$$

$$Q'_1 = 2 \bmod 3,$$

$$Q'_2 = (105/5)^{5-2} \bmod 5 = (21)^3 \bmod 5 = 9261 \bmod 5 = (1852 \cdot 5 + 1) \bmod 5$$

$$Q'_2 = 1 \bmod 5$$

$$Q'_3 = (105/7)^{7-2} \bmod 7 = (15)^5 \bmod 7 = 759375 \bmod 7$$
$$= (108482 \cdot 7 + 1) \bmod 7$$

$$Q'_3 = 1 \bmod 7.$$

The result is

$$n = (35 \cdot 2 \cdot 2 + 21 \cdot 3 \cdot 1 + 15 \cdot 2 \cdot 1) \bmod 105 = 233$$
$$= (2 \cdot 105 + 23) \bmod 105$$
$$n = 23 \bmod 105.$$

Indeed,

$$23 = \begin{cases} 7 \cdot 3 + 2 \\ 4 \cdot 5 + 3 \\ 3 \cdot 7 + 2 \end{cases}$$

B.4 COMPUTER ARITHMETIC FOR LARGE INTEGERS

The number of bits in a computer word limits the range of integers that can be manipulated by a particular processor, as well as the accuracy of a floating point number. If we have n-bit internal registers, then we can only represent integers in the range $0 \leqslant i \leqslant 2^n - 1$. About half of this range is allocated to positive integers and half to negative integers. We use one bit for the sign. To represent a negative integer we use the *two's complement*. The largest positive integer that can be represented by a 32-bit processor is $2^{31} - 1 = 2,147,483,367$, a very large number, but certainly smaller than the United State's budget deficit expressed in dollars, the number of atoms in the universe, or the time in seconds since the extinction of the dinosaurs.

The problem we address now is how to do arithmetic on large integers given the physical limitation imposed by the architecture of a processor. Number

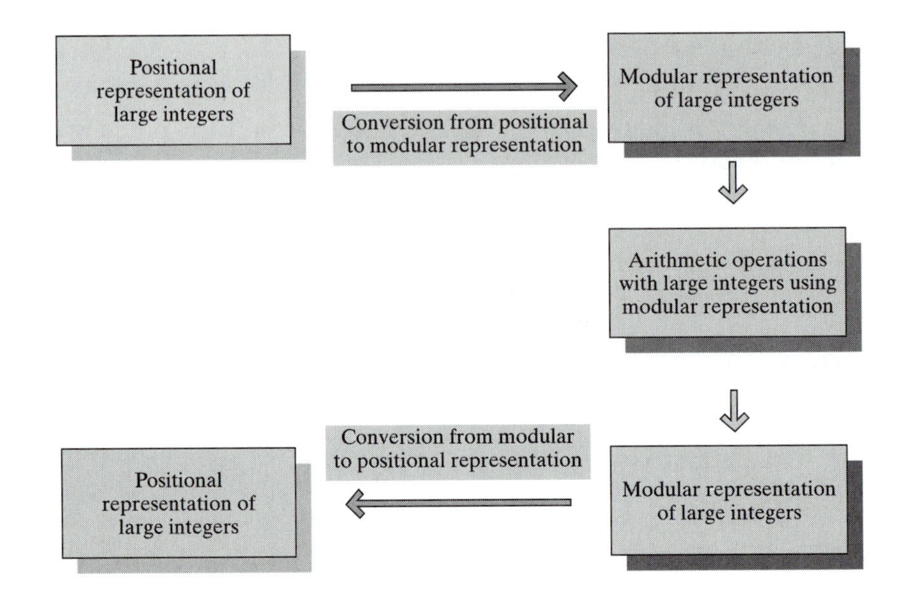

FIGURE B.1: The schematics of performing integer arithmetic on large integers based upon number theory. The positional representation of integers is first converted to a modular representation, then the arithmetic operations are performed using this modular representation, and finally the results are converted back to positional representation.

theory provides an elegant solution to the problem, as shown in Figure B.1. Recall that the "modulo m" relation, with m a positive integer, partitions the set of positive integers into m equivalence classes, namely integers whose residues, or remainders, when divided by m are $0, 1, 2, \ldots, m - 1$. For example, $113 \bmod 3 = 1234567892 \bmod 3 = 2$.

Now consider a set of "moduli" that *contain no common factors*, $p_1, p_2, \ldots, p_q$. Given an integer n, we can divide n by these moduli and compute the corresponding residues:

$$r_1 = n \bmod p_1, \quad r_2 = n \bmod p_2, \quad \ldots \quad r_{q-1} = n \bmod p_{q-1}, \quad r_q = n \bmod p_q.$$

This is a reversible process and we can always compute n given the set $(r_1, r_2, \ldots, r_{q-1}, r_q)$ using the algorithm in Section B.3. Thus, the set $(r_1, r_2, \ldots, r_{q-1}, r_q)$ can be thought of as another *internal representation of the positive integer n*. It turns out that this internal representation allows us to perform addition, subtraction, and multiplication of large, positive integers in a very convenient manner. It is easy to see that if we have two integers a and b with the internal representation $a = (a_1, a_2, \ldots, a_{q-1}, a_q)$ and $b = (b_1, b_2, \ldots, b_{q-1}, b_q)$ then:

$$\begin{aligned} a + b &= (a_1, a_2, \ldots, a_{q-1}, a_q) + (b_1, b_2, \ldots, b_{q-1}, b_q) \\ &= [(a_1 + b_1) \bmod p_1, (a_2 + b_2) \bmod p_2, \ldots, \\ &\quad (a_{q-1} + b_{q-1}) \bmod p_{q-1}, (a_q + b_q) \bmod p_q] \end{aligned}$$

$$
\begin{aligned}
a - b &= (a_1, a_2, \dots, a_{q-1}, a_q) - (b_1, b_2, \dots, b_{q-1}, b_q) \\
&= [(a_1 - b_1) \bmod p_1, (a_2 - b_2) \bmod p_2, \dots, \\
&\quad (a_{q-1} - b_{q-1}) \bmod p_{q-1}, (a_q - b_q) \bmod p_q] \\
a \cdot b &= (a_1, a_2, \dots, a_{q-1}, a_q) \cdot (b_1, b_2, \dots, b_{q-1}, b_q) \\
&= [(a_1 \cdot b_1) \bmod p_1, (a_2 \cdot b_2) \bmod p_2, \dots, \\
&\quad (a_{q-1} \cdot b_{q-1}) \bmod p_{q-1}, (a_q \cdot b_q) \bmod p_q].
\end{aligned}
$$

The modular representation of integers leads to relatively simple hardware implementation. You may recall from Chapter 4 that the classical one full bit adder must include a circuit to compute the *CarryOut* and must deal with the *CarryIn*. This is no longer necessary because the modular representation is *carry-free.*

There are also some disadvantages of using modular arithmetic. For example, given the modular representation, it is difficult to test if a number is positive or negative, to detect an overflow, or to perform integer division. It is also difficult to test if $a > b$. The idea of using modular arithmetic for the hardware implementation is attributed by D. Knuth [84] and others [133], to A. Svoboda and M. Valach.

APPENDIX C
Walsh-Hadamard Transform

Hadamard matrices and the Walsh-Hadamard Transform are used to describe quantum circuits and to define special classes of linear codes. In this section, we define the Hadamard matrix, discuss its properties and introduce the Fast Hadamard Transform.

C.1 HADAMARD MATRICES

A Hadamard matrix of order n, $H_n = [h_{ij}], 1 \leqslant i \leqslant n, 1 \leqslant j \leqslant n$, is an $n \times n$ matrix with h_{ij} either $+1$ or -1. The row vectors $\{h_1, h_2, \ldots, h_n\}$ of the matrix H_n, are pairwise orthogonal

$$h_k \cdot h_l = 0 \quad \forall (k, l) \in \{1, n\}.$$

A Hadamard matrix of order $2n$ is

$$H_{2n} = \begin{pmatrix} H_n & H_n \\ H_n & -H_n \end{pmatrix}.$$

A Hadamard matrix H_n has the following properties:

1. The product of a Hadamard matrix and its transpose satisfies the following relations
$$H_n H_n^T = nI_n \quad H_n^T H_n = nI_n.$$
2. The exchange of rows or columns transforms one Hadamard matrix into another one.
3. The multiplication of rows or columns by -1 transforms one Hadamard matrix into another one.
4. If $n = 2^{q-1}$ then H_{2n} can be expressed as the tensor product of q matrices of size 2×2
$$H_{2n} = H_2 \otimes H_2 \otimes \ldots H_2 \otimes H_2.$$

Recall that the tensor product of the $n \times n$ matrix $A = [a_{ij}]$ with the $m \times m$ matrix $B = [b_{rs}]$ is the $(m \cdot n) \times (m \cdot n)$ matrix $C = [c_{pq}]$ with $c_{pq} = a_{ij} \cdot b_{rs}$ where i, j, r, s, p, q are the only integers which satisfy

$$p = i \cdot m + r \quad \text{and} \quad q = j \cdot m + s.$$

In other words, C is the matrix obtained by replacing the element a_{ij} of matrix A by the $m \times m$ matrix $a_{ij} \cdot B$.

We now prove property (1). H_n and H_n^T can be written as

$$H_n = \begin{pmatrix} h_1 \\ h_2 \\ \dots \\ h_n \end{pmatrix}, \quad H_n^T = \begin{pmatrix} h_1^T \, h_2^T \, \dots \, h_n^T \end{pmatrix}$$

with $h_1, h_2, \dots, h_n$ the row vectors of H_n.
Then

$$H_n H_n^T = \begin{pmatrix} h_1 \cdot h_1 & h_1 \cdot h_2 & \dots & h_1 \cdot h_n \\ h_2 \cdot h_1 & h_2 \cdot h_2 & \dots & h_2 \cdot h_n \\ \dots & \dots & \dots & \dots \\ h_n \cdot h_1 & h_n \cdot h_2 & \dots & h_n \cdot h_n \end{pmatrix} = \begin{pmatrix} n & 0 & \dots & 0 \\ 0 & n & \dots & 0 \\ \dots & \dots & \dots & \dots \\ 0 & 0 & \dots & n \end{pmatrix} = n I_n.$$

Now multiply the previous equation with H_n^{-1}

$$H_n^{-1} H_n H_n^T = n H_n^{-1} I_n \quad \Rightarrow \quad H_n^T = n H_n^{-1}$$
$$H_n^T H_n = n H_n^{-1} H_n \quad \Rightarrow \quad H_n^T H_n = n I_n.$$

Properties (2), (3), and (4) follow immediately.

Given that $H_1 = [1]$, the Hadamard matrices of order 2, 4, and 8 are:

$$H_2 = \begin{pmatrix} 1 & 1 \\ 1 & -1 \end{pmatrix}, \quad H_4 = \begin{pmatrix} 1 & 1 & 1 & 1 \\ 1 & -1 & 1 & -1 \\ 1 & 1 & -1 & -1 \\ 1 & -1 & -1 & 1 \end{pmatrix},$$

$$H_8 = \begin{pmatrix} 1 & 1 & 1 & 1 & 1 & 1 & 1 & 1 \\ 1 & -1 & 1 & -1 & 1 & -1 & 1 & -1 \\ 1 & 1 & -1 & -1 & 1 & 1 & -1 & -1 \\ 1 & -1 & -1 & 1 & 1 & -1 & -1 & 1 \\ 1 & 1 & 1 & 1 & -1 & -1 & -1 & -1 \\ 1 & -1 & 1 & -1 & -1 & 1 & -1 & 1 \\ 1 & 1 & -1 & -1 & -1 & -1 & 1 & 1 \\ 1 & -1 & -1 & 1 & -1 & 1 & 1 & -1 \end{pmatrix}.$$

C.2 WALSH-HADAMARD TRANSFORM

Consider all binary q-tuples, $\mathbf{b}_1, \mathbf{b}_2, \dots, \mathbf{b}_{2^q}$. The *proper ordering*, π_q, of binary q-tuples is defined recursively as:

$$\begin{aligned} \pi_1 &= [0, 1], \\ \text{if } \pi_i &= [\mathbf{b}_1, \mathbf{b}_2, \dots, \mathbf{b}_{2^i}] \\ \text{then } \pi_{i+1} &= [\mathbf{b}_1 0, \mathbf{b}_2 0, \dots, \mathbf{b}_{2^i} 0, \mathbf{b}_1 1, \mathbf{b}_2 1, \dots, \mathbf{b}_{2^i} 1] \quad \forall i \in \{1, q-1\}. \end{aligned}$$

For example, given that $\pi_1 = [0, 1]$ then we have

$$\pi_2 = [00, 10, 01, 11]$$
$$\pi_3 = [000, 100, 010, 110, 001, 101, 011, 111]$$
$$\pi_4 = [0000, 1000, 0100, 1100, 0010, 1010, 0110, 1110, 0001, 1001, 0101, 1101, 0011, 1011, 0111, 1111].$$

Let $n = 2^q$ and let the n q-tuples under the proper order be

$$\pi_q = [\mathbf{u}_0, \mathbf{u}_1, \ldots, \mathbf{u}_{n-1}].$$

Let us enforce the convention that the leftmost bit of $\mathbf{u}_i$ is the least significant bit of the integer represented by $\mathbf{u}_i$ as a q-tuple. For example, consider the following mappings of 4-tuples to integers:

$$0000 \mapsto 0, \quad 1000 \mapsto 1, \quad 0100 \mapsto 2, \quad 1100 \mapsto 3, \quad 0010 \mapsto 4, \quad 1010 \mapsto 5$$
$$\ldots 0111 \mapsto 14, \quad 1111 \mapsto 15.$$

The matrix $H = [h_{ij}]$ with

$$h_{ij} = (-1)^{\mathbf{u_i} \cdot \mathbf{u_j}} \quad \forall (i, j) \in \{0, n - 1\}$$

where $u(i)$ and $u(j)$ are members of

$$\pi_q = [\mathbf{u}_0, \mathbf{u}_1, \ldots, \mathbf{u}_{n-1}]$$

is the Hadamard matrix of order $n = 2^q$. Observe that the rows and the columns of the matrix H are numbered from 0 to $n - 1$.

As an example consider the case $q = 3$. It is easy to see that the matrix elements of the first row, h_{0i}, and the first column, h_{i0}, are all $+1$ because $\mathbf{u}_0 \cdot \mathbf{u}_i = \mathbf{u}_i \cdot \mathbf{u}_0 = 0, \forall i \in \{1, 7\}$ and $(-1)^0 = +1$. It is also easy to see that the diagonal element $h_{ii} = -1$ when the weight of i (the number of 1s in the binary representation of integer i) is odd and $h_{ii} = +1$ when the weight of i is even. Individual calculation of $h_{ij}, \; i \neq j$ are also trivial. For example, $h_{32} = (-1)^{(110) \cdot (010)} = (-1)^1 = -1$. Indeed,

$$(110) \cdot (010) = 1 \cdot 0 + 1 \cdot 1 + 0 \cdot 0 = 1.$$

This shows that $H = H_3$.

Let $\mathbf{c} = (c_0 c_1 c_2 \ldots c_i \ldots c_{2^q - 1})$ be a binary 2^q-tuple, $c_i = 0, 1$ and B_q a $q \times 2^q$ matrix whose columns are all 2^q possible q-tuples $\mathbf{b}_i$. We define $\mathbf{c}(\mathbf{b})$ to be the component of $\mathbf{c}$ selected by the q-vector $\mathbf{b}$ according to the matrix B. The binary value $\mathbf{c}(\mathbf{b})$ can be either 0 or 1, depending upon the value of the corresponding element of the binary 2^q-tuple $\mathbf{c}$.

For example, let $q = 3$ and B_3 be given by

$$B_3 = \begin{pmatrix} 1 & 0 & 0 & 1 & 0 & 1 & 1 & 0 \\ 0 & 1 & 0 & 1 & 1 & 0 & 1 & 0 \\ 0 & 0 & 1 & 0 & 1 & 1 & 1 & 0 \end{pmatrix}.$$

Let $\mathbf{c} = (0\,1\,1\,1\,1\,0\,0\,1)$. Then $\mathbf{c}(1\,1\,0) = 1$ because $(1\,1\,0)$ is the 4th column of B and the 4th element of $\mathbf{c}$ is 1. Similarly $\mathbf{c}(1\,1\,1) = 0$ because $(1\,1\,1)$ is the 7th column of B and the 7th element of $\mathbf{c}$ is 0.

We can define a vector $\mathbf{R}(\mathbf{c})$ whose components $R_i(\mathbf{c})$ are $(-1)^{\mathbf{c}(\mathbf{b})}$ and their value can be either $+1$, or -1. In our example,

$$\mathbf{R}(\mathbf{c}) = (+1 \;\; -1 \;\; -1 \;\; -1 \;\; -1 \;\; +1 \;\; +1 \;\; -1).$$

Indeed:

$$\begin{aligned}
R_1(\mathbf{c}) &= (-1)^{\mathbf{c}(\mathbf{100})} = (-1)^0 = +1 \\
R_2(\mathbf{c}) &= (-1)^{\mathbf{c}(\mathbf{010})} = (-1)^1 = -1 \\
R_3(\mathbf{c}) &= (-1)^{\mathbf{c}(\mathbf{001})} = (-1)^1 = -1 \\
R_4(\mathbf{c}) &= (-1)^{\mathbf{c}(\mathbf{110})} = (-1)^1 = -1 \\
R_5(\mathbf{c}) &= (-1)^{\mathbf{c}(\mathbf{011})} = (-1)^1 = -1 \\
R_6(\mathbf{c}) &= (-1)^{\mathbf{c}(\mathbf{101})} = (-1)^0 = +1 \\
R_7(\mathbf{c}) &= (-1)^{\mathbf{c}(\mathbf{111})} = (-1)^0 = +1 \\
R_8(\mathbf{c}) &= (-1)^{\mathbf{c}(\mathbf{000})} = (-1)^1 = -1.
\end{aligned}$$

Let $\mathbf{d}$ be a binary q-tuple and let $\mathbf{c}$ be a binary 2^q-tuple. Let $\mathbf{R}(\mathbf{c}) = (-1)^{\mathbf{c}(\mathbf{b})}$ be a 2^q-tuple with entries either $+1$, or -1 as defined earlier. Then the *Walsh-Hadamard Transform* of $\mathbf{R}(\mathbf{c})$ is

$$\hat{\mathbf{R}}(\mathbf{d}) = \sum_{\mathbf{b} \in B_q} (-1)^{\mathbf{d} \cdot \mathbf{b}} \mathbf{R}(\mathbf{c})$$

or

$$\hat{\mathbf{R}}(\mathbf{d}) = \sum_{\mathbf{b} \in B_q} (-1)^{\mathbf{d} \cdot \mathbf{b} + \mathbf{c}(\mathbf{b})}.$$

For example, let $q = 3$, $\mathbf{d} = (1\,1\,1)^T$ and $\mathbf{c} = (0\,1\,1\,1\,1\,0\,0\,1)$. Then

$$\hat{\mathbf{R}}(1\,1\,1) = \sum_{\mathbf{b} \in B_3} (-1)^{(1\,1\,1) \cdot \mathbf{b} + \mathbf{c}(\mathbf{b})}$$

$$\begin{aligned}\hat{\mathbf{R}}(1\,1\,1) &= (-1)^{(1\,1\,1)\cdot(1\,0\,0)+\mathbf{c}(1\,0\,0)} + (-1)^{(1\,1\,1)\cdot(0\,1\,0)+\mathbf{c}(0\,1\,0)} \\ &\quad + (-1)^{(1\,1\,1)\cdot(0\,0\,1)+\mathbf{c}(0\,0\,1)} + (-1)^{(1\,1\,1)\cdot(1\,1\,0)+\mathbf{c}(1\,1\,0)} \\ &\quad + (-1)^{(1\,1\,1)\cdot(0\,1\,1)+\mathbf{c}(0\,1\,1)} + (-1)^{(1\,1\,1)\cdot(1\,0\,1)+\mathbf{c}(1\,0\,1)} \\ &\quad + (-1)^{(1\,1\,1)\cdot(1\,1\,1)+\mathbf{c}(1\,1\,1)} + (-1)^{(1\,1\,1)\cdot(0\,0\,0)+\mathbf{c}(0\,0\,0)} \\ &= (-1)^{1+0} + (-1)^{1+1} + (-1)^{1+1} + (-1)^{0+1} + (-1)^{0+1} \\ &\quad + (-1)^{0+0} + (-1)^{1+0} + (-1)^{0+1} = -2.\end{aligned}$$

Given the binary vector

$$\mathbf{t} = \mathbf{c} + \sum_{i=1}^{q} d_i v_i$$

with $\mathbf{d} = (d_1 d_2 \dots d_q)^T$ a binary q-tuple and v_i the i-th row of B_q, then $\hat{\mathbf{R}}$ is the number of 0s minus the number of 1s in $\mathbf{t}$.

C.3 THE FAST HADAMARD TRANSFORM

Orthogonal transforms such as Fourier, Hartley, Hadamard, or Haar can be computed using the Cooley and Tukey successive doubling method. Here we discuss the Fast Hadamard Transform.

If q is a positive integer and $M_{2^q}^{(i)} = I_{2^{q-i}} \otimes H_2 \otimes I_{2^{i-1}}$ with H_2 the Hadamard matrix, $I_{2^{q-i}}$ the identity matrix of size $2^{q-i} \times 2^{q-i}$, and $I_{2^{i-1}}$ the identity matrix of size $2^{i-1} \times 2^{i-1}$, then

$$H(2^q) = M_{2^q}^{(1)} M_{2^q}^{(2)} \dots M_{2^q}^{(q)}.$$

We prove this identity by induction. For $q = 1$ we need to prove that $H(2) = M_2^{(1)}$. I_n is the $n \times n$ identity matrix. By definition,

$$M_2^{(1)} = I_{2^{1-1}} \otimes H_2 \otimes I_{2^{1-1}} = H_2.$$

Assume that this is true for $q = k$ and consider the case $q = k + 1$. For $q = k$ we have:

$$H(2^k) = M_{2^k}^{(1)} M_{2^k}^{(2)} \dots M_{2^k}^{(k)}.$$

Now for $q = k + 1$ we have to prove that:

$$H(2^{k+1}) = M_{2^{k+1}}^{(1)} M_{2^{k+1}}^{(2)} \dots M_{2^{k+1}}^{(k)} M_{2^{k+1}}^{(k+1)}.$$

Then

$$M_{2^{k+1}}^{(i)} = I_{2^{k+1-i}} \otimes H_2 \otimes I_{2^{i-1}} = I_2 \otimes I_{2^{k-i}} \otimes H_2 \otimes I_{2^{i-1}} = I_2 \otimes M_{2^k}^{(i)}$$

Thus,

$$H(2^{k+1}) = (I_2 \otimes M_{2^k}^{(1)})(I_2 \otimes M_{2^k}^{(2)}) \dots (I_2 \otimes M_{2^k}^{(k)}) M_{2^{k+1}}^{(k+1)}.$$

We know from Chapter 2 that the tensor product of square matrices V, W, X, Y has the following property $(V \otimes W)(X \otimes Y) = VX \otimes WY$. Applying this property repeatedly after substituting $H(2^k)$ for $M_{2^k}^{(1)} M_{2^k}^{(2)} \ldots M_{2^k}^{(k)}$ we get

$$H(2^{k+1}) = (I_2 \otimes M_{2^k}^{(1)} M_{2^k}^{(2)} \ldots M_{2^k}^{(k)})(M_{2^{k+1}}^{(k+1)}) = (I_2 \otimes H(2^k) M_{2^{k+1}}^{(k+1)}).$$

But from the definition of $M_{2^{k+1}}^{(k+1)}$, we see that

$$M_{2^{k+1}}^{(k+1)} = H_2 \otimes I_{2^k}.$$

Thus,

$$\begin{aligned} H(2^{k+1}) &= (I_2 \otimes H(2^k))(H_2 \otimes I_{2^k}) = (I_2 H_2) \otimes (H(2^k) I_{2^k}) \\ &= H_2 \otimes H(2^k) = H(2^{k+1}). \end{aligned}$$

Compared with the traditional Hadamard Transform, the Fast Hadamard Transform allows a speedup $S_{FHT/HT}$ given by

$$S_{FHT/HT} = \frac{2^{q+1} - 1}{3q}.$$

For example, for $q = 16$ we have $S_{FHT/HT} = 2^{17}/48 \approx 128,000/48 \approx 2667$.

Given $\mathbf{R}(\mathbf{c}) = (+1, -1, -1, -1, -1, +1, +1, -1)$ let us compute $\hat{\mathbf{R}} = \mathbf{R}H$ with $H = M_8^1 M_8^2 M_8^3$. First we calculate M_8^1, M_8^2, and M_8^3 as follows:

$$M_8^1 = I_4 \otimes H_2 \otimes I_2 = \begin{pmatrix} 1 & 0 & 0 & 0 \\ 0 & 1 & 0 & 0 \\ 0 & 0 & 1 & 0 \\ 0 & 0 & 0 & 1 \end{pmatrix} \otimes \begin{pmatrix} 1 & 1 \\ 1 & -1 \end{pmatrix}.$$

Thus,

$$M_8^1 = \begin{pmatrix} 1 & 1 & 0 & 0 & 0 & 0 & 0 & 0 \\ 1 & -1 & 0 & 0 & 0 & 0 & 0 & 0 \\ 0 & 0 & 1 & 1 & 0 & 0 & 0 & 0 \\ 0 & 0 & 1 & -1 & 0 & 0 & 0 & 0 \\ 0 & 0 & 0 & 0 & 1 & 1 & 0 & 0 \\ 0 & 0 & 0 & 0 & 1 & -1 & 0 & 0 \\ 0 & 0 & 0 & 0 & 0 & 0 & 1 & 1 \\ 0 & 0 & 0 & 0 & 0 & 0 & 1 & -1 \end{pmatrix}.$$

Then

$$\begin{aligned} M_8^2 = I_2 \otimes H_2 \otimes I_2 &= \begin{pmatrix} 1 & 0 \\ 0 & 1 \end{pmatrix} \otimes \begin{pmatrix} 1 & 1 \\ 1 & -1 \end{pmatrix} \otimes \begin{pmatrix} 1 & 0 \\ 0 & 1 \end{pmatrix} \\ &= \begin{pmatrix} 1 & 1 & 0 & 0 \\ 1 & -1 & 0 & 0 \\ 0 & 0 & 1 & 1 \\ 0 & 0 & 1 & -1 \end{pmatrix} \otimes \begin{pmatrix} 1 & 0 \\ 0 & 1 \end{pmatrix}. \end{aligned}$$

Thus,

$$M_8^2 = \begin{pmatrix} 1 & 0 & 1 & 0 & 0 & 0 & 0 & 0 \\ 0 & 1 & 0 & 1 & 0 & 0 & 0 & 0 \\ 1 & 0 & -1 & 0 & 0 & 0 & 0 & 0 \\ 0 & 1 & 0 & -1 & 0 & 0 & 0 & 0 \\ 0 & 0 & 0 & 0 & 1 & 0 & 1 & 0 \\ 0 & 0 & 0 & 0 & 0 & 1 & 0 & 1 \\ 0 & 0 & 0 & 0 & 1 & 0 & -1 & 0 \\ 0 & 0 & 0 & 0 & 0 & 1 & 0 & -1 \end{pmatrix}.$$

Finally,

$$M_8^3 = I_1 \otimes H_2 \otimes I_4 = \begin{pmatrix} 1 & 1 \\ 1 & -1 \end{pmatrix} \otimes \begin{pmatrix} 1 & 0 & 0 & 0 \\ 0 & 1 & 0 & 0 \\ 0 & 0 & 1 & 0 \\ 0 & 0 & 0 & 1 \end{pmatrix}.$$

Thus,

$$M_8^3 = \begin{pmatrix} 1 & 0 & 0 & 0 & 1 & 0 & 0 & 0 \\ 0 & 1 & 0 & 0 & 0 & 1 & 0 & 0 \\ 0 & 0 & 1 & 0 & 0 & 0 & 1 & 0 \\ 0 & 0 & 0 & 1 & 0 & 0 & 0 & 1 \\ 1 & 0 & 0 & 0 & -1 & 0 & 0 & 0 \\ 0 & 1 & 0 & 0 & 0 & -1 & 0 & 0 \\ 0 & 0 & 1 & 0 & 0 & 0 & -1 & 0 \\ 0 & 0 & 0 & 1 & 0 & 0 & 0 & -1 \end{pmatrix}.$$

Then

$$\mathbf{R}M_8^1 = (+1, -1, -1, -1, -1, +1, +1, -1) \begin{pmatrix} 1 & 1 & 0 & 0 & 0 & 0 & 0 & 0 \\ 1 & -1 & 0 & 0 & 0 & 0 & 0 & 0 \\ 0 & 0 & 1 & 1 & 0 & 0 & 0 & 0 \\ 0 & 0 & 1 & -1 & 0 & 0 & 0 & 0 \\ 0 & 0 & 0 & 0 & 1 & 1 & 0 & 0 \\ 0 & 0 & 0 & 0 & 1 & -1 & 0 & 0 \\ 0 & 0 & 0 & 0 & 0 & 0 & 1 & 1 \\ 0 & 0 & 0 & 0 & 0 & 0 & 1 & -1 \end{pmatrix},$$

or

$$\mathbf{R}M_8^1 = (0, +2, -2, 0, 0, -2, 0, +2).$$

Next

$$(\mathbf{R}M_8^1)M_8^2 = (0, +2, -2, 0, 0, -2, 0, +2) \begin{pmatrix} 1 & 0 & 1 & 0 & 0 & 0 & 0 & 0 \\ 0 & 1 & 0 & 1 & 0 & 0 & 0 & 0 \\ 1 & 0 & -1 & 0 & 0 & 0 & 0 & 0 \\ 0 & 1 & 0 & -1 & 0 & 0 & 0 & 0 \\ 0 & 0 & 0 & 0 & 1 & 0 & 1 & 0 \\ 0 & 0 & 0 & 0 & 0 & 1 & 0 & 1 \\ 0 & 0 & 0 & 0 & 1 & 0 & -1 & 0 \\ 0 & 0 & 0 & 0 & 0 & 1 & 0 & -1 \end{pmatrix},$$

or

$$(\mathbf{R}M_8^1)M_8^2 = (-2, +2, +2, +2, 0, 0, 0, -4).$$

Finally,

$$(\mathbf{R}M_8^1M_8^2)M_8^3 = (-2, +2, +2, +2, 0, 0, 0, -4)\begin{pmatrix} 1 & 0 & 0 & 0 & 1 & 0 & 0 & 0 \\ 0 & 1 & 0 & 0 & 0 & 1 & 0 & 0 \\ 0 & 0 & 1 & 0 & 0 & 0 & 1 & 0 \\ 0 & 0 & 0 & 1 & 0 & 0 & 0 & 1 \\ 1 & 0 & 0 & 0 & -1 & 0 & 0 & 0 \\ 0 & 1 & 0 & 0 & 0 & -1 & 0 & 0 \\ 0 & 0 & 1 & 0 & 0 & 0 & -1 & 0 \\ 0 & 0 & 0 & 1 & 0 & 0 & 0 & -1 \end{pmatrix},$$

or

$$\hat{\mathbf{R}} = (\mathbf{R}M_8^1M_8^2)M_8^3 = (-2, +2, +2, -2, -2, +2, +2, +6).$$

APPENDIX D

Fourier Transform and Fourier Series

We review a few important facts about a widely used transformation of real and complex valued functions—the Fourier series and the Fourier Transform. Fourier Transforms are used in signal processing, image processing, and in a fairly large number of other scientific and engineering applications. The reader is encouraged to consult a reference text such as [86] for an in-depth treatment of the subject.

In case of signal processing the *Fourier analysis* allows us to construct the *spectrum* of the signal, and to study the properties of the signal in the frequency domain. For example, it is trivial to filter a signal in the frequency domain or, more generally, to compute the response of a linear circuit to the signal. *Fourier synthesis* allows us to transform back from the frequency domain to the time domain.

The Fourier Transform is often used to identify patterns in a set of data. Instead of computing the correlation of two functions in the time domain, $\int g(t)h(t - T)dt$, we simply compute the product of their Fourier Transforms $\mathcal{F}[g(t)]$ and $\mathcal{F}[h(t)]$ and study the maxima of the product in the Fourier Transform domain.

Given a real valued function $f(t)$ and a complex valued function $\mathcal{F} : \mathbb{R} \mapsto \mathbb{C}$, the Fourier Transform of $f(t)$ is the function

$$\mathcal{F}[f(t)] = \overline{f}(s) = \frac{1}{\sqrt{2\pi}} \int_{-\infty}^{+\infty} e^{-ist} f(t)dt$$

whenever the above integral is convergent. When $f(t)$ is a square integrable function, also called an $L2$ function, the convergence of the integral is assured, and then $\overline{f}(s)$ is also an $L2$ function. Moreover, any $L2$ function is the Fourier Transform of an $L2$ function, thus the Fourier Transform provides a bijective correspondence between $L2$ functions.

The *inverse Fourier Transform* of the function $\overline{f}(s)$ is

$$\mathcal{F}^{-1}[\overline{f}(s)] = f(t) = \frac{1}{\sqrt{2\pi}} \int_{-\infty}^{+\infty} e^{ist} \overline{f}(s)ds.$$

An important property of the Fourier Transform is that

$$\mathcal{F}(f \circ g) = \sqrt{2\pi}\, \mathcal{F}(f)\mathcal{F}(g)$$

with $f \circ g$ being the convolution of f and g.

A function $f(t)$ is said to be *periodic* with period T if $f(t + T) = f(t)$, where T is a positive constant. The least value of $T > 0$ is called the *period* of $f(t)$, $1/T$ is called the *fundamental frequency* of $f(t)$, and $\omega = 2\pi/T$ is the *angular frequency*.

A function $f(t)$ with period T has no Fourier Transform unless it is zero everywhere $f(t) \equiv 0$. However, if the integrals

$$c_n(\omega) = \frac{1}{T} \int_{-\frac{T}{2}}^{+\frac{T}{2}} f(t) e^{-in\omega t} dt$$

with n an integer are defined, then $f(t)$ has a series

$$\sum_{n=-\infty}^{+\infty} c_n(\omega) e^{in\omega t}$$

called the *Fourier series* which for a large class of functions $f(t)$ is convergent on any compact interval. This convergence is ensured if $f(t)$ satisfies the *Strong Dirichlet Conditions*. This means that in one interval of length T (and in any other such interval because $f(t)$ is periodic with period T) the function $f(t)$ has derivatives at all, but finitely many, points. If the function f has real values then

$$c_n = \bar{c}_n.$$

In this case,

$$f(t) = \frac{a_0}{2} + \sum_{n=1}^{\infty} (a_n \cos n\omega t + b_n \sin n\omega t)$$

with

$$c_n = a_n - ib_n.$$

In practice, we sometimes deal with discrete rather than continuous functions. For example, a digital image may consist of an $n \times m$ array of pixels. In this case instead of the Fourier Transform we use the Discrete Fourier Transform (DFT) of a set on N real or complex numbers.

Let v and w be two vectors in an N-dimensional complex vector space, $v, w \in \mathbb{C}^N$, $v = (v_0, v_1, v_2, \ldots, v_{N-1})$ and $w = (w_0, w_1, w_2, \ldots, w_{N-1})$. We restrict our discussion to the case when $N = 2^n$. DFT maps v into w as follows

$$w_k \equiv \frac{1}{\sqrt{N}} \sum_{j=0}^{N-1} v_j e^{2\pi i jk/N}.$$

In addition to one-dimensional Fourier Transforms, we sometimes use higher-dimensional Fourier Transforms for two-, three-, or higher-dimensional objects.

Bibliography

1. E. S. Abers. *Quantum Mechanics.* Prentice Hall, Upper Saddle River, NJ, 2003.
2. A. D. Aczel. *Entanglement: The Greatest Mystery in Physics.* Four Walls Eight Windows Publishing House, New York, NY, 2001.
3. M. Agrawal, N. Kayal, and N. Saxena. *PRIMES is in P.* http://www.cse.iitk.ac.in, 2002.
4. D. Z. Albert. *Quantum Mechanics and Experience.* Harvard University Press, Cambridge, MA, 1992.
5. A. Barenco, C. H. Bennett, R. Cleve, D. P. DiVincenzo, N. Margolus, P. Shor, T. Sleator, J. Smolin, and H. Weinfurter. *Elementary Gates for Quantum Computation.* Preprint, http://arxiv.org/archive/quant-ph/9503016 v1, March 1995.
6. J. S. Bell. "On the Einstein-Podolsky-Rosen Paradox." *Physics,* 1:195–200, 1964.
7. J. S. Bell. *Speakable and Unspeakable in Quantum Mechanics: Collected Papers on Quantum Philosophy.* Cambridge University Press, Cambridge, 1987.
8. P. Benioff. "The Computer as a Physical System: A Microscopic Quantum Mechanical Hamiltonian Model of Computers as Represented by Turing Machines." *J. Stat. Phys.*, 22:563–591, 1980.
9. P. Benioff. "Quantum Mechanical Models of Turing Machines that Dissipate no Energy." *Physical Review Letters*, 48:1581–1584, 1982.
10. P. Benioff. "Quantum Mechanical Models of Turing Machines." *J. Stat. Phys.*, 29:515–546, 1982.
11. C. H. Bennett. "Logical Reversibility of Computation." *IBM Journal of Research and Development*, 17:525–535, 1973.
12. C. H. Bennett. "The Thermodynamics of Computation—A Review." *International Journal of Theoretical Physics*, 21:905–928, 1982.
13. C. H. Bennett and G. Brassard. "Quantum Cryptography: Public Key Distribution and Coin Tossing." *Proc. IEEE Conf. on Computers, Systems, and Signal Processing,* 175–179, IEEE Press, 1984.
14. C. H. Bennett, G. Brassard, C. Crépeau, R. Jozsa, A. Peres, and W. K. Wooters. "Teleporting an Unknown State via Dual Classical and Einstein-Podolsky-Rosen Channels." *Physical Review Letters*, 70(13):1895–1899, 1993.
15. C. H. Bennett. "Quantum Information and Computation." *Physics Today*, 24–30, October 1995.
16. C. H. Bennett and P.W. Shor. "Quantum Information Theory." *IEEE Trans. on Information Theory*, 44(6):2724–2742, 1998.
17. C. H. Bennett, P. W. Shor, J. A. Smolin, and A. V. Thapliyal. "Entanglement-Assisted Capacity of a Quantum Channel and the Reverse Shannon Theorem." *IEEE Trans. on Information Theory*, 48(10):2637–2655, 2002.
18. C. H. Bennett, T. Mor, and J. A. Smolin. *The Parity Bit in Quantum Cryptography.* arXiv.quant-ph/9604040, July 5, 2002.

19. E. Bernstein and U. Vazirani. "Quantum Complexity Theory." *Proc. 25th ACM Symp. of Theory of Computing*, ACM Press, NY, 11–20, 1993.

20. E. Bernstein and U. Vazirani. "Quantum Complexity Theory." *SIAM J. Computing, 26*: 1411–1473, 1997.

21. A. Berthiaume and G. Brassard. "The Quantum Challenge to Structural Complexity Theory." *Proc. 7-th Annual Conf. on Structure in Complexity Theory*, 132–137, IEEE Press, Los Alamitos, CA, 1992.

22. A. Berthiaume and G. Brassard. "Oracle Quantum Computing." *Proc. Workshop on Physics of Computation*, 195–199, IEEE Press, Los Alamitos, CA, 1992.

23. G. Birkhoff and S. Mac Lane. *A Survey of Modern Algebra.* Macmillan Publishing, New York, NY, 1965.

24. D. Bohm. *Quantum Theory.* Prentice Hall, Upper Saddle, NJ, 1951.

25. M. Born. "The Statistical Interpretations of Quantum Mechanics." *Nobel Lectures, Physics 1942–1962*, 256–267. December 11, 1954. See also: http://www.nobel.se/physics/laureates/1954/born-lecture.html.

26. D. Bouwmester, A. Ekert, and A. Zeilinger, Editors. *The Physics of Quantum Information*, Springer Verlag, Heidelberg, 2003.

27. G. K. Brennen, C. M. Caves, P. S. Jessen, and I. H. Deutsch. "Quantum Logic Gates in Optical Lattices." *Physical Review Letters*, 82(5):1060–1063, 1999.

28. L. de Broglie. "The Wave Nature of the Electron." *Nobel Lectures, Physics 1922–1941*, 244–256. December 12, 1929. See also: http://www.nobel.se/physics/laureates/1929/broglie-lecture.html.

29. J. Brown. *The Quest for the Quantum Computer.* Simon and Schuster, New York, NY, 1999.

30. A. W. Burks, H. H. Goldstine, and J. von Neumann. "Preliminary Discussion of the Logical Design of an Electronic Computer Instrument." *Report to the US Army Ordnance Department*, 1946. Also in: *Papers of John von Neumann*. W. Asprey and A. W. Burks, Editors, 97–146, MIT Press, Cambridge, MA, 1987.

31. A. R. Calderbank and P. W. Shor. "Good Quantum Error-Correcting Codes Exist." *Physical Review A*, 54(42):1098–1105, 1996.

32. A. M. Childs and I. L. Chuang. "Universal Quantum Computation with Two-Level Trapped Ions." *Physical Review A*, 63(1), 012306, 2001.

33. J. I. Cirac and P. Zoller. "Quantum Computation with Cold Trapped Ions." *Physical Review Letters*, 74(20):4091–4094, 1995.

34. J. I. Cirac and P. Zoller. "A Scalable Quantum Computer with Ions in an Array of Microtraps." *Nature*, 404:579–581, 2000.

35. R. Cleve, A. Ekert, L. Henderson, C. Macchiavello, and M. Mosca. *On Quantum Algorithms.* Preprint, http://arxiv.org/archive/quant-ph/99036061 v1, 1999.

36. J. W. Cooley and J. W. Tukey. "An Algorithm for the Machine Calculation of Complex Fourier Series." *Math. Comp.*, 19: 297–301, 1965.

37. D. G. Cory, A. F. Fahmy, and T. F. Havel. "Nuclear Magnetic Resonance Spectroscopy: An Experimentally Accessible Paradigm for Quantum Computing." *Proc. PhysComp96*, T. Toffoli, M. Biafore, and J. Leao, Editors. New England Complex Systems Institute, 87–91, (1996).

38. D. G. Cory, A. F. Fahmy, and T. F. Havel. "Ensemble Quantum Computing by NMR Spectroscopy." *Proc. Nat. Acad. Sci.*, 94(5):1634, 1997.

39. D. G. Cory, M. D. Price, and T. F. Havel. "Nuclear Magnetic Resonance Spectroscopy: An Experimentally Accessible Paradigm for Quantum Computing." *Physica D*, 120:82–101, 1998.

40. T. M. Cover and J. A. Thomas. *Elements of Information Theory.* Wiley Series in Telecommunications. John Wiley & Sons, New York, NY, 1991.

41. W. van Dam. "A Universal Quantum Cellular Automaton." *Proc. PhysComp96*, T. Toffoli, M. Biafore, and J. Leao, Editors. New England Complex Systems Institute, 323–331, (1996).

42. D. Deutsch. "Quantum Theory, the Church-Turing Principle and the Universal Quantum Computer." *Proc. R. Soc. London A*, 400:97–117, 1985.

43. D. Deutsch and R. Jozsa. "Rapid Solution of Problems by Quantum Computations." *Proc. R. Soc. London A*, 439:553–558, 1992.

44. D. Deutsch. *The Fabric of Reality.* Penguin Books, New York, NY, 1997.

45. P. A. M. Dirac. "Theory of Electrons and Positrons." *Nobel Lecture*, December 12, 1933. See also: http://www.nobel.se/physics/laureates/1933/dirac-lecture.html.

46. P. A. M. Dirac. *The Principles of Quantum Mechanics.* Fourth Edition. Sec. 2, 4–7, Clarendon Press, Oxford, 1967.

47. D. P. DiVincenzo. "The Physical Implementation of Quantum Computation." *Fortschritte der Physik*, 48(9-11):771–783, 2000.

48. M. I. Dyakonov. *Quantum Computing: A View from the Enemy Camp.* Preprint, http://arxiv.org/archive/quant-ph/0110326 v1, October 2001.

49. A. Einstein. "Elektrodynamik bewegter Körper." *Annalen der Physik*, 891:921, 1905.

50. A. Einstein, B. Podolsky, and N. Rosen. "Can Quantum-Mechanical Description of Physical Reality Be Considered Complete?" *Physical Review*, 47:777, 1935.

51. A. Ekert and R. Jozsa. "Quantum Algorithms: Entanglement Enhanced Information Processing." *Proc. R. Soc. London A*, 356(1743): 1769–1782, 1998. See also: http://arxiv.org/archive/quant-ph/9803072 v1, November 2000.

52. A. Ekert, P. Hayden, and H. Inamori. *Basic Concepts in Quantum Computing.* Preprint, http://arxiv.org/archive/quant-ph/0011013 v1, November 2000.

53. R. P. Feynman, R. B. Leighton, and M. Sands. *The Feynman Lectures on Physics, Volumes 1,2, and 3.* Addison-Wesley, Reading, MA, 1977.

54. R. P. Feynman. "Simulating Physics with Computers." *Int. J. Theoret. Phys.* 21:467–488, 1982.

55. R. P. Feynman. "Quantum Mechanical Computers." *Found. Phys.* 16:507–531, 1986.

56. R. P. Feynman. *QED—The Strange Theory of Light and Matter.* Princeton University Press, Princeton, NJ, 1985.

57. R. P. Feynman. *Lectures on Computation.* Addison-Wesley, Reading, MA, 1996.

58. E. Fredkin. "Digital Machines: An Informational Process Based on Reversible Universal Cellular Automata." *Physica D*, 45:254–270, 1990. See also: http://digitalphilosophy.org/dm_paper.htm.

59. H. Fritzsch. *An Equation that Changed the World.* The University of Chicago Press, Chicago, IL, 2004.

60. W. Fulton and J. Harris. *Representation Theory—A First Course.* Third Edition. Springer-Verlag, Heidelberg, 1997.

61. P. Galison. *Einstein's Clocks, Poincare's Maps.* W. W. Norton & Co. Publishing House, London, 2003.

62. I. M. Gelfand. *Lectures on Linear Algebra.* Dover Publications, New York, NY, 1989.

63. N. A. Gershenfeld and I. L. Chuang. "Bulk Spin-Resonance Quantum Computation." *Science*, 275(5298):350–356, 1997.

64. D. Gottesman. "An Introduction to Quantum Error Correction." *Proc. Symp. in Applied Math*, Preprint, http://arxiv.org/archive/quant-ph/00040072 v1, April 2000.

65. I. S. Gradshteyn and I. M. Ryzhik. *Table of Integrals, Series, and Products.* Academic Press, Orlando, FL, 1980.

66. R. B. Griffiths and C.-S. Niu. "Semiclassical Fourier Transform for Quantum Computation." *Phys. Rev.* 76: 3228–3231, 1996. See also: http://arxiv.org/archive/quant-ph/9511007 v1, November 1995.

67. L. K. Grover. "A Fast Quantum Algorithm for Database Search." *Proc. ACM Symp. on Theory of Computing*, ACM Press, NY, 212–219, 1996.

68. L. K. Grover. "Quantum Mechanics Helps in Searching for a Needle in a Haystack." *Phys. Rev. Lett.* 78: 325–328, 1997.

69. L. K. Grover. "A Framework for Fast Quantum Mechanical Algorithms." *Proc. Symp. on Theory of Computing*, ACM Press, NY, 53–62, 1998.

70. Y. Hardey and W.-H. Steeb. *Classical and Quantum Computing.* Birkhäuser, Boston, MA, 2001.

71. S. W. Hawking. *A Brief History of Time.* Bantam Books, New York, NY, 1988.

72. W. Heisenberg. "The Development of Quantum Mechanics." *Nobel Lectures, Physics 1922–1942*, 290–301, December 11, 1933.

73. D. Jaksch, H.-J. Briegel, J. I. Cirac, and P. Zoller. "Entanglement of Atoms via Cold Controlled Collisions." *Physical Review Letters*, 82(9):1975–1978, 1999.

74. G. Johnson. *A Shortcut through Time.* Alfred A. Knopf, New York, NY, 2003.

75. R. Jozsa. "Entanglement and Quantum Computation." *Geometric Issues in the Foundations of Science.* S. Hugget, L. Mason, K.P. Tod, S.T. Tsou, and N. M. J. Woodhouse, Editors. Oxford University Press, 1997. See also: http://arxiv.org/archive/quant-ph/9707034 v1, 1997.

76. R. Jozsa. *Quantum Algorithms and the Fourier Transform.* Preprint, http://arxiv.org/archive/quant-ph/9707033 v1, 1997.

77. R. Jozsa. *Searching in Grover's Algorithm.* Preprint, http://arxiv.org/archive/quant-ph/9901021 v1, 1999.

78. R. Jozsa. *Quantum Factoring, Discrete Logarithms, and the Hidden Subgroup Problem.* Preprint, http://arxiv.org/archive/quant-ph/0012084 v1, 2000.

79. R. Jozsa. *Illustrating the Concept of Quantum Information.* Preprint, http://arxiv.org/archive/quant-ph/0305114 v1, 2003.

80. B. E. King, C. S. Wood, C. J. Myatt, Q. A. Turchette, D. Leibfried, W. M. Itano, C. Monroe, and D. J. Wineland. "Cooling the Collective Motion of Trapped Ions to Initialize a Quantum Register." *Physical Review Letters*, 81:1525–1528, 1998.

81. A. Yu. Kitaev. *Quantum Measurements and the Abelian Stabilizer Problem.* Preprint, http://arxiv.org/archive/quant-ph/9511026 v1, 1995.

82. E. Knill, R. Laflame, H. Barnum, D. Dalvit, J. Dziarmaga, J. Gubernatis, L. Gurvis, G. Ortiz, L. Viola, and W. H. Zurek. "Quantum Information Processing." *Kluwer Encyclopedia of Mathematics, Supplement III*, 2002.

83. D. Knuth. *The Art of Computer Programming. Vol. 1*, Second Edition, Addison-Wesley, Reading, MA, 1969.

84. D. Knuth. *The Art of Computer Programming. Vol. 2*, Second Edition, Addison-Wesley, Reading, MA, 1981.

85. A. N. Kolmogorov and S. V. Fomin. *Elements of the Theory of Functions and Functional Analysis.* Dover Publications, Mineola, NY, 1999.

86. E. Kreyszig. *Advanced Engineering Mathematics.* John Wiley & Sons, New York, NY, 1998.

87. R. Landauer. "Irreversibility and Heat Generation in the Computing Process." *IBM Journal of Research and Development*, 5:182–192, 1961.

88. S. Lloyd. "A Potentially Realizable Quantum Computer." *Science*, 261:1569–1571, 1993.

89. Y. Manin. *Classical Computing, Quantum Computing, and Shor's Algorithm.* Talk at the Bourbaki Seminar, June 1999. Preprint, http://arxiv.org/archive/quant-ph/9903008 v1, March 1999.

90. S. McCartney. *ENIAC; The Triumphs and Tragedies of the World's First Computer.* Walker and Company Publishing House, New York, NY, 1999.

91. L. Meitner and O.R. Frisch. "Disintegration of Uranium by Neutrons: A New Type of Nuclear Reaction." *Nature*, 143: 239-240, 1939.

92. E. Mertzbacher. *Quantum Mechanics.* Third Edition. John Wiley & Sons, New York, NY, 1998.

93. G. J. Milburn. *Schrödinger's Machines.* Perseus Books, Cambridge, MA, 1998.

94. G. J. Milburn. *The Feynman Processor.* W.H. Freeman and Company, New York, NY, 1996.

95. C. Monroe, D. M. Meekhof, B. E. King, W. M. Itano, and D. J. Wineland. "Demonstration of a Fundamental Quantum Logic Gate." *Physical Review Letters*, 74(25):4714–4718, 1995.

96. C. Monroe, D. Leibfried, B. E. King, D. M. Meekhof, W. M. Itano, and D. J. Wineland. "Simplified Quantum Logic with Trapped Ions." *Physical Review A*, 55:R2489–2491, 1997.

97. M. Mosca and A. Ekert. *The Hidden Subgroup Problem and Eigenvalue Estimation on a Quantum Computer.* Preprint, http://arxiv.org/archive/quant-ph/9903071 v1, May 1999.

98. M. A. Nielsen and I. L. Chuang. *Quantum Computing and Quantum Information.* Cambridge University Press, 2000.

99. B. W. Ogburn and J. Preskill. "Topological Quantum Computation." *Lecture Notes in Computer Science*, 1509:341–359, Springer-Verlag, Heidelberg, 1999.

100. R. Omnès. *The Interpretation of Quantum Mechanics.* Princeton Series in Physics. Princeton University Press, Princeton, NJ, 1994.

101. C. M. Papadimitriou. *Computational Complexity.* Addison-Wesley, Reading, MA, 1994.

102. D. A. Patterson and J. L. Hennessy. *Computer Organization and Design; The Hardware/Software Interface.* Second Edition. Morgan Kaufmann, San Francisco, CA, 1998.

103. W. Pauli. "Exclusion Principle and Quantum Mechanics." *Nobel Lectures, Physics 1942–1962*, 27–43, December 13, 1946. See also: http://www.nobel.se/physics/laureates/1945/pauli-lecture.html.

104. A. O. Pittinger. *An Introduction to Quantum Algorithms.* Birkhäuser, Boston, MA, 1999.

105. M. K. E. L. Planck. "The Genesis and Present State of Development of the Quantum Theory." *Nobel Lectures, Physics 1901–1922.* June 2, 1920. See also: http://www.nobel.se/physics/laureates/1918/planck-lecture.html.

106. S. Popescu. *Bell's Inequalities and Density Matrices. Revealing "Hidden" Nonlocality.* Preprint, http://arxiv.org/archive/quant-ph/9502005 v1, February 1995.

107. J. F. Poyatos, J. I. Cirac, and P. Zoller. "Quantum Gates with "Hot" Trapped Ions." *Physical Review Letters*, 81(6):1322–1325, 1998.

108. J. F. Poyatos, J. I. Cirac, and P. Zoller. "Schemes of Quantum Computations with Trapped Ions." *Fortschritte der Physik*, 48(9-11):785–799, 2000.

109. J. Preskill. *Fault Tolerant Quantum Computation.* Preprint, http://arxiv.org/archive/quant-ph/9712048 v1, December 1997.

110. J. Preskill. *Lecture Notes for Physics 229: Quantum Information and Computing.* California Institute of Technology, 1998.

111. J. Preskill. *Quantum Clock Synchronization and Quantum Error Correction.* Preprint, http://arxiv.org/archive/quant-ph/0010098 v1, October 2000.

112. E. Rieffel and W. Polak. "An Introduction to Quantum Computing for Non-Physicists." *ACM Computing Surveys*, 32(3):300–335, 2000.

113. E. Schrödinger. "The Fundamental Idea of Wave Mechanics." *Nobel Lectures, Physics 1922–1941*, 305–314. December 12, 1933.

114. E. Schrödinger. "The Present Situation in Quantum Mechanics." *Die Naturwissenschaften*, 23:807–812; 823–828; 944-849, 1935. See also *Proceedings of the Cambridge Philosophical Society*, 31:555–563, 1935 and 32:446–452, 1936.

115. B. Schumacher. "Quantum Coding." *Physical Review A*, 51(4):2738–2747, 1995.

116. C. E. Shannon. "Communication in the Presence of Noise." *Proceedings of the IRE*, 37:10–21, 1949.

117. C. E. Shannon. "Certain Results in Coding Theory for Noisy Channels." *Information and Control*, 1(1):6–25, 1957.

118. C. E. Shannon and W. Weaver. *A Mathematical Theory of Communication.* University of Illinois Press, Urbana, IL, 1963.

119. P. W. Shor. "Algorithms for Quantum Computation: Discrete Log and Factoring." *Proc. 35 Annual Symp. on Foundations of Computer Science*, 124–134, IEEE Press, Piscataway, NJ, 1994.

120. P. W. Shor. "Scheme for Reducing Decoherence in Quantum Computer Memory." *Physical Review A*, 52(4):2493–2496, 1995.

121. P. W. Shor. *Polynomial-Time Algorithms for Prime Factorization and Discrete Logarithms on a Quantum Computer.* Preprint, http://arxiv.org/archive/quant-ph/9508027 v2, January 1996.

122. P. W. Shor. "Fault-Tolerant Quantum Computation." *37th Ann. Symp. on Foundations of Computer Science*, 56–65, IEEE Press, Piscataway, NJ, 1996.

123. P. W. Shor. "Polynomial-Time Algorithms for Prime Factorization and Discrete Logarithms on a Quantum Computer." *SIAM J. Computing,* 26: 1484–1509, 1997.

124. P. W. Shor and J. Preskill. *Simple Proof of Security of the BB84 Quantum Key Distribution Protocol.* http://www.arXiv.org/quant-ph/0003004, May 2000.

125. P. W. Shor. *Introduction to Quantum Algorithms.* Preprint, http://www.arXiv/quant-ph/0005003, July 2001.

126. P. W. Shor. "Why Haven't More Quantum Algorithms Been Found?" *Journal of the ACM*, 50(1): 87–90, 2003.

127. D. R. Simon. "On the Power of Quantum Computation." *SIAM J. Computing,* 26: 1474–1483, 1997.

128. L. Sklar. *Philosophy of Physics.* Westview Press, Boulder, CO, 1992.

129. A. Sørensen and K. Mølmer. "Quantum Computation with Ions in Thermal Motion." *Physical Review Letters*, 82(9):1971–1974, 1999.

130. A. Steane. *The Ion Trap Quantum Information Processor.* Preprint, http://arxiv.org/archive/quant-ph/9608011 v2, August 1996.

131. A. Steane. "Quantum Computing." *Reports Prog. Phys.*, 61, 117, 1998. See also: http://arxiv.org/archive/quant-ph/97080222 v2, September 1997.

132. O. Stern. "The Method of Molecular Rays." *Nobel Lectures, Physics 1942–1961*, 8–16. December 12, 1946. See also: http://www.nobel.se/physics/laureates/1943/stern-lecture.html.

133. N. S. Szabo and R. I. Tanaka. *Residue Arithmetic and Its Applications to Computer Technology.* McGraw-Hill, New York, NY, 1967.

134. L. Szilard. "Über die Entropieverminderung in einem Thermodynamichen System bei Eingriffen Intelligenter Wesen." *Zeitschrifft für Physik,* 53:840–856, 1929.

135. A. J. Thomasian. *The Structure of Probability Theory with Applications.* McGraw-Hill New York, NY, 1969.

136. A. C. de la Tore, A. Daleo, and I. Garcia-Mata. *The Photon-Box Bohr-Einstein Debate Demythologized.* Preprint, http://arxiv.org/archive/quant-ph/9910040 v1, October 1999.

137. Q. A. Turchette, C. S. Wood, B. E. King, C. J. Myatt, D. Leibfried, W. M. Itano, C. Monroe, and D. J. Wineland. "Deterministic Entanglement of Two Ions." *Physical Review Letters,* 81:3631–3634, 1998.

138. A. M. Turing. "On Computable Numbers with an Application to the Entscheidungsproblem." *Proc. London Math. Soc. 2*, 42:230, 1936.

139. L. M. K. Vandersypen, M. Steffen, G. Breyta, C. S. Yannoni, M. H. Sherwood, and I. S. Chuang. "Experimental Realization of Shor's Quantum Factoring Algorithm Using Nuclear Magnetic Resonance." *Nature*, 414: 883–887, 2001. See also: http://arxiv.org/archive/quant-ph/0112176 v1, December, 2001.

140. S. A. Vanstone and P. C. van Oorschot. *An Introduction to Error Correcting Codes with Applications.* Kluwer Academic Publishers, Boston, MA, 1987.

141. J. von Neumann. "Fourth University of Illinois Lecture." *Theory of Self-Reproduced Automata*, A. W. Burks, Editor. 66, University of Illinois Press, Urbana, IL, 1966.

142. J. von Neumann. *Mathematical Foundations of Quantum Mechanics.* Trans. R. T. Bayer. Princeton University Press, Princeton, NJ, 1955.

143. D. Wells. *The Penguin Book of Curious and Interesting Puzzles.* Penguin Books, 1992.

144. D. J. Wineland, M. Barrett, J. Britton, J. Chiaverini, B. DeMarco, W. M. Itano, B. Jelencovic, C. Langer, D. Leibfried, V. Meyer, T. Rosenband, and T. Schätz. *Quantum Information Processing with Trapped Ions.* Preprint, http://arxiv.org/archive/quant-ph/0212079 v2, March 2003.

145. W. K. Wootters and W. H. Zurek. "A Single Quantum Cannot Be Cloned." *Nature*, 299:802–803, 1982.

146. *Conference on Lasers and Electron Optics.* http://optics.org/articles/news/9/6/3/1.

147. *Semiconductor Industry Association Roadmap 2000-2001.* http://public.itrs.net.

Glossary

Abelian group Algebraic structure; a group with a commutative binary operation. See also *group*.

adiabatic Term frequently used in thermodynamics meaning without transfer of heat (without change in temperature). For example, if we open the valve of a gas canister, the gas rushes out and expands without having the time to equalize its temperature with the environment. The rushing gas feels cool.

adjoint matrix $A^\dagger$ the adjoint of matrix $A = [a_{ij}]$ is obtained by transposing A, and then, taking the complex conjugate of each of its elements,

$$A^\dagger = [a^*_{ji}].$$

The order of the two operations can be reversed. See also *dual operator*.

algorithm A formal description of a computational process; a set of elementary steps necessary to carry out a computation. Ultimately, an algorithm should be translated into a set of instructions to be executed by a computing device. Such instructions include logical operations performed upon a set of operands in registers, as well as reading and writing the contents of the memory.

alphabet Set of symbols accepted as input by an automaton.

amplitude, probability amplitude Given a set of basis or logical states $| i\rangle$, a quantum system may be in a superposition of these states, $| \psi\rangle = \sum_i \alpha_i | i\rangle$. The complex numbers α_i are called amplitudes or probability amplitudes.

ancillas Auxiliary qubits used to assist in a computation.

anti-commutator of two operators Given two operators $\mathbf{A}$ and $\mathbf{B}$, the *anti-commutator* of the two operators is

$$\{\mathbf{A}, \mathbf{B}\} = \mathbf{AB} + \mathbf{BA}.$$

$\mathbf{A}$ anti-commutes with $\mathbf{B}$ if

$$\{\mathbf{A}, \mathbf{B}\} = 0.$$

automorphism An isomorphism of a group onto itself. See also *isomorphism* and *homeomorphism*.

Banach space A vector space which is complete and has a norm.

basis states of a quantum system Linearly independent, complete subset of orthonormal state vectors in the vector space of the quantum system.

basis states of a qubit Two orthonormal states such as $| 0\rangle$ and $| 1\rangle$.

Bayes rule Probability rule. Assume that it is known that event $\mathcal{A}$ occurred, but it is not known which one of the set of mutually exclusive and collectively exhaustive events

$\mathcal{B}_1, \mathcal{B}_2, \dots, \mathcal{B}_n$ has subsequently occurred. Then the conditional probability that one of these events, $\mathcal{B}_j$, occurs, given that $\mathcal{A}$ occurs is

$$P(\mathcal{B}_j \mid \mathcal{A}) = \frac{P(\mathcal{A} \mid \mathcal{B}_j)P(\mathcal{B}_j)}{\sum_i P(\mathcal{A} \mid \mathcal{B}_i)P(\mathcal{B}_i)}$$

with $P(\mathcal{B}_j \mid \mathcal{A})$ as the *a posteriori probability*.

BB84 A quantum key distribution protocol proposed by Charles Bennett and Gilles Brassard in 1984.

beam splitter An optical device (*e.g.*, a half-silvered mirror) that splits an incoming beam of photons into a reflected and a transmitted beam. The color of the light (thus, its wavelength) is not altered by a beam splitter: a behavior consistent with a wave.

Bell's Inequality A constraint on the sum of averages of measured observables, based on assumptions of *local realism* (*i.e.*, of locality and of existence of hidden variables). Quantum mechanics predicts a value for the sum of averages of observables which is in violation of Bell's Inequality. Consider two photons with the same polarization, moving apart from one another at the speed of light. "If the coherence of the system can be maintained, a measurement made at one location will assign a single definite state to the system, which should be verified by any subsequent measurements, no matter where they occur. Whether the first photon passes through the polarizer or is absorbed completely determines the polarization of both photons from that point onward. The polarizer can be thought of as asking the system *are you polarized at angle θ?* and accepting only a yes/no answer, which the system is then obliged to repeat consistently thereafter. In the Copenhagen interpretation, nothing can be said about the angle of polarization of the photons before the measurement. The angle of polarization has no objective reality until it is measured, and the measurement result is not predetermined by anything, it is intrinsically random. After it is measured, then the polarization state of both photons is completely known and will not change. In a *hidden variable theory*, there are variables that determine the real polarization angle of the photons; these variables have definite values from the moment the photons are formed, and they determine what the result of the polarization measurement will be. Let us make the polarization measurement on the second particle at a different angle from our measurement on the first. Now the two photons have different probabilities of passing through their respective polarizers. We can expect that sometimes photon A will pass through its polarizer, but photon B will be absorbed by its polarizer. We call such events errors, in the sense that there is a disagreement between the measurements of the two polarizers. Suppose the polarizer at A is rotated by an angle θ relative to the polarizer at B. A certain number of errors $\mathcal{E}$ are produced. Now what happens if we rotate the B polarizer through an angle θ in the opposite direction, so that the total angle between them is 2θ? We expect more errors, but how much more? Twice as many? Bell demonstrated that the error rate with an angle 2θ must be less than or equal to twice the error rate at angle θ: $\mathcal{E}(2\theta) \leqslant 2\mathcal{E}(\theta)$, provided that the photons have a definite polarization (as postulated by the hidden variable theories) and that the A photon cannot affect the B photon's state instantaneously (this is the requirement of locality)" (see `http://www.telp.com/philosophy/qw3.htm`). When we obtain two contradictory results using two different theoretical models, it means that one of the models is wrong. Then we must turn to experiments to determine which one is wrong. In this case, the experiments show that Bell's Inequality is violated by quantum systems.

Bell, John Stewart British physicist (1928–1990). His work led to the possibility of exploring seemingly philosophical questions in quantum mechanics, such as the nature of reality, directly through experiments. Bell started from locality and argued for the existence of deterministic hidden variables. He considered measurements of spin components along arbitrary directions corresponding to each of the two EPR particles (see the *EPR paradox*). Bell calculated what happened when the measurement direction was kept constant for one particle and varied for the other. He was able to show that the behavior predicted by quantum theory could not be duplicated by a hidden variable theory if the hidden variables acted locally. As shown by Bell and others, local realist theories (*i.e.*, theories with hidden variables) satisfy a so-called *Bell's Inequality*. This is a constraint on the relationship between the joint probability densities of the signals recorded in the two wings of the apparatus; it involves the four distinct cases that may be obtained by having two settings in each wing. Quantum theory, on the other hand, does not obey Bell's Inequality. In this way, Bell had opened up the possibility of experimental philosophy: the study of what are normally thought of as philosophical issues in experiments.

Bell states Four distinct quantum states of a two-particle system with a very strong coupling of the individual states of the particles. These states form a normal basis

$$| \beta_{00} \rangle = \frac{| 00 \rangle + | 11 \rangle}{\sqrt{2}},$$

$$| \beta_{01} \rangle = \frac{| 01 \rangle + | 10 \rangle}{\sqrt{2}},$$

$$| \beta_{10} \rangle = \frac{| 00 \rangle - | 11 \rangle}{\sqrt{2}},$$

$$| \beta_{11} \rangle = \frac{| 01 \rangle - | 10 \rangle}{\sqrt{2}}.$$

binary alphabet An alphabet with only two symbols usually denoted as 0 and 1.

binary symmetric channel Abstraction for a communication channel. The input and output alphabets consist of two symbols only. There is a well-defined mapping of the two symbols in the input alphabet to the two symbols of the output alphabet. A *noiseless binary symmetric channel* maps a 0 at the input into a 0 at the output and a 1 into a 1. A *noisy symmetric channel* maps a 0 into a 1, and a 1 into a 0 with probability p; an input symbol is mapped into itself with probability $1 - p$.

bit The basic unit of classical information. It can take one of two values, 0 or 1.

black box A computational device whose inner structure is unknown and which implements a set of operations. An experiment with a black box, called a *query*, consists of supplying an input and observing the output. The smallest number of queries necessary to determine the operation is called the *query complexity*.

Bloch sphere A qubit is represented as a vector **r** from the origin to a point on the sphere with a radius of one, the so-called Bloch sphere. θ is the angle of the vector **r** with the z axis, and φ is the angle of the projection of the vector in the xy plane with the x axis; γ has no observable effect. The state of a qubit $|\psi\rangle$ can then be expressed using three real numbers, θ, φ, and γ

$$|\psi\rangle = e^{i\gamma}\left[\cos\frac{\theta}{2} + e^{i\varphi}\sin\frac{\theta}{2}\right] = \alpha_0 \,|\, 0\rangle + \alpha_1 \,|\, 1\rangle$$

with

$$\alpha_0 = e^{i\gamma}\cos\frac{\theta}{2} \quad \text{and} \quad \alpha_1 = e^{i\gamma}e^{i\varphi}\sin\frac{\theta}{2}.$$

In this representation, $e^{i\gamma}$ is an overall phase factor that is not observable, and thus, it generally is ignored.

block code A code where a group of information symbols is encoded into a fixed-length code word by adding a set of parity check or redundancy symbols.

Bool, George British mathematician (1815–1864). Best known for his contribution to symbolic logic (Boolean algebra) but also active in probability theory, algebra, analysis, and differential equations. Bool founded symbolic logic. He lived, taught, and is buried in Cork City, Ireland.

Boolean algebra Boolean algebra is a set B with two binary operations, $\bigcup$ and $\bigcap$, that are commutative, associative, distributes each over the other, plus a unary operation $\neg$. The identity elements $\Phi, U \in B$ satisfy the relations $b \bigcup \Phi = b$, $b \bigcap U = b$, $b \bigcup \neg b = U$, and $b \bigcap \neg b = \Phi$ for all elements $b \in B$. One interpretation of Boolean algebra is the collection of subsets of a fixed set X. We take $\bigcup, \bigcap, \neg, \Phi$, and U to be set union, set intersection, complementation, the empty set and the set X, respectively. Equality here means the usual equality of sets. Another interpretation is the calculus of propositions in symbolic logic. Here we take $\bigcup, \bigcap, \neg, \Phi$, and U to be disjunction, conjunction, negation, a fixed contradiction and a fixed tautology, respectively. In this setting, equality means logical equivalence.

Bounded-error Probabilistic Polynomial Time (BPP) Term used in computational complexity theory to describe a class of probabilistic algorithms. BPP consists of polynomial time algorithms which produce the correct result with a probability at least 2/3 (or other values strictly between 1/2 and 1). We may get the wrong answer when we run such an algorithm once, but by repeatedly running the algorithm and then taking a majority vote, the probability of obtaining the correct answer can be made arbitrarily close to 1, while maintaining the polynomial running time. See also *polynomial algorithms.*

Bounded-error Quantum Polynomial (BQP) Term used in computational complexity theory to describe a class of probabilistic algorithms. BQP functions are all functions with the domain being the set of binary strings computable by uniform quantum circuits whose number of gates is polynomial in the number of input qubits, and which give the correct answer at least 2/3 of the time.

bosons Quantum particles with an integral spin, such as photons and mesons. The spin quantum number of bosons can be $s = 1$, $s = -1$, or $s = 0$. The spin of a composite system of bosons is a multiple of ± 1 See also *fermions.*

cardinality of a finite set Let G be a finite set. The cardinality of G, denoted as $\mid G \mid$ represents the number of elements in the set.

Cauchy sequence A sequence $\{a_n\}$ is Cauchy if for any $\epsilon > 0$ there exists $N \in \mathbb{R}$, such that $|| a_k - a_r || < \epsilon$ for $k, r > N$.

cellular automaton (CA) A cellular automaton is a discrete and deterministic system made up of cells like the points in a lattice. It follows a simple digital rule. A well known CA is the game of life invented by John Conway.

certificate A document containing the public key of an entity, signed by an authorized party.

channel capacity Maximum data rate through a communication channel.

checksum An error detection method. The sender of a message typically performs a 1s complement sum over all the bytes of a protocol data unit and appends it to the message. The receiver recomputes the sum and compares it with the one in the message and decides that there is no error if the two agree.

Chinese remainder theorem Let $p_1, p_2, \ldots, p_i, \ldots, p_q$ be positive integers that are pairwise coprime (*i.e.*, $\gcd(p_i, p_j) = 1$, if $i \neq j$). Let $P = p_1 \cdot p_2 \ldots \cdot p_i \ldots \cdot p_q$ and let $r_1, r_2, \ldots, r_i, \ldots, r_q$ be integers. Then there is exactly one integer n such that

$$0 \leqslant n \leqslant P \quad \text{and} \quad n = r_i \bmod p_i \quad 1 \leqslant i \leqslant q.$$

The theorem provides an answer to the following problem posed sometime during the fourth century A.D. by the Chinese philosopher Sun Tsu Suan-Ching: "There are certain things whose number is unknown. Repeatedly divided by 3 the remainder is 2, by 5 the remainder is 3, and by 7 the remainder is 2. What is the number?" Similar problems appear in the Hindu text Brahma-Sphuta-Siddhanta in the sixth century A.D.

Church-Turing principle Every function which can be regarded as computable can be computed by a universal computing machine.

ciphertext Plain text encoded with a secret key.

circuit complexity The smallest number of gates necessary to implement an operation on a fixed number of qubits.

closed quantum system An idealization of a system of quantum particles. The system is assumed to be isolated and the interaction of the particles with the environment is non-existent. In reality, we can only construct quantum systems with a very weak interaction with the environment.

`CNOT`, `controlled-NOT` **gate** A two-qubit gate. It has as input a control qubit and a target qubit. The control input is transferred directly to the control output of the gate. The target output is equal to the target input if the control input is $\mid 0 \rangle$, and it is flipped if the control input is not $\mid 0 \rangle$.

code In coding theory, the set of all valid code words. The code is known to sender and receiver; if the message received is not a code word, then the receiver decides that an error has occurred. See also *code word*.

code word An n-tuple constructed by adding r parity check bits to k information symbols to support error correction, error detection, or both. Here $n = r + k$.

coding theory Study of error-correcting and error-detecting codes.

coherent states Minimum uncertainty states. For such states, $\Delta(X)\Delta(P_X) = \hbar/2$, rather than $\Delta(X)\Delta(P_X) \geqslant \hbar/2$.

communication channel A communication channel is an abstraction for the medium used by two entities to communicate with each other.

commutator of two operators Given two operators **A** and **B**, the *commutator* of the two operators is $[\mathbf{A}, \mathbf{B}] = \mathbf{A}\,\mathbf{B} - \mathbf{B}\,\mathbf{A}$. **A** commutes with **B** if $[\mathbf{A}, \mathbf{B}] = 0$.

complete vector space A vector space is complete if any Cauchy sequence is convergent. See also *Cauchy sequence*.

compression Encoding a data stream to reduce its redundancy and the amount of data transferred.

conjugate matrix The conjugate A^* of a matrix $A = [a_{ij}]$ is obtained by taking the complex conjugate of each element, $A^* = [a_{ij}^*]$.

conservative logic gate A logic gate that conserves the number of 1s at its input. For example, the `Fredkin` gate is a conservative gate.

Copenhagen interpretation The Copenhagen school is the name given to the group of theoreticians who shared the views of Niels Bohr regarding quantum mechanics interpretations. In the Copenhagen interpretation, nothing can be said about a property of a quantum system before the property is measured. Such a property has no objective reality until it is measured, and the measurement result is not predetermined by anything; it is intrinsically random. After it is measured, then the property is completely known and will not change. Through a measurement one cannot learn anything about the past state of the system; the measurement can only provide information about the future state of the system.

coset of a group For a subgroup H of a group G, $H \subset G$, and an element $x \in G$; define xH to be the set $\{xh : h \in H\}$ and Hx to be the set $\{hx : h \in H\}$. A subset of G of the form xH for some $x \in G$ is said to be a *left coset of H* and a subset of the form Hx is said to be a *right coset* of H. For any subgroup H, we can define an equivalence relation ($\equiv$) by $x \equiv y$ if $x = yh$ for some $h \in H$. The equivalence classes of this equivalence relation are exactly the left cosets of H, and an element $x \in G$ is in the *equivalence class* xH. Thus, the left cosets of H form a partition of G. Any two left cosets of H have the same cardinality, and in particular, every coset of H has the same cardinality as $eH = H$, where e is the identity element. Thus, the cardinality of any left coset of H has cardinality equal to the order of H. The same results are true of the right cosets of G. One can prove that the set of left cosets of H has the same cardinality as the set of right cosets of H.

cyclic redundancy check (CRC) Error-detecting code; the parity check symbols are computed over the characters of the message and are then appended to the packet by the networking hardware.

de Broglie's Equation Links the momentum, p, of a particle to the wavelength, λ, of the wave associated with that particle and Planck's constant h

$$p = \frac{h}{\lambda}.$$

de Broglie, Louis Victor Pierre Raymond, duc A great French physicist, de Broglie was born August 15, 1892 in Dieppe and died March 19, 1987 in Paris. In his Nobel Prize lecture in 1929, de Broglie explained the background of the ideas in his doctoral thesis "Thirty years ago, physics was divided into two camps: ... the physics of matter, based on the concepts of particles and atoms, which were supposed to obey the laws of classical Newtonian mechanics, and the physics of radiation, based on the idea of wave propagation in a hypothetical continuous medium, the luminous and electromagnetic ether. But these two systems of physics could not remain detached from each other: they had to be united by the formulation of a theory of exchanges of energy between matter and radiation. ... In the attempt to bring the two systems of physics together, conclusions were in fact reached which were neither correct nor even admissible when applied to the energy equilibrium between matter and radiation. ... Planck ... assumed ... that a light source ... emits its radiation in equal and finite quantities—in quanta. The success of Planck's ideas has been accompanied by serious consequences. If light is emitted in quanta, must it not, once emitted, possess a corpuscular structure? ... Jeans and Poincaré [showed] that if the motion of the material particles in a source of light took place according to the laws of classical mechanics, then the correct law of black-body radiation, Planck's law, could not be obtained. ... Thus, I arrived at the following general idea which has guided my researches: for matter, just as much as for radiation, in particular light, we must introduce at one and the same time the corpuscle concept and the wave concept. In other words, in both cases we must assume the existence of corpuscles accompanied by waves. But corpuscles and waves cannot be independent, since, according to Bohr, they are complementary to each other; consequently it must be possible to establish a certain parallelism between the motion of a corpuscle and the propagation of the wave which is associated with it."

decoding The process of restoring encoded data to its original format.

decoherence The destruction of the superposition of pure quantum states due to the interaction of the quantum system with the environment.

decompression The process of restoring compressed data to its original format in case of lossless compression or to a format very close to the original one in case of lossy compression.

decryption The process of recovering encrypted data; the reverse of encryption.

de Morgan laws If x and y are Boolean variables, then

$$\overline{x + y} = \overline{x} \times \overline{y} \quad \text{and} \quad \overline{x \times y} = \overline{x} + \overline{y}$$

where $+$ and $\times$ represent the Boolean operations AND and OR, respectively, and $\overline{x}$ is the negation of x.

dense coding Encoding multiple bits into a single qubit.

density matrix of a quantum system The density matrix or operator is a simplifying notation to represent pure or mixed states. For a pure state $| \varphi\rangle$, the density operator is $\rho = | \varphi\rangle\langle\varphi |$; for mixed states, it is a probabilistic expression $\rho = \sum_i \mu_i | \varphi_i\rangle\langle\varphi_i |$ with $\sum_i \mu_i = 1$, and with more than one $\mu_i \neq 0$.

Deutsch, David Physicist born in Haifa, Israel, and educated at Cambridge and Oxford Universities in the UK. Member of the Quantum Computation and Cryptography group at Clarendon Laboratory at Oxford University, author of the theory of parallel universes.

Deutsch's principle Extends the Church-Turing principle. Every finitely realizable physical system can be perfectly simulated by a universal computing machine operating by finite means.

diameter of the electron According to high-energy electron-electron scattering, the diameter of the electron is $<10^{-18}$ m.

Dirac, Paul Adrien Maurice British mathematician (1902–1984). His work has been concerned with the mathematical and theoretical aspects of quantum mechanics. He used a noncommutative algebra for calculating atomic properties leading to the relativistic theory of the electron (1928) and the theory of holes (1930). This latter theory required the existence of a positive particle having the same mass and charge as the known (negative) electron. This particle, the positron, was discovered experimentally at a later date (1932) by C. D. Anderson. The importance of Dirac's work lies essentially in his famous wave equation, which introduced special relativity into Schrödinger's equation. In 1932, he became Lucasian Professor of Mathematics at Cambridge. In 1933, Dirac shared the Nobel Prize for physics with Schrödinger.

Dirac's `ket` and `bra` notation Notation used in quantum mechanics for state vectors. An n-dimensional `ket` vector $|\psi\rangle$ can be expressed as a linear combination of the orthonormal `ket` vectors $|0\rangle, |1\rangle, \ldots, |i\rangle, \ldots, |n-1\rangle$

$$|\psi\rangle = \alpha_0 |0\rangle + \alpha_1 |1\rangle + \ldots + \alpha_i |i\rangle + \ldots + \alpha_{n-1} |n-1\rangle$$

where $\alpha_0, \alpha_1, \ldots, \alpha_i, \ldots, \alpha_{n-1}$ are complex numbers. In matrix representation, the `ket` vector is expressed as the column matrix

$$|\psi\rangle = \begin{pmatrix} \alpha_0 \\ \alpha_1 \\ \vdots \\ \alpha_i \\ \vdots \\ \alpha_{n-1} \end{pmatrix}.$$

For each `ket` vector $|\psi\rangle$, there is a *dual*: the `bra` vector denoted by $\langle\psi|$. The `bra` and `ket` vectors are related by Hermitian conjugation

$$|\psi\rangle = (\langle\psi|)^\dagger, \quad \langle\psi| = (|\psi\rangle)^\dagger.$$

The `bra` vector $\langle\psi_a|$, the *dual* of the `ket`, is expressed as a linear combination of the orthonormal `bra` vectors $|0\rangle, |1\rangle, \ldots, |i\rangle, \ldots, |n-1\rangle$

$$\langle\psi| = \alpha_0^*\langle 0| + \alpha_1^*\langle 1| + \ldots + \alpha_i^*\langle i| + \ldots + \langle n-1|$$

where $\alpha_0^*, \alpha_1^*, \ldots, \alpha_i^*, \ldots, \alpha_{n-1}^*$ are the complex conjugates of $\alpha_0, \alpha_1, \ldots, \alpha_i, \ldots, \alpha_{n-1}$. The dual `bra` vector is expressed as the row matrix

$$\langle\psi| = \begin{pmatrix} \alpha_0^* & \alpha_1^* & \ldots & \alpha_i^* & \ldots & \alpha_{n-1}^* \end{pmatrix}.$$

distinguishable states of a quantum system Two states of a quantum system are distinguishable if they are orthogonal. If two states are distinguishable, then a measurement exists that guarantees to determine which one of the two states the system is in.

dual operator/matrix $A^\dagger$, the dual or the adjoint of a matrix A, is obtained by transposing the matrix and then taking the complex conjugate of each element. The order of the two operations can be reversed. See also *adjoint matrix*.

efficient computation A computation is efficient if it requires, at most, polynomial resources as a function of the size of the input. For example, if a computation returns the value of the function $f(x)$, with x as a bit string of length n, then the computation of $f(x)$ is efficient if the number of steps it requires is bounded by n^k for some k.

eigenstate In quantum mechanics, an *eigenvector* of an observable (Hermitian operator). See also *eigenvector*.

eigenvalue A scalar, λ_i, associated with an *eigenvector*, $\mid \psi_i\rangle$, of a linear operator (observable), $\mathbf{O}$, as

$$\mathbf{O} \mid \psi_i\rangle = \lambda_i \mid \psi_i\rangle.$$

In quantum mechanics, the eigenvalues of an operator represent those values of the corresponding observable that have non-zero probability of occurring. The set of all the eigenvalues is called the operator (matrix) *spectrum*.

eigenvector A state vector, $\mid \psi_i\rangle$, is an eigenvector of a linear operator (observable), $\mathbf{O}$, if when operated on by the operator the result is a scalar multiple of itself as

$$\mathbf{O} \mid \psi_i\rangle = \lambda_i \mid \psi_i\rangle.$$

The scalar λ_i is called the *eigenvalue* associated with the eigenvector. If $\mid \psi_i\rangle$ is an eigenvector with eigenvalue λ_i, then any non-zero multiple of $\mid \psi_i\rangle$ is also an eigenvector with eigenvalue λ_i. If the set of state vectors $\mid \psi_0\rangle, \mid \psi_1\rangle, \ldots, \mid \psi_{n-1}\rangle$ are eigenvectors to *different* eigenvalues $\lambda_0, \lambda_1, \ldots, \lambda_{n-1}$, then the state vectors $\mid \psi_0\rangle, \mid \psi_1\rangle, \ldots, \mid \psi_{n-1}\rangle$ are necessarily linear independent.

electromagnetic field An electromagnetic field consists of an electric and a magnetic field perpendicular to each other and oscillating in a plane perpendicular to the direction of propagation of the electromagnetic wave.

entanglement The translation of the German term *Verschränkung* used by Schrödinger, who was the first to recognize this quantum effect. It means that a two-particle quantum system is in a state that cannot be written as a tensor product of the states of the individual particles. The two quantum particles share a joint state, and it is not possible to describe one of the particles in isolation.

entropy Measure of the uncertainty, or the degree of disorder, in a system.

entscheidungsproblem German term for the "decision problem" posed by Hilbert: "Could there exist, at least in principle, a definite method or process, by which it could be decided whether any mathematical assertion was provable?"

EPR paradox Experiment proposed by Einstein, Podolsky, and Rosen in 1935 to show that quantum mechanics is not a complete theory. This version of the Einstein-Podolsky-Rosen (EPR) paradox is due to David Bohm. Consider a *spin*-0 particle that decays into two *spin*-1/2 particles. The z component of the spin, s_z, of `particle 1` has two eigenstates, $\alpha_{+1/2}$ and $\alpha_{-1/2}$. Similarly, `particle 2` has two eigenstates, $\beta_{+1/2}$ and $\beta_{-1/2}$. The combined state vector of the two-particle system is $y = \{(1/2)[\alpha_{+1/2}\beta_{-1/2} - \alpha_{-1/2}\beta_{+1/2}]\}$. The minus sign corresponds to the fact that the total spin is zero. This state vector is said to be entangled; that is, the plus state for `particle 1`, $\alpha_{+1/2}$, is always associated with the minus state of `particle 2`, $\beta_{-1/2}$, and vice versa. However, in the orthodox view of quantum theory, neither spin has a specific value of s_z until a measurement is made. According to von Neumann, a measurement of s_z on `particle 1` causes the state vector to collapse to either $y = \alpha_{+1/2}\beta_{-1/2}$ or $y = \alpha_{-1/2}\beta_{+1/2}$. Therefore, if the s_z of `particle 1` is measured to be $+\hbar/2$, then that of `particle 2` will be $-\hbar/2$, and vice versa. Both spins now have specific values of s_z. That is, the measurement has an instantaneous effect on the spin of `particle 1` and the spin of `particle 2`, even though the particles spatially are separated. But locality insists that a measurement on the spin of `particle 1` cannot have an instantaneous effect on the spin of `particle 2`, because nothing can travel faster than the speed of light. If one wishes to retain locality, one must dispute the orthodox view that the individual spins do not have values of s_z.

equivalence class Given a set S and an equivalence relation ($\sim$) over S, an equivalence class is a subset H of S with the form

$$H \subset S = \{x \in S \mid x \sim a\}$$

with $a \in S$. This equivalence class is usually denoted as $[a]$ and consists of those elements of S which are equivalent to a.

equivalence relation Equivalence relations are often used to group together objects that are similar in some sense. An equivalence relation on a set S is a binary relation on S that is reflexive, symmetric, and transitive. If the equivalence relation is denoted as $\sim$, $\forall(a, b, c) \in S$:

1. Reflexivity: $a \sim a$.
2. Symmetry: if $a \sim b$ then $b \sim a$.
3. Transitivity: if $a \sim b$ and $b \sim c$ then $a \sim c$.

ergodic process Stochastic process for which time averages and set averages are equal to one another.

ergodicity An attribute of stochastic systems; generally, a system that tends in probability to a limiting form that is independent of the initial conditions.

error-correcting code Code allowing the receiver to reconstruct the code word sent in the presence of transmission errors.

error-detecting code Code allowing the receiver to detect transmission errors.

Euclid of Alexandria The most prominent mathematician of antiquity best known for his treatise on mathematics, *The Elements*. Euclid was born about 325 B.C. and died about 265 B.C. in Alexandria.

Euclid's algorithm Algorithm to compute the greatest common divisor of two integers, m and n. The algorithm was most likely known a few hundred years before appearing around the year 300 B.C., in Book Seven of Euclid's *The Elements*.

Euler, Leonhard A great mathematician (1707–1783) born in Bâle, Switzerland. Euler created a good deal of analysis, and revised almost all of the branches of pure mathematics known at that time, filling up the details, adding proofs, and arranging the whole in a consistent form. In 1748, he wrote *Introductio in Analysin Infinitorum*, an introduction to pure analytical mathematics, and in 1755, the *Institutiones Calculi Differentialis*.

Euler's formulae

$$\sin\theta = \frac{1}{2i}(e^{i\theta} - e^{-i\theta}), \qquad \cos\theta = \frac{1}{2}(e^{i\theta} + e^{-i\theta}).$$

Euler's function The function $\varphi(n)$ denotes the number of positive integers m, where $m \leq n$, which are relatively prime to n (*i.e.*, $\gcd(m, n) = 1$).

Euler's number Euler's number, $e = 2.71828...$, is the only positive real number with the property that the function $f(x) = e^x$ is the same as its derivative.

Euler's theorem Given any positive integer n, let the positive integer a be relatively prime to n. Then $a^{\varphi(n)} \equiv 1 \pmod{n}$. Here $\varphi(n)$ denotes the number of positive integers m, where $m \leq n$, which are relatively prime to n (*i.e.*, $\gcd(m, n) = 1$).

factoring an integer To factor an integer N means to write N as a product of prime numbers $N = p \times q_1 \times q_2 \ldots \times q_n$.

Fast Fourier Transform (FFT) FFT is an algorithm proposed by J. W. Cooley and J. W. Tukey in 1965, which reduces the number of operations to compute the Fourier Transform from $2n^2$ to $2n \log_2 n$. A similar idea is used for the Quantum Fourier Transform. The algorithm decomposes the transformation for $n = 2^m$, recursively, into two transforms of length $n/2$ using the identity

$$\begin{aligned}\sum_{j=0}^{n-1} a_j\, e^{-i2\pi jk/n} &= \sum_{j=0}^{n/2-1} a_{2j}\, e^{-i2\pi(2j)k/n} + \sum_{j=0}^{n/2-1} a_{2j+1}\, e^{-i2\pi(2j+1)k/n} \\ &= \sum_{j=0}^{n/2-1} a_j^{even} e^{-i2\pi jk/(n/2)} + e^{-i2\pi k/n} \sum_{j=0}^{n/2-1} a_j^{odd} e^{-i2\pi jk/(n/2)}.\end{aligned}$$

fermions Quantum particles with a spin multiple of one-half (*e.g.*, electrons, protons, and neutrons). The spin quantum number of fermions can be $s = 1/2, s = -1/2$. The spin of a complex system of fermions is a multiple of $\pm 1/2$. See also *bosons*.

Feynman, Richard Phyllis American physicist (1918–1988). He received his doctorate from Princeton in 1942 under J. A. Wheeler (he was also advised by E. Wigner) and worked on the atomic bomb project at Princeton University (1941–1942) and then at Los Alamos (1943–1945). Feynman's main contribution was to quantum mechanics. He introduced diagrams (now called Feynman diagrams) that are graphic analogues of the mathematical expressions needed to describe the behavior of systems of interacting particles. He was awarded the Nobel Prize in 1965, jointly with Schwinger and Tomonoga, for fundamental work in quantum electrodynamics and physics of elementary particles. His later work led to the current theory of quarks, fundamental in pushing forward an understanding of particle physics. He made significant contributions to the field of quantum computing as well. According to his obituary published in the Boston Globe, "He was widely known for his insatiable curiosity, gentle wit, brilliant mind and playful temperament."

field Algebraic structure consisting of a set F equipped with two binary operations, addition and multiplication, such that:

1. Under addition, F is an Abelian group with identity (or neutral element) 0;
2. Under multiplication, the non-zero elements form an Abelian group; and
3. The distributive law holds: $a(b + c) = ab + ac, \ \forall(a, b, c) \in F$.

Fredkin, Edward Professor of Computer Science at Carnegie-Mellon University. Sometime around 1960, he suggested that the universe is some kind of computational device, a highly parallel computational machine known as cellular automaton (CA). This idea was first published in a scientific journal in 1990 [58].

Fredkin gate A three-qubit gate with two target inputs, a, b, and one control input, c, and three outputs, a', b', and $c' = c$. When the control input is not set, $c = 0$, and the target inputs appear unchanged at the output, $a' = a$ and $b' = b$. When the control is set, $c = 1$, and the target inputs are swapped, $a' = b$ and $b' = a$. The control input is always transferred to the output unchanged. There are both classical and quantum versions of the `Fredkin` gate.

frequency Characteristic of a periodic signal, the number of cycles per unit of time. It is measured in Hertz (Hz) or cycles per second.

fundamental theorem of algebra Any polynomial equation of degree n has n solutions.

fundamental theorem of arithmetic Any positive integer $n > 1$ can be expressed as a product of prime numbers with unique non-negative exponents as

$$n = 2^{n_2} \cdot 3^{n_3} \cdot 5^{n_5} \cdot \ldots = \prod_{p \text{ prime}} p^{n_p}.$$

Galois field A finite field with q elements is also called a *Galois field*, and it is denoted as $GF(q)$. In particular, $\mathbb{Z}_p$ is denoted as $GF(p)$. It is easy to prove that a finite field has p^n elements with p a prime number.

game of life A game invented by John Conway with the following rules: given a rectangular $2D$ lattice (like a checkerboard) with two states per cell, one (live) or zero (dead). At each tick of a clock, the value at each cell: (a) if zero (dead) becomes one (birth) if exactly three of its nearest eight neighbors are ones (live); (b) remains one (survival) if two or three neighbors are ones (live); or (c) becomes zero (death) or stays zero (dead) in all other cases. The number of neighbors that are ones (live) is based on the value of the cells before the rule is applied.

gate (quantum gate) Building block of a quantum circuit. A physical system capable of transforming one or more qubits.

Gedanken experiment "Gedanken" is the German word for "thought." A thought experiment enables us to prove or disprove a conjecture when the experiment enabling us to study the physical phenomena is not feasible. We construct the result of the thought experiment according to a set of assumptions and a model of the system: all based upon universally accepted laws of physics.

greatest common divisor Given two integers, m and n, their *greatest common divisor*, denoted as $\gcd(m, n)$, is the largest integer that divides both m and n.

group Algebraic structure consisting of a set closed under a single-valued binary operation "·", called multiplication, which satisfies three conditions:

1. It is associative;
2. It has a neutral or identity element; and
3. Each element has an inverse under multiplication.

See also *Abelian group*.

Hadamard gate The H gate describes a unitary quantum "fair coin flip" performed upon a single qubit. For example, it transforms an input qubit in state $|0\rangle$ into a superposition state $(|0\rangle + |1\rangle)/\sqrt{2}$, or $(|0\rangle - |1\rangle)/\sqrt{2}$. The transfer matrix of a `Hadamard` gate is

$$H = \frac{1}{\sqrt{2}}\begin{pmatrix} 1 & 1 \\ 1 & -1 \end{pmatrix}.$$

Hadamard, Jacques French mathematician (1865–1963) who proved the prime number theorem, developed Hadamard matrices, and did work on the calculus of variations.

Hamiltonian Similar to classical mechanics, the Hamilton function, $H(q, p)$, represents the energy of a quantum system expressed in terms of dynamical variables, such as position q and momentum p.

Hamming bound The minimum number of parity check bits necessary to construct a code with a certain error-correction capability (*e.g.*, a code capable to correct all single-bit errors).

Hamming distance The number of positions where two code words differ.

Hamming, Richard Wesley An American mathematician (1915–1998) with a B.S. in 1937 from the University of Chicago and Ph.D. in mathematics in 1942 from the University of Illinois at Urbana-Champaign with a thesis on "Some Problems in the Boundary Value Theory of Linear Differential Equations." In 1945, he joined the Manhattan Project, and one year later started working for Bell Laboratories. At Bell Labs, he worked with Claude Shannon and John Tukey. Hamming is best known for his work on error-detecting and error-correcting codes. His fundamental paper on this topic appeared in 1950.

Heisenberg, Werner German physicist; the founder of quantum mechanics (1901–1976). Heisenberg and his fellow student Pauli started their study of theoretical physics under Sommerfeld in 1920. In 1924–1925, he worked with Niels Bohr at the University of Copenhagen. In 1925, Heisenberg formulated matrix mechanics, the first coherent mathematical version of quantum mechanics. Matrix mechanics was further developed in 1926 in a paper co-authored with Born and Jordan. He is perhaps best known for the *Uncertainty Principle*, discovered in 1927. In 1928, Heisenberg published *The Physical Principles of Quantum Theory*. In 1932, he was awarded the Nobel Prize in physics for the creation of quantum mechanics. During the Second World War, Heisenberg headed the unsuccessful German nuclear weapons project. In 1946, he was appointed director of the Max Planck Institute for Physics and Astrophysics at Göttingen.

Heisenberg's Uncertainty Principle Intrinsic property of the quantum systems: the precise knowledge of some basic physical properties such as position and momentum is simply forbidden. The *Uncertainty Principle* states that the uncertainty in determining the position ΔX and the uncertainty in determining the momentum ΔP_X at position X are constrained by the inequality:

$$\Delta X \, \Delta P_X \geq \frac{\hbar}{2}$$

where $\hbar = h/2\pi$ is a modified form of Planck's constant.

Hermitian operator or matrix A linear operator or a matrix A with the property that $A = A^{\dagger}$, where $A^{\dagger}$ is the dual, or the adjoint of A. See also *adjoint matrix*.

Hertz The unit to measure the frequency of a periodic phenomena, abbreviated as Hz. Named after the great German physicist Heinrich Hertz. 1 Hz = 1 cycle/second. Multiples of this unit are 1 KHz = 1000 cycles/second, 1 MHz = 10^6 cycles/second, 1 GHz = 10^9 cycles/second, etc.

Hertz, Heinrich A German physicist (1857–1894); the first to demonstrate experimentally the production and detection of Maxwell electromagnetic waves. The photoelectric effect was first discovered accidentally in 1887 by Hertz, while carrying on investigations on the electromagnetic waves.

hidden variable theory A theory based upon the assumption that there is a variable or variables that determine the real properties of quantum particles. These variables have definite values from the moment the particle is created, and they determine the result of the measurement performed on that property of the quantum particle.

Hilbert, David Eminent German mathematician (1862–1943). In 1895, he was appointed to the chair of mathematics at the University of Göttingen, and he remained there for the rest of his career. Hilbert's first work was on invariant theory, and, in 1888, he proved his famous basis theorem. He published *Grundlagen der Geometrie* in 1899, putting geometry in a formal axiomatic setting. He delivered the speech "The Problems of Mathematics" at the Second International Congress of Mathematicians in Paris, challenging mathematicians to solve fundamental questions such as: continuum hypothesis, the well ordering of the reals, Goldbach conjecture, the transcendence of powers of algebraic numbers, the Riemann hypothesis, and the extension of the Dirichlet principle. Hilbert's work in integral equations led to the research in functional analysis and established the basis for his work on infinite-dimensional space, later called Hilbert space.

Hilbert space, n-dimensional An n-dimensional complex vector space with an inner product and thus, with a norm denoted as $\mathcal{H}_n$. An n-dimensional Hilbert space is isomorphic with $\mathbb{C}^n$. Each vector in $\mathcal{H}_n$ can be thought of as a column vector with n complex components. The norm of a vector $a \in \mathcal{H}_n$ is $|| \, a \, ||^2 = \langle a, a \rangle$ with $\langle a, a \rangle$ as the inner product. See also *inner product in an* n*-dimensional Hilbert space*.

homomorphism An isomorphism of a topological space. See also *isomorphism*.

impure states of a single qubit Impure, or mixed states, are represented by points inside the Bloch sphere. This implies that the trace of the square of their density matrix is less than one; $\mathrm{Tr}(\rho^2) < 1$. See also *pure states of a single qubit.*

index of a subgroup Let G be a group with cardinality $\mid G \mid$, and $H \subset G$ be a subgroup of G with cardinality $\mid H \mid$. Then the index of the subgroup H is

$$I_{G/H} = \frac{\mid G \mid}{\mid H \mid}.$$

information A form of matter that can be communicated, recorded, and computed with, regardless of the physical embodiment. For example, information written on a magnetic media such as a CD can be transferred to another media [e.g., paper (it may take a large stack of paper to transfer all the information on a CD), spoken word, or music]. We distinguish between deterministic, probabilistic, and quantum information. Each type of information is characterized by specific *information units.* For example, deterministic information is characterized by bits (0 or 1).

inner product in an n-dimensional Hilbert space The inner product of two vectors $a = (a_1, a_2, \ldots, a_i, \ldots, a_n)^T$ and $b = (b_1, b_2, \ldots, b_j, \ldots, b_n)^T$, $a, b \in \mathcal{H}_n$ is denoted as $\langle a, b \rangle$. Then $\langle a, b \rangle = a^\dagger b = \sum_i a_i^* b_i$, with a_i^* the complex conjugate of a_i. In Dirac's notation, the inner product of two vectors $\mid \psi_a \rangle, \mid \psi_b \rangle \in \mathcal{H}_n$ is written as $\langle \psi_a \mid \psi_b \rangle$. See also *inner product of two vectors.*

inner product of two vectors If $\mathcal{A} = F^n$ is an n-dimensional vector space over the field of complex numbers, $F = \mathbb{C}$, then the scalar product of two vectors is also called the inner product of the two vectors. See also *scalar product.*

integer factorization Decomposition of a positive integer N into a product of prime numbers. A very important, practical problem because the security of widely used cryptographic protocols is based upon the conjectured difficulty of large integers factorization.

irreversible/non-invertible gate A gate characterized by the fact that, knowing the output, we cannot determine the input for all possible combinations of input values. The irreversibility of classical gates, other than NOT, means that there is an irretrievable loss of information, and this has very serious consequences regarding the energy consumption of classical gates.

isomorphism The word derives from the Greek *iso*, meaning "equal" and *morphosis*, meaning "to form" or "to shape." Formally, an isomorphism is bijective morphism. Informally, an isomorphism is a map which preserves sets and relations among elements. A space isomorphism is a vector space in which addition and scalar multiplication are preserved. Two groups, G_1 and G_2 with binary operators "+" and "×", are isomorphic if there exists a map, $f : G_1 \mapsto G_2$, which satisfies the relation $f(x + y) = f(x) \times f(y)$. An isomorphism preserves the identities and inverses of a group. See also *automorphism and homomorphism.*

isothermal Term frequently used in thermodynamics: meaning at a constant temperature.

key distribution Mechanism for distribution of cryptographic keys; in particular, of public keys.

Lagrange's theorem If H is a subgroup of a finite group G, then $| H |$ divides $| G |$.

Landauer's Principle Landauer's Principle traces the energy consumption in a computation to the act of erasing information. When a computer erases one bit of information, the amount of energy dissipated into the environment is at least $k_B T \ln(2)$, with k_B (Boltzmann's constant) and T (the temperature of the environment). An equivalent formulation: the entropy of the environment increases by at least $k_B \ln(2)$ when one bit of information is erased.

latency The time needed for an activity to complete.

least common multiple Given two integers, m and n, their *least common multiple* denoted as $\text{lcm}(m, n)$ is the smallest positive integer that is a multiple of both m and n.

light A form of electromagnetic radiation.

linear operator A linear operator $\mathbf{A}$ between two vector spaces $\mathcal{A}$ and $\mathcal{B}$ over the field F is any function from $\mathcal{A}$ to $\mathcal{B}$, $\mathbf{A} : \mathcal{A} \mapsto \mathcal{B}$, linear in its inputs

$$\mathbf{A}\left(\sum_i c_i \alpha_i\right) = \sum_i c_i \mathbf{A}(\alpha_i)$$

with $\alpha \in \mathcal{A}$, $c_i \in F$, and $A(\alpha) \in \mathcal{B}$.

L2 function A function $f(x)$ is square integrable, or $L2$ function, if

$$\int_{-\infty}^{+\infty} | f(x) |^2 \, dx$$

is finite.

Manhattan Project A United States government research project to produce an atomic bomb. It was called the Manhattan Project because the first research had been done at Columbia University in Manhattan. The main research and development activity was later transferred to Los Alamos, in New Mexico, site of the Los Alamos National Laboratory.

matrix exponentiation If β is a real number and if matrix A is such that $A^2 = I$, then

$$e^{i\beta A} = \cos(\beta)I + i\sin(\beta)A.$$

matrix in a Hilbert space An $n \times m$ matrix A is regarded as a linear operator from an n-dimensional Hilbert space, $\mathcal{H}_n$, to an m-dimensional Hilbert space, $\mathcal{H}_m$, namely

$$A : \mathcal{H}_n \longrightarrow \mathcal{H}_m.$$

maximum likelihood decoding Decoding strategy when a received n-tuple is decoded into the code word to minimize the probability of errors.

Maxwell's demon In 1871, Maxwell proposed the following "thought" experiment. Imagine the molecules of a gas in a cylinder divided in two by a wall which has a slit; the slit is covered with a door controlled by a little demon. The demon examines every molecule of gas and determines its velocity; those of high velocity on the left side are let to migrate to the right side and those with low velocity on the right side are allowed to migrate to the left side. As a result of these measurements, the demon separates hot from cold in blatant violation of the second law of thermodynamics.

measurement of a quantum system The process which makes a connection between the quantum and the classical worlds; generally considered as an irreversible operation which destroys quantum information about an observable (property) of a quantum system and replaces it with classical information. In quantum mechanics, the measurement of an observable of a quantum system (such as momentum, energy, or spin) is associated with a Hermitian operator, $\mathbf{A}$, on the Hilbert space of state vectors of the system. If v is an eigenvector of $\mathbf{A}$ with eigenvalue λ, then measuring the system in a state described by state vector v will always give the result λ. If the state vector is not an eigenvector of $\mathbf{A}$, the measurement process forces the system to jump (collapse) randomly to a state corresponding to a state vector v_i, an eigenvector of $\mathbf{A}$. The result of the measurement is the corresponding eigenvalue of $\mathbf{A}$, λ_i. See also *projective measurement.*

minimum feature size Chips are generally categorized by their "minimum size feature"—the width of the smallest line or gap actually constructed in the manufacturing process. This size continually is getting smaller, as the chipmakers seek to pack more circuitry into the same space.

mixed state A quantum system whose state $\mid \psi\rangle$ is not known precisely is said to be in a mixed state. A mixed state is a superposition of different pure states; the system is in a state $\mid \varphi_i\rangle$ with probability p_i. The density operator of a quantum system in a mixed state is $\rho = \sum_i p_i \mid \varphi_i\rangle\langle\varphi_i \mid$. The trace of the density operator is $\mathrm{Tr}(\rho^2) < 1$. A mixed state is also called an *impure state.* See also *pure state of a single qubit.*

model of a physical system Abstraction used to study the properties of a system. For example, a Turing machine is a high-level model capturing essential properties of a computer. A model captures only the "relevant" properties of the system. Here "relevant" means: important for a particular type of study the model is designed for. We cannot draw any conclusion regarding the heat dissipation of a computer from the Turing machine model, because the model does not take into account physical characteristics of a computer, such as the energy required for an elementary operation. Sometimes, we construct a scale model of a physical system. Such a model allows us to study some properties of the system without actually building it. A scale model can be used to perform experiments that are unfeasible, very difficult, or very costly to carry out. For example, the model of an airplane wing allows us to draw conclusions about its lift-off properties.

modular representation of integers Consider a set of "moduli" that contain no common factors $p_1, p_2, \ldots, p_q$. Given an integer n, we can divide n by these moduli and compute the corresponding residues

$$r_1 = m \bmod p_1, \quad r_2 = n \bmod p_2, \ \ldots, \ r_{q-1} = n \bmod p_{q-1}, \quad r_q = n \bmod p_q.$$

The set $(r_1, r_2, \ldots, r_{q-1}, r_q)$ provides an internal representation of the positive integer n, which allows us to perform addition, subtraction, and multiplication of large, positive integers in a very convenient manner. The modular representation is *carry-free* and leads to relatively simple hardware implementation. Yet, given the modular representation, it is difficult to test if a number is positive or negative, to detect overflow, or to perform integer division. It is also difficult to test if $a > b$.

Moivre's formulae Give the absolute value and the argument of a product of complex numbers z and z':

$$| z \cdot z' | = | z | \cdot | z' |, \qquad \arg(z \cdot z') = \arg(z) + \arg(z').$$

norm of a vector in a Hilbert space A non-negative function which measures the "length" of a vector. If $| \psi\rangle \in \mathcal{H}_n$, then the norm, $\| \; | \psi\rangle \; \|$, can be computed from the inner product of the state vector with itself, $\| \; | \psi\rangle \; \|^2 = | \langle \psi | \psi\rangle |$.

normal operator in a Hilbert space An operator $\mathbf{U} \in \mathcal{H}_n$ with the property that $[\mathbf{U}, \mathbf{U}^\dagger] = \mathbf{U}\mathbf{U}^\dagger - \mathbf{U}^\dagger\mathbf{U} = 0$.

normal unitary basis of an n-dimensional Hilbert space, $\mathcal{H}_n$ A set of n vectors, $| \psi_0\rangle, | \psi_2\rangle, \ldots, | \psi_i\rangle, \ldots, | \psi_{n-1}\rangle$, where each vector has the norm (or length) equal to one, $| \; | \psi_0\rangle \; | = | \; | \psi_1\rangle \; | = \ldots = | \; | \psi_{n-1}\rangle \; | = 1$, and any two of them are orthogonal, $\langle \psi_i | \psi_j\rangle = 0$, for $(i \neq j)$.

normalization condition in quantum mechanics Requirement that the square of the modulus of the projections of a state vector on the orthonormal basis sum is equal to 1. This translates into the condition that the sum of probabilities of all possible outcomes of a measurement of a quantum system must be equal to 1.

***n*-tuple** Vector with n components. Each component is a symbol from an input alphabet; for example, a binary 4-tuple is a vector with four components, each one being either 0 or 1.

numerical simulation (exact/approximate) of a physical system New methodology of science which became available after the introduction of the stored program computer in mid 1940s. We use a computer to study the behavior of a physical system under different conditions. For numerical simulation, we first construct a model of the physical system, then we design a program that reflects the essential properties of the physical system captured by the model. The results of simulation are only as good as the model is. In 1982, Richard Feynman argued that in traditional numerical simulations, such as weather forecasting or aerodynamic calculations, computers model physical reality only *approximately*. He advanced the idea that physics was computational, and that a quantum computer could do an *exact simulation* of a physical system, even of a quantum system. Numerical simulation complements the two traditional scientific methods: theoretical modelling and experiments.

observable A physical property of a quantum system that can be measured by an external observer. Mathematically, each observable X has an associated operator $\mathcal{M}_X$.

operator A function used to transform the state of a system.

oracle models An oracle implements a computational model and solves a problem for which there is no efficient solution. Oracle models are used to compare the power of two computational models. For example, in 1994, D. Simon showed that a quantum computer with a specific oracle $\mathcal{O}$ efficiently could solve problems that classical computers having access to the same oracle $\mathcal{O}$ could not solve efficiently. See also *efficient computation*.

order of an integer r modulo N Given two integers r and N, the order of r modulo N is the smallest integer k such that $r^k = 1 \bmod N$ with two additional conditions:

1. $r^1 \neq 1 \bmod N$, and
2. $(r^{k-1} + r^{k-2} + \ldots + r^2 + r + 1) \neq 1 \bmod N$.

orthogonal state vectors Two vectors, $| \psi_a \rangle$ and $| \psi_b \rangle$ in $\mathcal{H}_n$, are *orthogonal*, and we write $| \psi_a \rangle \perp | \psi_b \rangle$ if their inner product is zero:

$$\langle \psi_a | \psi_b \rangle = 0 \implies | \psi_a \rangle \perp | \psi_b \rangle.$$

orthonormal basis in a Hilbert space, $\mathcal{H}_n$ A basis consisting of a set of n orthonormal vectors such as $| 0 \rangle, \ | 1 \rangle, \ \ldots, \ | i \rangle, \ \ldots, \ | n - 1 \rangle$. Any two vectors from the set are orthogonal, and all have a norm equal to 1. See also *normal unitary basis of an* n-*dimensional Hilbert space,* $\mathcal{H}_n$.

outer product in a Hilbert space, H_n The *outer product* of a `ket` vector and a `bra` vector, $| \psi_a \rangle \langle \psi_b |$, is a *linear operator*. For example, in $\mathcal{H}_4$, we have

$$| \psi_a \rangle \langle \psi_b | = \begin{pmatrix} \alpha_0 \\ \alpha_1 \\ \alpha_2 \\ \alpha_3 \end{pmatrix} \begin{pmatrix} \beta_0^* & \beta_1^* & \beta_2^* & \beta_3^* \end{pmatrix} = \begin{pmatrix} \alpha_0\beta_0^* & \alpha_0\beta_1^* & \alpha_0\beta_2^* & \alpha_0\beta_3^* \\ \alpha_1\beta_0^* & \alpha_1\beta_1^* & \alpha_1\beta_2^* & \alpha_1\beta_3^* \\ \alpha_2\beta_0^* & \alpha_2\beta_1^* & \alpha_2\beta_2^* & \alpha_2\beta_3^* \\ \alpha_3\beta_0^* & \alpha_3\beta_1^* & \alpha_3\beta_2^* & \alpha_3\beta_3^* \end{pmatrix}.$$

parity check symbols Symbols added to a message to increase the redundancy and support error-correcting and/or error-detecting capabilities of a code.

Pauli matrices Two by two matrices describing transformations (rotations) of a single qubit. They are defined as $\sigma_i^\dagger = \sigma_i, \quad i = (1, 2, 3)$. They are Hermitian, they square to unity, $\sigma_1^2 = \sigma_2^2 = \sigma_3^2 = 1$, and they satisfy the relation $\sigma_1\sigma_2 = i\sigma_3$ (this is also true for a cyclic permutation of indices), as well as the relation $\sigma_1\sigma_2 + \sigma_2\sigma_1 = 0$ (this is also true for a cyclic permutation of indices).

Pauli, Wolfgang Ernst Austrian-born physicist (1900–1958). In 1924, Pauli proposed a quantum spin number for electrons. Best known for his (Pauli) exclusion principle, proposed in 1925, which states that no two electrons in an atom can have the same four quantum numbers. He predicted mathematically, in 1931, that conservation laws required the existence of a new particle which he proposed to call the "neutron." In 1933, he published his prediction and he made the claim that the particle had zero mass. The particle, which we now know as the neutron, has a non-zero mass and was discovered by Chadwick in 1932. Pauli's particle was named the "neutrino" by Fermi in 1934, and at that time, he correctly stated that it was not a constituent of the nucleus of an atom. The neutrino was later found experimentally.

Pauli's Exclusion Principle Consider a pair of electrons on the same orbital around the nucleus of an atom. *Pauli's Exclusion Principle* dictates that the two electrons cannot be in identical states, including the spin. They must have their spins oriented in opposite directions, because they already share three quantum numbers. If, as a result of an experiment, one of the electrons is made to change the orientation of its spin, then a simultaneous measurement of the other finds it in a state with opposite spin.

photoelectric effect When light shines on a negatively charged metal plate, the surface emits electrons. This phenomena was explained by Albert Einstein in 1905 based upon Plank's quantum theory. Einstein got the Nobel Prize in 1921 for explaining the photoelectric effect.

photon From the Greek word "photos" meaning light. A light particle.

polarization filter A partially transparent material that transmits light of a particular polarization.

polarization measurements Polarization is measured by passing a photon through a polarizer. If the polarizer axis is oriented parallel to the polarization of the photon, the photon passes through unimpeded. If it is oriented perpendicular to the polarization of the photon, the photon is absorbed. At an intermediate angle, the photon will have a certain probability of being transmitted.

polarization of light As an *electromagnetic radiation*, light consists of an electric and a magnetic field perpendicular to each other and, at the same time, perpendicular to the direction the energy is transported by the electromagnetic (light) wave. The electric field oscillates in a plane perpendicular to the direction of light and the way the electric field vector travels in this plane defines the polarization of the light. When the electric field oscillates along a straight line, we say that the light is *linearly polarized*. When the end of the electric field vector moves along an ellipse, the light is *elliptically polarized*. When the end of the electric field vector moves around a circle, the light is *circularly polarized*. If the light comes toward us and the end of the electric field vector moves around in a counterclockwise direction, we say that the light has *right-hand polarization*; if the end of the electric field vector moves in a clockwise direction, we say that the light has *left-hand polarization*.

polynomial time algorithm Term used in computational complexity theory to describe algorithms which require a number of steps, $T(n)$, bounded by a polynomial, with n the size of the input. The class of polynomial time algorithms is denoted by $\mathcal{P}$. The class of polynomials which require a number of steps not bounded by a polynomial is denoted by $\mathcal{NP}$.

prime number A positive integer p is prime if its only factors are 1 and itself.

program An algorithm expressed in a programming language understood by a physical computer.

projection operator, projector The outer product of a unit vector with itself as

$$| \Psi_a \rangle \langle \Psi_a | \; = \; \mathbf{P}_a.$$

It has the defining property: $\mathbf{P}_a^2 = \mathbf{P}_a$. A complete set of orthogonal projectors in $\mathcal{H}_n$ is a set $\{\mathbf{P}_0, \mathbf{P}_1, \dots, \mathbf{P}_i, \dots, \mathbf{P}_{m-1}\}$ such that

$$\sum_{i=0}^{m-1} \mathbf{P}_i = 1.$$

projective measurement A projective measurement is characterized by a set of projectors, $\mathcal{M}_i$, such that $\sum_i \mathcal{M}_i = I$ and $\mathcal{M}_i \mathcal{M}_j = \delta_{ij} \mathcal{M}_i$. The outcome of the measurement is the one with the index i associated with $\mathcal{M}_i$. The probability of outcome i for a system in state $| \psi\rangle$ is $p_i = | \mathcal{M}_i | \psi_i\rangle |^2$. Given the outcome i, the quantum state "collapses" to the state $M_i | \psi_i\rangle / \sqrt{p_i}$.

proper factor The positive integer p is a proper factor of N if another integer q exists such that $N = pq$, and $p \neq 1$, $p \neq N$.

public key cryptography Communicating entities have both a *private* and a *public* key. A secure message is sent to an entity E by encrypting it with E's *public key*; E decrypts the message with its own *private key*.

pure state of a single qubit Pure states of a single qubit are represented by points on the Bloch sphere. This implies that the trace of the square of their density matrix is one, $\mathrm{Tr}(\rho^2) = 1$. See also *mixed state* or *impure states of a single qubit* .

quantum Latin word meaning "some defined quantity." In physics, it is used with the same meaning as *discrete* in mathematics.

quantum communication channel Physical media allowing two parties to exchange quantum information. For example, an optic fiber allowing photons with a certain polarization to circulate from a source to a destination.

quantum computer Device able to store and transform quantum information.

quantum computing New discipline emerging during the last three decades of the twentieth century.

quantum information Information based upon quantum mechanics. The information is encoded as a property of a quantum particle (*e.g.*, the spin of an electron or the polarization of a photon).

quantum mechanics Discipline of modern physics founded by Heisenberg.

quantum parallelism Term capturing the fact that a quantum computer can manipulate an exponential set of inputs simultaneously. For example, consider a function $f(x)$ having as argument a binary vector $x = (x_1, x_2, \ldots, x_l)$ of length $l = 2^n$. A quantum computer can evaluate all $f(x_1), f(x_2), \ldots, f(x_l)$ simultaneously.

quantum particle Atomic or sub-atomic particle obeying the laws of quantum mechanics.

qubit Quantum bit. Mathematical abstraction for a quantum system capable of storing one bit of information.

ray in a Hilbert space A mathematical abstraction that exhibits only direction. It can be represented as a straight line through the origin of the coordinate system.

relatively prime integers A set of n integers, $\{m_1, m_2, \ldots, m_n\}$, are said to be *relatively prime* if they do not have any common factors.

relativity postulate Formulated by Albert Einstein in his paper "On the Electrodynamics of Moving Bodies" published in *Annalen der Physik* in 1905. "There is no way to tell which unaccelerated reference frames is *truly* at rest."

reversible physical process A physical process is said to be reversible if it can evolve forwards as well as backwards in time. A reversible system can always be forced back to its original state from a new state reached during an evolutionary process. Reversibility is a fundamental property of nature.

RSA algorithm Public key encryption algorithm named after its inventors: Rivest, Shamir, and Adleman.

scalar product of two vectors Given an n-dimensional vector space $\mathcal{A} = F^n$ over the field F, the scalar product of $\alpha, \beta \in \mathcal{A}$ is denoted by (α, β). The scalar product is a bilinear map with the property $(\alpha, \beta) = (\beta, \alpha)$. If $F = \mathbb{C}$, the field of complex numbers, then the scalar product is also called the inner product.

Schrödinger, Erwin Austrian born physicist (1887–1961). He is one of the founders of quantum physics and has made significant contributions to statistical mechanics and the general theory of relativity. Schrödinger transferred the idea of a wave associated to a particle, predicted by de Broglie, to Bohr's atomic model. In 1927, he showed the mathematic equivalence of his equation $\mathbf{H}\Psi = \mathbf{E}\Psi$ and Heisenberg's matrix mechanics. That same year he became Max Planck's successor for the theoretical physics chair at the University of Berlin. He was awarded the Nobel Prize for Physics in 1933 (together with Paul Dirac) for his contributions to the development of quantum mechanics.

Schrödinger's Equation Equation introduced by Erwin Schrödinger for the wave function, $\Psi_n(q)$, of a *stationary state* of energy, E_n, of an atom in terms of the Hamiltonian function, $H(q, p)$ where

$$H\left(q, \frac{h}{2\pi i}\frac{\partial}{\partial q}\right)\Psi_n(q) = E_n \Psi_n(q).$$

The dynamics of the wave function is governed by the following partial differential equation:

$$i\frac{h}{2\pi}\frac{\partial}{\partial t}\Psi(q,t) = H\left(q, \frac{h}{2\pi i}\frac{\partial}{\partial q}\right)\Psi(q).$$

Schwartz inequality Inequality satisfied by any state vectors $(\mid \psi_a\rangle, \mid \psi_b\rangle) \in \mathcal{H}_n$ where

$$\langle \psi_a \mid \psi_a\rangle\langle \psi_b \mid \psi_b\rangle \geq \mid \langle \psi_a \mid \psi_b\rangle \mid^2 .$$

second law of thermodynamics The entropy of a system is a non-decreasing function of time.

Shannon, Claude Elwood American mathematician and electrical engineer (1916–2001); founder of the modern information theory. He graduated from the University of Michigan in 1936 with bachelor's degrees in mathematics and electrical engineering. In 1940, he earned both a master's degree in electrical engineering and a Ph.D. in mathematics from the Massachusetts Institute of Technology (MIT). Shannon joined the mathematics department at Bell Labs in 1941 and remained affiliated with the Labs until 1972. He became a permanent member of the faculty at MIT in 1958 and a professor emeritus in 1978. In 1948, Shannon published his landmark work *A Mathematical Theory of Communication*, and in 1949, he published the *Communication Theory of Secrecy Systems*, generally credited with transforming cryptography from an art to a science.

Shannon's entropy The *entropy* of a random variable X, with a probability density function $p_X(x)$, is a positive real number, such as

$$H(X) = -\sum_x p_X(x) \times \log_2 p_X(x).$$

Shannon's theorem Relates the channel capacity for a noisy channel to the signal-to-noise ratio and the bandwidth of the noiseless channel.

singlet electron state The anti-symmetric state of a pair of electrons with *anti-parallel* spins $1/\sqrt{2}(|\uparrow\downarrow\rangle - |\downarrow\uparrow\rangle)$. The electrons have different spin quantum numbers $+1/2$ and $-1/2$, and the total spin of the state is zero. See also *triplet electron state.*

spectral decomposition of a normal operator in a Hilbert space In $\mathcal{H}_n$, every normal operator $\mathbf{N}$ has n eigenvectors $| n_i\rangle$, and, correspondingly, n eigenvalues λ_i, $1 \leqslant i \leqslant n$. If $\mathbf{P}_i$ are the projectors corresponding to these eigenvectors, $\mathbf{P}_i = | n_i\rangle\langle n_i |$, then the operator $\mathbf{N}$ has the spectral decomposition of

$$\mathbf{N} = \sum_i \lambda_i \mathbf{P}_i.$$

spin The observable associated with the intrinsic rotation of the electron is the intrinsic angular momentum, also called the *spin angular momentum*. The "spin" is the quantum number characterizing the intrinsic angular momentum of the electron and, for that matter, of other quantum particles. There are two types of particles:

1. *bosons*, particles whose spin quantum number can be $s = 1$, $s = -1$, or $s = 0$ (*e.g.*, photons and mesons), and
2. *fermions*, particles whose spin quantum number can be $s = 1/2$ or $s = -1/2$ (*e.g.*, electrons, protons, and neutrons).

superposition probability rule If an event in quantum mechanics may occur in two or more indistinguishable ways, then the probability amplitude of the event is the sum of the probability amplitudes of each case considered separately.

superposition state of a quantum system If the states $| 0\rangle, | 1\rangle, \ldots, | n-1\rangle$ of a quantum system are distinguishable and if the complex numbers α_i satisfy the condition $\sum_i | \alpha_i |^2 = 1$, then the state $\sum_i \alpha_i | i\rangle$ is a valid quantum state called a superposition state. See also *basis states of a quantum system.*

Stern-Gerlach experiment Experiment revealing the spin of quantum systems.

system A physical entity that can find itself in a set of states. For example, the state of a computer is determined by the contents of the memory, of internal registers, of external storage devices, and so on.

teleportation In a science fiction context, making an object or person disintegrate in one place and have it reembodied as the same object or person somewhere else. In the context of quantum information theory: "a way to scan out part of the information from an object A, which one wishes to teleport, while causing the remaining, unscanned, part of the information to pass, via the Einstein-Podolsky-Rosen effect, into another object C, which has never been in contact with A. Later, by applying to C a treatment depending on the scanned-out information, it is possible to maneuver C into exactly the same state as A was in before it was scanned" (see `http://www.research.ibm.com/quantuminfo/teleportation`). In this process, the original state is destroyed.

tensor product of two Hilbert spaces The tensor product of $\mathcal{H}_n$ and $\mathcal{H}_m$ is $\mathcal{H}_n \otimes \mathcal{H}_m = \mathcal{H}_{nm}$ when, given that $(e_0, e_1, \ldots, e_{n-1})$ is an orthonormal basis for $\mathcal{H}_n$

and $(f_0, f_1, \ldots, f_{n-1})$ is an orthonormal basis for $\mathcal{H}_m$, then
$(e_0 \otimes f_0, e_1 \otimes f_1, \ldots, e_{n-1} \otimes f_{n-1})$ is an orthonormal basis for $\mathcal{H}_{nm}$.

tensor product of two linear operators or matrices in a Hilbert space The tensor product of the $n \times n$ matrix $A = [a_{ij}]$ with the $m \times m$ matrix $B = [b_{rs}]$ is the $(m \cdot n) \times (m \cdot n)$ matrix $C = [c_{pq}]$ with $c_{pq} = a_{ij} \cdot b_{rs}$, where i, j, r, s, p, q are the only integers which satisfy

$$p = i \cdot m + r \quad \text{and} \quad q = j \cdot m + s.$$

In other words, C is the matrix obtained by replacing the element a_{ij} of matrix A by the $m \times m$ matrix $a_{ij} \cdot B$. For example, the tensor product of vectors (a, b) and (c, d) is the vector:

$$\begin{pmatrix} a \\ b \end{pmatrix} \otimes \begin{pmatrix} c \\ d \end{pmatrix} = \begin{pmatrix} ac \\ ad \\ bc \\ bd \end{pmatrix}.$$

tensor product of two vector spaces The *tensor product*, $\mathcal{A} \otimes \mathcal{B}$, of two vector spaces, $\mathcal{A}$ and $\mathcal{B}$, over the same field, F, is the dual $\mathcal{G}(\mathcal{A}, \mathcal{B})^*$ of the space $\mathcal{G}(\mathcal{A}, \mathcal{B})$ of bilinear functions from $\mathcal{A}$ and $\mathcal{B}$ to F.

thermodynamic entropy The thermodynamic entropy of a gas, S, quantifies the notion that a gas is a statistical ensemble, and it measures the randomness (or the degree of disorder) of the ensemble. The entropy is larger when the vectors describing the individual movements of the molecules of gas are in a higher state of disorder than when all of them are well organized and moving in the same direction with the same speed, such as

$$S = k_B \ln \Omega,$$

where k_B is Boltzmann's constant, $k_B = 1.3807 \times 10^{-23}$ Joules per degree Kelvin, and Ω is the number of microstates. A version of this equation is engraved on Ludwig Boltzmann's tombstone.

thermodynamics Thermodynamics is the study of energy, its ability to carry out work, and the conversion between various forms of energy, such as the internal energies of a system, heat, and work. Thermodynamic laws are derived from statistical mechanics. C. P. Snow summarized the three laws of thermodynamics as follows:

1. You cannot win. Matter and energy are conserved, thus you cannot get something for nothing.
2. You cannot break even. You cannot return to the same energy state because there is always an increase in disorder. The entropy always increases.
3. You cannot get out of the game. Absolute zero is unattainable.

Toffoli gate A three-qubit gate with two control inputs, a and b, and one target input, c. The outputs are: $a' = a$, $b' = b$ and c'. When $c = 1$, then
$c' = 1 \oplus (a \text{ AND } b) = \text{NOT}(a \text{ AND } b)$, otherwise $c' = c$. The `Toffoli` gate is a universal gate, and it is reversible. There are both classical and quantum versions of the `Toffoli` gate.

Toffoli, Tommaso Professor of electrical engineering at Boston College.

trace of a linear operator The trace of an operator is the trace of the matrix associated with the operator.

trace of a matrix The trace of a matrix A is the sum of its diagonal elements, and it is denoted as $\text{Tr}(A)$.

transpose matrix A matrix whose rows are the columns of the original matrix (the rows of the original matrix become the columns of the transpose).

triplet electron state The state of a pair of electrons with *parallel* spins, $|\uparrow\uparrow\rangle$ or $|\downarrow\downarrow\rangle$, or in a symmetric superposition of anti-parallel spins, $(1/\sqrt{2})(|\uparrow\downarrow\rangle + |\downarrow\uparrow\rangle)$. The total spin of a triplet state is $+1$.

truth table A method to characterize the output of a logic circuit. One constructs a table with one entry for every possible combination of input values.

Turing, Alan Mathison British mathematician (1912–1954) regarded as the founder of the modern computer science. He entered King's College at Oxford in 1931 and graduated in 1935 with a degree in mathematics. He was elected a fellow of King's College at Cambridge in 1935. In 1936, he published "On Computable Numbers, with an Application to the Entscheidungsproblem." Turing defined a computable number as a real number whose decimal expansion could be produced by a Turing machine starting with a blank tape. In 1939, Turing started to work full-time at the Government Code and Cypher School at Bletchley Park. Together with W. G. Welchman, Turing developed a machine in late 1940, which was able to decipher all of the messages sent by the German Enigma encryption machines of the Luftwaffe (the German air force during the Second World War). After the war, Turing was invited by the National Physical Laboratory in London to design a computer. In March 1946, he submitted a report proposing the Automatic Computing Engine (ACE), an original design for a modern computer. In 1948, he moved to Manchester, and in 1950, he published a paper, "Computing Machinery and Intelligence in Mind" where he proposed the Turing Test. This is the test used today to answer the question of whether a computer can be intelligent. Alan Turing was elected a Fellow of the Royal Society of London in 1951.

Turing machine An abstract machine which moves from one state to another using a precise, finite set of rules (given by a finite table), depending on a single symbol it reads from a tape. An ordinary Turing machine has a control unit and a read-write head; it performs a sequence of read-write-shift operations on an infinite tape divided into squares. The dynamics of the Turing machine is described by quintuples (A, T, A', T', σ) of the form $AT \mapsto T' \sigma A'$. The significance of this notation is that when the control unit is in state A and the symbol currently scanned by the read-write head is T, then the machine will first write T' in place of T and then shift left one square, right one square, or remain on the same square, depending upon the value of σ ($\sigma = -, +, 0$, respectively). The new state of the control unit will be A'. An n-tape Turing machine is one where T, T', and σ of each quintuple are themselves n-tuples.

uniform family of (quantum) circuits A set of (quantum) circuits with one circuit for each number of bits. If a Turing machine can compute a function F in T(n) number of steps for an input of size n, then there is a family of uniform circuits such that the circuit with n input bits has at most $cf(n)^2$ elements, with c depending upon the complexity of the Turing machine.

unitary matrix A matrix $A = [a_{ij}]$ with complex elements; $a_{ij} \in \mathbb{C}$ is said to be unitary if $A^\dagger A = I$. Here $A^\dagger$ is the *adjoint* of A, a matrix obtained from A by first constructing A^T, the *transpose* of A, and then taking the complex conjugate of each element (or by first taking the complex conjugate of each element and then transposing the matrix). The determinant of a unitary matrix is 1.

unitary operator A linear operator $\mathbf{A}$ on a Hilbert space that preserves the inner product, thus the distance. See also *unitary matrix*.

universal set of quantum gates A set of gates with the property that there exists a network of them capable to implement every single unitary operation.

vector space An algebraic structure consisting of:

1. An Abelian group $(V, +)$ whose elements are called "vectors" and whose binary operation "+" is called *addition*;
2. A field F of numbers [either $\mathbb{R}$ (the real numbers) or $\mathbb{C}$ (the complex numbers)], whose elements are called "scalars";
3. An operation called "multiplication with scalars" and denoted by "·", which associates to any scalar $c \in F$ and vector $\alpha \in \mathcal{A}$ a new vector $c \cdot \alpha \in \mathcal{A}$.

wave function Function describing the state of a stationary system and the evolution in time of a non-stationary system. See also *Schrödinger's equation*.

Young's double-slit experiment Experiment revealing the interference phenomena related to the wave-like behavior of light (photons).

Index